T0176639

Practical Power Plant Engineering

Practical Power Plant Engineering

A Guide for Early Career Engineers

Zark Bedalov
Vancouver
BC, CA

Registered Office
John Wiley & Sons, Inc., 111 River Street, Hoboken, NJ 07030, USA

Editorial Office
111 River Street, Hoboken, NJ 07030, USA

For details of our global editorial offices, customer services, and more information about Wiley products visit us at www.wiley.com.

Wiley also publishes its books in a variety of electronic formats and by print-on-demand. Some content that appears in standard print versions of this book may not be available in other formats.

Library of Congress Cataloging-in-Publication Data
Names: Bedalov, Zark, author.
Title: Practical power plant engineering : a guide for early career
   engineers / Zark Bedalov, Vancouver BC, CA.
Description: Hoboken, NJ, USA : Wiley, 2020. | Includes bibliographical
   references and index.
Identifiers: LCCN 2019027657 (print) | LCCN 2019027658 (ebook) | ISBN
   9781119534945 (hardback) | ISBN 9781119534983 (adobe pdf) | ISBN
   9781119534990 (epub)
Subjects: LCSH: Electric power-plants.
Classification: LCC TK1191 .B43 2020 (print) | LCC TK1191 (ebook) | DDC
   621.31/21–dc23
LC record available at https://lccn.loc.gov/2019027657
LC ebook record available at https://lccn.loc.gov/2019027658

Cover Design: Wiley
Cover Images: Courtesy of Zark Bedalov; Background © Martin Capek/Shutterstock

Set in 10/12pt WarnockPro by SPi Global, Chennai, India

Printed in the United States of America

V10016650_010820

Contents

Preface – Why This Book? *vii*
Acknowledgments *xv*
About the Author *xvii*

1   Plant from Design to Commissioning *1*

2   Plant Key One-Line Diagram *31*

3   Switching Equipment *75*

4   Designing Plant Layout *107*

5   System Grounding *121*

6   Site and Equipment Grounding *137*

7   Plant Lighting *157*

8   DC System, UPS *179*

9   Plant Power Distribution *191*

10   Insulation Coordination, Lightning Protection *209*

11   Voltage and Phasing Standards *239*

12   Cables and Supporting Equipment *253*

13   Power Factor Correction *285*

14   Motor Selection *303*

15   **Variable Frequency Drives (VFDs) and Harmonics**   *321*

16   **Relay Protection and Coordination**   *341*

17   **Plant Automation and Data Networking**   *379*

18   **Generation**   *407*

19   **Power Dispatch and Control**   *441*

20   **Diesel Engine Generator Plant and Standby Power**   *461*

21   **Reliability Considerations and Calculations**   *475*

22   **Fire Protection**   *495*

23   **Corrosion, Cathodic Protection**   *517*

24   **Brief Equipment Specifications and Data Sheets**   *531*

25   **Solar Power**   *567*

26   **Wind Power**   *599*

      Index   *643*

Preface – Why This Book?

This book is a result of 50 years of practical experience from working in a number of industries with ever-changing technologies and by associating with many experienced engineers; electrical and other engineering backgrounds.

Starting as an engineer is not easy. You are facing a big transition. I'm certain this book will help get you through the most critical phase of your development as an electrical engineer and make you the confident and knowledgeable professional that you wanted to be when you decided to be an engineer.

There are a lot of books on the market explaining the theory of electrical engineering, but there are no books on practical engineering and experience. There used to be the old Westinghouse (now ABB) TD (blue) Book and Donald Beeman, General Electric Co. 1955: Industrial Power Systems Handbook, both of which I have proudly used as a young engineer. Both books now seem to be largely outdated. Computers have taken over much of the handmade calculations.

The information contained in this book is by no means all encompassing. An attempt to present the entire subject of practical electrical engineering would be impractical. However, this book does present guidelines to provide the reader with a fundamental knowledge sufficient to understand the concepts and methods of practical design and equipment selection and operations.

The first hint of a book came in Venezuela. After three years on a job with a local engineering company heading the engineering department, I decided to move on. The boss called me with a special request, saying: "Please stay on for another 3 months and write a book on how to do electrical engineering. You seem to do this work with a lot of common sense. I thank you for your help in leading and teaching our younger engineers. So, stay on, please."

The above dialog happened three years after seeing the movie Papillion (1978) filmed in Venezuela with Steve McQueen. The day after the movie, I fell into a snowy ditch somewhere north of Toronto. I had to leave the car and walk alone in the snow for a couple of hours in a total whiteout. A day later I spoke to my wife, we were going to Venezuela. Both of us loved the tropics and had it enough of Canadian winters.

Einstein: Theory is when everything is known but nothing works.

Experience is when everything works but no one knows why.

When we join theory and experience nothing works and no one knows why.

We quit our jobs, sold everything, and went to Caracas. "How smart was that," I heard it many times? Once in Caracas, I left my resume with six major engineering companies and then we went to a beach. Two weeks later, we returned to the Hotel Sabana Grande in Caracas. The owner said that I had many calls. I had five job interviews and took a job with a company that had a contract to build a 4×400 MW power plant. They badly needed an experienced electrical engineer. At that time, I had about 10 years of experience with a great company called Shawinigan Engineering from Montreal, Canada. That company was later taken over by SNC-Lavalin, Inc., my last employer.

Three years later, after the plant was built, I told my Venezuelan boss that I enjoyed it greatly, but I gotta be moving on. I moved on to Riyadh, Saudi Arabia. It was 1981. It seemed I was at the right place at the right time. There was so much going on in Saudia. At that time, some large generation existed in the Eastern province for oil production and barely in the cities of Riyadh and Jeddah. We began the electrification of the country in a major way. After Saudia, I went to several other international posts with companies like Fluor and Bechtel. Finally, I ended up with SNC-Lavalin for the past 17 years as a commissioning engineer. That makes it a total of 50 years as a lead design and commissioning engineer for power plants, heavy industrial plants, and power systems. Of that, 10 years were as an independent engineer on my own.

In the years after Venezuela, I often lectured younger engineers on many engineering issues and had discussions with companies to create a manual that would help their electrical engineers to follow and practice good engineering. It took a while. Finally, in 2015, I agreed to do a book. It took me one and half years to complete a draft copy. Now, it is here in your hands.

As an experienced electrical engineer I have noted huge obstacles young engineers were facing to become experienced engineers. I'm not talking about civil or mechanical, but electrical engineers. Let me explain. For mechanical engineers, everything is visible. Here's a pump, pipe, valve, filter, and strainer. All of it, recognizable objects. What's on the drawing is what you see in the real life. Open a valve and water or oil flows. You see it, hear it. If it leaks, you see it and you replace a gasket or clean a clogged filter.

Electrical engineers, however, in the same environment face an invisible world. Some call it "The mystery world". You may be able to recognize a few pieces of equipment from the drawings, but this is not what matters. In the electrical engineering, it is what you don't see that matters. If something goes wrong, you don't know which way to look. There are no electrons anywhere to be seen? Where do you start? Well, the first several years will be difficult, but with some experience and guidance, you start seeing the invisible. One young engineer told me that he came to his first job interview ready to solve a bunch of differential equations, but all that school teaching didn't seem to matter.

It is clear that mechanical engineers have a head start in the plants and the plant designs. Right off the bat, they are confident about themselves, of what they are seeing, doing, and learning. They will eventually become project managers and will boss the electrical engineers. I have seen this over and over again, anywhere in the world. Young mechanical engineers talk job immediately with confidence and are liked by the bosses because they talk the same language. Meanwhile electrical engineers are fearful to ask questions or suggest anything. They struggle for years. The bosses seem don't know how to talk to them.

So, while our mechanical engineering colleagues confidently talk about the things they do, and advance in their experience and carriers, we the electrical engineers appear shy and aimless, struggling in the world that has no resemblance to what we studied so diligently for many years.

Even the language is different. Here come buzzwords. Everyone uses buzzwords and most of them you don't understand. If that is bad enough, it gets even worse. Young mechanical engineers appear to be smarter. They seem to be learning faster every day in their *visual* world. As they say, a picture is worth 1000 words. On the other hand, a young electrical engineer gets very little visual information and thus retains less. Mechanical engineers are immediately immersed into the overall (big) picture of the plant, while the electrical engineers are pinned down to look at details.

Without an experienced engineer to explain things and to guide him, a young electrical engineer is lost. It would help if he only knew the questions. Not even that. He goes home after work and wonders: What is the reason for having me there? Will I ever be useful?

Let me give you an idea what happens on your first day on the job. You graduated from a difficult faculty of electrical engineering. It was tough and struggle, but you studied hard and endured, and felt you were on the top of the world. The world is yours. What a great feeling of accomplishment and exuberation that you can do anything.

Then you start looking for a job, and soon realize that the world is not all yours. The employers are not looking at your grades but at your experience, of which you have not much to show for. You cannot choose your job and will be happy to take anything that comes along. Finally, after three months of job searching, an engineering company was willing to give you a try.

You will be working on a new project. On the first day, you are introduced to your colleagues from all the disciplines and given a lot of drawings and reports to read. The material is mostly process and mechanical, to give you an orientation of what the project is all about. You were also told to talk to everyone and ask any questions you might have regarding the material you were reading as well as to acquaint yourself with the things the others were doing.

Then after a month of "doing nothing" your lead electrical engineer gives you for instance a couple of tender documents for a 10 MVA transformer just received from the bidders. One is for a transformer with 8% impedance and the other with 9.5%. The first one is more expensive than the other. The Lead tells you to evaluate the cost benefit of one over the other and if you have any questions feel free to ask. Since you were a junior engineer and need a bit of a help, he reminds you that the larger impedance causes more Watt and Var losses and higher voltage drop, while the lower impedance allows for higher fault level on the downstream bus, which may force the project to use more expensive equipment.

Wow, what now? That day back at home you look through your text books and find nothing relevant to help you out. Well, of course not. The text books tell you about the transformers and the transformation in general, but nothing specific for a particular application. That may be the last time you looked at your school books.

This actually is your first day at work. Remember that exuberating feeling when you graduated? You could do anything? Well, your Lead lowered you down to the real world. Now you feel hopeless and lacking confidence. You start asking questions all around and gradually acquire some knowledge but you are still far away from being able to decide which transformer to recommend. Fortunately, your Lead had already made that decision. Of course, he wouldn't let his junior engineer to decide on such an important matter. He just wanted to test you on how you think, how you formulate your questions, and how you deal with the engineers around you.

Welcome to the job. It'll be tough and it'll take time. All of us have started like this. You'll be doing fine if you immerse yourself into the project and start building up your practical experience over several years of working with experienced engineers on a variety of projects. This also includes those of other disciplines to learn what is important to them and how to select the electrical equipment to drive and automate their equipment. This real world book will help you get there.

This book is a result of 50 years of design and field engineering by experienced engineers and teaching others to do the same. As an experienced engineer with acquired practical knowledge, I'm ready to share it with the new coming engineers and lead them through a transition for which there is no blueprint or book, until now. This book provides useful information as a reference guide for all the electrical engineers. It fills the gap between the Academia and being an experienced engineer. If you read this book, you will learn a half of it you need to know and all the proper questions you should ask.

Hopefully this book will spawn others to write books. Your first job is a step into the open, away from your school. As soon as you start reading it, you realize this is a different world and it won't be easy. I agree, it won't be easy, but this unique book in your hands will give you a kick start, help you interact with other engineers and understand what is going on in the design office and in the field around you.

Why not searching on Google? Yes, there is plenty of this stuff and hundreds of answers on the internet. Well, if you only knew what you were looking for and had knowledge to properly assess it for your application? Without proper feedback, you don't know what is right and what is wrong and how to resolve doubts. "The Internet often seems to be a source of befuddlement rather than enlightenment," as Gregg Easterbrook eloquently put it in his outstanding book "Sonic Boom." This book gets straight to the point of what you need to know.

It's not an easy task to cover all the electrical engineering activities into a single readable 500+ page book. Many chapters would require a book by itself. The goal was to summarize the engineering activities and to direct the reader onto the right path and base from which he (she) can build experience needed to make proper engineering decisions. Everything in it, this author has experienced and then confirmed through commissioning and discussions with other engineers.

The theory is essential. It forms your basic knowledge fundamentals. The fact is this; our professors teach us to become professors. That's fair enough. The best students in our class became professors. An engineer you become with practical experience by associating with other engineers, facing multiple engineering applications and problems, making mistakes and reaching accomplishments.

Recently, I spoke to a professor about Variable frequency drives (VFDs), Chapter 15. I was telling him how I use them to regulate the plant flows on demand so I can employ smaller storage tanks, etc., while he was talking about flux vectors inside the rectifiers. "I'm not trying to make rectifiers. I'm just applying them for various useful plant applications," I told him? That's the difference. Because of this issue, many engineering schools are changing. Nowadays, students are forced to work between the semesters. Students are telling me that it's a hard go, as it is not easy to land summer jobs as unfinished engineers.

If you happen to get a job with a manufacturer, your life may be a bit easier. You will be trained for a specific job to work on some electrical equipment, such as improving a lightning arrester, rectifier, or a grounding switch. Soon, you will notice that designing a piece of electrical equipment is mostly of making it smaller, cooler, and with different materials. Then, you also realize that the job is 10% electrical and 90% mechanical engineering, and start wondering: "Is that it?" Well, maybe you'll like it. I didn't.

I graduated with a diploma on power transformers. My first job for two years was mostly how to make better cooling for transformers. I worked on hollow conductors for cooling water passing through them. There was nothing electrical about that. Why didn't they hire a mechanical engineer to do that, I wondered? On the other hand, if you get a job to design power systems for various plants it's a different story. It's an electrical story.

So, between you and me, I had enough of mechanical engineering and them taking advantage of us and bossing us around by saying; "I was not smart enough, so I went into mechanical engineering." I heard that line a lot. With this book, I want to even out the playing field and help you young electrical engineers stand your ground, be productive, and contribute almost immediately.

One of my first job interviews was at Toronto Hydro. An engineer showed me a picture and asked me if I knew what it was? I saw six bushings and said that it was a transformer. I was wrong. It was a high voltage oil circuit breaker. I failed that job interview. I should have known that a breaker had six identical insulators. Transformer has 3 + 3 unequal bushings.

I moved on, looking for my first job in Canada and ended up at Pinkerton Glass for a job interview. A secretary gave me a test sheet to fill in before meeting an engineer. The sheet said "Practical test for electricians." I filled it up as best as I could, guessing on a half of it. I failed that one too. None of those famous differential equations could have helped me. It was so bad; the Engineer didn't even waste his time to see me.

Many years later I was already an experienced engineer. Our company, Fluor, had a project with the Xerox Corp. in NY State. As a lead electrical engineer I was invited to visit the plant and scope the work for adding a new ink toner line to the existing plant. We started touring this large plant. As an electrical engineer I prefer to look first at the plant overall one line diagram. This is the Chapter 2 in this book. Having acquired the big picture, then I visit the plant and observe it from my electrical perspective. Well, anyway, we started touring as soon as I got there. The mechanical engineer, my tour guide, looked at me and said: "You are an electrical engineer, right?"

"Yes, any problem with that," I answered jokingly?

"No nothing, but, let me pass one by you. Here we have a problem," he started talking. "We've been struggling for 2 years now with it. Occasionally, we have light flickers in the plant. It happens suddenly and then nothing for a few days and then again. The whole plant flickers and then everything is back to normal. Do you have any suggestion what that might be," he asked?

"I really don't know. It can be anything," I answered. "Does it happen at night or day, high load, low load," I inquired? Suddenly, I realized I was in an invisible world of electrical engineering.

"Well, I agree," the plant engineer said. "It's unpredictable; anytime. We checked it with the local utility. They said that it must be something internal within our plant as they don't experience flickers on their system."

"If I were you I would look at the main transformer since the whole plant is flickering. It may be coming from there," I said. "Otherwise, you may have to shut down the plant and megger all the major electrical equipment, starting from the incoming transformer."

"Hmm, I'll mention it to my electrician," he answered.

So much for that, I thought. We continued touring and got into a rather noisy room. My guide pointed to their 2000 HP, 5 kV compressor, the biggest drive in the plant. I came closer to the compressor and spotted a drop of oil below its big cabinet on which it read: "Surge Pack." I told him that there was no reason for the oil to be on the floor here and suggested that if he wouldn't mind we open the cabinet.

"Don't worry about the oil, I'll call our cleaning staff to clean it up," my tour guide mentioned it somewhat embarrassed. I insisted and we opened the cabinet. Inside it I saw more oil, obviously leaking from the surge capacitor. I turned to him and said: "This may be your source of flicker. The capacitor is leaking and occasionally breaks down and creates a brief short circuit to discharge itself."

Weeks later, he called me and thanked me for the discovery. They replaced the capacitor and, thank goodness, resolved the issue. I wrote back to him, "We were lucky to be in the room before the cleaning lady had a chance to remove the oil drop. Please don't fire that lady."

In the invisible electrical world, you often have to be lucky.

Read this book and practice it. If you have read it and understood 80% of it, you don't need more schooling, though it would help. Not even calculus. I have nothing against math. I was pretty good at it. Nowadays, since the use of computers for power system studies, the highest level of math I have used was $\sqrt{3}$ and $\cos \phi$ on my calculator. True, I have been using "per unit" a lot in my conservative short circuit fault calculations, though. Most of all, I needed know how to prepare estimates for various project options and calculate percentage power losses or voltage drops in power lines. And, of course, it was always important to be able to answer questions right on the spot during the project meetings with the project owners.

This book will not make you an expert on any of these subjects. For that you will need a lot of experience and hopefully some good mentoring. But it will give you a good start and capability to discuss the subject with some confidence and ask good questions during your job interviews. With this knowledge, you can start your job from a solid base, rather than starting from nothing.

Hopefully, this book will point to you the path on how engineers think in planning and resolving the problems and the basic elements of the engineering considerations; scope of work – big picture, engineering tasks, economics (cost of equipment and production), reliability, and automation requirements.

A few more notes.

As a junior engineer, I grasped from other engineers the concept of looking at the big picture, and what matters, while leaving other things for later. These are likely to change anyway, so why bother thinking of them now. Looking at a big picture means developing a design criteria for all parts of the project right at the start, such as, determining the short circuit levels (by computer) at various plant busses, allowable voltage drops, outage contingencies, big motor start, etc. Then, I develop overall key one line diagram. Once I have done that and have it on paper, I have my base and reference and talking points for everything that fits inside. It's not frozen in concrete. I can adapt it as I go along. If you do this once, you can replicate it again and again on other projects, based on the clients' requirements and new requests, capacity of production and levels of automation and security demanded.

Most employers do look for individuals who talk in terms of looking forward and grasping the whole concept of the project. One day at SNC-Lavalin, I had a chance to see my file they made after my job interview and it read, "Impressive view of looking at big picture."

Also, as a beginner engineer, I was once interviewed by an American company. In their books of potential candidates, next to my name was written: "Capable of looking and seeing a bigger picture. Hire him, when opportunity comes up." I heard that little jewel from a head hunter who called me up and offered me other positions. Head hunters dig out their information from their contacts in the HR (Human Resources) of the companies they deal with. I was not born with it. I looked up to and learned it from other engineers who had impressed me.

When you do your work, share it with others. Be a team worker to be able to succeed in this type of work. There are other disciplines involved on the project and often you will find out something that will help you make a proper decision or make a modification you didn't notice before. So, be a team player. Then, when you are done with an assignment, tell your boss that you are done. He loves to hear that. If you are holding it back, soon it will be noticed. Bosses like to hold onto their best producers and keep them busy.

When you are finished with your assignment, tell your lead engineer to give you feedback. Don't be afraid of his review. Don't work in isolation. Share the ownership of your assignment with him. Involve him as you work on it. He will appreciate that. Besides, he is likely far more experienced than you are and experience is what counts in electrical engineering. He may have a different approach for certain parts. Broaden your thinking, evaluate his input and implement it if beneficial and more economical.

Always show interest. I was always curious. As a kid I wanted to be an engineer. Being a medical doctor also crossed my mind. Later on I realized that being a doctor would have been a mistake. I was not made for it. I function by logical thinking. I don't want to learn anything by heart. I learn and gain experience through logical interaction. Most of it was by observing and listening to others. Being interested puts you in front of the events. Many a time I was sent to investigate a situation about which I had not a clue. Nobody else on the project had a clue either. So, others rejected it and declined to be involved. I always went for it, tried my best and most of the time I found solutions.

Enjoy engineering. This has been your biggest investment so far. Go for it.

Vancouver, BC, Canada
16 June 2019

Zark Bedalov

Acknowledgments

The book was initially prepared for lectures for students at the University of Electrical Engineering, Zagreb, Croatia. Thank you Dean Prof. Mislav Grgic and ProDean Prof. Marko Delimar.

This author thanks all the contributors, mainly other engineers who helped in creation of this book, either directly or through discussions and suggestions.

I thank SNC Lavalin for sending me on interesting projects and assignments to share experience with knowledgeable engineers from the various equipment suppliers, owners' engineers and project managers, some of who insisted on me to write this book.

Eddie Fung of SNC Lavalin for major contributions in all parts of the book.

Wes Yale of TNC Power Systems for grammar and technical edits and contributions to Chapter 12.

Jose Galindo of Galmea Consultants, Spain for their contribution on the use of Waste Heat, Chapter 20.

University Professor Martin Jadric for contributions to Chapters 14 and 15.

Ken Thomson, P.Eng of SNC Lavalin for the exchange of ideas.

Nancy Walczak of SNC Lavalin who taught me how to put the illustrations together.

GrafikaPlus.hr for preparations of the illustrations.

Khusrau Baghaei of SNC Lavalin for contribution to Chapter 17.

Sorin Segal. P. Eng. Technical editing.

Bob McNair, P. Eng. Technical editing.

Michael McBurnie, P. Eng. Technical editing on Chapter 15 and 23.

Mohd Akram, Abdullah, Eng. TNB, Malaysia, Power plant commissioning and managing

Sumit Verma, CV Shahin Kumar and Dhaneesh Kc of Alstom (GE) India for great technical exchanges on a recent project in Malaysia.

And many more.

Let us not forget those engineers who I have worked with and who have left significant impact on my early career.

Andy Sturton, Shawinigan Engineering	Relay Protection, Grounding
Jim Erath, Shawinigan Engineering	Power System Engineering
Ernie Siddel, Canadian Atomic Energy	Reliability
Farid P. Dawalibi, Shawinigan Engineering	Grounding

The first three I knew as experienced electrical engineers who readily shared their valuable experience on me and other young engineers. The forth one was a junior engineer just like me when we worked on some power projects. He took great interest in the grounding issues on the projects. Then he wrote the first computer program to calculate the substation grounding requirements. If one searches on the net anything to do with grounding, his name will appear as one of the greatest authorities in this area, including on so many IEEE papers.

About the Author

Zark Bedalov explains that as a kid he was always intrigued by electricity. He would turn a light switch on and off for hours and observed the light to come on. If the switch sparked, that intrigued him even more. There was no one there to explain to him what was going on in the wires. In his early school days, he was punished for smashing a capacitor to pieces for trying to  see what was inside. A lot of greasy paper was inside.

He graduated at University of Electrical Engineering, Zagreb, Croatia, in 1965. Also, attended some master degree studies at the University of Toronto.

His first job was in a factory for power transformers. Soon he realized that making a transformer is 90% mechanical and 10% electrical. That was not what he was expecting. So he skipped over the border to Wiena, Austria, and arranged for immigration to Canada. Finding a job in Canada was not easy, in particular if you are an engineer. Being an engineer comes with responsibility and that requires "Canadian Experience." It took him about six months to start as a draftsman on mining projects. Had a lot of support from many senior engineers and three years later was certified a "Professional Engineer."

From thereon, Zark was in his domain and in demand. Early on, he changed companies every three to four years to learn more. He worked for almost 50 years on large projects, power plants, and heavy industries, all around the world, employed by major international engineering companies, such as Bechtel, Fluor, Atomic Energy of Canada, SNC Lavalin, and often independent, teaching along the way and enjoying the work and life. Now retired, he writes on electricity and teaches young engineers on how the electricity makes the factories and power plants function.

1

Plant from Design to Commissioning

CHAPTER MENU

1.1 Planning, 1
    1.1.1 Plant Design Procedure, 3
    1.1.2 Codes and Standards, 4
1.2 Project Development, 5
    1.2.1 Type of Project, 5
    1.2.2 Conceptual Design for Feasibility Study, 5
    1.2.3 Detailed Design, 6
    1.2.4 Engineering Documents, 8
    1.2.5 Equipment Specifications and Data Sheets, 9
    1.2.6 Equipment Numbering, 11
    1.2.7 Load List, 13
    1.2.8 Generated Cable List, 13
    1.2.9 Schematic/Wiring Diagrams, 14
1.3 Precommissioning and Commissioning, 17
    1.3.1 Precommissioning, 17
    1.3.2 Commissioning, 19
    1.3.3 Reliability Run, 23
    1.3.4 Power Plant Grid Tests, 24
    1.3.5 Commissioning Reports, 25
1.4 Project Economics, 26
    1.4.1 Budget Estimate, 26
    1.4.2 Levelized Cost of Energy (LCOE), 27
    1.4.3 Marginal Cost of Energy, 29
    1.4.4 Profitability of an Industrial Plant, 29
Reference, 30

1.1 Planning

The electrical power distribution systems have to be designed to fit the plant electrical requirements. The power systems must be well planned, considering the technological process, cost, reliability, maintenance, control, operating flexibility, and future growth. Furthermore, the undertaking must take into account

Practical Power Plant Engineering: A Guide for Early Career Engineers, First Edition. Zark Bedalov.
© 2020 John Wiley & Sons, Inc. Published 2020 by John Wiley & Sons, Inc.

the safety of people and equipment, continuity of power supply, installation, and operating costs.

Electrical engineers' responsibility is to prepare design criteria and single-line diagrams, power system studies, calculate fault currents, locate load centers within the plant, estimate load diversity, select the grounding system, define the routes of overhead lines, prepare plant layouts, and develop the electrical protection system, all of it to suit the plant location and the prevailing standards. Furthermore, he/she must procure the equipment and participate in the plant construction and commissioning.

The basic concept for a single-line diagram representing the *power plant* power distribution is generally established by the utilities. The main engineering effort is on implementing the power system around the generating units. Depending on the generator unit MW size, a decision will be made on having generator breakers next to the generators or employing high voltage (HV) breakers instead, in the switchyard to serve as the unit breakers for the generator/transformer groups. That is one of the most significant factors that define the overall concept of the diagram. The power plant station service generally uses less than 5% of power of the generator MW rating, thus, the one-line diagram for a power plant is relatively simple in comparison to the industrial plants (see Chapter 18 for more details).

One-line diagrams for industrial plants vary significantly from industry to industry. The load is fully distributed around the various operating activities, such as crushing, grinding, mixing, drying, pumping, batching, each of which requires a considerable engineering effort and decision making process to arrive at an optimal economic diagram that can be scaled and readily expanded in the future.

This book is written in 26 chapters to cover all the technical aspects of electrical engineering and to transfer practical experience onto young electrical engineers. In order to present it in a meaningful way, the book explains the technical details around a fictitious, though realistic power plant and industrial projects. An industrial project offers a greater variety of requirements and lends itself better for practical analysis.

This analysis can be applied to other plants that use similar electrical equipment, such as transformers, motors, generators, variable frequency drives (VFDs), cables, switchgear, overhead lines, fire protection, control systems, grounding, lighting, etc.

The project is commenced by an investor (company) who have decided to build a power or an industrial plant (cement factory, steel manufacturing, oil refinery, wood mill, plastic cups, fruit canning, etc.) on a particular location for a particular operating capacity (produced MW, tons of cement, tons of steel, tons of paper, tons fruit, etc.).

The investor company had already prepared a rough estimate proposal for a project with a simple budget estimate of ±40% accuracy and had received

Project development steps

Owner	Initial feasibility study → 40% Estimate → Bank
Engineer/ owner	Conceptual design → 20% Estimate for bank loan
	Review of process alternatives Eng. drawings: flow, P&ID, one lines, based on projections procurement of major long lead equipment
Engineer	Detailed design based on actual procurement engineering specifications and drawings procurement: mechanical, electrical and control site construction of infrastructure
Engineer/ contractor/ engineer	Project construction Civil → mechanical, Electrical, controls Precommissioning: individual equipment commissioning: system by system
Contractor/ owner	Release to production ownership and production

Figure 1.1 Project development.

positive indications of financing from a bank to develop a feasibility study and a more detailed cost estimate. Figure 1.1 shows the steps of the project development.

From the electrical power system perspective, the first step is to review the project flow diagrams produced by mechanical engineers and on that basis prepare electrical design criteria and develop a **key one-line diagram** (see Chapter 2). The key one-line diagram will envelop all the process facilities within the plant starting from the power source down to the individual equipment users and services. This is followed by preparing a (±20%) budget cost estimate as part of the conceptual design inclusive of the cost for engineering and construction and then present it to the bank to secure a loan.

1.1.1 Plant Design Procedure

Plant design is a joint effort by multiple engineering disciplines: process, mechanical, civil, electrical, architectural, structural, estimating, scheduling, procurement, document controls, and project management. Every department is doing its work in strict coordination with others to insure everyone is "on the same page" and that nothing falls "between the cracks." Lead engineers of all the disciplines are on the email circulation of everything that is happening on the project to insure all the design decisions and project changes are being communicated and implemented, both "vertically and horizontally" through the project organization chart. Regular meetings are held on a weekly basis to assess the progress, critical path schedule, any design changes, manpower shortfalls, and any delays in design, procurement, fabrication, and installation and their impact on the project schedule. Design options and temporary measures are being reviewed to overcome the delays in the equipment deliveries and/or equipment failures.

1.1.2 Codes and Standards

The publications listed below form a part of this book. Each publication shall be the latest revision and addendum in effect at the time of issue of contract and design specifications, unless noted otherwise.

ANSI	American National Standards Institute
CSA	Canadian Standards Association
IEEE	Institute of Electrical and Electronic Engineers
IEC	International Electro-technical Commission
ISA	Instrument Society of USA
ISO	International Standards Organization
NACE	The Worldwide Corrosion Authority
NEMA	National Electrical Manufacturer's Association
NEC	National Electrical Code
NFPA	National Fire Protection Association
UL	Underwriters Laboratories of USA

Standards, codes, and guidelines listed above and referenced in every chapter of this book are widely used by the engineers in the industry, both as directives and guides. When working in another country, the local standards must also be applicable. This book refers to and often presents data courtesy of these engineering standards, which are considered one of the major sources of guidelines and good engineering practices for engineers.

The standards and codes are extremely important. Actually, there are three parts to your success as an experienced engineer:

(1) Understanding the applicable standards and knowing how and where to apply them.
(2) Referring to the suppliers' equipment catalogs and reviewing the graphs and performance data sheets to determine the proper equipment ratings and supplies for your specific applications.
(3) Building experience by field reviews of the equipment performance and its related hardware.

1.2 Project Development

1.2.1 Type of Project

A typical project referred to in this book is an ore-bearing property, owned by an investor, which according to the exploration figures contains a large ore body of Cu/Zn ore. This can equally be a mining property of coal, silver, and gold, or it may be a brewery batch plant or a large harbor development.

On the power industry side, this may be a hydro development project, for which a catchment area is defined and dammed to create a head and estimate the flow of water that can be controlled from the area. The electrical part of the project, though large, is relatively small in comparison with the huge civil infrastructure required to be built. Utilities typically take ownership of these large projects.

1.2.2 Conceptual Design for Feasibility Study

If the project gets a go ahead by the investment partners, the investors or their bank will provide initial funding to engage an engineering company to investigate the project a bit closer. There are a number of different engagements possible between the owner and the engineer that can be employed. That may be a subject of another book.

The project team with all the engineering disciplines has been given an assignment to develop conceptual drawings and budget estimate. The conceptual design includes plant layouts, load flow diagrams, electrical design criteria for all the electrical activities and equipment, a key one-line diagram, and a reasonable accurate capital cost estimate to be presented as a "bankable feasibility study." This document will also serve as a basis for the future detail engineering design to build the project, should the project proceed further.

The design calculations in this conceptual study will use the system and equipment characteristics from previous projects similar to this one to generate the design parameters for the new project. The budget estimate is obtained from the various budget quotations, previous projects, and earlier work done in the country of the project with their labor rates.

During this three to six months of engineering phase, some major long lead mechanical and electrical equipment may be ordered on the basis of a possibility of cancelling the orders if things don't "pan out." The electrical lead equipment may include the main transformers and HV switching equipment.

The flow diagrams show the flow of the ore, through the various plant processes, including additions of other ingredients: water, heat, and fuel to process it and take it to the final product. In this case, the final products are bars of

Cu and Zn. Where there is Cu, there is a chance of minor percentage of gold, while silver is never far off from a deposit of Zn.

From the electrical design perspective, design criteria, major cable routing, and one-line diagrams will define the shape of the plant power distribution and other aspects of the electrical equipment and plant operation.

1.2.3 Detailed Design

Detail design will follow the conceptual phase. This "detailed" phase may last anywhere from 9 to 18 months for an industrial plant or two to four years for a power plant, depending on the type of plant. The conceptual drawings will be reworked and expanded. Procurement phase will commence by preparing the purchase specifications to specify equipment performance requirements and also to make the interface diagrams to tie up with the related mechanical equipment. Often, the electrical design may have to wait a while for the mechanical design to near its completion and their suppliers' drawings are in hands to determine the electrical ratings and the interfacing connections needed.

The first effort will be to update the electrical plant design criteria and the key one-line diagram from the conceptual design phase. These two items are your two big pictures, and the foundations for everything else you plan to build on.

System studies: The detailed design will present final one-line diagrams with the actual impedances and equipment characteristics. It will use the data based on the results from power system studies: load flows, motor start, voltage drops, phase and ground short circuits, arc flash, insulation coordination, and step and touch potential. The studies will determine more precise and factual system characteristics and prove that the selected equipment ratings conform to the requirements set out in the design criteria. For these calculations, we will use the software from various system houses, such as Easy Power, ETAP, Cyme, and others. The plant data will be laid out on a computer and let the computer do the math. Not only that, the computer teaches you the power system functioning. One can introduce changes and alternatives and then observe the impact of the changes on the power system performance. It allows you to select the optimal solutions.

Interfaces: At this time, schematic and wiring diagrams for all the motors and valves, cable lists, and plant layouts will be prepared.

One of my bosses once told me: "Project usually fails at the interfaces." He was right. The projects require a huge effort by many personnel working on the project, ranging from secretaries to the managers. Possibilities of errors are ever-present. The interface changes may be due to a late design modification initiated by other engineering departments. If not well communicated and reconfirmed, the changes may not get on the drawings. This applies also for the communications between the engineering departments, the suppliers, and fabricators. If the equipment arrives to site with incorrect connections, it will lead to a lot of confusion on site, "throwing blame around of who said what, and so on." This is where the experience comes in from working on large projects and by recognizing how the equipment is supposed to work and how it relates to the other equipment. Experienced engineers would notice problems if incorrect drawings cross their desks.

Every discipline can use approximations, add (+) or delete (−) a few inches or feet here and there on the drawings. The electrical engineers have no such a benefit. We have to produce drawings that match the equipment perfectly. Electrical drawings show several hundreds of thousands of wires, power, and controls interfacing between the various electrical and mechanical equipment. The only grace we get is that we can bend the plant cables around in the cable trays.

You may have done your job to perfection, but unfortunately, when you come to the construction site, you may face some disappointments. You will notice the supplier's actual equipment does not match the drawings you received to prepare your diagrams from. The suppliers have just got confused and sent you drawings they had engineered for a previous customer, or they had made changes but failed to inform you.

Do not panic now. This is something to get used to. It happens. Once the wires are connected, you may notice different problems stemming from errors, suppliers' incorrect designs, and of course, the wiring errors. This is where precommissioning and commissioning comes into play to make sure everything is properly tested and made to work as intended.

Everyone can make mistakes. Let us be honest about it. Even mechanical engineers can make a mistake here and there. But there is nothing like what the electrical engineers face. Thousands and thousands of wires are laid out in the field, and each one must find its proper place or it may turn out to be a major mistake and error, which will have to be troubleshooted later during the plant commissioning. Fortunately, with the advances in technology, a half of wiring in the modern plant is now replaced by communication cables, coax, a pair of wires, etc., carrying thousands of signals which can be shaped and configured as part of the plant control system. But that is another story. That certainly is a wiring relief, but our problems will now likely resurface in the software during commissioning (see Chapter 17).

You as an electrical engineer will prepare or work on the following drawings and documents:

- Equipment and installation specifications.
- *System studies*: Load flow for voltage drops, short circuits for the equipment ratings, large motor starts, and relay coordination.
- One-line diagrams.
- Design criteria.

- Layouts for electrical equipment, lighting, cable trays, load Lists, cable schedules and terminations, embedded grounding, equipment grounding, lightning, and power corridors.
- Prepare schematic and wiring diagrams for each motor, valve, and feeder,
- Review of civil, mechanical, and instrumentation drawings.
- Review of suppliers' drawings, and more.

That is a lot. A project of this magnitude may require thousands of electrical drawings and hundreds of documents.

1.2.3.1 Cost of Change

At the project meetings, you will note that the design is still open to changes. With the design criteria and key single-line diagram in hand, you will be discussing with your civil, process, and mechanical counterparts on what is possible and reasonable and what is not, and what will cost "an arm and a leg" and what may be a more reasonable option. The developer may suddenly decide to add another process line in the plant, which may stretch your "almost finished" power distribution system or completely change it. Remember, a plant change on paper is 10 times less costly than doing it during construction (Figure 1.2).

1.2.4 Engineering Documents

During the plant design, an electrical engineer with his team of designers must prepare the following documents:

Drawings: Drawings are being prepared for the specific electrical equipment and as layouts for the equipment installations. The former are included with corresponding equipment specifications, while the latter are part of the construction (installation) specifications. The drawings are to be prepared by experienced designers with a help and under the supervision of a lead engineer.

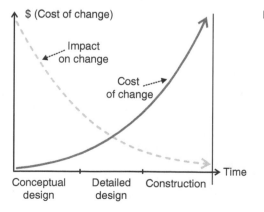

Figure 1.2 Cost of change.

Initially, the drawings are prepared as preliminary and issued to various suppliers for tendering purposes. Once a winning bidder is selected, the preliminary drawings are updated and finalized based on the fabrication drawings received from the selected supplier and finally issued for construction. The drawings must be marked with appropriate revisions as they are being revised and issued.

There are different methods of marking the drawing revisions. Here's one. The preliminary issues of the drawings are labeled with revisions, Rev. A, B, C, ... or PA, PB, PC, ... The final drawings for construction are marked as Rev. 0, 1, 2, 3. Minor changes not affecting the contents or performance may be modified without raising the revision number.

Reports: During the project, many situations are encountered where the engineer is required to prepare official reports to evaluate various options and make recommendations of possible changes and improvements to the project. The changes may be due to the project cost reductions, technological changes, or changes to the site or operating conditions.

Coordination with other engineering disciplines: The electrical engineer must also review the mechanical, process, and civil engineering drawings to familiarize themselves with the buildings and mechanical equipment, as well as to insure the mechanical equipment includes appropriate electrical parameters specific for the project.

1.2.5 Equipment Specifications and Data Sheets

These documents will be prepared by the lead engineers for the electrical equipment, such as transformers, motor control centers (MCCs), VFDs, switchgear, etc. Revisions to these documents may follow the same procedures as identified for the drawings. Following a receipt of the tenders from the suppliers, the engineer prepares technical tender evaluations with appropriate conclusions, recommendations, and specific conditions for purchasing the equipment. As part of the award of contract, the specifications and data sheets are updated to match that of what was agreed on "as purchased" (see Chapter 24 for some specification details and data sheets).

A typical small or big project requires a number of specifications with data sheets to be written. The specifications define the equipment performance requirements and workmanship. The data sheets cover the specific equipment rating requirements. The specification for a particular piece of equipment can be updated from project to project with some minor changes. It is the data sheet that changes in a big way as the application and ratings may be completely different from project to project.

Hopefully, the new specifications will be similar to those of your previous projects. Often, one can change the project name and the spec number and then revise the data sheet to suit the equipment you need for your new project.

Try not to repeat yourself in the documents. Sooner than you think, someone will call and ask you: "What do you want: 1000 A breaker written in the specification or 1200 A breaker listed in the Data Sheet." If you want to talk about the breaker in the specification, just note: "For the ratings, refer to the Data Sheet."

From project to project, try to maintain the same ID number for the same design product, if the project permits it. For instance:

Specification and data sheet, respectively, for MCCs on project ABC:
   ABC – xxx – TS31 – DS31
Specification and data sheet, respectively, for MCCs on project XYZ:
   XYZ – xxx – TS31 – DS31

Try to group the documents for the type of equipment and services. Leave some gaps as there are differences in scope from project to project. When you are dealing with equipment like MCCs located in various different parts of a plant, write a common spec with several data sheets added to it for different areas.

Here is a list of specifications from a recent project in Minnesota on a 55 MW power plant using turkey litter as fuel:

(1) Electrical contribution to mechanical engineers' specifications.

TS01	Electrical requirements for mechanical equipment
TS02	Electrical requirements for 480 V motors up to 200 kW
TS03	Electrical requirements for medium voltage (MV) motors over 200 kW

(2) Main power distribution

TS11	Switchyard equipment and hardware
TS13	Large transformers
TS14	Standby diesel generator
TS15	Relay protection panels
TS17	13.8 kV transformers
TS21	13.8 kV switchgear

(3) Plant equipment

TS23	MV motor controllers
TS24	MV VFDs

TS31	480 V MCCs
TS32	480 V VFDs
TS33	Unit substations and low voltage (LV) switchgear
TS35	Station battery and chargers
TS36	Uninterruptible power supply (UPS) equipment and panels
TS37	Rigid bus ducts
TS38	Cable bus ducts
TS39	Lighting and distribution panels
TS40	Power and control cables
TS41	Plant heat tracing panels and hardware

(4) Services and plant installation

TS43	Plant CCTV
TS44	Plant public address
TS45	Plant telephones and data
TS51	Plant fire detection and suppression system
TS52	Plant heat tracing
TS54	Overhead distribution lines
TS55	Switchyard installation
TS57	Plant installation

1.2.6 Equipment Numbering

There are a number of methods on how to number the equipment. Some clients may have their own numbering system for all of their projects. The numbering is generally done by the mechanical department, except for the purely electrical equipment, which is numbered by the electrical group. The most popular numbering systems are the intuitive systems, as follows:

Example: **20 PU 007**; Area 20; **PU**mp sequence number: 007. The pump number is also given to the associated motor.

The plant process areas may be given the specific area designations:

00: For a general site, including the main substation
10: Crushing
20: Conveying
30: Milling
40: Flotation, etc.

For the equipment, assign intuitive designations as follows:

PU, pump; SP, sump pump; TK, tank; CR, crane; AG, agitator; VF, vent fan; RF, roof fan; LP, lighting panel; HR, heater; CV, conveyor; TR, transformer; CY, cyclone; etc.

The sequence is from 001 to 999. The sequence numbering may restart from 001 for each area.

The numbers do not have to be consecutive. The sequence number 001 may start from the basement, 101 on the first level, etc., to better describe the equipment location.

Be consistent and make sure the pumps are sequenced in pairs: 001/002, 103/104, the odd number is, for instance, on the left side approaching the motors.

Cables are generally numbered and defined by the loads and not the MCC sources. Some companies insist that cable numbers include indications of the source (the MCC bucket) and the destination (load). Standardize control cables for the common hardware such as pushbuttons by using consistent cable C suffixes, programmable logic controller (PLC) connections. For instance, the cables for the pump 40 PU 003 are labeled:

40 PU 003 P: For the motor power cable.
40 PU 003 C1: For the control cable to PLC equipment.
40 PU 003 C2: For the control cable to the Push Button station.
40 PU 003 C3: For the control cable to the field level switch or other sensor.

Make sure the numbering is consistent for all plant motors. This will allow for easier understanding of the plant cabling and purposes. For instance, establish that cable C1 for all the plant motors is always the cable from the motor starter going to PLC. If that cable is not used, that number is skipped for that particular drive. Push button station always occupies cable C2, etc. This consistent method of numbering will help you computerize the cable lists and to sort the lists to specific equipment.

The equipment numbers are needed to create load lists, which will later be expanded to create the list of schematics and cable lists. For instance by adding to the list of loads: the load names, voltage, kW rating, MCC, and bucket, a computer program can create a list of cables, select the cable sizes, and assign "From – To" destinations without any manual input.

If you add a typical schematic diagram type for each drive or feeder to the load list, the computer program will assign and fill in the cables and cable details on the drawings for each particular drive/feeder without any human input.

1.2.7 Load List

Based on the data given in the load list (Table 1.1), the computer calculates the cable size, type, length, and routing (see Table 1.2).

Adopt minimum cable sizes: #12 American wire gauge (AWG) for power and #16 AWG for controls.

Should the kW rating change and a new kW rating be entered into the list, the cable size will be computer-updated automatically when the cable list is regenerated.

The same computer program generates and prints the schematic and wiring diagrams automatically adding to the drawing of all the specific data including the cable data, input/output (I/O) data, and wiring terminations.

1.2.8 Generated Cable List

Table 1.1 Load list (simplified).

ID	Name	Volt	kW	MCC	In service	Type
40 PU 003	Feed Pump, Water Tank 1	460	15	MCC11	1	T1
40 PU 004	Feed Pump, Water Tank 1	460	15	MCC11	SB	T1
30 AG 006	Agitator, Tank 6	460	10	MCC03	1	T1
30 CR 001	Crane, 60 Tone, Main Hall	460	100	MCC03	1	F1
10 RF 003	Crusher Roof Fan	460	1	MCC 01	1	T3

In service: 1 = operating, SB = on standby.
Type: this is the designation of the type of operating circuit to be implemented.

Table 1.2 Cable list (simplified).

Cable ID	Cable size	kW	Type/class	From	To	Length
40 PU 003 P	1-3/c #10 AWG,	20	Teck 600V	MCC11	PU Motor	120
40 PU 003 C1	1-5/c #16 AWG,		Teck 300V	MCC11	PLC 2	15
40 PU 003 C2	1-3/c #14 AWG,		Teck 300V	MCC11	PB Stn	120
40 PU 004 P	1-3/c #10AWG,	20	Teck 600V	MCC11	PU Motor	120
40 PU 004 C1	1-5/c #16 AWG,		Teck 300V	MCC11	PLC 2	15
40 PU 004 C2	1-3/c #14 AWG,		Teck 300V	MCC11	PBStn	120
40 PU 004 C3	1-2/c #14 AWG,		Teck 300V	MCC11	TH1	130
30 AG 005 P	1-3/c #12AWG,	15	Teck 600V	MCC05	PU Motor	85
30 AG 005 C1	1-5/c #16 AWG,		Teck 300V	MCC05	PLC 3	20
30 AG 005 C2	1-3/c #14 AWG,		Teck 300V	MCC05	PB Stn	85
10 SP 003 P	1-3/c # 12 AWG,	5	Teck 600V	MCC01	PU Motor	60
10 SP 003 C2	1-3/c #14 AWG,		Teck 300V	MCC01	PB Stn	60

1.2.9 Schematic/Wiring Diagrams

In the last 20 years, the schematic and wiring diagrams have changed in the following ways. This author has participated on all three of them:

(1) Relay logic diagram. That is past. We will not dwell on this wiring approach any more.
(2) Connection to PLC I/O (see Figure 1.3).
(3) Connection to distributed control system (DCS) communication modules (see Figure 1.4).

It is not known whether there is a forth step around the corner, but it could be said that each advance has brought us considerable progress and simplification to the design of the schematic/wiring diagrams.

Specific operating logic for each motor and valve is no longer needed to be shown in the diagrams as it used to be with the relay logic diagrams. The circuits (Figure 1.3) were drawn and wired uniformly into DCS or PLC I/O cards, while the specific logic to each drive is developed as software. Though this method

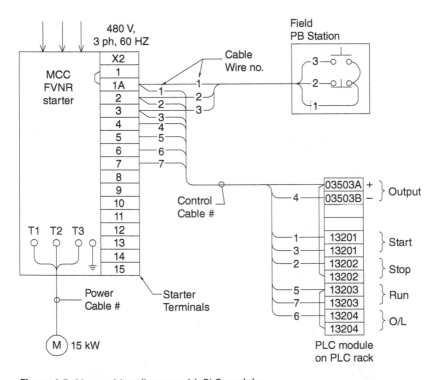

Figure 1.3 Motor wiring diagram with PLC module.

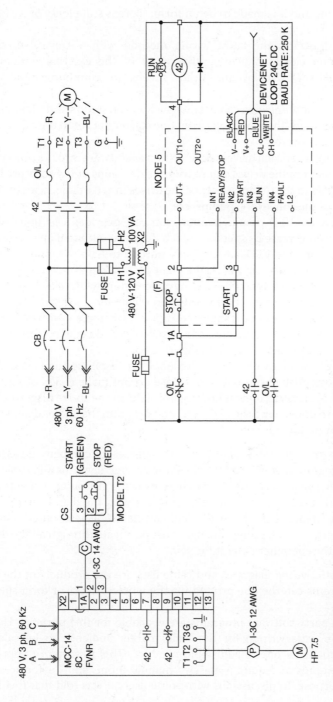

Figure 1.4 Motor wiring diagram to DeviceNet communication module.

needs less wiring and it is made to be uniform, there is still plenty of wiring to be done.

Figure 1.4 illustrates the latest wiring method with a DeviceNet communication loop clearly indicating that most of the external wiring has disappeared. For DeviceNet and other means of communications (see Chapter 17).

Evidently, this motor needs a cable for its pushbutton station and a DeviceNet loop to a DN module. The loop loops from one starter bucket to the other. That is it. The rest is software.

The specific logic for each drive is now written as software into the processors to receive status from the drives and to feed the decisions of the software logic back to the inputs to start/stop/sequence the drives in accordance with the flow requirements of the conveyors, pumps, etc.

Or if the drive is a VFD-operated motor, software provides a set point to the drive to increase/decrease its speed to match the plant output at any particular moment. Therefore, the VFDs are not only needed to help the motor to start softly but to also continuously adjust the plant production of a certain product in the plant operation. This could never have been done with relay logic.

What is the difference between a wiring and schematic diagram?

A wiring diagram of a motor is shown in Figure 1.3, complete with all the cabinet terminations. A schematic diagram is the stretched version of the wiring diagram (Figure 1.4a) and is shown in Figure 1.4b.

As a result of the innovations, the site labor for installing the field control wiring has substantially been reduced. Please do not make a sigh of relief, as yet. Though the operating logic is no longer visible on the above diagrams, you will now have to understand the PLC logic and program the ladder diagrams to make the plant motors function like an orchestra.

Computer program: It is desirable that the engineering company develops a software program that will create schematic/wiring diagrams and cable lists directly from the project load list database by using attributes that automatically get filled on the typical model drawings with the data sourced from the load list. Manual entry to these documents is the biggest source of errors on the project. A small project change must permeate through all the documents. Let the computer enter it for you.

The schematic/wiring diagrams and cable lists are the products of the load list. The diagrams can then be printed for the whole project or for a specific area or MCC.

Some third-party software programs are available for this purpose. Unfortunately, these were written for the wide market audience to be saleable to every company as one unadjustable product. This third-party approach unfortunately tends to require a massive manual input, and for that reason, defeats the purpose. In discussions with some users, I was told that the input is overwhelmingly manual and leads to erroneous inputs. It was not efficient, and the software was abandoned.

This author has developed its own program on FoxBASE for that purpose. It is updated for every new project to be project-specific resulting in minimal manual entry, mostly for cable lengths.

1.3 Precommissioning and Commissioning

Commissioning of an industrial plant is a bit simpler. The plant operating system can be broken down into smaller subsystems, such as crushing, milling, which could be precommissioned and commissioned totally independently.

In the power plants, the generating units are large operating blocks, which are tested one at a time along with the water or fuel paths (input) and the electricity path to HV switchyard (output), as well as the unit services and operating controls all at the same time. Station services are commissioned separately.

1.3.1 Precommissioning

Precommissioning and commissioning of an industrial plant or a power plant are different activities. They must be approached differently in particular if the plants are fully automated. Precommissioning is testing of the equipment such as switchgear, MCC, VFD, or transformers on an individual basis in an energized state, but totally disconnected from the other operating equipment.

Secondary injection: First, the switchgear is meggered and high-pot tested. Then, the protective relays are tested by secondary injection (simulation) to trip the breakers due to overloads or undervoltages according to the protective relay setting sheets for each breaker. Protective current transformers (CTs) and potential transformers (PTs) circuits are fed to the tester to simulate the operating state. The secondary injection is performed by a three-phase tester, shown in Figure 1.5.

Circuit breakers: Each breaker in the switchgear can be tested for functioning in its drawn-in (connected), test and withdrawn position. The switchgear is

Figure 1.5 Three-phase tester.

not energized, but the circuit breakers can be operated because the 125 V_{dc} control circuits are energized to allow the breakers to function. Furthermore, there may be an additional control circuit at 24 V_{ac} or V_{dc} used for remote operation and signaling to and from the plant control system.

In each of the three positions, the switchgear and the circuit breaker leave its mark.

In the withdrawn position, one can test the breaker to charge the activating spring and to open/close without affecting the other breakers in the assembly. The breaker test position is a half-drawn-out position. In this position, one can fully test breaker in all aspects of control and interlocks, but without affecting the other parts of switchgear assembly.

In the connected position, the breaker can be fully tested provided the incoming and the tie breakers are locked and held in a withdrawn position. This test position is very useful in the commissioning (energized) phase of the plant testing that follows the precommissioning.

Similar precommissioning activities are carried out on MCCs for each motor or feeder circuit to enable the assembly to be energized and to power motors and feeders for further tests. Each motor is being bumped for its rotation to match that of the pumps or conveyor travels, etc. For this activity, the motors are decoupled from the pumps.

Furthermore, the motor branch circuit breakers are also pretested to establish their minimum instantaneous protection settings to suit the motor inrush currents (see Chapter 3 for details).

Wiring: During the precommissioning, a lot of simulation will be required to be performed to test the equipment and cable wiring. This includes jumpering the contacts and injecting volts or currents from other sources to command the operation of the switchgear breakers or MCC starters. All the wiring and schematics of the field devices and hard wired safety interlocks must be verified.

Wiring diagrams used to be checked during the precommissioning stage too, but these diagrams are now becoming a rarity and often obsolete. As mentioned earlier, wiring diagrams have been greatly simplified by using the communication links, such as Ethernet, DeviceNet, Modbus. The present schematics have all the terminals marked just the same as the wiring diagrams used earlier. It seems to be a trend now. Perhaps not in the industrial projects yet, but, certainly, it is a trend in the large power plants.

You may then ask, how do you make cable terminations if you do not have wiring diagrams? That is a very good question. Well, what many contractors now use are the cable tabulation lists showing the terminations from the terminals shown on the equipment A to the terminals on equipment B, but without giving any significance to each wire. The wiremen can swiftly terminate the wires as listed and let someone else think if the list was right or wrong. As a result, not all the signaling is being precommissioned. Some parts of it may be rung out, but not precommissioned to verify the interlocks. It is left to the commissioning group to test it and prove it on the equipment performance basis. Again here, this approach refers to the large power plants and not industrial projects.

Transformer oil: Transformers oil is tested for its dielectric strength several days prior to energizing. New oil should have a strength of >65 kV/cm, while older oils must demonstrate the insulating strength of >60 kV/cm. If the strength is lower than those desired, the transformer oil must be purified by the heating and filtering equipment to exhaust the moisture before energizing. Oil samples will be taken from the transformer during installation a week before energizing.

Figure 1.6 presents an actual handover chart of a power plant from construction, through precommissioning, commissioning, and reliability run (RR) to operation and ownership transfer.

The precommissioning checks on a larger piece of equipment, for instance, a large hydroelectric generator, is a relatively complex endeavor. A large number of interlocks must be simulated, much of them from the software. Some precommissioning can be done and must be done, such as unit trip logic and emergency stops and safety trips. In order to make the simulations more manageable, control functions are usually broken down into a number of sequential steps. These critical offline tests are performed before the online tests are attempted in order to minimize any unforeseen inadvertent operation. During this process, the unit is tested through a restricted logic to allow the checks on part of the logic and then proceed to the next ever larger step. The rest is left for commissioning. Since there are too many interlocks to be dealt with, there is also a fear that some of the simulation jumpers may be forgotten and left behind. Jumpered contacts left behind hide unreal bypassed conditions.

1.3.2 Commissioning

Commissioning is often called wet commissioning. It is an engineering activity that follows the precommissioning phase, often called dry commissioning. Commissioning is an engineering activity dedicated to testing the plant or part of the plant, in a fully energized state of applied electricity, pressure, heat, steam, and water with all the auxiliary systems in service.

During the days of relay logic and manual plant control, the commissioning was not that extensive. Nowadays, the plants have become fully automated and generally unmanned, having inputs from numerous field devices, start, stop, overload, ready, local/remote (Loc/Rem), breaker position, trip, analogs, etc. To diagnose the plant directly on the operator's screen rather than by pulling, simulating, and jumpering, the wires in the panels takes a lot of coordinated effort in many parts of the plant at the same time.

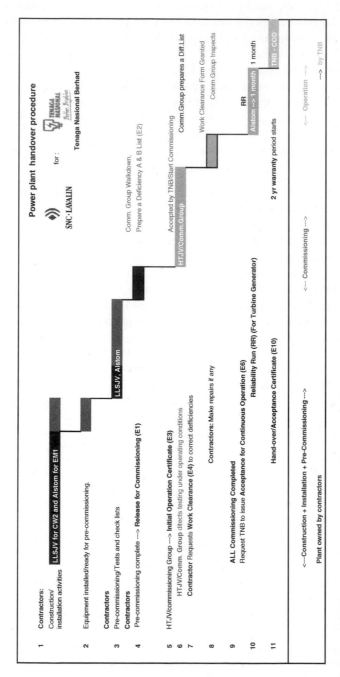

Figure 1.6 Plant handover procedure. Source: Courtesy of SNC-Lavalin.

Only a fully energized condition offers a controlled testing environment that allows all the logic elements to be properly commissioned and proven. This work in a noisy environment requires good communications between all the parties. Often, the cell phones do not work due to the noise interference. On the last project, the proper communication was only achieved after the house telephones operating on fiber optic cables were installed.

Primary injection: Commissioning of large power plants carries considerable risks and safety issues. This is because commissioning is mostly performed under the primary injection when the equipment is fully energized. The operating voltage is applied, and the currents are flowing through the cables and breakers. This flow can now be monitored on the real front panel instruments and/or relays. The flow of current (power) can be changed by loading the feeders and circuit breakers.

While most of the precommissioning and troubleshooting is performed from the schematic diagrams, just like always has been, the commissioning is performed based on the operational requirements. Commissioning as mentioned above is energizing and testing of the groups of equipment working together under active pressure of oil, water, steam, air, or electricity, with a minimum of simulation. This is the first time the equipment and the control system will face fuel, water, air, oil, and electricity. Believe me, this makes a lot of difference in the equipment behavior, particularly the instrumentation. The instrumentation finally gets to be checked for flow, speed, voltage, pressure, etc. Some of the instruments will now prove to be faulty, poorly calibrated, or incorrectly selected, or perhaps incorrectly wired or leaking.

The precommissioning of the power plant control system is mostly skipped in favor of the commissioning as mentioned earlier when the generator is energized in a controlled manner and tested.

Most of the wiring or control logic (95%) is expected to be correct, but a small part (5%) may be incorrect. The operator's screen may not indicate an error, but it will show what is not functioning or functioning incorrectly. It is not an easy task to comprehend as multiple malfunctions may be a result of a single wrong input. But which? Where to start? The logic interlock strings have to be investigated by observing the schematics, by removing the wired and software interlocks, one interlock at a time to zoom in on the possible targets or "usual suspects."

What are the interlocks? These are the "permissives" that enable or disable another drive. This may be a case of set of pumps of which one only is allowed to run at any time. Safety Interlocks are hard wired into the electrical circuits as in Figure 1.7. Process interlocks are entered into software as shown on the right.

Sometimes, missing wiring must be added, wires reversed, corrected, removed, and jumpered to make it correct according to the schematics.

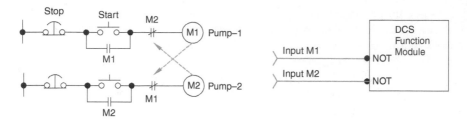

Figure 1.7 Plant interlocks, hard wired and software.

And if it still does not work, then the schematic must be corrected and the circuit rewired. Often, an interposing relay may not be seating properly. There are so many ways to fail?

The wiring tabulations, mentioned earlier, are generally not being updated. The schematics are updated in red and issued "As Built." The commissioning approach, though relatively cumbersome, avoids all the simulations to be done and saves the time as the work is concentrated on fixing faulty wiring and wrong logic lines only, and not on all the wiring and all the logic. This approach may work providing the unit was being fully tested under all the operating conditions of loading, ramping, and stopping to expose the functioning of all the instrumentation, the transmitters, and indicating instruments. The transmitters may work correctly for certain loading and operating sequences, but rapid changes in the load may expose their weaknesses.

In my judgment, the following are typical commissioning problems:

(1) *Mechanical*: Rare. They tend to be visible, but take longer (days) to fix and replace with spare parts if available.
(2) *Electrical wiring*: Due to wrong wiring, wrong schematics, inappropriate contact seating. This is diagnosed by going over the schematics and fixed within hours.
(3) Control logic software errors. This is diagnosed and modified within hours. Often, several attempts are tried.
(4) Nonresponsive devices (level switches, analogs, pressure switches, transmitters, etc.). The devices may be faulty or inappropriate for the application. These problems are relatively common and difficult to deal with because of their unpredictable and intermittent behavior. These issues may take several days to be discovered, repaired, and the fix confirmed when repaired.

Here is an example of unpredictable nature of certain field devices. A large hydro generator sometimes fails to transit from the synchronous condenser mode to the generator mode when requested. Having spent days to review the field devices that control this activity, the focus shifts toward three-level switches operating in a tube controlled by an orifice. The switches may work well for days, but then suddenly fail to provide a transition signal. After days of trying and figuring out many situations, it is finally correctly concluded that the water in the tube is too turbulent at certain conditions and water pressures and affects the operation of the switches. Several new orifice sizes were tried and eventually a proper size is selected and placed. The trial period to prove a particular orifice may take weeks. Also, the same orifice size used on Unit 1 may not be the right one for Unit 2, due to varied operating conditions.

Naturally, all the changes implemented on Unit 1 in the wiring, control logic, and unresponsive field devices are immediately updated on the next unit if it is available, thus, insuring faster commissioning and release to operation.

In an industrial plant, the commissioning is based on a system-by-system basis grinding, flotation, conveying, leaching, etc. Switchgear, transformers, MCCs, and motors are tested together under partially energized conditions. Each system is semiautonomous and gradually energized to prove it operates in proper sequence for starting and shutting down under specific conditions. The systems are being commissioned until the whole plant is made operational and ready for starting up from the control room and it is correctly represented on the control system monitors.

Some say that commissioning is the slowest game in town. It is 90% logistics and preparation to position right people to the right places, keep them focused on the task with proper tools, and to prepare the safe environment and operating conditions for the test. The actual test may take only a few minutes to show a change in state on the instruments, or that something had moved or rotated. Conveyors start moving in sequence, grinders start grinding, and pumps start pumping and filling the tanks. Level switches regulate the speed of the drives to prove the set points established by the control system. The automation starts directing the process.

For the commissioning to work well, one has to have a good cooperation of the mechanical, process, and controls engineers and excellent means of communications in the field. Why are the electrical engineers most often assigned to be the commissioning engineers? Well, I think because the electrical engineering is the hardest part of the project to understand. Electricity is the blood flow of the plant. It connects all the pieces in an invisible way that experienced electrical engineers can understand.

1.3.3 Reliability Run

RR is also shown in Figure 1.6. This is a big commissioning event as the unit enters the operation and starts producing electricity. Typically, RR is applied to power plants but not to the industrial plants. RR lasts anywhere from 7 to 30 days. A period of 30 days is the most common RR period. For the plant to enter an RR, it means it has been fully commissioned and proven to be operational, with some minor listed deficiencies that do not affect the production.

These can be fixed during the RR or later. Successful RR leads to ownership transfer and a start of a one to two years warranty period.

During an RR, the owner's operators take over the plant operation in the presence of the suppliers in supervisory role. This is also a phase of practical training for the operators. The operators follow the directives of the dispatch center and load the machines accordingly in terms of MW and MVARs. The intent is to operate and expose the plant to all the operating modes and transitions without restrictions and as often as possible. As more and more automation and supervision is added to the plants the commissioning and RR tests get more and more involved.

Each plant owner may dictate different constraints for RR. The rules may also depend on the unit performance during the commissioning. If the plant has been failing often, owners may impose more stringent conditions. In general, the unit must operate 30 days, 24 hours a day without a major failure that would cause the unit to reduce its capability to carry load. If that happens, the RR is restarted from the beginning. For instance, a failure of a pump with a successful automatic transition to the healthy pump will not be a cause to stop the RR, but considered a successful operating action.

It is much easier to commission a plant if you had followed it through the design and construction into commissioning. Sometimes a commissioning engineer may be invited to do a commissioning on a plant that he/she may not be familiar with. There were a number of cases like this. One of difficult ones was in Lahore, Pakistan, where we were invited to conduct a commissioning and RR test after it has failed in 12 earlier attempts over a period of two years. It was a 5×30 MW thermal plant.

During the RR, the plant was supposed to operate flawlessly for seven days without a single alarm while being tested under all the operating conditions. In addition, there was a specific test during the RR while all the plant units were running, called "Islanded Test." An unexpected three-phase fault is arranged on the HV line, connected to the plant. To pass the test, the plant had to separate itself from the grid, shutting down four units, while one unit was supposed to be left running and maintaining the plant station service load.

Well, this time, it worked well, and the plant passed the test. We knew that this time, we had to take a different approach. We modified the protective relay settings and readjusted the governor transfer functions. Every short circuit is different, and the units are required to conform accordingly. Perhaps we were just plain lucky. It is hard to tell. This time the protective relays operated selectively and the new governor settings acted correctly.

1.3.4 Power Plant Grid Tests

The power plant or a unit connected to a HV grid is often subjected to a number of grid tests that must be completed to confirm its compatibility with the grid and to operate from the dispatch center. Here is a list of some of the grid tests:

- Load dispatch following capability
- Reactive power dispatch follow
- Isochronous capability (droop)
- Dispatch MW ramp rate
- Fast ramp MW response
- Synchronous condenser mode switchovers
- Primary/secondary high-frequency response
- Under-frequency trip setting check (simulation)
- Loss of station service AC supply – switchover to standby power
- Black start, synchronizing on islanded grid dead bus
- House load operation on generator power for two hours

1.3.5 Commissioning Reports

Most commissioning engineers observe the testing in a passive way, and at the end of the test, they ask for the commissioning reports by the contractors. In their preprepared reports, you will never find a hick up or any failure noted. So, why bother having the reports, when everything is 100% perfect. In fact their reports are already done before they even start testing. Everything is Yes, Yes, OK, OK! There are tests that indeed go smoothly. Faults do happen and often due to multiple reasons that can be either fixed quickly or postponed and fixed later. That has to be recorded, as well as all the temporary deviations due to missing components, etc.

We do our own commissioning or precommissioning report for reasons noted below. We do it directly on our PCs as the events evolve. We sit with the test engineers at the end of a large desk in front of the main control panel and observe and write:

- We do not make a story. We write the action.
- The testers pay more attention to their work in our presence. They cannot hide the actual facts.
- As active people, we can ask for repeat tests or test it from a different perspective or condition.
- Our report is chronological and timed as it happens. One never knows what and when something unusual will happen. It does happen and often. One wants to capture the moment and to include all the background details on what was in action before the occurrences of a failure.
- It includes the failures as a record, why and how they were resolved and made to be OK.
- Later on during the operation when something fails, one can look into the commissioning reports and figure out if this is a recurrence of the same fault and likely to be expected to happen again.

1.4 Project Economics

1.4.1 Budget Estimate

A typical project passes through a number of development phases, starting from an initial estimate, conceptual design for feasibility study, detailed design, construction, and start up.

Let us name some typical projects:

- a power plant: hydro, diesel generation, gas combined cycle, etc., and
- an industrial project that may be an ore exploitation process, a factory of detergents, potato chips, and any other similar facility.

Utilities typically take care of the large power projects of this nature. They make a budget estimate of the project cost and evaluate it against the revenue based on the kWh to be sold to 'the consumers. The projected cost will include the initial capital for the equipment, materials, and labor over the years of construction, cost of money and plant maintenances, and operation (see Section 1.4.2)

The overall cost must include the transmission line from the power plant to the major switchyard. In case of a hydro project, the transmission lines are inevitably long and at higher transmission voltages. Diesel generation plants, on the other hand, are generally built for specific consumers in remote isolated areas. This may be a case of a mine up North needing 10–20 MW of power, for which the cost of building a transmission line would be rather prohibitive, in particular, if one evaluates it on the basis of the cost of km of line per MW delivered.

It is not only the utilities that are involved with power generation. There are small power producers called Independent Power Producers (IPPs) for generating anywhere from 2 to 100 MW. They generally do not get involved with power distribution and readily sell all the power to the local utilities.

The cost of fuel over the years of plant exploitation is always the most prevalent factor in the evaluation of a project development. This is where the hydro facility jumps ahead in spite of its huge cost of civil infrastructure. A gas fired plant will certainly be less costly on the basis of the initial cost of the plant and transmission line, but it is the cost of fuel per kWh produced that matters over the life of the plant. A diesel generation plant up North may be built at a low initial cost, but the cost of fuel is high, due to the additional cost of trucking and barging it to the site and storing it there for a full year. But do not forget that the usage of waste heat from the diesel engines can lift the plant efficiencies and reduce the cost of fuel usage.

An evaluation of a power plant also depends on the daily/weekly operating cycle. For hydro, the idea is to maximize the water usage, in particular during the rainy seasons to avoid spilling. Not many power hydro plants are built with a 100% operating capacity factor. A hydro project operating at 50% capacity factor is common, while I have seen hydro plants built for a projected 12% annual capacity factor. In other words, a plant of 100 MW installed capacity with a projected 12% capacity factor can produce, based on the estimated water availability, only 12 MW on a daily average for the whole year. It produces power when it has enough water drawing it down to the minimum operating level (MOL). Producing power below the MOL would be highly inefficient use of water resources and may make it difficult to recover back to the higher more efficient operating levels. Such plants with low capacity factors may be operating as base load generation during the rainy seasons and also as peaking plants because of their quick start capability when the marginal energy cost is the highest.

On the other hand, a fossil-fuel operating plant costs a lot more to operate and is likely to be used for peaking duties in a daily cycle only. Therefore, the overall economics of building a fossil-fuel operating plant in an area, which includes a mixture of different types of generation, must be estimated on the basis of its low operating hours.

While the power plant projects are built with the highest quality of equipment and redundancy, intended for 40 years of operation, an industrial plant may be built for a shorter duration of, say 10 years.

Levelized Cost Of Electricity (LCOE) is one of the yardsticks the owner's accountants use to compare the energy options for power plants. The formulae of totalizing the lifetime cost of production against the lifetime revenue are quite complex summations, using discount rates, inflation, and present worth accounting. Let the accountant work his figures. You as an engineer should understand the math behind it and offer technical options that may reduce the operating and initial costs to make the project more feasible.

1.4.2 Levelized Cost of Energy (LCOE)

Figure 1.8 [1] shows the relative costs in $/kWh for a number of generating options over the plant lifetime.

For instance, everything including a large hydro generation costs $0.04/kWh compared to an offshore wind farm priced at $0.19/kWh during the lifetime of the plant. The graph also shows the range of the cost for other alternatives. Clearly, the utility may have a hard time selling the wind power in this situation.

$$LCOE = \sum_{(1-n)} \text{total lifetime cost (\$)} / \sum_{(1-n)} \text{total lifetime}$$

$$\text{energy production (MWh)} \quad (1.1)$$

The units of LCOE are money/energy (usually $/MWh or c/kWh[1]).

---

1 If you want to convert between the two, it is handy to remember that 1c/kWh = 10$/MWh.

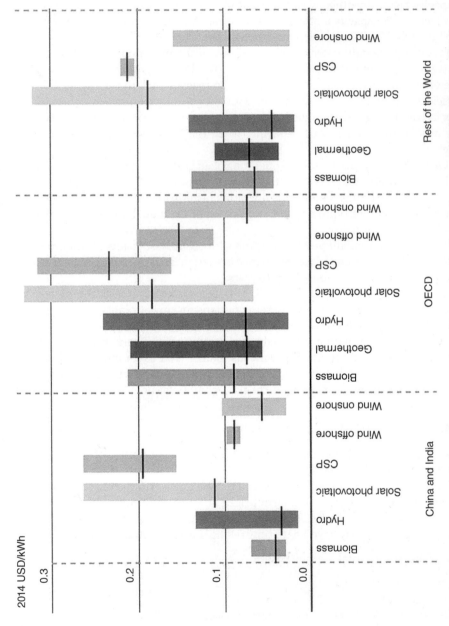

Figure 1.8 Typical LCOE cost ranges and weighted averages for electricity generation. Source: Courtesy of IRENA Publications (2014).

As the name suggests, the "money" part of this equation consists of costs: specifically a summing up of all the costs spent over the whole lifetime (from year 1 to year n) of the project. So this is the money spent building a power plant – at the start, capital expenditure (capex), and a long list of operational expenditure (opex), for example fuel, maintenance and repairs, land lease, insurance, tax, and interest on bank loans. If you can sell your system for something at the end of the project lifetime, you can knock this residual value off your list of costs.

Some projects, such as solar PV and hydro, will involve considerable up-front capex, but followed by years of very low operating cost. Others, such as a gas-fired power plant, will see the majority of spending over the project lifetime during the operating years of burning fuel.

Indeed, levelized cost analysis is all about comparing different energy systems with very different cost structures on a "fairer," long-term basis. Comparing the two examples mentioned earlier, the revenue of the solar PV will depend on how sunny it is and by how much the output of the solar panels degrades over time. On the other hand, the revenue for the gas plant will depend on the capacity factor and how often it runs. The later will be more difficult to predict though, as it will depend on the interplay between electricity sale and gas purchase prices.

1.4.3 Marginal Cost of Energy

The LCOE cost can be considered the average cost of a particular type of generating source. Utilities are interested in the marginal energy cost also. It is the cost experienced by the utilities for the last (peaking) kWh of electricity produced and sold. The marginal cost is highly variable and could vary throughout the day between negative pricing when there is over generation and could increase to hundreds of $/MWh when the demand is high and supply is low. The marginal cost determines the ranking of the type of generating source that will be dispatched. Those with the lowest marginal costs are the first ones to be energized to meet the demand, while the plants with the highest marginal costs are the last to be brought on line.

For example, a wind or solar power plant has no fuel cost and relatively low O&M costs. It yields the lowest marginal energy provided when the sun is shining and the wind is blowing. There is a big difference in production cost whether the plant is generating 1 or 100 MW. Similarly, a gas turbine plant also has low marginal cost if the gas price is low, which it is right now (2017).

1.4.4 Profitability of an Industrial Plant

What is the profitability of an industrial plant? The investor is typically interested in the initial capex and a quick construction schedule to insure a quick loan repayment from the plant operating cost. The investors like to use a simple formula called a **"payback time."** The investor is typically looking at a maximum of five-year payback plan to repay the cost of the initial plant (capex) with the sale of the product, ore, or other merchandize and enjoy a loan and cost-free life thereafter. Naturally, the economics are highly dependent on the commodity prices of the materials produced. Once the project is initiated, the cost estimate follows a more detailed approach.

Reference

1 IRENA: International Renewable Energy Agency (2014). LCOE, Levelized Cost of Energy.

2

Plant Key One-Line Diagram

CHAPTER MENU

2.1   One-Line Diagrams, 32
      2.1.1   What Is the One-Line Diagram, or Single-Line Diagram?, 32
2.2   The Electrical Project, 34
2.3   Site Conditions, 36
      2.3.1   Source of Power, 36
      2.3.2   Ambient Derating Factors, 36
      2.3.3   Reliability Criteria, 37
2.4   Connection to Power Utility, 38
      2.4.1   Source Impedance, 40
      2.4.2   Line Conductor, 41
      2.4.3   HV Circuit Breaker Fault Interrupting, 43
      2.4.4   Double or Single Incomer Connection, 43
      2.4.5   Utility Generating Capacity, 45
      2.4.6   Firm Capacity, 45
      2.4.7   Line Protection, 45
      2.4.8   Lightning, 45
2.5   Main Plant Substation, 45
2.6   Load Site Placement, 47
      2.6.1   Crushing, 48
      2.6.2   Grinding and Conveying, 48
2.7   The Key One-Line Diagram, 50
      2.7.1   Load Investigation, 50
      2.7.2   Connected Load – Operating Load, 51
      2.7.3   Voltage Level Selection, 53
      2.7.4   Switchgear Breaker Ratings, 54
      2.7.5   Single Incomer Substation for a Small Plant, 56
      2.7.6   13.8 or 33 kV Switchgear for a Larger Plant, 57
      2.7.7   Transformer Connections: Cable, Cable Bus, or Bus Duct?, 58
      2.7.8   Medium Voltage Switchgear and Controllers (4.16 kV), 59
      2.7.9   Low Voltage Service Voltage, 60
      2.7.10 Bus Tie Breaker Switching, 61
      2.7.11 Plant Transformation, 63
      2.7.12 Voltage Regulation, 64
      2.7.13 Overhead Distribution Lines, 66

Practical Power Plant Engineering: A Guide for Early Career Engineers, First Edition. Zark Bedalov.
© 2020 John Wiley & Sons, Inc. Published 2020 by John Wiley & Sons, Inc.

2.8   Transformer System Grounding, 67
 2.8.1   Transmission Level, 67
 2.8.2   MV Systems, 68
 2.8.3   LV Systems, 68
 2.8.4   Generator Neutrals, 68
2.9   Transformer Winding Configurations and Phasing, 68
2.10 Standby Power, 70
2.11 Insulation Coordination, 71
 2.11.1 Substation Shielding, 73
2.12 Plant Control System, 74
2.13 Fire Protection, 74
Reference, 74

2.1   One-Line Diagrams

Designing a **Key one-line diagram** is the most important task in the development of an electrical system for a power or an industrial plant. This diagram is a result of the key decisions made by the engineers working on the project. This book devotes significant time in explaining the electrical components, which are fundamental in building the functional electrical one-line diagram.

The one-line diagram represents the electrical power distribution formed to suit the technological process for the proposed project (see Chapter 1). The electrical engineers must focus on acquiring information on the type of process, load magnitude, load centers, quality and availability of power, power loss tolerance, and required plant reliability.

The *key* one-line diagram prepared at the initial stage of design will be conceptual in nature. It will encompass the other one-line diagrams for the specific parts of the plant. It will serve for discussions, cost estimates, and to offer the other design team engineers a basis for their equipment selections. Figure 2.1 is not a "key" diagram, but a part of a plant one line diagram.

The design procedure in this chapter is described in light detail to arouse interest of the electrical engineers in the design and operation of electrical systems for industrial manufacturing and power plants. More clarifying details related to the specific equipment specifications, applications, and reasons for their selection can be found in the chapters that follow.

2.1.1   What Is the One-Line Diagram, or Single-Line Diagram?

Mechanical engineers have their "flow diagrams." Electrical engineers have their single-line diagrams showing the electrical power flows and plant overall integration. As the name implies, it is the principal electrical diagram or our big picture of the plant or specific part of the plant, whereby the three phases are represented in a simplified single-line form. The diagrams show all the

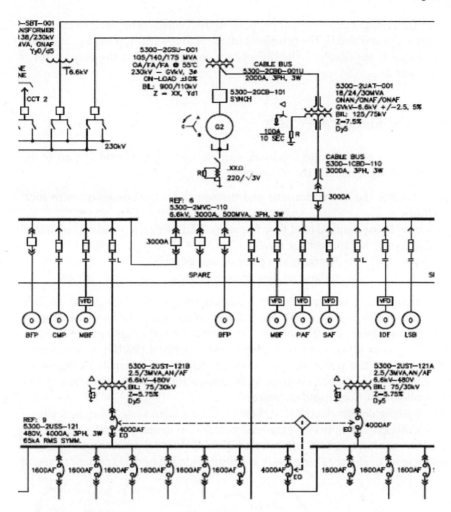

Figure 2.1 Part of a plant one line diagram.

major transformers, loads, circuit breakers, and cables or line connections, including the ratings: kW (HP), MVA, V, A, AWG (mm^2), leading to the major plant equipment. The key diagram includes references to the partial more detailed one-line diagrams for the specific process areas. One medium size industrial plant may have 20 individual one-line diagrams starting from the key one-line diagram down to the individual MCC 480 V (400 V) diagrams.

Decisions must be made on the main switchyard, the number of incoming transformers, and the selection of the plant busbar voltages for distribution of power to the major load centers for large and small loads and primary and secondary power lines to remote plant facilities.

Note: The voltages and frequency applied in this book will be those of the North American standards. The principles of calculations and application used here are equally applicable to the IEC system voltages used in the other parts of the world.

2.2 The Electrical Project

The activities presented in this book, some of it in this chapter as part of the electrical design, include the following:

- Determine the site conditions and discuss the interconnection with local Utility.
- Review of mechanical load flow diagrams, P&D diagrams and establish load (kW) estimates and voltage levels.
- Prepare one-line diagrams and plant design criteria.
- Conduct system studies and determine electrical equipment ratings.

As the starting point, the mechanical engineers will develop 10–20 flow diagrams of the plant process. A small part of the ore handling flow diagram is shown in Figure 2.2. The process (instrumentation) engineers will develop 30–40 process piping and instrumentation diagrams (P&ID) to instrument and automate the plant, as shown on a small P&ID segment in Figure 2.3 for a feed pump. The P&ID gives us the indications on how the plant will be controlled, monitored, and operated.

As a young design electrical engineer you have been assigned to be a part of a multidiscipline engineering team responsible to develop a project estimated to consume about 30 MW of power, or 37.5 MVA at 0.8 power factor (pf). The design team of electrical engineers and draftsmen led by an experienced senior

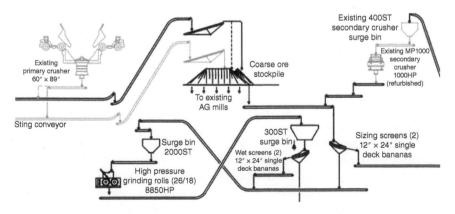

Figure 2.2 Part of plant flow diagram.

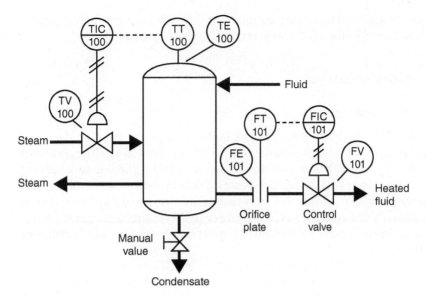

Figure 2.3 A part of a P&ID diagram.

lead electrical engineer is responsible to design the electrical power distribution system and procure the electrical equipment to power up and control the plant equipment. The plant distribution system will follow the national standards and voltages for 60 Hz (50 Hz) frequency as applicable to the location of the project.

The principal operating item in this facility and the biggest electrical load is a large 10 MVA semi autogenous grinding (SAG) mill operated as a variable speed drive cyclo-converter. It receives the ore from the crushers, as 10–20 cm chunks of raw material and reduces it to 1–2 cm large gravel. The mill speed is regulated in accordance with the hardness of the ore, which may vary on a daily basis. This material is then conveyed to ball mills, which pulverize the product to allow it to be mixed with water and pumped around as slurry through the rest of the plant. The slurry will be then subjected to some chemical treatment processes to separate the metal from the ore.

What is the operating basis for this project? The owner will look at operating the plant 24 hours a day to maximize his early output in order to quickly pay off the loan to the financial institution. The only way to do that is to run the facility around the clock in two 12-hour shifts. This operating regime is also favored by the local utility, as it allows them to run more generation around the clock as less costly base load and flatten the load cycle.

A housing complex for 100 people will be built at a site about 3 km away from the main process plant for the workers to build and operate the plant. This is an area away from the plant crushers. Operators are happy to work long shifts seven days a week on a basis of three weeks in and one week out. The housing complex (camp) will be a continuation of a construction camp for 300 people, which will be used for a full year prior to the start of production.

2.3 Site Conditions

2.3.1 Source of Power

Let us assume, the local utility has just built a new power plant on the coast, about 120 km away from the ore deposit and have extended a 230 kV transmission line to a city 50 km away from the deposit. This is the line to which the plant will be connected to. The line passes 20 km by the proposed mining site. To further simplify the matters, we will assume that the utility has sufficient spare capacity and is happy to furnish power to the new facility. This is a fortunate situation as it makes it feasible to import the power instead of generating it on its own.

The plant load will be relatively constant with ±10% variability. Utilities love constant load, which they can supply as a base load. The base load energy is less costly to produce in $/MWh.

The plant's electrical distribution system must operate in a stable manner within the prescribed tolerances of voltage and frequency as stipulated by the standards, in spite of the load variations. The load may be subject to changes, both MW and MVAR, caused by the operating cycle and duty of the plant large motors.

The plant owner must determine, based on the history of operation of the generating plant, if the source of power is reliable enough to meet the plant requirements. The plant process can tolerate short power outages without detrimental effects, but longer outages would be a concern with respect to the economies of the plant production.

Studies will have to be made to find out if a wind farm or a solar plant could be economical and possibly developed in the vicinity to supplement the imported power.

2.3.2 Ambient Derating Factors

The electrical equipment will be operating at an altitude of 1700 m. The equipment shall be derated for the altitude in accordance with the applicable ANSI C57.40 and IEC 282-1.2 standards. The following derating factors are applicable for the 1700 m site altitude:

Voltage: 0.93
Current: 0.99

The voltage derating will be factored by the suppliers of the equipment, such as switchgear, transformer bushings, to ensure the equipment insulation is designed for lower air density at the specified altitude. The current (ampacity factor) derating for the plant cables is not significant, well within our conservative plant selection estimates.

Applicable ANSI or IEC standards for ampacity derating factors will be used for the power cables buried or installed in multiple duct banks, also discussed in Chapter 12.

The other site ambient conditions, such as road conditions, minimum/maximum temperatures, humidity, rainfall, number of lightning days, are normally included in the tender document for the vendors to design the equipment accordingly.

2.3.3 Reliability Criteria

Before we start putting things together, let us clarify the design reliability criteria, discussed in Chapter 21.

In order to ensure adequate availability and continuous plant operation, the power distribution system must be designed to tolerate and override certain failures of the equipment. Generally, the system will be designed for "a single contingency failure" of the principal distribution equipment. This is sometimes called "a single outage contingency." In other words, the design will fully cover for a single failure of one major piece of equipment, such as transformer, pump, motor, but not for a simultaneous failure of two pieces of similar major equipment.

Therefore, each pump system, which is considered a critical primary part of the operating plant shall include $2 \times 100\%$ units. In some cases, for larger pumps, $3 \times 50\%$ pumps could be used. For sump pumps, roof fans, heating, ventilating air conditioning (HVAC), etc., which are considered the plant auxiliary services, there shall be no immediate substitutes. The switchgear busbars from the plant distribution transformers will be bussed together through bus tie breakers to allow for feeding the plant loads from a single transformer, in case of an outage of one transformer. Failure of some smaller distribution transformer may be tolerated by reconnecting the load to alternate sources of power supply.

Recovery from power outages will be either by having spares cable connected or piped, or by switching capability to feed power from alternate sources such as closing the bus tie circuit breaker.

No contingency consideration will be given to the failures of power cables, lines, or pipes, which can be replaced or fixed relatively quickly, except for the high voltage (HV) single conductor cables used in power plants for 138 kV and higher, where one additional spare phase is added and laid out next to the operating cables. Therefore, during the plant design, one may ask: "What if...?" But not: "What if..., and if...?"

2.4 Connection to Power Utility

The investors will likely not be interested in building this industrial facility unless they have spoken to the utility and were assured that there is available generating capacity to power the project.

Let us talk to the Utility to acquire the information we need to build our plant power distribution system. Here are some of the issues to be clarified by the Utility engineers:

- Power agreement: firm or interruptible
- Tariffs, for power demand and time of use
- Line voltage and its daily and weekly profile
- Frequency off-limits
- Power factor tariffs and penalties
- Source impedance, inclusive of the transmission line impedance (conductors)
- Double- or single-circuit incoming transmission line
- Generating capacity, firm power, how many units are available, and their ratings
- Method of line protection
- Lightning level (number of lightning days/yr) in the area

Let us review each of the aforementioned issues:

Power agreement: The plant owner will sign a power agreement with the Utility. If the plant needs power 24 hours a day, every day, the owner will look at signing a power agreement for uninterruptible firm power supply, if available. An interruptible power supply allows the utility to occasionally cut the power supply in a specific amount or in total. Naturally, this contract comes with a lower tariff. The plant owner will likely insist on an uninterruptible power supply (UPS).

Tariffs: In addition to the nominal charge for kilowatt-hour consumed, the utility will likely have additional tariffs as demand charge, peak load, and reactive power hour consumption (MVARh consumption). The demand is the load averaged over a specified time (15 minutes, 30 minutes, or 1 hour) in kW or KVAR. The peak load may be the maximum instantaneous load or a maximum average load over a designated period of time. The reactive energy charge may be applicable for the load operating at <95% power factor at the point of interconnection (POI).

Voltage operating range: For this plant, 69, 138, and 230 kV voltages can be used. In this case, 230 kV is available and preferred. We have to determine the percentage range of voltage oscillations received from the utility and how stable it is. The next thing is to decide if our plant will need an automatic on-load tap changers (OnLTCs) on the main incoming power transformers, or simple off-load tap changers (OffLTCs). See a typical transformer in Figure 2.4.

Figure 2.4 Large oil filled transformer.

Assume a 20% adder to the cost of a transformer with an automatic tap changer. On the other hand, an automatic LTC can, in addition to keeping the plant voltage constant, better regulate the flow of power and save us some money in penalties charged by the utility for the low plant power factor.

If the voltage swings are large, OffLTC changers may not be able to provide a manageable operating solution. They can regulate the plant voltage manually to a preselected percentage tap, while the OnLTCs manage the plant voltage automatically and linearly to the full tap range of ±10% on the primary winding.

If OffLTCs are employed, the operator will have to shut down the plant in order to change the taps, if desired. Naturally, manual tap changes cannot be performed on a regular daily basis. Voltage at night may be higher than during the day. So once you set the taps, that is it. You may be forced to change the tap settings again if the operating conditions alter.

Suppose, you have decided that your incoming transformers are to be rated 230 to 13.8 kV. Also, you were informed that the incoming voltage from the utility varies from 215 to 245 kV, but most likely toward the lower range (see Chapter 24 for more details on transformers).

Let us examine in Table 2.1 the voltage range at the secondary side of the transformers with OffLTCs for the various taps and voltage swings:

Transformer voltage: 230 to 13.8 kV
Taps on primary side range: ±10% in 2.5% steps.

Based on the aforementioned, a choice would be to operate the transformer with OffLTC at −2.5% taps for the primary voltage range of 215–245 kV. Negative taps on the transformer primary winding are the taps of choice used for boosting the plant 13.8 kV voltage. On the other hand, the OnLTC, if used, will maintain voltage relatively steady in smaller tap movements within the full tap range.

Table 2.1 Secondary voltage for primary grid voltage.

Grid voltage	215 kV	230 kV	245 kV
Taps set at:			
−5.0%	13.54	14.48	~~15.42~~
−2.5%	12.98	14.14	15.05
0	12.89	13.8	14.69
+2.5%	12.57	13.45	14.32
+5.0%	~~12.25~~	13.10	13.95

Most of the industrial plants would purchase transformers with OffLTCs. In this case, based on the discussions with the utility and due to the expected significant variations in the day/night voltage profile, we would prefer transformers with automatic OnLTCs to make sure we have a stable voltage in the plant at all times.

The plant voltage profile is not determined solely by the utility but also by the plant motor load. Plant reactive MVAR load will likely have to be partly drawn from the utility, as explained in Chapter 13.

The smaller plant transformers, which distribute power to lower voltages, will generally have (±5%) OffLTC tap changers. Taps for each transformer will have to be set to obtain the most comfortable voltage profile throughout the plant during the normal plant operation and for large motor starting. This can be determined by a computer **load flow study** and confirmed during the plant operation. With the choice of **OnLTC** on our main transformers, we can consider that our plant distribution voltage will be relatively constant at 13.8 kV at all times, irrespective on what the utility throws at us.

The typical voltage drop criteria to be considered in the design of the plant distribution system is <15% for large motor starting, and <3% for large motor while running.

2.4.1 Source Impedance

This is the system *subtransient* impedance Z'' representing the generating capacity of the utility at the POI. It also includes the impedance of the interconnecting transmission line. The source impedance is derived from the short-circuit level at the plant as advised by the utility. The figure given will likely be based on a present and future generation planned by the utility. This value will be used as the base for determining the interrupting ratings of the plant circuit breakers that connect to the transmission line and the voltage regulation and capability of the plant large motors to start properly.

We have to determine the source impedance for two different extreme cases, the maximum and the minimum values, as follows:

- The maximum source impedance (minimum fault level) when the utility is operating on light load with a minimum generating capacity connected to the grid. This source impedance will be used for voltage regulation calculations and large motor starting duty. If the supply network is weak (low short-circuit level), soft, or variable frequency starting may be required for starting large motors in order to satisfy the utility flicker requirement and to minimize the impact on other nearby customers connected to the grid.
- The minimum source impedance (maximum fault level) is when the utility is operating on high load with maximum generating capacity. This impedance will be used to determine the short-circuit interrupting duty of the plant circuit breakers.

For our system studies and calculations, we will use $MVA_b = 30\,MVA$ figure as our per unit MVA base. This is the *base* rating of our main incoming transformers: 30/40 MVA, 230 to 13.8 kV.

2.4.2 Line Conductor

We received the conductor (name) information from the utility. It will be a single Hawk ACSR (aluminum conductor steel reinforced) conductor. The overhead conductors are symbolically called by the names of birds. The data for the Hawk conductor can be obtained from online sources. The best source of data for the conductors is the old T&D Westinghouse handbook, from which we find the following data:

Hawk
Type: ACSR, 477 kcmil
Stranding: 26/7
Ampacity: 660 A
Resistance: 0.135 Ω/km
Inductive reactance: 0.24 Ω/km
Capacitive reactance: 0.188 Ω/km

The Hawk line carrying capacity is well in excess of our plant requirements. Utilities like to build lines with sufficient capacity for future expansions. The maximum expected current from the plant at 230 kV is 100 A at 40 MVA. The line length from the plant to the local utility source of generation is estimated at 120 km.

➤ Calculate the line parameters in pu for the system studies:

$$R_{line} = 0.135\ \Omega/km \times 120\ km = 16.2\ \Omega$$

Now, calculate line characteristics in pu

$$R_{\text{line pu}} = R_\Omega \times \frac{\text{MVA}_b}{\text{kV}^2} \text{ or } R_\Omega \times \frac{\text{kVA}_b}{1000 \times \text{kV}^2}$$

$$= 16.2 \times \frac{30}{230^2} = 0.009 \text{ pu or } 0.9\% \tag{2.1}$$

$$Xl_{\text{line pu}} = 0.24 \times 120 \times \frac{30}{230^2} = 0.016 \text{ pu}$$

$$= 1.6\% \text{ Line inductive reactance in pu}$$

$$Xc_{\text{line pu}} = 0.188 \times 120 \times \frac{30}{230^2} = 0.0125 \text{ pu}$$

$$= 1.25\% \text{ Line capacitive reactance in pu}$$

➤ Convert line characteristics (pu) from one MVA base to another:

$$\text{From MVA}_{b1} \text{ to MVA}_{b2} \rightarrow R_{\text{pu b2}} = R_{\text{pu b1}} \times \frac{\text{MVA}_{b2}}{\text{MVA}_{b1}} \tag{2.2}$$

$$\text{From MVA}_{b2} \text{ to MVA}_{b1} \rightarrow R_{\text{pu b1}} = R_{\text{pu b2}} \times \frac{\text{MVA}_{b1}}{\text{MVA}_{b2}} \tag{2.3}$$

➤ Convert line characteristics (Ω) from one system voltage kV base to another:

$$\text{From } V_{b1} \text{ to } V_{b2} \rightarrow R_{\Omega\, b2} = R_{\Omega\, b1} \times \frac{V_{b2}^2}{V_{b1}^2} \tag{2.4}$$

$$\text{From } V_{b2} \text{ to } V_{b1} \rightarrow R_{\Omega\, b1} = R_{\Omega\, b2} \times \frac{V_{b1}^2}{V_{b2}^2} \tag{2.5}$$

The utility "informed" us that fault level at our plant bus, projected for the future with possible expansion is **10 kA at 230 kV**. This is calculated approximately as: $\text{MVA}_{sc} = 4000 \text{ MVA}$.

Therefore, we can now calculate the **source impedance** at our 230 kV bus as follows:

$$X_{s\,\text{pu}} = \frac{\text{MVA}_b}{\text{MVA}_{sc}} = \frac{30}{4000} = 0.0075 \text{ pu or } 0.75\%$$

Or pu source impedance if kA fault interrupting current from the source (power utility) is given:

$$X_{s\,\text{pu}} = \frac{\text{MVA}_b}{\sqrt{3} \times \text{kA} \times \text{kV}} = \frac{30}{1.73 \times 10 \times 230} = 0.0075 \text{ pu, or } 0.75\%$$

This source impedance will be used in our computer studies to represent the utility at our plant point of interface. It is a conservative value that will provide plenty of margin in our calculation.

However, for our quick hand calculations and for our interrupting ratings of the switchgear, we will rate our equipment on a conservative basis of an infinite fault level from the utility (zero source impedance). Therefore, the utility can now be called an Infinite bus at the point of interface with our plant.

If we now calculate the fault MVA_{sc} from the Utility on our selected MVA_b base, it is:

$$MVA_{sc} = \frac{MVA_b}{Z_s} = \frac{30}{0} = \text{Infinite MVA}$$

2.4.3   HV Circuit Breaker Fault Interrupting

Now we can determine the incoming 230 kV breaker as follows: It has to have interrupting capacity of at least 4000 A. We select 3 phase, 245 kV, 1200 A, 40 kA fault interrupting, basic impulse level BIL: 1050 kV peak (see Figure 2.5).

Why 40 kA? This may sound too excessive for our requirements, but these are the ratings at the low end for the 230 kV equipment.

The transformer differential protection scheme and metering would need current transformers (CTs) on the grid side of the HV breaker.

2.4.4   Double or Single Incomer Connection

The 30/40 MVA plant can be connected to the incoming utility transmission line with one or two transformers. This will depend on the plant reliability

Figure 2.5 245 kV circuit breaker.

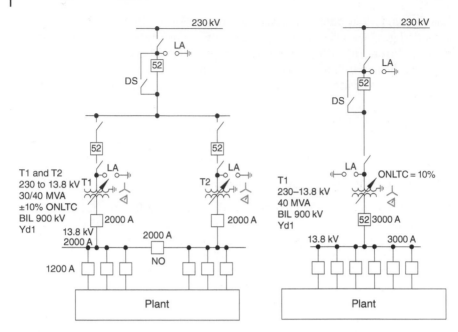

Figure 2.6 Double and single incomer diagram.

requirements. Earlier we determined that the plant must meet the "single contingency criteria." Therefore, we conclude that the plant will be connected to the grid with two transformers (two incomers). Each transformer must be capable of carrying the full load of the plant. Figure 2.6 shows the substations diagrams with two and single incomers for the same plant. A switchyard with a single HV breaker would not meet our reliability criteria of full redundancy for the plant operation. This is mainly because a major transformer failure may cause a total plant shutdown for an extended period of time.

The transformers in our 30/40 MVA plant will be required to share the plant load, but each will also be capable of carrying the full load of the plant on their ONAF cooling in case one transformer fails.

Double power entry will be more reliable, though considerably more costly. A substation with two transformers, in addition to more HV switches will need three incoming HV circuit breakers to feed the two transformers. This substation will also require considerably more space.

For a smaller plant of up to 10 MVA, a single HV incomer may be considered acceptable as the cost of the HV breakers and two transformers might be excessive in comparison with the cost of the plant. In fact, a single incomer switching yard may include nothing more than a single H frame pole structure.

To support a smaller plant, several diesel generators (DGs) can be brought to site in trailers to temporarily replace a faulty transformer.

2.4.5 Utility Generating Capacity

This information we require to be able to determine if our plant will need some supplementary firm generation at site and also if we would be able to expand the plant in the future in case our ore body miraculously doubles up. We will look into a possibility of having a solar plant or a wind farm to augment our power sources. The solar and wind resources cannot be counted in as firm capacity, but simply as a source of power to displace the fuel consumption or import of power (see Chapters 25 and 26).

2.4.6 Firm Capacity

This term is used in generation to determine the overall available MW capacity not considering one unit (single outage failure); therefore, a power plant with two units of 50 MW each has a firm capacity of one unit (50 MW). That is one generator unit out of two units, or two out of three units. Therefore, a firm capacity of a generating plant is the available generation MW capacity not counting one unit which is held as spare. Firm capacity of a transmission line is defined as one out of two circuits. A single circuit line has no firm capacity.

2.4.7 Line Protection

Utilities use specific protective relays and have definite strategies for the line protection: two to three zone distance, negative sequence, etc. The utility will likely ask us to match the protection relaying at our end to that of their end and to coordinate the settings between the two ends. The line protection will likely be by pilot differential relaying with fiber optic communications.

2.4.8 Lightning

We can look at the historic data of the lightning days/year in the area. The plant may be in a desert environment with a low isokeraunic level or in some mountainous region of high lightning intensity. Hopefully, we can obtain this data from the utility and design the switchyard and the overhead lines accordingly with appropriate overhead shielding, grounding, and lightning arrester protection (see Chapter 10 for more details).

2.5 Main Plant Substation

Question? What is the difference between a switchyard and a substation? Is there a precise definition of one and the other? I have never heard one. In my circles, we called a facility with HV switches, circuit breakers, and transformers a substation. A switchyard was the same, but without transformers.

The main substation on this project will contain a number of major items of equipment: transformers, HV switches, HV circuit breakers, arresters, and protection panels as well as medium voltage (MV) switchgear connected to the low voltage side of the transformers. Specifications must be urgently written for the long lead items. In this substation, the transformers should be given a priority. Large transformers are long lead items, requiring 12—18 month delivery plus the procurement time. Transformers rated up to 10 MVA can be obtained within nine months. Often, large transformers may be purchased ahead of time with a provision of being cancelled if the project is not approved to proceed. To purchase the transformers, we need know: plant load, future load, voltages, and method of cooling.

Based on the projected load estimate let us assume the main transformers will be oil immersed/forced air cooled, as follows: two (2) 230 to 13.8 kV, 30/40 MVA, YNd1, ONAN/ONAF/ONAF(prov.), BIL 900 kV, 55 °C rise at 40 °C ambient. We will explain the details later.

For our plant, each transformer must carry 30/40 MVA on ONAN/ONAF/ONAF cooling. Each stage of fan cooling adds about 15% capacity to the base rating. In addition, we can specify the transformer to have a 55/65 °C temperature rise allowance. A transformer rated at 55 °C rise has about 10% spare MVA reserve over a transformer rated 65 °C rise of the same MVA rating. Obviously, a 55 °C transformer is built to a more efficient cooling design.

We choose a single ONAF cooling stage, including a provision (prov.) for adding additional stage of cooling fans if necessary in the future.

Remember, the transformer base rating is 30 MVA. This value will be used in the study calculations.

What are the designations for the transformer cooling?

ONAN: Oil Natural Air Natural → Without fans.
ONAF: Oil Natural Air Forced → With fans.

The transformers will be furnished with a conservator tank and all the standard auxiliaries. We noted the winding configuration as YNd1. Star Primary, HV Neutral solidly (effectively) grounded, Delta secondary, lagging 30° in counter clockwise (CCW) convention. The most popular winding configurations in the industry for high voltage (HV) transformers at 230 kV and above are Yd1 and Yd11.

Transformers up to 10 MVA can be ordered as "sealed tank design," without a conservator.

The main and plant oil immersed transformers will be placed outside, next to the plant buildings in their independent vaults with oil containment basins. The walls of the vault will be fire rated (see Chapter 4 for the NFPA guidelines). The main transformers will be provided with the Deluge water mist protection system (see Chapter 22).

Furthermore, we decided that the transformers would have OnLTC in the range of ±10%. As we are 120 km away from the power plant, the voltage may not be stable. Daily 24-hour load cycle will vary from hour to hour. The automatic tap changers will give us some form of voltage stability for the plant operation. We may also need some additional means of voltage regulation within the plant to enable us to further improve the voltage profile through power factor correction. To meet this requirement, additional capacitors and reactors may be considered (see Chapter 13).

2.6   Load Site Placement

At this conceptual phase of the engineering design, we have to determine the locations of loads within the plant as well as their kW ratings to determine the major power routes of the distribution system. Depending on the kW load magnitudes, we will determine the corresponding voltages for the distribution equipment.

Loads (motor and feeders) up to 200 kW	480 V, Motor voltage: 460 V
Loads (motors with VFDs) up to 500 kW	480 V, Motor voltage: 460 V
Loads above 200 kW	4.16 kV, Motor voltage: 4000 V

We will obtain the load data from our mechanical engineers. Roughly, we expect that the plant-connected load will be about 50 MW. The mining load is typically a motor load with 0.8 pf, but with power factor correction, we will get it over 0.9. This MW figure is assessed generally from the plant flow diagrams, based on the hardness of the ore, raw material processed, and product produced. In their flow diagrams, mechanical engineers may suggest a kW (HP) rating figure for each motor. However, the final kW (HP) ratings will be taken from the actual bids received from the suppliers. The tendency seems to be that suppliers more often overestimate rather than underestimate the load.

From the plant layout drawings, we determine the locations of the groups of loads: mining, crushing, conveying, grinding, process plant, tailings, and camp. The load centers will generally follow the flow of the ore. These locations may be kilometers apart from each other. The biggest groups of loads will be in the grinding and the process plant for conveying, pumping, agitation, and floatation. The camp, crushing, and tailings will be away from the process plant.

The power load centers may include either 480 V loads only, or both: 4.160 and 480 V.

Typical 4.16 kV controllers for large motors are rated at 400 A. Controllers rated 800 A are also available, but rarely used. Let us calculate the current of a

2000 kW, pf = 0.8 motor and verify if a 4.16 kV, 400 A controller can operate it safely:

$$\frac{2000}{1.73 \times 0.8 \times 4.16} = 347 \text{ A} < 400 \text{ A. OK!}$$

The power will be fed from the main substation to the plant load centers by 15 kV overhead lines or medium voltage (MV) cables, depending on the relative location of the main substation. In North America, the 13.8 and 4.16 kV voltages are often called 15 and 5 kV, respectively.

2.6.1 Crushing

This facility may be close to the grinding plant, but at a certain distance, to limit the dust spreading to the rest of the plant. Also a large (seven-day) ore stockpile will be placed between the crushing station and the grinding plant. Crushing usually operates a single 12-hour shift, while the rest of the plant marches on two shifts. A load center will be required to feed the 4 kV and 460 V motors. We will use 1.5 MVA transformers (13.2 kV/480 V or 4.16 kV/480 V) with MV controllers for motors >200 kW and 480 V MCC starters for motors rated ≤200 kW.

2.6.2 Grinding and Conveying

This is the location of the largest load, the 8 MW, (10 MVA) cyclo-converter motor. This load is too big for 4.16 kV and must be powered at 13.8 kV. It will have its own power feeder fed directly from the main 13.8 kV switchgear located at the main substation. For this variable frequency drive (VFD), the main power supply is converted to DC voltage and then back to AC voltage of varied frequency to operate at variable speed to suit the ore quality and hardness. This large piece of equipment is a complete package. It includes a number of smaller auxiliary loads, such as MCC, fans, lube pumps, lighting, and heating, all of them fed from the same power source feeder.

The grinding facility will also need a 4.16 kV load center for MV motors (Ball Mills) and a number of load centers for a large number of low voltage (LV) motors and drives. Drives mentioned here are typically referred to as the VFD operated motors. Two grinding ball mills may even use synchronous MV motors, which by controlling their excitations may help us improve the plant power factor (see Chapter 13).

The MV load center with 5 kV switchgear and MV motor controllers will be fed from two 13.2 to 4.16 kV, 10/15 MVA, Dy11, ONAN/ONAF, oil type transformers, which will be placed outdoors adjacent to the plant and have a joint provision for oil containment.

Oil containment is a concern in the plant due to a fire hazard. Dry transformers for feeding smaller loads are preferred due to their flexibility to be placed closer to the loads and for being less of a fire hazard. However, since dry transformers are commercially built only up to 3 MVA and at voltages up to 13.8 kV, oil immersed transformers will be ordered for this application. To conform to the NFPA guidelines, a fire-rated blast wall will be provided between the two transformers (see Chapter 4 for details).

In addition, this plant will need several load centers for the LV motors and drives. This can be covered by a number of unit substations with 2/3 MVA, 13.2 kV to 480 V transformers, 480 V switchgear and 480 V MCCs. For explanation, why 13.2 kV is used and not 13.8 kV (see Chapter 11).

Process (pumping, flotation): This facility will need several load centers for LV motors and drives. These loads will be fed from a couple of unit substations with 2/3 MVA, 13.2 kV to 480 V transformers. Each unit substation will include a 480 V switchgear and 480 V MCCs for feeding the LV loads.

Camp: The camp is typically located 3–5 km away from the main substation. An overhead distribution line will be erected to the Camp directly from the main substation. This line may include several other 15 kV circuits that may be going in the same direction. The camp load center will include a 500 kVA, 13.2 kV/ 480 V load center with LV distribution boards. Additionally, the load center will be furnished with one or two small emergency standby diesel generating units connected to the main load center via an automatic transfer switch (ATS) to feed the selected camp essential loads. Two small generators are recommended specifically for the camp due to the remote and challenging living environment.

Tailings: Tailing pumping station is 6 km away. The facility will require a 15 kV overhead line to feed a small 13.8 kV/480 V load center.

Water supply pumping: The manufacturing and process equipment will need fresh water supply either from the natural sources or by trucking and storing in the tanks near the process plant.

Mining: This load may be supplied by a portable DG at an open-pit mine location.

Auxiliary system: Add HVAC, lighting, cranes, sumps, eat tracing, etc., to the load estimate. These items generally are not shown on the process flow diagrams.

Plant standby generator: A standby DG 1 MW, will be connected to the 13.8 kV bus at the main substation to supply power to the plant essential load. It will automatically be energized and loaded in case of a total loss of the main supply. This may happen due to problems at the utility or a loss of the 230 kV line. The standby generator will supply power to 30% of lighting, heating, some pumping, and other essential and life critical services. It will not provide power to the process, but it will maintain the charge of the plant DC and control systems (see Section 2.10).

2.7 The Key One-Line Diagram

2.7.1 Load Investigation

An industrial plant of this magnitude will have approximately 400–500 motors and electrical feeders. Obviously, we cannot extract all these motors and feeders from the flow diagrams, but as the design progresses, and the major equipment is procured, the number of motors and drives generally increases and the load grows. Having gone through the load investigation and the load placement within the plant, we have finally arrived at a point of drawing an overall plant diagram, the **key one-line diagram** as shown in Figure 2.7. We will call the feeders drawn above the 13.8 kV drawing line as the incomers and those below the line will be called the plant feeders.

Now we can assign the loads (MW) to each area of the plant and to specific MCCs, separated by different voltage categories, MV and LV. The presentation

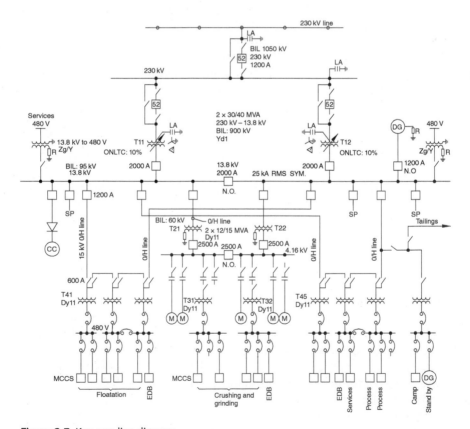

Figure 2.7 Key one-line diagram.

of the diagram is relatively simplistic, made for the book only. A more comprehensive drawing would be required if prepared for the real project.

The main 13.8 kV switchgear for the aforementioned project will require about 11 feeder breakers to feed various 4.16 kV and 480 V load centers, three incomers, and a bus tie breaker for a total of 16 breaker and switchgear cells. You will note that the 480 V load centers are looped together to two 13.8 kV breakers from either side of the 13.8 kV bus to enhance reliability of supply of the load centers.

2.7.2 Connected Load – Operating Load

The overall estimate for the **connected load** (Table 2.2) as part of the conceptual design broken down at voltage levels is as follows:

Of course, the figure given earlier is not the actual operating load. This is yet to be determined. We will arrive at that by using diversity and other load factors.

Having participated in the design of many projects, we have yet to know the actual operating load in relation to the connected load. It surely depends on the type of plant. We would expect that a batch plant would have a higher percentage compared to a mining plant. During the design, it is evident that the connected load far exceeds the expected operating load and far exceeds the power distribution transformer capability. The connected load must be "treated" with several diversity and load factors.

During the commissioning, we do observe the operating currents of the individual drives and note that the actual load factors are considerably lower than projected. This is partly due to the fact that mechanical engineers and suppliers continually oversize the plant motors due to their fears and rounding up to the higher frame sizes to insure flows are adequate and in line with expectations. We have yet to hear an owner complaining that we have overloaded the system, or that we have oversized the system. There always seems to be "a lot of fat" in the system.

Table 2.2 Connected load.

Voltage	MW
13.8 kV	9.5
4.16 kV	15
480 V	20
Cable and equipment losses	0.5
Growth (10%)	5
Connected load	**50**

Table 2.3 Load list.

ID	Name	kW	Volt	Serv.	LF	UF	(DF)	Load
20AG01	Tank#1 Agitator	11	480	1	0.7	1	0.7	7.7
15LT03	Lighting transformer	15	480	1	0.5	1	0.5	7.5
20PU21	Tank #1 Supply pump	30	480	1	0.8	1	0.8	24
20PU22	Tank #1 Supply pump	30	480	0	0.8	1	0.5	0
15FA11	Crusher building fan	30	480	1	0.6	1	0.6	18
15CR01	Crusher building crane	50	480	1	0.4	0.1	0.05	2.5
21BM01	Ball mill	2000	4160	1	0.7	0.6	0.7	840
21BM02	Ball mill	2000	4160	1	0.7	0.6	0.7	840
15WR01	Welding receptacle	60	480	1	0.5	0.1	0.05	3.0
	Connected	**4226**				**Operating**		**1742.7**

Abbreviations: ID, identification number; kW, connected load; Serv., unit in service (1) or on standby (0); LF, load factor (<1); UF, utilization factor (<1); DF, calculated diversity factor (Serv *LF*UF).

We were all told to size the plant motors to make them operate at not less than 80% of their nominal ratings. This is to save on the cost of motors as well as to operate them at their highest efficiency and thus reducing the cost of energy. My personal feeling is that when all is well and done, the operating load will be around 50–60% of the connected load.

Occasionally, it does happen that a motor is undersized and must be replaced. But, more often, the motors are oversized and operate with a low load factor. That scenario does not seem to concern the owners. Falling short certainly is a bad news.

Here, we go. Let us determine the projected operating load. First, we prepare the complete "Load List." A partial example of the list is given in Table 2.3 for several plant loads.

Based on the small sample, the operating load is 41.2% of the connected total load. This is just an engineering guess on the paper, for now.

One can create his own tabulation to suit the particular plant and define the load factors for its own reasonable comfort suited to a type of the plant. In this case, we used the factors for the peaking duty: peaking operation for one hour duration. The load in the table is calculated by multiplying the connected load (kW) with the diversity factor (DF).

It was mentioned earlier that the process plant will be operating 24 hours, while the crushers will be working one 12-hour shift only. Most of the maintenance and admin will be shut down too. Since we are concerned with the maximum plant load over one hour, the crushing plant is included. A welding receptacle or a crane or similar loads, for instance, have low utilization factors as maintenance work is not a continuous work.

Based on the load estimate for the 4.16 kV load, the plant transformers will be 13.2 to 4.16 kV, 12/15 MVA, Dyn1, ONAN/ONAF, BIL 110 kV, 55 °C rise at 30 or 40 °C ambient, depending on the environment. If the plant is in the Northern region, the ambient temperature can be <40 °C.

2.7.3   Voltage Level Selection

In our earlier discussion, we have proposed the voltage levels for the plant. Here, we will confirm the plant voltage levels for the primary and secondary plant distribution and the plant loads. The primary switchgear (Figure 2.8) is located at the main substation several kilometers away from the process plant with convenient routes to feed all the plant buildings with overhead distribution lines.

For the primary distribution, we can apply either 13.8 or 20, or even 33 kV voltages. The switching equipment and cables at 20 and 33 kV are considerably more expensive compared to 13.8 kV equipment. However, higher voltages may be needed if the distances are over 10 km and if there is a need for transfer of larger blocks of power. Since the project requires a large number of short feeders, it does not seem to be cost-effective to distribute minor load at higher voltages.

The distances to the plant load centers are not large (<6 km) and 13.8 kV can be employed without significant voltage drops (<5%), which is to be verified by system studies.

Therefore, 13.8 kV voltage seems to be the most appropriate for the primary distribution throughout the site for this 40 MVA plant.

Figure 2.8 Plant MV switchgear.

We have already noted that a single transformer must be able to carry the full plant load of 40 MVA for an emergency condition of having one transformer out of service. The maximum incoming current without overloading from a 30/40 MVA transformer is 1675 A at 13.8 kV and 700 A at 33 kV. The standard 13.8 kV switchgear breakers go up to 4000 A. Forced air or water cooled breakers up to 5000 A are also available, but considered less reliable. Since we do not wish to have water complications in the switchgear, we can limit ourselves to lower breaker ampacities.

A good engineering practice dictates that breakers can be loaded to up to 80% of their nominal ratings. For instance 960 A for a 1200 A frame breaker. However, in case of an emergency like having a failure of one incoming transformer, the incoming breaker can be loaded closer to its nameplate rating.

Since the maximum incoming current at 13.8 kV is expected to be 1675 A (40 MVA), we can choose 2000 A incoming breakers. The breaker maximum loading is calculated to be 84% of the expected maximum rated current. The switchgear bus will be of the same rating, as the incoming breakers.

The most common types of breakers at 4.16 and 13.8 kV are vacuum type breakers.

2.7.4 Switchgear Breaker Ratings

The breakers must be selected based on their continuous capability and short-circuit interrupting duty. The interrupting rating of the switchgear breakers will be based on the short-circuit fault contributions from the source and the plant motor load to a fault on its bus. For the calculations of fault contributions on the 13.8 kV side of the main transformers, we will use conservative values and ignore the system source impedance, i.e. we will assume it to be zero, as noted earlier.

Let us assume the impedance of the main transformers is 9%.

For the selection of the switchgear breakers it is required to determine the following:

- *Voltage*: It must be the most economical voltage, at which we can transmit power throughout the plant.
- *BIL level*: It goes with the selected voltage and the method of system grounding as a standard. For 13.8 kV switchgear it is 95 kV peak for indoors and 110 kV peak for outdoors.
- *Breaker continuous rating*: It is usually the minimum frame size that can carry the load with 20% spare.
- *Breaker interrupting kA capacity*: It is based on the calculation of *the bus fault for the worst-case scenario*, with at least 20% spare.
- *Bus continuous and interrupting rating*: It is determined in a similar manner as for the breakers.

Let us proceed with the basic calculations:

➤ We will calculate short-circuit fault levels on the MVA basis and then convert it to kA (see Chapter 3 for the breaker ratings on kAic [kA interrupting capacity] basis).

The worst-case short-circuit scenario for the 13.8 kV switchgear bus is one with the plant operating with a single main transformer and the 13.8 kV bus tie breaker is closed. The interrupting rating for the 13.8 kV switchgear is available at 25, 40, 50, and 63 kAic r.m.s. symmetrical.

➤ Contribution from the grid is based on infinite bus criteria (source impedance = 0). Therefore, the contribution through the transformer impedance is $MVA_{sc} = 30/0.09 = 333$ MVA.
➤ Assume 60% of the plant-operating load is motor load (0.60×40 MVA = 24 MVA).

The motor impedance $X'' = 17\%$ (0.17 pu) on the motor kW base, as per ANSI standards.

Motor fault contribution can be then calculated as follows: $24/0.17 = 141$ MVA. The cycloconverter is not taken into account as a short-circuit fault contributor. It is considered a DC load. VFD-operated loads are not included either. If this work were being done on a computer as a system study, the 4 kV motors would be entered individually, while the motor load on the 480 V buses would be entered as grouped loads, one for each 480 V bus.

➤ Therefore, the total fault at the 13.8 kV busbar is $333 + 141 = 474$ MVA.
➤ Add a 25% margin to this figure: 1.25×474 MVA $= 592$ MVA$_{sc}$, or 24.8 kAsc.
➤ Select switchgear and breaker rating: 40 kAic r.m.s. symmetrical.

Based on the aforementioned, we conservatively determine, without taking into account the cables and line impedances, that 13.8 kV switchgear (breakers and bus) should have interrupting rating of 40 kAic.

Furthermore, we determine the 13.8 kV feeder breakers, which are delivering power to the plant can be sized either 800 or 1200 A continuous rating as applicable for the various plant load centers. In Europe and Asia, it is common to use a mixture of breaker size on the same bus. In North America, the breakers tend to be of the same frame size for better exchangeability and fewer spare parts.

We now wish to determine the interrupting rating of the 4.16 kV plant switchgear. The impedance of the plant transformers is assumed to be 7% on their 12 MVA basis. We will also include the impedance of the main (grid) transformers of 9% on a 30 MVA base. All the transformer impedances must be converted on a common base, in this case 30 MVA, as follows:

$$X_{pu} = 0.07 \times \frac{30}{12} = 0.175 \text{ pu } (17.5\%), \quad \dots \text{based on Eq. (2.3).}$$

We calculate the interrupting rating of the 4.16 kV circuit breakers on 30 MVA basis:

Fault contribution from the source: $\text{MVA}_{\text{sc}} = \frac{30\,\text{MVA}}{0.09 + 0.175} = 113\,\text{MVA}.$

Fault contribution from 5 MVA motor load: $\frac{5}{0.17} = 29\,\text{MVA}_{\text{sc}},$

It totals to about 142 MVA, or 19.7 kA$_{\text{sc}}$ on the 4.16 kV bus. By adding a margin factor of 1.20–1.25 to the breakers and the 4.16 kV switchgear, we conservatively select 40 kAic r.m.s. symmetrical.

The BIL insulation level rating for this type 4.16 kV indoor switchgear is 60 kVpeak.

The preceding calculations were based on the fact that the main transformers were not operating parallelly. The transformer pairs are of equal design and construction and with approximately equal impedances. If we allow the parallel operation, the voltage profile would improve throughout the plant. Large plant motors would likely start without any difficulty. So why do not we operate the plant with the transformers in parallel?

If we allow a parallel operation, the fault contribution from the source to the 13.8 kV switchgear would come from both the main transformers, as follows:

$$\text{MVA}_{\text{sc}} = 2 \times \frac{30\,\text{MVA}}{0.09} = 666\,\text{MVA}$$

Therefore, with the motor contribution added, the fault level would considerably increase. The interrupting rating required for our 13.8 kV switchgear for this application would be a step higher, which may be cost excessive for 13.8 kV. If you observe the one-line diagram, motor contribution from the 480 V buses to faults on their own buses would be increased. However, the contribution to the other buses would be minor due to the cable, O/H line and transformer impedances between the 480 V bus and the fault located at any other bus.

The continuous ratings for the bus tie and the incomer breakers will be determined further in Section 2.7.6.

2.7.5 Single Incomer Substation for a Small Plant

Let us review a small plant having a 5/7 MVA, ONAN/ONAF, 138 to 4.16 kV transformer. A single incoming transformer would suffice. This plant may have some 4.16 kV and some 480 V loads. The 138 kV primary side may include a 600 A fused load interrupter with arcing horns to allow for switching the magnetizing current of the unloaded transformer. Except for switching the magnetizing current, while the LV side is kept open, the switch is not allowed to be operated. The fuse serves for short-circuit protection only. The maximum incomer current on the transformer 4.16 kV secondary side is 970 A < 80% of 1200 A frame breaker for continuous rating.

A 1200 A rated breaker would be appropriate for this application. The incomer is fed to a single 4.16 kV switchgear 1200 A bus.

2.7.6 13.8 or 33 kV Switchgear for a Larger Plant

There is a correlation between the rating of the switchgear incoming breakers and the incoming transformers ratings.

For an even larger plant primary distribution system, the use of 33 kV would be likely, with the transformer incoming and bus tie breakers rated at 2000 A. These 33 kV breakers can accommodate 60/80 MVA, ONAN/ONAF incoming transformers at 80% of their breaker nameplate ratings. The switchgear bus can also be selected as 2000 A, while the plant feeders to the individual plant load centers will be any of those available: 1200, 1600, or 2000 A breakers. In this instance, we can standardize on a continuous rating of 1200 A, which can be loaded up to 960 A (55 MVA at 33 kV), i.e. 80% of the breaker nominal rating and 100% of the breaker nameplate rating for a temporary emergency usage.

Here is the switchgear for a large plant either at 13.8 or 33 kV, metal-clad or 33 kV gas insulated switchgear (GIS) with vertical sections having single breaker positions. In summary, the two switchgear assemblies are rated as follows:

• Nominal voltage	13.8 kV	33 kV
• BIL, indoors/outdoors	95 kV/110 kV	170 kV
• Phasing	3 ph, 3 w	3 ph, 3 w
• Interrupting capacity	40 kA r.m.s. symmetrical	31 kA r.m.s. symmetrical
• Construction	Metal-clad	Metal-clad or GIS
• Incoming and tie breakers	2000 A	2000 A
• Normal operation, incomers	37.6 MVA at 80%	72 MVA at 80% rating
• Maximum loading, incomers	47 MVA at 100%	92 MVA at 100% rating
• Feeder breakers	1200 A	800, 1200 A

The breaker 2000 A rating for the incoming and tie breakers at the 13.8 kV switchgear is more than adequate for both normal and emergency duty. The branch feeders on the switchgear are usually positioned in such a way to minimize the power flow across the bus tie breaker.

Evidently, 13.8 kV voltage can be used for the plants up to 50 MVA. Certainly, 20 or 33 kV voltage can be equally applied and used for this 40 MVA plant as well as in even larger industrial facilities as the primary plant distribution voltage. The economies of these options would have to be worked out during the design phase of the project, when the largest plant loads are confirmed.

2.7.7   Transformer Connections: Cable, Cable Bus, or Bus Duct?

Let us now return to our 30/40 MVA plant. We confirm that the primary distribution will be at 13.8 kV, secondary at 4.16 kV, and LV at 480 V. Therefore, the primary 13.8 kV circuits will be brought into each process building: crushing, milling, floatation, and processing. Additional feeders will be needed for remote tailings and camp.

230 kV incoming to main 30/40 MVA transformers: The transformers will be located outdoors as part of the main switchyard. The HV connections to the transformer bushings will be by bare conductor drops from the switchyard overhead buses.

13.8 kV connections from the main transformers to 13.8 kV switchgear: The plant must be capable of working with a single transformer outage on its ONAF rating at 40 MVA → 1675 A. The transformers will be placed adjacent to the switchgear building with the LV side facing the 13.8 kV switchgear. This will be a short straight run by a three-phase 2000 A cable bus into the transformer top bushings. Use 133% rated cables for ungrounded systems. A cable bus installation is far simpler, more flexible, and less costly than a rigid three-phase bus duct.

HV connection to 12/15 MVA, 13.8 to 4.16 kV transformers: The HV connection will be rated at 650 A for 15 MVA ONAF rating. This can be accomplished by a drop from a 15 kV overhead line pole as a buried cable connection of 2–500 kcmil (2–250 mm^2) per phase, 133% cable insulation for ungrounded systems, into the transformer HV cable boxes. Lightning arresters are required at transition and the overhead line to cables.

LV connection to 12/15 MVA, 13.8 to 4.16 kV transformers: A cable bus from the transformers to the switchgear: 2100 A, 3 ph for 15 MVA, 133% cable insulation for ungrounded systems, with top entry from the transformer bushings into 4.16 kV switchgear, also as top entry.

HV connection to 2/3 MVA, load center transformers at 13.8 kV: 150 A cable drop from overhead line, 133% cable insulation for ungrounded systems to the HV load interrupter on the dry type transformers. Lightning arresters are required at the cable/line transitions.

At 4.16 kV: 400 A cable from 4.16 kV switchgear, 133% cable insulation for ungrounded systems. Load interrupter is not required if the load center is in the same room as the 4.16 kV switchgear. Lightning arresters may not be required. Those arrester used on the transformer HV side are considered also effective for the LV side. However, arresters are inexpensive at this voltage level. It will not hurt to install them in here.

LV connection at 480 V from 2/3 MVA transformers to LV Switchgear: This connection is assumed to be butted directly from the transformers to the LV (480 V) switchgear buses.

Overcurrent Settings for the Incomers: An electrical inspector in USA will likely verify and insure that the interrupting device (breaker) always has a lower setting than the cable it is protecting. In the case of the 13.8 kV incomers, even though the cable bus may have a rating of 2000 A and it connects to a 2000 A frame breaker, the breaker would likely have the protective relay overcurrent trip set lower than the cable bus rating, for instance 1700 A.

Or to be even more selective, the relay settings of the incomers could be arranged to have two groups of settings: 900 and 1700 A. The Group 1 setting of 900 A is for the two incomer breaker operation and as soon as one transformer fails and the full 13.8 kV load is transferred to one breaker only, the relay setting change would be initiated to Group 2 (1700 A) for the 40 MVA transformer rating by the breaker status interlock of the failed breaker. Similarly, the same can be arranged for the 4.16 kV incomers.

2.7.8   Medium Voltage Switchgear and Controllers (4.16 kV)

The 4.16 kV connected load is estimated to be around 15 MW at 0.8 pf (18.7 MVA). The 4.16 kV loads are critical loads for the plant production, requiring adequate availability of power supply. To establish the highest level of availability of the plant, the MV load board will be supplied from two 13.8 kV feeders from the main substation. The MV (5 kV) control assembly will be used to feed the plant MV motors rated over 200–2000 kW and 2/3 MVA, 4.16 kV to 480 V unit substations (Figure 2.9). The 4.16 kV loads will be controlled by the MV controllers, employing MV-fused 400 A contactors.

The MV motor control assembly will be located close to the plant. It will have two incoming 4.16 kV switchgear feeder cells and a tie breaker. Each incomer breaker will be rated 2500 A to be able to carry the full MV load from a single incoming transformer of 12/15 MVA transformers.

The 4.16 kV motor control board is a metal enclosed assembly, NEMA class E2 (fused) with integral 120 V_{ac} control transformers. The four motor controller cells on each side will be attached to the regular 4.16 kV metal-clad switchgear in the middle. The incoming breakers will automatically interact with the tie breaker on a 2 out of 3 switching principle to carry the full load in case of a transformer outage.

• Nominal voltage	4.16 kV, 60 Hz, 3 ph, 3 w
• BIL	60 kV
• Switchgear, interrupting capacity	40 kA r.m.s. symmetrical
• MV controllers, interrupting	40 kA r.m.s. symmetrical
• Assembly	Metal-clad/metal enclosed
• Incoming and tie breakers	2500 A, vacuum type
• Motor controllers (contactors)	400 A, fused, for up to 2000 kW maximum, NEMA E2

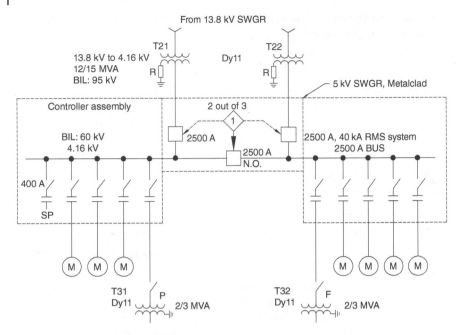

Figure 2.9 4.16 kV motor controller assembly.

The incomers and tie breakers are assembled in their metal-clad, single cell enclosures. Motor and feeder controllers can be stacked up as two units per vertical section. The controllers for feeding the unit substation feeders are of latched type to ensure they are immediately energized and restored following a restoration from a power failure. The unit substation radial overhead feeders are protected by GE Multilin feeder protective relays, or equal, capable of automatic one shot reclosing. Each motor controller includes a set of medium voltage fuses, a vacuum contactor, and GE Multilin motor protector relay or equal. Fuses for the motors are of R-rated type.

Fuses for the transformer and outgoing feeders are of current limiting L-rated type (see Chapter 3).

All the breakers and controllers are provided with means of operating from local and remote positions. The means of communication for the breaker operation, interlocks, and status are by Ethernet from the control room (see Chapter 17).

2.7.9 Low Voltage Service Voltage

A simplified one-line diagram of one of the LV switchgear is shown in Figure 2.10. It is mostly used to distribute power to plant MCCs. The switchgear breakers are rated from 800 to 3200 A frame sizes. The breaker

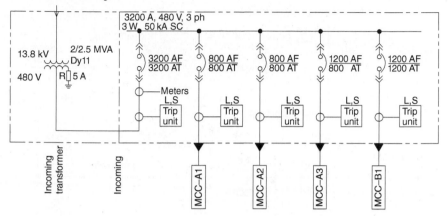

Figure 2.10 Part of LV switchgear.

trip unit can be lower than the frame sizes to suit the load. This is shown as 800AF/600AT, as frame and trip rating for the breaker. L and S are the characteristics for the long and short breaker sensor trip element, respectively.

The common LV switchgear voltages are as follows:

In USA: 480 V, 3 ph, 60 Hz. The lowest voltage used is 120 V, 1 ph.

Canada: 600 V, 3 ph, 60 Hz. The lowest voltage is 120 V, 1 ph.

Europe, Asia, South America, Australia: 400 V, 3 ph, 50 Hz. The lowest plant voltage is 231 V, 1 ph, which is the line to ground voltage from 400 V, 3 ph.

These voltages are used to feed low voltage, three-phase motors up to 200 kW, and auxiliary feeders.

Smaller motors are fed at 1 ph, 120 V in USA/Canada, or 231 V in the IEC countries.

Larger motors up to 500 kW can also be fed at LV (400–600 V), if controlled by VFDs or SoftStarts (see Chapter 15).

2.7.10 Bus Tie Breaker Switching

2.7.10.1 Incoming Transformer Failure

According to the overall key one-line diagram, a part of which we repeat here in Figure 2.11, the incoming transformer T11 feeds Bus A, while the T12 feeds the Bus B of the main 13.8 kV switchgear. The switchgear bus tie breaker is held open during the normal plant operation and its operating control switch on its front panel is held on Loc/*Rem* position.

The incoming breaker transfer switching can be arranged by the plant control system (automatic) or hard wire logic (manual) on the switchgear. The bus tie breaker is not allowed to be closed while both incomers are closed. Once either

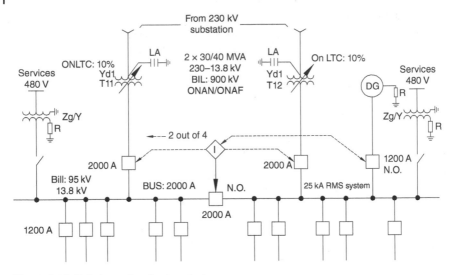

Figure 2.11 Switchgear breaker interlocks.

incoming circuit breaker opens, the tie breaker closes immediately behind, if its control switch is also placed on Loc/**Rem**. Generally, all the main breakers should have their switches on placed on Loc/*Rem* position.

In case of a failure of one of the transformers, the faulty transformer (Assume: T12 on B bus) will be isolated by being tripped automatically or manually (locally). Once the incoming breaker for the transformer T12 opens, a signal is sent to fully isolate T12 by tripping its HV breaker. This will be followed by closing the 13.8 kV bus tie breaker.

Please note, all the B buses from 13.8 kV down to 480 V are temporarily shut down following the isolation of the faulty T12 transformer. When the tie breaker closes, the healthy transformer (T11) now feeds both Buses A and B. The control system now starts restoring the power (incoming and tie breakers) sequentially from the top bus to the lowest bus in that order to operate the plant on the single transformer. The whole process of power restoration is completed within several seconds.

This automatic switching process explained above is more applicable to the power plants, which may have three or four buses operated from 13.8 kV down to 480 V. The industrial plants have fewer plant bus levels, and the switching following a transformer failure may be manual.

The actual switching is arranged on the "Break before Make" principle. This is also called a "Dead Transfer." Therefore, the two incomers + tie breakers operate on 2 out of 3 closed principle. However, since we are also planning to connect the standby DG to the Bus B in case of a total outage, the operating logic must accommodate an operating condition of 2 out of 4 breakers closed. Should there be a total power outage, both switchgear incomer breakers are opened, a signal is sent to the standby DG to start and close onto the 13.8 kV bus. That done, the tie breaker closes next (2 out of 4 conditions) and the essential power restoration commences across the whole 13.8 kV bus and the plant.

The breaker operating and closing coils as well as the protection and trip circuits operate on $125\,V_{dc}$ circuits fed from the station battery. By having the DC system available, the main breakers can be operated during a total plant outage.

The plant restoration following a total blackout and DG operation follows by the appearance of voltage on the HV side of the main transformers. HV breakers are closed and a proper safe moment is awaited to initiate the restoration of the overall power distribution system.

Switching and restoring the service when the faulty transformer is brought back follows a similar procedure in reverse order. The incomer B is switched ON, on its HV side. This initiates the tie breaker T to open first, followed by closing the incomer breaker B. During this restoration to normal operation, Bus B again passes through a "dead" transition, and its feeder breakers and contactors are turned OFF during a brief transition. Latched feeder breakers are restored immediately, while the other loads must be restored by the plant control system by reviewing the permissive logic of each circuit. If the switchgear breaker cell control switches are left on **Loc**/Rem, they must be turned to Remote to allow the automatic reintegration of the lost services to proceed.

2.7.10.2 Switching from Control Room

Automatic switching from the control room is possible only if the breaker Loc/Rem switches are held on "Remote." All the breakers and controllers are provided with means of operating from a local and remote position. The breakers that are left on Loc position (most likely intentionally) will not be restored. The means of communication from the plant control system to the switchgear is by Ethernet (see Chapter 17).

Breaker Loc/Rem switch operation:

- Loc (Local) means manual.
- Rem (Remote) means automatic from control room.

2.7.11 Plant Transformation

Every effort shall be made to minimize the number of transformations from the grid to the loads. Unnecessary transformers add to the voltage drops as well as to the cost of the plant in the transformers and additional associated switching and protection equipment.

For this plant, we can use two or maximum three levels of transformation, as follows:

Grid → 230 kV–T1–13.8 kV → 13.2 kV–T2–4.16 kV → 4.0 kV Motors, or

Grid → 230 kV–T1–13.8 kV → 13.2 kV–T2–0.48 kV → 0.46 kV   Motors   and feeders

Grid → 230 kV–T1–13.8 kV → 13.2 kV–T2–4.16 kV → 4.16 kV–T3–0.48 kV → 0.46 kV Loads

Grid → 230 kV–T1–13.8 kV → 13.2 kV–T2–0.48 kV → 0.48 kV–T3–0.48 kV → 0.48 kV Ltg. Panels[1]

2.7.12 Voltage Regulation

2.7.12.1 Voltage Regulation ΔV

The full range of load regulation of a power transformer is the change in secondary voltage Vs, expressed in percentage of rated Vs for a specified power factor. This occurs when the rated MVA output at a specified power factor is reduced to zero, with Vp maintained constant. One can use these approximate formulae to calculate ΔV at any operating load and power factor:

$$\Delta V_{pu} \text{ at rated or any load: } \Delta V_{pu} = RP_{load} + XQ_{load} \text{ all in pu} \qquad (2.6)$$

For instance: Transformer: 30 MVA as a $MVA_b = 1$ pu, having $Z = 9\%$ (0.09 pu) impedance
on 30 MVA base, efficiency 99.5% at rated load.

➤ Total loss $= (1 - 0.995) \times 30$ MVA $= 150$ kW. The total loss includes no load and load losses.

With this information, we can calculate transformer resistance R, followed by calculating the reactance X. After that, we calculate the voltage regulation.

Transformer resistance R is calculated from the load losses. Since we do not have that figure we will assume that the *load loss* is equal to 80% of *total loss*. The rest is the *no load* loss.

Assume operating load of 25 MVA (MVA load pu $= 0.833$ pu) at power factor pf $= 0.85$.

➤ R or Load loss in $\% = 0.8 \times \frac{150}{30,000} \, 100 = 0.4\%$. Z is given as 9%
➤ X is calculated from: $\sqrt{Z^2 - R^2} = \sqrt{9^2 - 0.4^2} = 8.99\%$
➤ $P_{load} = MVA_{pu} \times \cos\varphi = 0.833 \times 0.85 = 0.71$ pu
➤ $Q_{load} = MVA_{pu} \times \sin\varphi = 0.833 \times 0.52 = 0.43$ pu
➤ $\Delta V_{pu} = (0.004 \times 0.71 + 0.0899 \times 0.43) = 0.0414$ pu or 4.1 %. From Eq. (2.6).

2.7.12.2 Motor Start Voltage Drop

The plant will include a number of large motors, for which **cable sizing** must be verified with respect to the voltage drop during its start. During the motor

---

1 For a specific reason that is explained in Chapter 7, we will add additional isolation transformers for the lighting systems.

start, the voltage at the motor terminals must be >85% of the motor nominal voltage. Here is a quick check without going through a computer study.

For selecting the power cable for the motor we use National Electrical Code (NEC) for ampacities of Cu and Al cables [1]. Most of the engineers would have this booklet (Code) on their desks and use it for the various engineering activities ranging from the switchgear and cable selection to fire protection regulations. A similar Code is also available in Canada.

Let us calculate voltage drop for a motor rated as follows:

Motor: 100 kW, 0.85 pf, 480 V
Motor sub-transient Impedance: $Z_m = 17\%$,
Power cable length: 100 m.

➢ Select the cable. Calculate motor nominal current: $I_n = \dfrac{100}{\sqrt{3}\,V\,pf} = 125\,A$
➢ Select the cable from NEC Ampacity table.

We look for a cable size for: $1.25 \times 125\,A = 155\,A$ (~25% margin was added).

➢ Cable selected from the code: 3c # 1/0, Cu, 90 °C, capable of carrying 170 A.

We calculate the voltage drop on the motor kW base (kWb): 100 kW = 1 pu. Motor impedance on the motor base if not known can be assumed as: $Z_m = 0.17$ pu as per ANSI.

Cable impedance Z_Ω for cable 3c #1/0 AWG = 0.035 Ω/100 m. Value was taken from relevant tables.

➢ Calculate impedance $Z_{c\,pu}$ on per unit (pu) value for 100 m cable:
➢ Cable impedance in pu: $Z_{c\,pu} = Z_\Omega \times \dfrac{kWb}{1000 \times kV^2} = 0.035 \times \dfrac{100}{1000 \times 0.480^2} = 0.015$ pu/100 m.
➢ Calculate voltage drop ratio corresponding to the motor/cable impedance ratio:

$$\frac{Z_{c\,pu}}{Z_m + Z_{c\,pu}} = \frac{0.17}{0.17 + 0.015} = 0.919\,pu = 91.8\%$$

➢ Therefore, the voltage % on motor terminals during the motor start is equals to 91.8% OK!
For motor cable 100 m long: $\Delta V = 8.2\% < 15\%$ allowed.

This calculation assumed the 480 V bus is operating at 100% voltage at the time the motor is initiated to start.

2.7.12.3 Conclusion
This motor operating with the proposed cable 3c #1/0 AWG is acceptable for the actual cable length of 100 m. If the distance is doubled to 200 m, a new heavier cable would have to be selected as the $\Delta V = 16.4\% > 15\%$ is over the operating limit.

2.7.13   Overhead Distribution Lines

The major part of the power distribution between the main substation and the remote areas and the main process plant will be by overhead distribution lines. The lines will follow the most direct routes or along the roads to facilitate ease of construction and maintenance. The 13.8 kV feeders will be laid out either as radial feeders or ring main loop feeders to connect several unit substations and other major loads in a loop fed from two different sources.

The MV distribution lines will use wood poles and wood cross-arms (see Figure 2.12). In tropics, use concrete poles and steel cross-arms (see Chapter 9). The distribution lines will use Hawk ACSR 466 kcmil (660 A), or Partridge ACSR 266 kcmil bare conductor having more than 470 A capacity in free air and high pulling strength. High capacity conductors are used not only to maintain low voltage drops but also for their tensile strength for pulling and to be able to implement longer spans between poles.

The lines will be protected from lightning by an overhead shield wire on the top. The shield wire, popularly called overhead power ground wire (OPGW), will also contain fiber optic single mode, multicore cable inside it. All the metallic pole hardware will be grounded to a pole ground wire, which will then be led to the pole grounding rods. Depending on the soil conditions, the grounding rods will be installed at every pole in the dry areas and every several poles in the moist, low soil resistive areas.

Figure 2.12 Distribution line.

2.8   Transformer System Grounding

Earlier in this section, we chose the main (grid) transformers to have YNd1 winding configurations. We did not explain the reasons for our specific choice. Let us clarify it. The method of system grounding is a critical issue of every electrical installation and it must be defined for each voltage level. The transformer winding configuration plays an essential role in this matter. The system grounding (equipment neutrals) has evolved over the years by using different methods of high resistance to no resistance grounding. In the last 20 years, the industry has basically standardized on the following methods:

• Transmission level, 69 kV and higher	Solid (zero resistance) → No fault limitation
• MV 3.3–33 kV	Resistance → Limitation to 100 A
• LV 400–600 V	Solid (zero resistance) → No Limitation
• Generators, 5 kV and higher	High resistance → Limitation to 5–10 A

The zero resistance (solid) grounding is now officially called "effective grounding."

In some parts of the world, the LV systems for manufacturing plants are left ungrounded or resistance grounded to allow the plant operation to continue, but to clear the fault at some convenient time. A phase to phase fault occurs if the plant receives another fault on a different phase. In that case, the affected part of the plant is tripped.

Brief reasoning for the above methods of grounding and standardization are given below. For further discussion on the above standardization (see Chapter 5).

2.8.1   Transmission Level

The HV and EHV equipment at this voltage level is expensive. The transformer γ/Δ winding configuration with solid (effective) grounding with no resistance on the primary winding allows the manufacturers to apply lower BIL levels for the HV windings and to lower the cost of transformers and other equipment. The line tower footing ground resistance must also be kept low to reduce the BIL level of the line, thus cutting the cost of towers while offering the use of lower clearances and shorter insulator strings.

2.8.2 MV Systems

Having established the grid transformation to be γ/Δ for the primary distribution, the 13.8 kV delta system becomes ungrounded. A stable and detectable grounding system is required for this project, to insure the neutral point will not be floating in case of a ground fault. This can be accomplished by grounding/station service transformers. These will be 300 kVA, 13.8 kV to 480 V, Zg/γ with a 100 A grounding resistor on the primary. The secondary neutral is grounded directly to ground without a resistor to render a four-wire system, suitable for the LV 277/480 V services.

The grounding/station service transformers are connected to the 13.8 kV buses with fused links.

The MV 13.8 to 4.16 kV transformers are Δ/γ. The LV neutral is resistance grounded to limit the fault current to about 100 A. The selected fault current is low enough to limit the damage caused by ground faults to the MV equipment and high enough to be used for selective relay protection and most importantly to limit the insulation overstressing due to spiking arcing ground faults caused by the system capacitances.

2.8.3 LV Systems

The LV systems will use 13.8 kV to 480 V, Δ/γ transformers. The LV (star) side will be solidly grounded with no intentional resistance. This will cause the ground fault currents to be as high as phase fault currents and can be operated by the phase current protection. In the mining industry, LV neutrals are resistance grounded as shown in Figure 2.10.

2.8.4 Generator Neutrals

This is addressed in more details in Chapter 5.

The standby DG may be grounded as the MV system to limit the fault current to 100 A.

2.9 Transformer Winding Configurations and Phasing

Based on the foregoing discussion, the plant distribution transformer selection will be as follows:

- *Grid transformers (2)*: 230 to 13.8 kV, YNd1, 30/40 MVA, ONAN/ONAF, oil Immersed, solidly grounded.
- *Process plant transformers (2)*: 13.2 to 4.16 kV, Dyn11, 12/15 MVA, ONAN/ONAF, oil immersed, 100 A grounding resistor.

Figure 2.13 Plant phasing diagram.

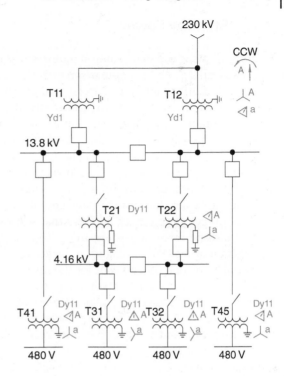

- *Plant transformers (2)*: 4.16 kV to 480 V, Dyn11, 2/3 MVA, ONAN/ ONAF, dry type, solidly grounded.
- *Plant transformers (7)*: 13.2 kV to 480 V, Dyn11, 1/1.5 MVA,

ONAN/ONAF, dry type, solidly grounded.

Note Figure 2.13, we have selected lagging (−30°) winding configuration for the grid transformers and leading (30°) for the plant transformers. This will make the secondary voltage (4.16 kV) of the second-level plant transformers in phase with the 230 kV grid voltage.

Furthermore, note that all the 480 V buses are not in phase. Those fed from the 13.8 kV buses are 30° apart from those fed from the 4.16 kV buses: for instance T41 and T31, respectively.

It is important to note that the 480 V buses from those transformers cannot be interconnected together. On our project in this book, this type of phasing may not be important, as the plant power distribution system is of radial nature without a need for synchronizing.

In other larger plants, where there is a requirement for synchronizing, it is essential to establish proper transformer winding configurations to suit. Engineers must learn how to assign correct transformer winding configuration and phasing orientation for the transformers on the electrical diagrams.

2.10   Standby Power

Each major plant will need some standby power (see Figure 2.14). Often, it is called "essential power" in comparison to the "emergency power," which usually comes from station batteries. The essential power is needed to ensure safety of the personnel and to provide for an orderly plant shutdown as well as quick power restoration and plant operation following a power outage. The standby power must be sized to feed all the critical (essential) loads: part of lighting, DC chargers, plant control system, UPS, heating, and heat tracing where necessary. Also, to power the sensitive process loads and products, which must not be let to harden or freeze up following a power failure.

As noted earlier, the critical path circuit breakers will be provided with latched contactors or circuit breakers that will remain turned on upon a loss of power to be ready to restore the power supply to the essential loads in anticipation of the DG starting up and feeding the essential load. All the other loads will be turned off and wait for the normal power to be restored and for the plant control system to gradually restore power to all the plant services. To insure, the standby power plant is not overloaded during a plant outage, the plant motor loads will drop out or be forced to shut down during a loss of normal power and be held open until restarted by the plant control system onto the normal power.

Standby units often are purchased as packaged standardized container units, with acoustic enclosures, fully wired, and ready for operation. The standby plant must be designed to be black start capable. The engines will be provided with DC batteries for starting.

The cost of the diesel equipment is proportional to its speed as much as it is to its power. Standby power typically uses 1200 or 1800 rpm, 1–1.5 MW units. The standby unit will be provided with provisions to be tested and

Figure 2.14 Standby generator.

synchronized/connected to the plant distribution system during the normal operation to prove its standby duty readiness on a weekly basis.

A loss of voltage for three seconds is a good criterion for initiating a start-up of standby units. Within 15 seconds, the unit will be running and be ready to "synchronize" on the plant dead bus and pick up the load. More on diesel engine generation in Chapter 20.

The plant control system must be operated on the emergency power (DC battery supply) to ensure proper power restoration is initiated after the normal power source is restored. The power restoration procedure (back from a failure) and switching off the standby generation in this type of plants is typically a manual procedure.

The principal loads requiring back up of essential power are

- Camp, including water supply (partial)
- DC chargers to power UPS and plant batteries
- Thickeners, rakes, and U/F pumps
- Instrument air
- Fire detection system
- Lighting (partial)

The standby generators will make use of the plant electrical infrastructure (distribution system) to deliver standby power to all parts of the plant.

The camp will include two 0.5 MW standby units operated at 480 V with synchronizing capability to feed the miscellaneous critical and essential loads. The emergency power will be tied directly to the camp distribution system via ATSs, which will respond to initiate the standby generation following a loss of the principal power received over the 13.8 kV overhead lines.

The camp standby gensets and the switchboard will be trailer-mounted, placed adjacent to the camp main switchboard.

2.11 Insulation Coordination

The electrical equipment must be designed with the insulation withstand capability to operate in the area subjected to lightning strikes and switching surges. Switching surges are not concerned with this plant, i.e. to the voltages 230 kV and lower. The main substation must be properly shielded and protected with appropriately rated lightning arresters (Figure 2.15) placed close to the electrical equipment and grounded by connecting to the transformer enclosure and base and then to the substation grounding grid, using the shortest path possible to ground.

This approach will also protect the transformers from lightning strikes that may be arriving over the overhead line. Similarly, lightning arresters must be installed at all the distribution lines where overhead lines merge with cables or

Figure 2.15 Lightning arrester.

Porcelain type 3EP3
(420 kV)

transformers. The transition points: cable to line, etc., are the reflection points for the lightning traveling waves (see Chapter 10 for insulation coordination).

The insulation level provided on any piece of apparatus, the transformers in particular, constitute an appreciable part of the cost. The equipment withstand capability is called basic insulation level or simply BIL, for instance 13.8 kV switchgear operating outdoors must be rated for BIL is equal to 110 kVpeak, while the same type of switchgear operating indoors is allowed to be rated 95 kVpeak.

The electrical equipment is, therefore, furnished with lightning arresters, which operate as valves to discharge (spark over) the overvoltage strikes to ground, while a smaller tolerable portion of overvoltage below the equipment BIL level continues to the equipment. The basic insulation level or BIL determines the dielectric strength of the apparatus and is expressed by peak value of the full wave withstand waveform as shown in Figure 2.16.

The voltage withstand capacity of all the equipment in an electrical substation or on an electrical transmission system must be determined based on its operating system voltage as shown in Table 2.4.

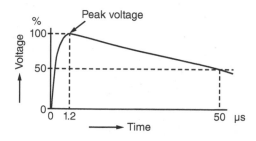

Figure 2.16 BIL impulse wave shape.

Table 2.4 System BILs.

System (kV)	BIL (kVpeak)
11	75
13.8	95/110
33	200
66	450
132	650/750
230	900/1050

2.11.1 Substation Shielding

The switchyard design for over voltage protection in addition to employing the protective equipment should also be shielded (protection coverage) to force the lightning strikes to hit the masts or shielding wires strung between the shielding masts, and not the operating equipment. Utilities typically have an area charted for flash density as Zones 1 and 2, which generally corresponds to the isokeraunic levels of the region and which can be applied on this project (see Chapter 10). Substations in Zone 2 area must be shielded with overhead shield wires or lightning masts.

Based on the criteria (Chapter 10) and the method of grounding, the equipment shall be protected by the lightning arresters having the following ratings (see Table 2.5):

Table 2.5 System lightning arresters.

	Grid	13 kV Transformer	O/H lines	4.16 kV Switchgear
Voltage (kV)	230	13.8	13.8	4.16
Highest voltage (kV)	245	15	15	5
Grounding	Solid	Resistance	Resistance	Resistance
BIL (kV)	900	95	110	60
MCOV[a] (kV)	150	15	15	—
Vr rating (kV)	188	18	18	—
Type (class)	Station	Line	Distr.	Not Req.[b]
Discharge current (kA)	10	5	5	—

a) It is the maximum continuous operating voltage r.m.s. (root mean square) voltage that can be continuously applied to the arrester terminals. The selection must consider the voltage regulation that may be applicable to the operating system.

b) The 5 kV switchgear is fed from a transformer which in turn is connected to a 13.8 kV O/H line. The line is protected by lightning arresters on both ends of the line. This protection coverage is considered satisfactory for the 5 kV switchgear.

2.12   Plant Control System

The plant control will be directed from the central control room, having several levels of controls from the production level to the supervisory and management level utilizing DCS and PLCs technologies and appropriate communication links: Ethernet, DeviceNet, ControlBus, as applicable (see Chapter 17 for details).

2.13   Fire Protection

The plant will be fully protected with fire detection and protection system, including sprinklers and hydrant systems (see Chapter 22 for details).

Reference

1 NEC (National Electric Code) – Ampacity Tables for Cables

3

Switching Equipment

CHAPTER MENU

3.1 MV Switchgear, 75
   3.1.1 Breaker Ratings, 76
   3.1.2 Switchgear "Constant MVA" or "Constant kA" Interrupting, 79
   3.1.3 Breaker Status Contacts, 80
   3.1.4 Switchgear Arc Flash Design, 81
3.2 Circuit Breakers, 84
3.3 MV MCC Motor Controllers, 85
   3.3.1 4.16 kV MCC Assembly, 85
   3.3.2 Motor Controllers, 86
   3.3.3 Key Interlocks, 88
3.4 LV Unit Substations, 89
   3.4.1 LV Distribution, 89
   3.4.2 Unit Substation, Equipment Features, 90
   3.4.3 LV Switchgear, 91
3.5 Motor Control Centers (MCCs), 94
   3.5.1 Assembly, 94
   3.5.2 Motor Starters, 95
   3.5.3 Branch Circuit Breakers for Feeders, 96
   3.5.4 Branch Circuit Breakers for Motors, 96
   3.5.5 MCC Interrupting Capacity, 97
   3.5.6 MCC Breaker Characteristics, 99
   3.5.7 Motor Starter Selection, 103
   3.5.8 Breaker and Contactor Operation, 103
   3.5.9 Arc Flash Modern MCC Design, 104
References, 104

3.1 MV Switchgear

Electrical switchgear (Figure 3.1) is one of the principal elements of the plant distribution system. It houses the switching devices, controls, metering, and protection. The main components of the switchgear are the medium voltage (MV) circuit breakers employed to distribute power to the major plant load

Practical Power Plant Engineering: A Guide for Early Career Engineers, First Edition. Zark Bedalov.
© 2020 John Wiley & Sons, Inc. Published 2020 by John Wiley & Sons, Inc.

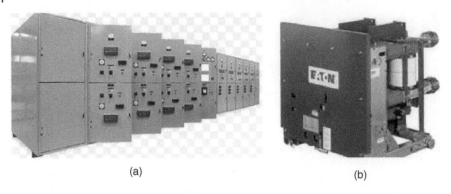

(a)   (b)

Figure 3.1 (a) MV switchgear and (b) circuit breaker.

centers. The MV switchgear [1] is built to 4.16, 6.6, 11, 13.8, 27.6, and 33 kV voltages. A 2.4 kV voltage can still be found, but it is presently on a way out.

For higher voltages up to 750 kV, gas insulated switchgear (GIS) is employed. GIS is several times more expensive than the open-type switchyard, but due to their compact construction with compressed SF6 gas, it takes considerably less space and may be mandatory in some locations. This type of switchgear is always provided with double parallel busbars.

Worldwide, ANSI/ CSA and IEC standards are applicable.

Here are some ratings from recent projects. The ratings may differ from supplier to supplier. The essential ratings are listed in Tables 3.1 and 3.2.

Metal-clad (MC) enclosure is defined by ANSI Std. 37.20.2-1996 as medium-voltage switchgear fully sectionalized having metallic compartments for controls and instruments, incoming and outgoing cables, insulated busses and draw-out electrically operated breakers. Each compartment is provided with grounded metal barriers. Rated maximum voltage levels for MC switchgear range from 4.76 to 38 kV with main bus continuous current ratings of 1200, 2000, 3000, and 4000 A.

The close and latch rating is the breakers ability to close in on a short circuit and stay closed for a period of 10 cycles. Based on the latest standards, the close and latch rating was changed from an RMS value to a peak value. The peak rating is 2.6 times the maximum kilo ampere short circuit rating.

3.1.1 Breaker Ratings

MV breaker frame ratings in North America tend to be standardized to three Amp values: 1200, 2000, and 3000 A. In Europe, 600 and 800 A ratings are also used. A typical switchgear assembly can have breakers of several different frames, particularly in Europe. In North America, it is common to have 1200 A breakers as the feeders, while larger 2000 A breakers are employed for the

Table 3.1 MV switchgear ratings, ANSI Standards.

Voltage (kV)	4.16	13.8
Max voltage (kV)	4.76	15
Power frequency withdstand (kV RMS)	19	36
BIL (kV peak)	60	95/ 110
Breaker rating (A)	1200, 2000, 3000	1200, 2000, 3000
Short circuit interrupting rating (kA RMS symmetrical)	25, 40, 50, 63	25, 40, 50, 63
Asymmetrical interrupting (total kA asym)	31, 49, 59, 80	31, 49, 59, 80
Close and latch current (kA peak)	65, 104, 130, 164	65, 104, 130, 164
Bus, self cooled (kA)	1.2, 3, 4	1.2, 3
Breaker type	Vacuum, SF6	Vacuum, SF6
Operating time (ms)	50 (3 cycles)	50 (3 cycles)
Enclosure, NEMA[a]	Metal-clad	Metal-clad
Application	Indoors	Indoors
Grounding switches	No	No
Control voltage V_{dc}	125	125
Arc Flash design	Type 2	Type 2

BIL, Basic Impulse Level.
a) Refer Chapter 24 for the NEMA and IEC IP standard enclosures.

incoming breakers and bus ties. The largest usable breaker at these voltages is 3000 and 4000 A as a special case.

There are number of issues to be dealt with in determining the main characteristics of the switchgear, as follows:

(1) Feeder breakers are selected on the basis of the intended load. A 1200 A breaker, for instance, can be selected for the load of up to 960 A, which is 80% of 1200 A.

(2) Switchgear and the breakers are rated for the interrupting duty kA_{sc}, at least 25% over the expected bus short circuit fault level in the worst case scenario.

(3) Bus tie breakers are usually selected to be equal to the bus rating. For instance, 2000 A bus tie breaker for a 2000 A bus. The bus rating is not selected on the basis of the total bus load. The load must be distributed along the bus to minimize the flow of current on the bus. The bus rating is rated for the current on the bus in case of one transformer carrying the "full" load. Again here, the bus load that matters is not the connected load, but the running (operating) load.

(4) The incoming breakers are selected on the basis of the MVA of the incoming supply transformers at their highest ratings. The breaker can be loaded up

Table 3.2 MV switchgear ratings, IEC Standards.

Rated voltage (kV)	12	36
Service voltage (kV)	11	33
Frequency (Hz)	50	50
Short circuit interrupting rating (kA RMS symmetrical)	25, 40, 50, 63	25, 40, 50, 63
Rated peak withstand (kA RMS)	63	63
Power freq. Withstand (kV)	28	70
BIL (kV$_p$)	75	170
Short time withstand (kA/3 s)	25	25
Main bus bar (A)	1250, Cu	1250, Cu
Breaker rating (A)	600–1250	600–1250
Breaker type	Vacuum	Vacuum
Enclosure (IP class)[a]	IP42	IP41
Application	Indoors	Indoors
Earthing switches (kA/3 s)	31.5	31.5
Control voltage (V_{dc})	110	110

a) Refer Chapter 24 for the NEMA and IEC IP standard enclosures.

Table 3.3 Breaker loading.

Maximum loading at 80% breaker rating	Maximum temperature loading at 100% (MVA)
$\sqrt{3} \times 4.16 \times 1200 = 5.75$ MVA	7.2
$\sqrt{3} \times 4.16 \times 2000 = 9.6$ MVA	12
$\sqrt{3} \times 4.16 \times 3000 = 14.4$ MVA	18
$\sqrt{3} \times 13.8 \times 1200 = 23$ MVA	28.6
$\sqrt{3} \times 13.8 \times 2000 = 38$ MVA	48.0
$\sqrt{3} \times 13.8 \times 3000 = 57$ MVA	71

to 80% of its capacity during normal continuous operation and up to 100% during an outage of the paired incoming transformer.

Let us look at the 4.16 and 13.8 kV switchgear and the maximum incoming transformer MVA rating for the breakers of 1200, 2000, and 3000 A in Table 3.3:
Once the total load is known and the transformer maximum MVA is determined, the incoming breakers and transformer ratings are selected in conjunction. The incoming breakers are rated to match the transformer ONAN ratings

at 80% of the breaker nominal rating. The 100% breaker rating is selected to match the maximum transformer cooling rating. This author has seen projects where the ratings of the breakers and transformers were not coordinated. The transformer ONAF ratings were considerably higher than the incoming breaker 100% rating?!? The transformer ONAF rating went unused.

Therefore, the bus tie and the incoming breaker sizing will depend on the MVA rating of the transformers at ONAN and ONAF loadings. In the afore-mentioned presentation, it is clear that a 13.8 kV plant distribution system with 2000 A incoming breakers can use 38 MVA base rated transformers and with fan cooling of up to 48 MVA. In case of a transformer failure, the remaining transformer and incoming breaker can be temporary loaded to nearly 48 MVA (2000 A). On "our project," 2000 A rated incoming breakers were also selected in consideration of the plant growth in the future.

3.1.2 Switchgear "Constant MVA" or "Constant kA" Interrupting

The switchgear interrupting duty is calculated based on the short circuit fault level on the bus assuming the source impedance to be zero (neglected), with some margin for safety and expansion. The MV switchgear interrupting rating is known in terms of "Constant MVA," or lately in "Constant kA."

In 1997, the ANSI C37.06 Standard "Preferred Ratings and Related Required Capabilities" was revised to acknowledge the fact that vacuum and SF6 breaker technologies, compared to the old air-magnetic and oil type circuit breakers, have nearly constant interrupting capabilities over a range equal to or less than the maximum rated voltage of the breaker. This change reflected the move to harmonize the IEC and ANSI/IEEE standards and present the ratings on the "constant kA" basis instead of the earlier rating of "constant MVA." In 2000, the ANSI C37.06 standard was changed to match the IEC short circuit ratings. The updated short circuit ratings were changed to 20, 25, 31.5, 40, 50, and 63 kA for the switchgear voltages from 5 to 38 kV [1].

The earlier standards rated the switchgear breakers on the "constant MVA" by incorporating voltage range factor $K > 1$ to accommodate for the breaker operating range. The new "constant kA" circuit breakers and switchgear with $K = 1$ factor simplify the selection of circuit breakers and switchgear, while also more accurately represent the true physics of modern vacuum interruption technology.

We will calculate the interrupting ratings for our main and plant switchgear with incoming transformers, as shown in Figure 2.7, as follows:

T_{11}: 30 MVA, 13.8 kV, $Z_T = 9\%$ (0.09 pu)
T_{21}: 12 MVA, 4.16 kV, $Z_T = 7\%$ (0.07 pu)
Constant MVA class = 1.25 × incoming transformer MVA/Z_T.
Constant kA class = 1.25 × incoming transformer MVA/Z_T converted to kA.
  Select kA rating.

We will select the switchgear rating on the constant kA basis;

For T_{11}: $kA = \frac{1.25 \times MVA}{\sqrt{3} \times kV \times Z_T} = \frac{1.25 \times 30}{\sqrt{3} \times 13.8 \times 0.09} = 17.5\,kA \rightarrow$ select 25 kA RMS symmetrical, interrupting capacity.

For T_{21}: $kA = \frac{1.25 \times MVA}{\sqrt{3} \times kV \times Z_T} = \frac{1.25 \times 12}{\sqrt{3} \times 4.16 \times 0.07} = 29.7\,kA \rightarrow$ select 40 kA RMS symmetrical, interrupting capacity.

1.25 factor is taken as a margin for the short circuit interrupting capacity.

3.1.3 Breaker Status Contacts

A very important feature of the switchgear breakers are the breaker status contacts "52a" and "52b." In their normal de-energized (disconnected) state, sometimes called a shelf state, 52a is normally open, while 52b is normally closed contact. When a breaker closes, its 52a contact closes, while 52b contact opens. Contact 52a follows the status of the breaker, while 52b does the opposite. Similarly, there are truck operating contacts, which indicate if the breaker is inserted in, connected, test, or withdrawn positions. These mechanically operating contacts are extremely important in creating the operating logic of the breakers, both within the control system as well as for the electrical interlocking in the relay protection schemes.

The breaker functioning interlocks can be tested (precommissioned) if the breaker is withdrawn into the "test" position. The breaker can be padlocked in all three positions, which is often done as a safety feature.

The switchgear can be a simple assembly with one incoming feeder and a number of outgoing feeders. However, a larger plant where higher plant availability is needed will require more complex switchgear; with two incoming feeder breakers and a bus tie breaker in the middle. In a switchgear assembly with a tie breaker, the tie breaker is typically held open. Each incoming feeder operates and feeds a half of the plant. This arrangement allows the plant to operate if one of the incoming feeders or transformers fails. The faulty breaker is tripped (opened) manually or automatically, followed by a manual or auto close of the bus tie breaker to transfer the plant load entirely to the healthy incomer. This arrangement is called a two out of three interlocking operation with "break before make" transition. Therefore, when the first incomer breaker opens, it immediately initiates the tie breaker to close. Only two out of three breakers can be held closed at any time. Should there be a total outage from the source; the bus tie breaker will not budge as both incoming breakers are down.

Functioning of this logic is assured by the breaker status permissive interlocks 52a and 52b contacts. The breaker switching logic is typically initiated by the undervoltage relays (27), which in this situation can be regular interposing relays with contact changing states at 100% and 0% voltage. The same transfer logic can be initiated by the plant control system, providing the breaker status contacts are fed to the control system.

3.1.4 Switchgear Arc Flash Design

The Arc Flash [2] can be initiated by ether accidental contact, underrated equipment, circuit breaker insufficient interrupting capacity, contamination or tracking over insulated surfaces, deterioration or corrosion of equipment or parts, as well as other causes. Most of the damage caused to the operating personnel occurs during repairs conducted on live switchgear; for instance due to the attempts to rack out the breaker while it is still energized (Figure 3.2).

In an arc flash incident, electrical energy is converted into heat, pressure, light, and radiation. The incident energy is expressed in cal/cm^2. The goal is to reduce the personnel exposure of the incident energy to $<1.2\,cal/cm^2$.

The principal objective is to design switchgear to be resistant to the arc flash through several protective safety and design measures. The principal objective is to design switchgear to be resistant to the arc flash through several protective safety and design measures. NFPA defines arc flash hazard as:

A dangerous condition associated with the release of energy is caused by an electric arc. NFPA 70E Article 110.8(B)(1) [3] specifically requires *Electrical Hazard Analysis* conducted within all areas of the electrical system that operates at 50 V or greater. The results of the electrical hazard analysis will determine the work practices, protection boundaries, personal protective equipment (PPE), and other procedures required to protect employees from arc-flash or contact with energized conductors.

PPE [3] refers to protective *clothing*, *helmets*, *goggles*, or other garments or equipment designed to protect the wearer's body from *injury*, Figure 3.3. The hazards addressed by protective equipment include physical, electrical, heat, chemicals, *biohazards*, and *airborne particulate matter*. Protective equipment may be worn for job-related *occupational safety and health* purposes, as well as for *sports* and other *recreational activities*. "Protective clothing" is applied to traditional categories of clothing, and "protective gear" applies to items such as pads, guards, shields, or masks and others.

Figure 3.2 Arc flash accident. Source: Courtesy of EEP.

Figure 3.3 Personal protective equipment.

The arc resistant switchgear design must ensure that the highest energy levels are directed to the rear of the structure and away from the operator. Arc-resistant switchgear in addition to providing maximum safety through circuit separation and prevention of access to live parts, must incorporate sealed joints, top-mounted pressure relief vents, reinforced hinges or latches, and "through the-door racking," thus minimizing exposure to harmful gases and significantly reducing the risk of injury to personnel in the event of an arc-flash event.

While testing switchgear circuit breakers in live environment, provide a safety feature or selective interlocking to be activated on the front or remote panel to lower the relay settings to the instantaneous level (zero delay). This reduces the amount of time current flows and the amount of arc flash energy (I^2t) the system encounters during fault conditions, resulting in improved personal protection and prolonged equipment life.

Transformers with increased impedance values drastically reduce the available arc-fault current.

Some form of arc flash protection has lately become mandatory on the electrical switching equipment, though arc flash design adds significantly to the cost of the equipment.

There has been a lot of searching in recent years to determine the level of protection required. So, how much of protection is needed? It depends on the customer. A utility will insist on the highest levels of protection, while the industrial plants will likely insist on level 1 or level 2 arc flash protection, as outlined in the following.

Recognition of arc flash issues and their resolution has gone further in North America then in Europe and Asia since year 2000. The North American standards recommend the following categories of the arc resistant switchgear;

(1) ANSI/IEEE C37.20.7 [4] defines switchgear arc resistance in two basic categories:

ANSI type 1: Arc resistance from the front of gear only

ANSI type 2: Arc resistance provided from the front, sides, and rear

A letter suffix may be added to either of these two types to further define the type of protection provided:

A: Basic design

B: Arc resistance is maintained even while opening designated low voltage compartments

C: Arc resistance is maintained even when opening designated adjacent compartments

D: Special designation that supplements the Type 1 designation but identifies additional arc resistance in certain structures

(2) EEMAC G14–11987, Canada [5]

Accessibility Type A: Switchgear with arc resistant construction at the front only.

Accessibility Type B: Switchgear with arc resistant construction at front, back, and sides.

Accessibility Type C: (Utility requirement) switchgear with arc resistant construction at front, back, and sides and between compartments within the same cell or adjacent cells.

IEEE test guide C37.20.7 Arc-resistant switchgear performance is defined by its accessibility type, as follows:

Type 1: Freely accessible front of the equipment only.

Type 2: Freely accessible exterior (front, back, and sides) of the equipment only.

The components of the proper arc resistant designed switchgear are as follows:

- Arc-resistant switchgear must be designed to channel the exhaust gas and to withstand effects of internal arcing faults up to its rated arc short circuit current and duration. The arc-withstand capability of the switchgear enclosure is achieved by use of reinforced heavier gauge steel where needed, smart latching of doors and covers, and top mounted built-in pressure relief system.
- The formed steel compartmental design to provide sealed joints under fault conditions. This prevents smoke and gas from escaping to other compartments, a condition that can occur with switchgear compartments designed with conventional flat bolted panels.
- Integral, pressure release flap vents mounted on top of each individual vertical section provide for controlled upward release of arc created overpressure, fire, smoke, gases, and molten material out of the assembly without affecting structural integrity and protect personnel. Since arc pressure is vented out

through the top of each individual vertical section, the equipment damage is confined to individual structures, minimizing damage to adjacent structures.

- The racking mechanism is mechanically interlocked with the compartment door such that the door cannot be opened until the circuit breaker is opened and racked out to the test/disconnect position. This interlocking ensures that the levering of the circuit breaker into or out from the connected position is done with compartment door closed and latched, with no exposure to potential arc flash.
- Provision of easy access and viewing ports on the door to allow operator to carry out all normal functions with the door closed and latched with no exposure to potential arc flash, including breaker racking, manual charging of closing springs, manual opening and closing of the circuit breaker, viewing of open/close status of the breaker main contacts, viewing of charged/discharged status of the closing springs, viewing of mechanical operations counter and breaker position.

3.2 Circuit Breakers

In principle [6], there are three types of circuit breakers used in the industry, according to the voltage and application, as follows:

(1) MV and high voltage (HV) equipment, feeders, and lines. These breakers are located on MV switchgear (3.3–38 kV), HV GIS and outdoor switchyards. They serve as high-speed switching devices but have no intelligence to initiate the breaker operations. The initiation to trip for overloads, short circuit faults, differential protection, directional faults, reverse power, etc., is provided by protective relays which are mounted on either the front doors of the switchgear or on separate protection panels. The relays are directly connected to the current transformers (CTs) and potential transformers (PTs). The trip characteristics are defined by the choice of protective relays selected and the settings of the relays and miscellaneous permissive inputs received from other equipment.

(2) Low voltage (LV) switchgear and motor control centers (MCCs) for *feeder* protection. A feeder may be a transformer or a feed to another switchboard. These breakers do not need inputs from protective relays. They have their own serial trip elements. The protection elements included within the breakers are *thermal and magnetic elements*; used for protection against current overloads and short circuits, respectively. The characteristic curves of the protective elements are adjustable for time and overcurrent values. The thermal element is located on the top of the curve, while the magnetic

element forms the bottom part of the protective characteristic on the Time–Current graphs.

(3) LV switchgear and MCCs for *Motor* protection. These breakers do not need inputs from protective relays. They have their own serial trip elements. The protection elements included within the breakers are *magnetic elements only*; used for protection against short circuits. The characteristic curve of the magnetic protective element is adjustable for time and overcurrent values. Often it is set to override the motor inrush current $>10I_n$, with instantaneous trip setting. The thermal overcurrent protection is provided by an additional protective element "overload," included as part of the associated motor starter (contactor). The overload element is often nonadjustable, but selected specifically for each motor.

3.3 MV MCC Motor Controllers

3.3.1 4.16 kV MCC Assembly

Motor Controller Assembly (Figure 3.4) is a significant component of the plant power distribution. It is utilized to supply power to MV motors and transformer feeders for LV substations. A 4.16 kV motor controller MCC is a metal enclosed assembly, housing motor controllers, NEMA class E2 (fused) with integral 120 V control transformers. Controller (400 A) units are fitted two units per vertical structure, as listed in the following. Special 800 A controller units are mounted as a single unit per structure.

• Nominal voltage	4.16 kV, 60 Hz
• BIL	60 kV
• Phasing	Three phase, three wire
• Interrupting capacity	40 kA$_{sc}$
• Construction	Metal-clad/metal enclosed
• Incoming and switchgear tie breakers	1200, 2000, 3000 A
• Motor controllers (contactors)	200–400 A for motors up to 2000 kW

The MV controller assembly generally includes some regular switchgear cells in their Metal clad (MC) enclosures with draw-out vacuum type circuit breakers 1200, 2000, 3000 A as incoming breakers and a tie breaker if the assembly is fed from two independent sources. The controllers used for the unit substation feeders are typically of latched type to insure they are immediately energized upon power restoration.

Figure 3.4 MV MCC controller assembly.

3.3.2 Motor Controllers

MV motors are typically controlled by fused controllers and contactors and in some cases by breakers. Figure 3.5 shows the protective curves of the contactor fuses against the motor starting characteristic. The MV motors, referred to, are rated over 200 kW up to 2000 kW. For motors rated over the aforementioned range, 800 A contactors are available, or higher voltages may be arranged from the primary power distribution; like 6.6 or 13.8 kV. Each motor controller includes a set of MV fuses, a vacuum contactor, and motor multifunction protective relay.

All the breakers and controllers are provided with means of operating from a local and remote position. The means of communication for the unit operation, interlocks, and status is by Ethernet IP.

"Why MV controllers? Why not use the regular switchgear breakers?" Yes, regular switchgear breakers can be used and are often used by utilities. The reason is that MV controllers are less costly, less bulky and above all, they are more suited for regular daily switching duty.

Figure 3.5 Motor protection chart.

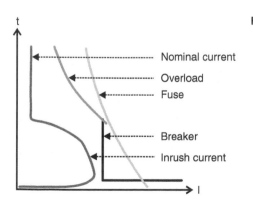

Figure 3.6 MV motor controller basic circuit.

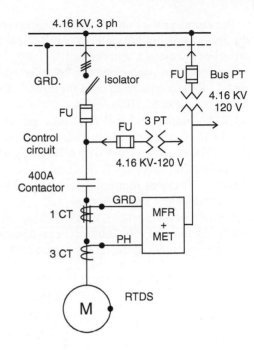

Table 3.4 MV contactor duty.

Contactor duty	4.16 kV		6.9 kV	
	400A	**800A**	**400A**	**800A**
Induction motors (kW)	2000	3750	2500	5000
Synchronous motors, (0.8 pf) (kW)	2000	3750	3375	6000
Synchronous motors, (1.0 pf) (kW)	2625	4500	4160	10 000
Transformers (kVA)	2500	4500	4000	6000
Capacitor banks (kVA$_r$)	2100	4800	3300	4800
Interrupting NEMA E2 (fused) (kA)	50	50	50	50

MV controller assemblies are available with the arc flash resistant design for Type 2B arc flash construction. Type 2B construction is defined as arc-resistant at front, back, and sides of the enclosure with the low voltage compartment door open.

MV controllers are available for induction motors (Figure 3.6), reversible motors, as well as for synchronous motors, either brushless or with static excitation with brushes, field forcing, and DC supply according to the following application Table 3.4. Reversible controllers (starters) need two contactors.

The motor protection adjustable characteristic must coordinate to override the motor starting characteristic and allow the fuse within the controller to interrupt in case of a short circuit.

The controller contactor fuses are selected based on the full load motor current I_n. R-class type fuses used for motors are made to suit the motor starting characteristic of $6I_n/10$ seconds.

NEMA Standards for R-rated MV power fuses require that they operate within 15–35 seconds when subjected to an RMS current 100 times the "R" rating. For example, a fuse with a 2R rating will open within 15–35 seconds on an applied current of 200 A (2×100).

E or L of current limiting type fuses are used in the controllers for feeders and transformers, selected at $1.4–1.5I_n$.

NEMA: For fuses rated ≤100 A, melting time shall be 300 seconds. For a current in the range of 200–240% of the continuous current rating of the fuse. Here are a few examples of the motor controllers:

Controller	4.16 kV	400 A
	I_n (minimum–maximum)	Fuse
Motors	18.7–31.1	70 – 2R
	46.8–62.3	130 – 4R
	93.6–137	200 – 9R
Feeders/transformers	7.2–10.7	15E
	18–21.4	30E
	28.7–36.7	50E
	71.5–89.3	125E

3.3.3 Key Interlocks

This mechanical safety feature is often used to interlock several circuit breakers or disconnect switches, or a cabinet door lock to a switch inside. The key operation engages the mechanical linkages between the breakers. For instance:

(1) A single key is provided and shared between two breakers. A breaker can be operated to close only when the key is inserted into it and turned. The key can be removed from the breaker only after the breaker is opened. The key is moved from one breaker to the other.
(2) Transformer incoming isolation switch. The operating sequence is as follows: the transformer HV isolation switch is operated first, followed by operating the breaker on the secondary side. A single key is used and applied first on the HV side to close the switch and energize the unloaded transformer. The key is then removed and inserted into the LV side breaker

to allow it to be closed. The same sequence, but in reverse order is arranged for the transformer de-energizing.

(3) Interlock on door key to electric switch. The cabinet door cannot be opened if the switch is closed. By opening the switch, it allows the key to open the cabinet door.

3.4 LV Unit Substations

3.4.1 LV Distribution

Unit substations, fed from higher voltage centrally located switchgear, are placed as close to the load as possible. They can be built and assembled as indoor or outdoor assemblies. The indoor model with dry type transformers is preferred, in particular in colder climates. Dry transformers do not need special foundations. A strong concrete floor or base is sufficient. Oil transformers require special foundations with oil containment spaces and vaults. In most cases, long bus ducts are needed to connect to their switchgear, while dry transformers can be directly coupled with the switchgear assemblies.

Unit substations are close coupled electrical assemblies of single or double ended type as required, fed from 4.16 up to 38 kV, and used to power the plant MCCs and LV loads. Figure 3.7 indicates the alternative configurations for feeding loads [5].

- *Radial*: The simplest form.
- *Primary selective*: This arrangement offers greater security of the incoming power supply.
- *Loop (ring main) double ended, selective*: This configuration is primary selective for a string of unit substations, allowing alternate supplies and isolation of a unit for maintenance.
- *Double ended*: It offers security of supply to the loads in case of a transformer failure. This arrangement can further be upgraded to include alternate power sources on the primary side.

Double ended unit substations are provided with LV switchgear, LV incoming, and bus tie breakers.

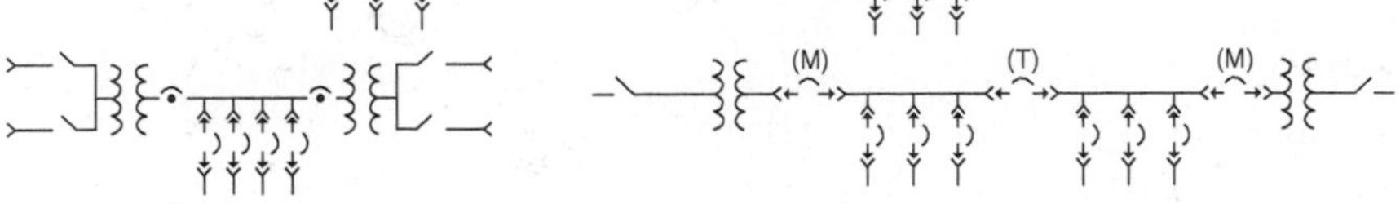

Figure 3.7 Unit substations for LV distribution.

A typical industrial unit substation may take a form of any of those shown in Chapter 9, including the Ring Main Loop diagram, depending on the plant layout and the power reliability requirements. Each unit substation generally comprises the following:

- One or two MV incoming section, complete with fused load break switch and lightning arrestors for each transformer. Load break switches are provided for the unit substations, which may be located away from and outside of the reach of the MV incoming switchgear.
- One or two dry- or oil-type transformers, copper windings, standard taps and with forced air cooling equipment. The most common primary MV voltages are 4.16, 6.6, 11, and 13.8 kV.
- LV (208, 400, 480, 600 V), three phase, three or four wire switchgear, with air break breakers equipped with current solid-state sensors. The breakers are electrically or manually operated.

3.4.2 Unit Substation, Equipment Features

The choice of transformer types are Delta/Star oil immersed for outdoor applications and dry type 300–3000 kVA made for ventilated areas. Since these unit substations are at the end of the distribution circuits, the transformer primary voltage is often selected 5% lower than nominal. For instance, 13.2 kV instead of 13.8 kV. The impedance of the dry-type transformers is usually standardized in the industry at 5.75%.

The Y secondary winding is mostly solidly grounded in power plants, in particular if unbalanced phase loading is expected. In some industrial plants, LV winding may be resistance grounded. Dry-type transformers having VPI (pressure impregnated polyester) Insulation or Cast fiberglass coil are rated in Table 3.5.

Table 3.5 Dry-type transformers.

Insulation type	VPI	Cast coil
Cost	Higher	–
kVA rating	300–3000	300–3000
Fan cooling	Yes	Yes
Temp. rise (°C)	Up to 150	Up to 115
Insulation class (°C)	H (220)	185
Primary voltage (kV)	3.3–34.5	3.3–34.5
BIL (kV crest)	Up to 200	Up to 200

Table 3.6 Load break switch ratings.

System voltage (kV)	BIL (kV)	Current (A)	Fault closing (kA Asym. Mom.)
5	60	600	40/64
5	60	1200	40/64
15	95	600	40/64
15	95	1200	40/64
27	125	600	40/64

The primary MV load break (interrupting) switches are fitted in their own cabinets close coupled and directly connected to the transformer primary bushings and lightning arresters.

The switch is provided with arcing horns to allow for switch closing and interrupting transformer magnetizing current. The door of the cabinet must have locking provision to disable door opening if the switch is energized. The MV switch must also be interlocked with the transformer LV incoming breaker as not to allow the switch to be operated if the LV breaker is closed. The switch is rated generally, as follows in Table 3.6.

A unit substation does not need an MV load interrupter switch if it is in proximity to the source breaker, but it can be included for convenience. This would be the case when the 13.8 kV switchgear with lockable breakers is in the same room as the 13.8 kV/480 V (600 V) unit substation.

The current limiting fuses are available, depending on the transformer ratings, as follows:

4.16 kV → 80E to 400E for 300 kVA to 2000 A, respectively.
13.8 kV → 20E to 125E for 300 kVA to 2000 A, respectively.

Lightning arresters for the transformer primaries are rated based on the method of system grounding of the incoming service. The MV voltage systems in the industrial and power plants are typically resistance grounded (see Chapter 5). Therefore, for the typical MV voltages, the arresters are rated above the nominal voltage.

3.4.3 LV Switchgear

LV switchgear as part of a Unit Substation (Figure 3.8) often called a Load Center for an industrial plant, is an assembly of vertical structures with draw-out breakers stacked three to four per vertical section, rated from 800 to 4000 A. LV switchgear is typically used for feeding MCCs.

(a)   (b)

Figure 3.8 (a) LV substation – double ended. (b) LV draw-out circuit breakers.

Higher ratings are possible, but rarely used. Molded case breakers can also be fitted into the assembly. Typically, each unit substation has a metering section with a multifunction meter and a separate Voltmeter and Ammeter. A typical double-ended assembly includes two incoming 2000–3000 A breakers and a 2000–3000 A tie breaker in the middle. The outgoing feeder breakers could be rated 800–1600 A. The incoming and the tie breakers are generally arranged to be electrically operated from a control room. In that case, they need be wired to operate on 125 V_{dc} with interposing relays to facilitate breaker opening and closing from remote location.

The LV switchgear is subjected to 50 kA fault levels and higher. The new trend is definitely to provide greater security against possible arc flash incidents and have all the feeder breakers furnished with remote switching capability. The breakers will also have to have a capability to reduce the arc flash energy by switching the breaker trip settings to the shortest time possible with a maintenance switch during the breaker maintenance.

LV switchgear breakers are equipped with the solid-state trip units, rated 800–4000 A.

Typical breaker operating curves are shown in Figure 3.9. It includes adjustable long, short, and instantaneous elements used for coordinating the breaker unit with the downstream protective devices on the system.

The trip unit rating may be different from that of the breaker frame rating. For instance, all the feeder breakers may be 1600 A frame, but the trip units may be selected as a mixture of 800, 1000, 1200, and 1600 A units. The trip unit includes the following adjustments:

(1) The long (pick-up) element can be selected from the choice of IEEE standard curves ranging from inversed to moderate. The pickup setting can be adjusted from 0.4 to 1.0I_n. I_n is the breaker trip unit rating. The setting adjustment is 70–120% of I_n for long pickup.
    The time dial is adjustable from 0.1 to 10 in 0.1 increments. The short-time element is adjustable from 100% I_n to 300% I_n and 0.1 to 0.5 seconds.

(2) Instantaneous element (three cycles) is usually optional and often disabled or set high if the breaker feeds an MCC. This is to allow the downstream

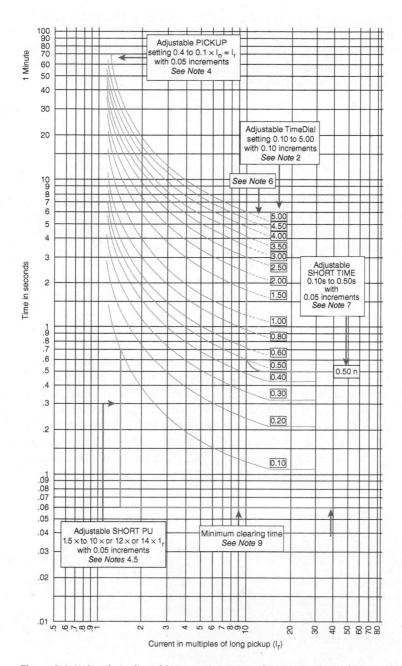

Figure 3.9 LV breaker adjustable operating curves Source: Courtesy of Eaton Corp.

protective devices to assume instantaneous settings in the device coordination process. The setting ranges from 400% I_n to 1000% I_n.

Also, the instantaneous setting may be automatically engaged and lowered down to a small manageable level during maintenance by activating a panel or remotely activated arc flash switch.

3.5 Motor Control Centers (MCCs)

3.5.1 Assembly

MCCs [7] are custom-made prewired low-voltage (230, 400, 480, and 575 V) assemblies of conveniently grouped control equipment, primarily used for the control of LV motors and feeder breakers. Small variable frequency drives (VFDs) can also be fitted into MCCs. MCC's are designed for feeding three phase motors of up to 200 kW (270 HP) and feeders of up to 500 kW.

MCCs (Figure 3.10) will be used on this "book project" for the crushing plant, grinding, plant ventilation, floatation cells, packaging, etc. The breakers are used for distributing power to motors, lighting panels or some smaller subpanels, which may operate specialized equipment that comes with its own controllers.

The MCCs conveniently incorporate all the motor control and power distribution needs while providing the ability to operate on one of many standard industrial communication networks such as DeviceNet, ProfiBus, EtherNet/IP, ModBus TCP, which in turn connects with the Plant PLC or Plant Control System DCS. Presently, the communication trend for the MCC starters is Ethernet IP. MCCs can communicate to the industry standard protocols and allow the plant operator to monitor and manage a production from the plant control room or some other remote location.

Figure 3.10 LV MCC assembly.

Horizontal bus is defined by the projected load on MCC, from 600 to 3000 A and vertical bus from 200 to 1200 A. MCC bus bracing is either 65 kA$_{ic}$ (kA interrupting asymmetrical current) standard or optional for up to 100 kA$_{ic}$. The MCC interrupting rating is determined based on the estimated three phase short circuit fault level, RMS symmetrical value.

The starters and feeder breakers fit into draw-out compartments. Size 5 starter will take a full MCC vertical section. Feeder breakers in MCCs go up to 500 kW, and VFDs up to 500 kW.

MCC is typically wired as NEMA Class II Type B, with all the individual starters prewired to terminal blocks as shown in the purchaser's typical schematic diagrams. Each starter at this plant is supplied with an individual control power transformer low voltage at one phase, 120 V_{ac} (220 V_{ac} in Europe), double phase fused primary and one secondary high rupturing capacity (HRC) fuse. The secondary winding of the control transformer is grounded to insure the control circuit trips in case the control circuit suffers a ground fault.

3.5.2 Motor Starters

Each MCC starter compartment includes the following components: circuit breaker, starter with overload, integral transformer, communication module, and terminations as shown in Figure 3.11.

The contactor is operated either manually via field pushbuttons or automatically through some prearranged operating logic written into the plant control software. The controller is mostly a Direct On Line (DOL) starter with an overload device. The starter comprises a circuit breaker, contactor, and a communication module. Most of the starters are DOL type Full Voltage Non Reversing (FVNR). There are also Full Voltage Reversing (FVR) starters, which include two contactors, for the drives that move in two directions. This is accomplished by reversing two phases on one of the contactors.

Figure 3.11 MCC starter bucket.

Also, the starter buckets may include two contactor combinations for two speed motors, but this is becoming a rarity due to the presence of Soft starts and VFDs (see Chapter 15).

Y/Δ starters are still in use, typically start on Star (reduced current and torque) and then switch to the Delta connection for higher torque.

3.5.3 Branch Circuit Breakers for Feeders

This concerns mainly the branch feeders for transformers, distribution panels, switchboards, and heaters. The breakers for feeders must have protective elements at both ends of the protective curve; thermal and magnetic. The thermal magnetic breaker has an adjustable curve for its thermal element and a fixed magnetic element for short circuits. The thermal part protects the feeders from overloads (top part of the protective curve on the diagram). It is adjustable from 70% to 120% of the trip unit nominal value. The magnetic part of the curve is more inversed as the feeder breaker does not deal with motor inrush currents.

The feeder breakers come in 125, 250, 400, and 1000 A frame sizes, depending on the suppliers. The interchangeable trip units or plugs are selected from a number of ratings suitable for the load protected. Let us say we have a load of 40 A maximum. For this scenario, we can select a breaker frame size 125 A with a trip unit of $1.25 \times 40 = 50$ A. On your one line diagram for this MCC, you would write the breaker as 50AT/125AF. This provides the trip and frame size designation.

3.5.4 Branch Circuit Breakers for Motors

As noted earlier, branch circuit breakers for the motor starters have magnetic element only as part of their protective characteristic curve. This element serves for the short circuit protection, but not for the thermal (overload) protection, see Figure 3.12. The overload protection is assumed by the starter overload element. The interrupting rating of the breakers is dependent on the incoming transformer and its impedance (Figure 3.12).

For instance, by assuming an incoming transformer of 2 MVA, 5.75% impedance, the fault MVA on 480 V bus can be calculated as $2/0.0575 = 34.8$ MVA, and the fault current as 42 kA RMS symmetrical. For this plant, we select 50 kA RMS symmetrical for all the MCCs. For plants needing higher interrupting ratings up to 100, 200 kA load limiters can be added to the breakers.

Though some MCCs may tolerate lower interrupting ratings in some parts of the plant, like those fed from 1 MVA transformers, my preference is to standardize on one rating throughout to avoid possible misplacement of breakers into wrong MCCs and uniformity of spares.

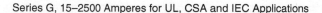

Series G, 15–2500 Amperes for UL, CSA and IEC Applications

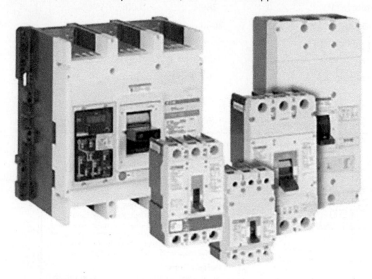

Figure 3.12 A family of circuit breakers [8].

What would happen if the fault level is greater than the interrupting capability of a breaker?

The breaker would be unable to trip on the fault and may cause an arc flash and serious damage and fire to MCC.

One electrical engineer, a colleague of mine told me that on the first job interview, he was asked how he would go about selecting a circuit breaker for a 50 HP motor? A very good interview question.

Let us summarize the selection of circuit breakers (Figure 3.13);

(1) *Application for feeder or motor*: Select breaker with thermal and magnetic characteristic for feeders or with magnetic characteristic only for motors.

(2) *Estimate three phase fault current*: Select breaker frame size for adequate interrupting rating. In a group of breakers like in MCC, establish a minimum frame size to be adequate for most of the breakers. This determines the range of the magnetic settings.

(3) *Calculate the load current and add 20–25% margin*: Select the breaker frame size and the trip element in it to match the calculated load current including rounding.

3.5.5 MCC Interrupting Capacity

One can conservatively (assuming source: infinite bus) calculate the short circuit level for the MCC and the breakers as

MCC system voltage (V)	480	600
Transformer feeding the MCC (kVA)	2000	2000
Transformer impedance (%)	5.75	5.75 (on transformer base rating)
Calculate interrupting (kA$_{ic}$)	41.8	33.5 (kA$_{ic} = \dfrac{kVA}{\sqrt{3} \times Z_{pu} \times V}$ … kA asymmetrical)
Select breaker interrupting (kA$_{ic}$)	50	50 (20–25% margin added)

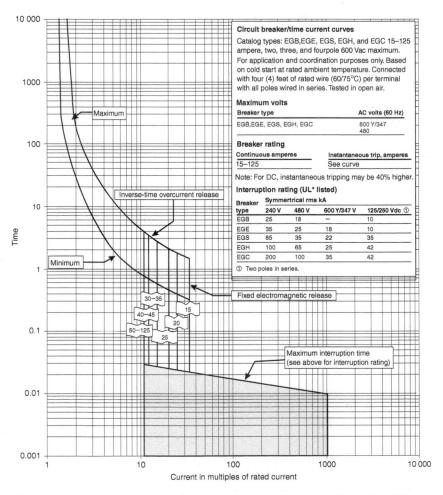

Figure 3.13 Breaker time – current characteristic. Source: Courtesy of Eaton Corp [7].

3.5.6   MCC Breaker Characteristics

As noted earlier, there are two breaker groups, also called molded case circuit breakers for their specific applications:

- Thermal-magnetic breakers for branch feeders
- Motor protectors with magnetic trips with additional current limiters: HMCP (Eaton's trade name).

For instance, one supplier has frame sizes for the *feeder breakers* (thermal-magnetic type), as listed below from frames E to R (125–2500 A). Each Frame can accommodate a large number of interchangeable trip units, often called "rating plugs." The breaker would be denoted on the one line diagrams as 30 AT/125 AF, representing a Frame 125 A breaker with a 30 A trip unit. One would presumably select this breaker for a feeder carrying a maximum 25 A load (80% of its trip rating).

For instance, Eaton Series G15-2500 A [8]: thermal Magnetic Type. UL489 and IEC 60947-2.

Frame	E – 15–125 A	Interrupting 35 kA symmetrical For trip units: 15, 20, 30, 40, …, 125 A
	J – 63–250 A	Interrupting 65 kA symmetrical at 480 V
	L – 250–630 A	Interrupting 65 kA symmetrical
	N – 400–1200 A	Interrupting 65 kA symmetrical
	R – 800–2500 A	Interrupting 65 kA symmetrical

The trip characteristics are fully adjustable for each trip unit on the thermal (overload) side, as shown in the breaker time–current characteristic chart (Figure 3.13).

The time current chart is presented in a logarithmical scale, whereby the current is presented in the multiples of the plug rating on the x axis. For instance, for a chosen plug of 25 A, 1 equals 25 A, 10 equals 250 A.

The adjustments to the plug are made on the overload side from minimum to maximum for the readings on the x axis. On the short circuit side, the magnetic setting is fixed as shown by the flags for the 15–125 A trip units. This chart shows a 15 A plug with a magnetic setting at 34 multiple of its rating, which is $34 \times 15 = 510$ A. For a 125 A plug, the maximum setting is $11 \times 125 = 1375$ A.

Motor protection circuit breakers (MPCBs) are specifically made for motor protection. Typical branch motor loads are protected by three-component starters, consisting of breaker, contactor and overload relay, or fuse, contactor and overload relay. The MPCB application-specific protection eliminates the need for motor overload relay found in the traditional three-component starter assembly. The branch motor load protection is simplified to an MPCB

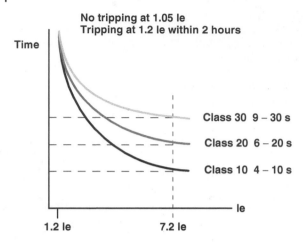

No tripping at 1.05 le
Tripping at 1.2 le within 2 hours

Figure 3.14 Electronic overload.

and contactor. For the overloads protection, the MCC starters use specific electronic trip units as seen in Figure 3.14.

For the short circuits, MPCB employs adjustable instantaneous protection.

The electronic trip unit provides typical motor overload relay functionality and branch circuit short-circuit protection against potential phase-to-phase or phase-to-ground faults. Overload protection is NEMA Class 10, 20, and 30 motor protection. The thermal memory feature is also added to prevent immediate motor restart after an overload trip, to allow the motor to cool down.

The time-current characteristic (Figure 3.15) illustrates Motor Circuit Protector (HMCP) type motor protector breaker adjustable instantaneous settings for the breakers from 3 to 100 A. For instance, a 30 A breaker can be adjusted from 80 to 320 A in 8 (A to F) steps.

The setting of the HMCP breaker magnetic element is best done in field by adjusting the breaker setting for the motor inrush current. The inrush current for new premium efficiency motors is expected to be $6.5-8I_n$, lasting for six to eight seconds. In many instances, the inrush current is even higher. Two identical motors of the same kW rating, coming from the same manufacturer may not have identical magnetic settings due to different inrush currents?!? Set it up for $7I_n$, then start the motor two times. If it stays on both times, keep the setting and write it in the book. If it does not, move the setting a step higher.

In the 1970s, the MCC branch short circuit protection was mainly performed by HRC fuses. Fuses were less costly than breakers and had a higher interrupting capability. Nowadays, circuit breakers have improved and increased their interrupting capability. An advantage of the breakers is that their protective curve is considerably more flexible and adjustable (selective) as shown in Figure 3.16.

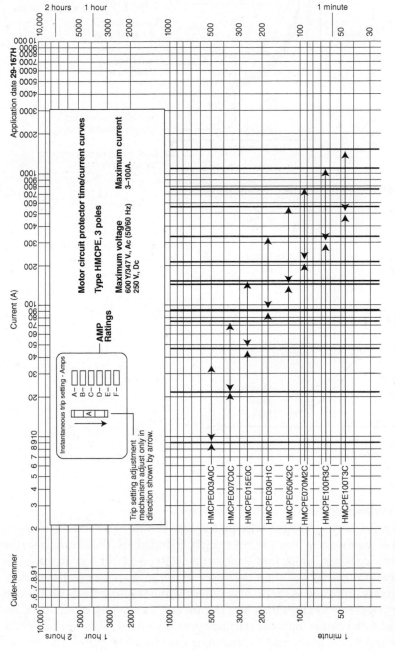

Figure 3.15 HMCP adjustable breakers. Source: Courtesy of Eaton Corp [7].

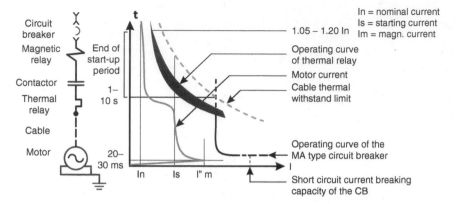

Figure 3.16 Typical motor protection with breakers.

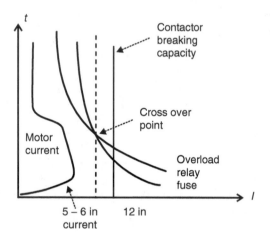

Figure 3.17 Motor protected by Fuse.

The overloads for motor starters nowadays are solid-state type or bimetal. The overload heater rating is selected for each motor based on the nominal current of the motor I_n plus 15–25% margin, rounded up. The allowance is the motor service factor, which for LV motors is 1.15 and 1.0 for 4 kV motors. It is the current the starter will allow the motor to run at as an overload condition.

Motors are protected as shown in the aforementioned diagrams with a circuit breaker (Figure 3.16) and with a fuse (Figure 3.17). Evidently, circuit breakers can be adjusted closer to motor starting curve than fuses. The overload curve cuts the fuse curve in the middle, assuming the top of the curve. The bottom part after the crossover point is fuse protection.

3.5.7 Motor Starter Selection

In most cases, the selected MCC vendor will select the MCC starters for motors and breakers for feeders, based on the list of motor and feeder loads provided to him. This would include the selection of the breaker type and size, contactor size, and the contactor overloads. You will provide to him the short circuit interrupting capacity requirement based on your calculations.

The starter Sizes 1–5 are selected based on the motor HP/kW ratings for 460 and 575 V MCCs, as shown in Table 3.7. Similarly, applicable tables are available for the 400 V, IEC voltage.

3.5.8 Breaker and Contactor Operation

Breakers and contactors (starters) have a different role in the operation of the plant following a plant outage. The feeder breakers will not shut down during a power loss unless they are tripped by some undervoltage relay. They will be immediately reenergized following a return of voltage.

Contactors, often called starters, with *integral* control $120V_{ac}$ transformers will trip following a loss of voltage. In fact they will shut down for a voltage dip below 80% and will remain de-energized upon a return of voltage. Upon voltage recovery, these contactors will have to be restarted, either manually or remotely as part of the plant restoration process.

Some contactors may be equipped with latching mechanisms to remain closed during a loss of voltage. These are called latched contactors. They are usually used for feeding transformers. Industrial plants usually use integral control transformers for the MCC starters. Power plants sometimes use DC power for the control of their starters. The contactors with controllers powered from a DC source will not shut down following a loss of plant voltage. The trend nowadays is changing toward the integral 120 V transformers to allow the power plant control system DCS to restart the starters following a programmed procedure.

Table 3.7 Starter selection.

Starter	Motor (HP)	Motor (kW)
Size 1	1–10	0.75–7.5
Size 2	15–25	11–19
Size 3	30–50	22.5–37.5
Size 4	60–100	45–75
Size 5	125–250	94–188

3.5.9 Arc Flash Modern MCC Design

New MCCs by some suppliers [9] include retracting stab technology that provides stab isolation, stab position indication, and unit interlock features to meet the arc flash requirements. These interlocks are designed to prevent users from opening unit doors or operating disconnects until stabs are properly connected or disconnected from the vertical bus. Remote racking accessory must be available to advance and retract the bus stabs from a distance of up to 8 m (25 ft), allowing the operator to be outside of the arc flash boundary.

An arc flash reduction maintenance system can be offered as a module that can be applied to the main breaker. It permits the operator to add an instantaneous trip setting to temporarily reduce the breaker's normal trip threshold during maintenance.

Arc Flash Containment: MCC assembly includes door and structure for Arc flash Type 2 accessibility rating, designed with arc-resistant features on its front, back, and sides. The doors, side sheets, and back sheets are made using 12-gauge steel. The non-arc-flash MCCs use 14-gauge steel. Isolation barriers between adjacent structures help to isolate blast energy within the MCC. In addition, 4-in. sections are added to the first and last structures of the lineup to increase the through-air spacing between the end of the horizontal bus assembly and the MCC side-sheet, reducing the likelihood for a potential fault to dead metal if an electrical arc spreads along the horizontal bus. Door latches and hinges are further strengthened to allow pressure-relief of internal energy that is released during an arc-fault event.

References

1 ANSI C84.1-1989 (1989). *Preferred voltage ratings for electric power systems and equipment (60 Hz)*. New York: American National Standards Institute. ANSI-C37.06–2009. AC high voltage circuit breakers rated on symmetrical current basis.

2 Eaton (2011) Arc flash reduction maintenance system. Online published information.

3 NFPA 70E (2015) Article 110.8(B)(1) Arc Flash; electrical hazard analysis.

4 C37.20.7-2007 (2007). IEEE guide for testing metal-enclosed switchgear for arc fault.

5 EEMAC (January 1987) G-14-1 Procedure for Testing Resistance of Metal-Clad Switchgear under Conditions of Arc Flash.

6 IEEE (1993). Recommended practice for electric power distribution for industrial plants, IEEE Standard 141–1993, December 1993.

7 Eaton (2016). Motor control centers – low voltage. Online catalog Information TB04300003E.
8 Eaton (2019). Molded case circuit breakers, HCMP, Publication TD00301007E.
9 Eaton (2015). Freedom FlashGuard, LV arc resistant MCCs. Technical Data and Specifications.

4

Designing Plant Layout

CHAPTER MENU

4.1 Plant Power Distribution Routes, 107
4.2 Underground Installations, 109
    4.2.1 Concrete Duct Banks, 109
    4.2.2 Soil Thermal Resistivity, 112
4.3 Plant Electrical Rooms, 112
4.4 Plant Design, 115
    4.4.1 Room Numbering, 115
    4.4.2 Cable Floor Openings, 115
    4.4.3 Process Plant Enclosures, 116
    4.4.4 NFPA Concerns for the Electrical Rooms, 116
4.5 Transformer Vault Design, 117
    4.5.1 NFPA Fire Guidelines, 117
    4.5.2 Oil Containment, 118
4.6 Plant Control Rooms, 119
References, 120

4.1 Plant Power Distribution Routes

The plant layout will in principle follow the process flow from the ore body to the final product output by going through the various specific process facilities. The electrical rooms will be located close to the load centers. Small plants would typically require a single electrical room, while the larger plants would need several electrical rooms to minimize the length of cables to the loads. The plant power distribution system will bring power at 13.8 or 4.16 kV in form of overhead lines and cables to the unit substations in the electrical rooms in the process areas.

Unit substations of up to 2/3 MVA will use dry transformers placed indoors within the electrical rooms, which will also contain 480 V motor control centers (MCCs), distribution boards (DBs), and lighting panels (LPs).

Practical Power Plant Engineering: A Guide for Early Career Engineers, First Edition. Zark Bedalov.
© 2020 John Wiley & Sons, Inc. Published 2020 by John Wiley & Sons, Inc.

An electrical engineer must be present during the initial phase of the plant layout design. The task of the engineer is to insure the electrical buildings and spaces for cable tray routing are properly laid out. The routes of the major overhead lines and power feeders must be well defined (see Chapter 9 for medium voltage (MV) distribution). Some cable corridors may be attached to the overland conveyor structures.

The future expansion for the electrical systems will generally manifest itself in making space provisions for the 230 kV and MV transmission/distribution line/cable corridors and spaces for additional substation transformers.

Buildings: In the tropics, plant buildings will be of open naturally ventilated construction, designed for tropical climate. Attention will be paid to proper channeling and drainage of heavy rainfalls.

In colder climate regions, the buildings will be properly ventilated and heated. The process and domestic piping inside and outside will be heat traced. Heat tracing may also be employed for de-icing under the concrete walkways and on the roof edges.

Permafrost concerns: The permafrost in the Northern regions may go down as deep as 3 m (see Figure 4.1). Digging and trenching in permafrost is extremely difficult. The active layer which melts during warmer months is about 0.3–0.4 m deep. Temperature of permafrost at 30 cm depth during the colder months may be around −6 °C, as shown in the graph, though the temperature outside may be −40 °C.

The graph presents a situation during a period of one year. This is important to know because the plant may include a number of water and fuel lines buried in shallow trenches within the active layer. They will have to be insulated, and electrically heat traced for −10 °C rather than −40 °C of the ambient as it is often thought. Even a pipe buried in snow will benefit from

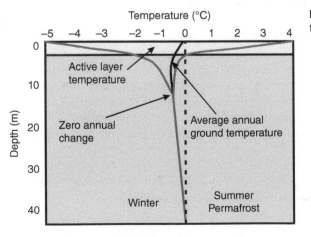

Figure 4.1 Permafrost temperature.

Figure 4.2 HV outdoor substation.

the snow coverage and will not be subjected to −40 °C. The building foundation design in the northern areas may also be a concern as permafrost may thaw under the buildings. In the Northern areas, preference is given to the designs with elevated structures.

Permafrost is a poor conductor for grounding. Grounding cables will likely be buried within the active layers of soil, and the grounding rods extended further down below 3 m depth.

Main substation: The main plant 230 kV substation (Figure 4.2) must be located relatively close to the plant load centers to reduce the length of cables and overhead lines to the specific site locations. The main substation will need an area of 100 m × 80 m, which includes a space for future expansion. The whole area is to be fenced. The substation is sized for the ultimate future plant capacity and the number of transformers. It includes the space for marshaling MV overhead distribution feeders to exit the substation at convenient locations to approach the plant load centers with minimum of crossings and interferences.

The plant MV power distribution will generally be installed in the following order of preference: overhead lines, cable trays, buried and lay in underground trenches. Where buried installation in ducts is planned, the cables will be placed in encased concrete duct banks for the road underground crossings, by observing all the installation standards prevailing in the country of the plant location. The cables buried in concrete ducts will be derated for lower ampacity in comparison with the overhead lines.

Cable ducts with appropriate space provisions will be erected for the ultimate phase of the plant to allow further installation of cables without additional trenching and power interruptions.

4.2 Underground Installations

4.2.1 Concrete Duct Banks

Duct banks carrying 5 kV or higher voltage cables are made of conduits of high density PVC, size 4 or 6 in. (100 or 150 mm) encased in concrete. Duct banks

Figure 4.3 Concrete PVC duct bank.

are required for all the cable laying and installation in the outdoor areas where there is a possibility of vehicular traffic. Figure 4.3 shows a precast concrete block with six PVC conduits.

Conduit risers from underground runs will extend a minimum 30 cm (12 in.) above grade in the open areas or above slabs to prevent floor liquids entering the conduits.

Risers out of the duct banks are of rigid galvanized steel conduit and are to be filled with sealing material. Risers must be sealed with appropriate electrical isolating foam.

Manholes are provided in the areas where conduits change direction and to allow for cable pulling every 100 m (330 ft). Some conduits must be left empty as spares for future. Concrete blocks with 20 or more conduits may be precast on site. NEMA has special tables to be used for derating the cable current carrying capacities when buried and or placed in concrete banks. A cable ampacity for a cable feeder must be judged based on its weakest section. In a fully composite run of overhead + buried + concrete duct bank, the concrete bank has the lowest ampacity. That section will govern the overall ampacity of the run. Manholes are built of precast sectional type concrete, 150 cm (60 in.) sides, designed to be drain proof and equipped with covers, pulling hooks, and cable supports.

The soil or concrete above the duct banks is colored in red to identify the buried locations.

Armored cables can be buried directly in ground or within PVC conduits in designated areas of no vehicular traffic. This applies for the roadway lighting in remote areas where the armored cables are directly buried along the roadways. Typical concrete duct bank installation is shown in Figure 4.4.

Electrical heating calculations need to be performed when large duct banks with significant amounts of conduits and cables are routed below grade in the

Figure 4.4 Cable duct installation.

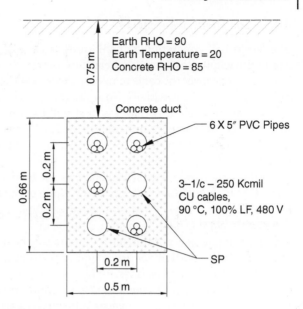

earth to determine if any additional restrictions on the conductors are required. The derating is based on many factors including but not limited to the following:

- Number and size of conduits and conductors,
- Configuration of the conduits and conductors,
- Spacing between the conduits in both the horizontal and vertical dimensions,
- Amount of earth above the conductors,
- The rho factor and the amount of the back fill material,
- Load factor of the conductors, over a certain critical period,
- The actual design load,
- Soil temperature, and
- Length of concrete bank.

For critical situations, computer software should be used to be able to vary the conditions and loadings and to arrive at acceptable solutions of maximum winter or summer loading, backfill quality, and duct bank composition. If the electrical conductors overheat past their rated use, the cable insulation could degrade to a point where a faulty condition may ensue. Concrete banks in the cities are becoming overloaded as the city density is increasing. In places where expansion is not feasible, the cities have undertaken the cooling arrangements of certain city duct banks to enhance their ampacities.

4.2.2 Soil Thermal Resistivity

The "electrical" concrete duct bank must cool itself through the surrounding soil and backfill. Resistance to heat flow between the cable and the ambient environment causes the cable temperature to rise. The heat transfer is measured by the soil thermal resistivity (rho) factor, measured in m°C/W. The higher the rho factor, likely it is that less heat will be transferred (see Figure 4.5).In principle, dense, compact, and moist soils and backfill conduct heat better than natural, loose, and dry soils. Air in soil increases thermal resistance for heat transfers.

A value often assumed for calculations for thermal resistivity [1] of soil in buried cable calculations is 0.9 m°C/W. None of the curves in Figure 4.5b ever get that low, even at very high soil densities. Typical density for a field soil that can sustain plant growth is around 1500 kg/m^3.

Organic material is not suitable for dissipating heat, no matter how dense it is made. Thermal resistivity of dry, granular materials, even when they are compacted to extreme density, is not usable for cable backfill. The air spaces control the flow of heat. Adding vegetation for example could result in significant soil drying, with potential consequences as discussed earlier. Clay soils in particular can crack on drying, resulting in development of air gaps. Every effort must be made to avoid this happening. Potential "hot spots" along the cable route (such as zones of well-drained sandy soils or vegetated areas that could lead to significant soil drying) should receive particular attention to ensure long-term success of any buried cable or cable bank installation. © *Reproduced by Permission of Gaylon Campbell.*

a. Adding water to a porous material drastically decreases its thermal resistance.
b. The thermal resistivity of a dry, porous material is strongly dependent on its density.

4.3 Plant Electrical Rooms

The electrical rooms must be designed to be large and comfortable for all the electrical equipment to allow for safe maintenance in front and behind the electrical cabinets and assemblies, and with some space for future expansions. The equipment will be laid out in such a way to insure that any piece of equipment can be withdrawn and maneuvered out through the equipment door. The electrical room is usually located at a central location near the electrical load centers and adjacent to the external transformers to reduce the length of the bus ducts

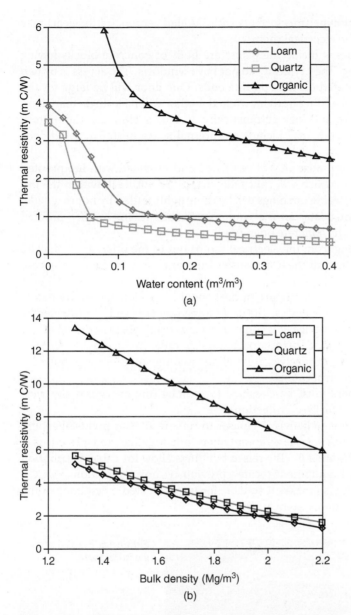

Figure 4.5 (a and b) Soil heat resistivity charts. Source: Courtesy of Decagon Devices [1].

and HV cables. Dry-type transformers with LV load centers (switchgear) may be located inside the electrical rooms.

The electrical rooms inside the plants are built of concrete blocks having two-hour fire rating. They typically do not have windows. The access is by two fire-rated doors located on the opposite ends. One door will be large for the equipment entries standing upright, while the other will be a single man door. The doors will close and lock automatically. In colder climates, the bottom part of the outside door may include an internal space heater to avoid door freeze ups.

The electrical room must be well ventilated and provided with fire proofed motorized dampers, which will shut following a fire outbreak within the fire protected zone. The cable openings will be thoroughly sealed by nonmagnetic, metallic cable gland plates in the equipment cabinets and by fire proofed expanding foam in the concrete openings.

Cable trays passing from the electrical rooms out to the other areas shall be cut off at the walls to limit the heat transfer from one room to the other in case of fire.

The room will include a mixture of heat, smoke, and ionization fire detectors and fire notification devices. The cable spreading area will be protected by sprinklers. In North America, it is common to use sprinklers over the cable tray banks in electrical rooms. This approach is not popular in the industry in Asia, where the use of more expensive gas flooding discharge methods is popular for the electrical and control rooms.

Hand extinguishers must be classified for combatting energized electrical fires (see Chapter 22 for Fire Protection).

The present trend is to deliver E-Houses to remote sites as preinstalled and prewired, with all the services, in containers put together and placed on a 1-m-high stands (Figure 4.6). The raised buildings allow for cable distribution under the electrical equipment located previously; unit substations, MCCs, panel boards, etc. It also makes it suitable for unobstructed assembling of the power plant distribution system at site.

Figure 4.6 Prefabricated electrical room.

4.4 Plant Design

4.4.1 Room Numbering

Each plant usually has several buildings and many rooms and offices. It is recommended to name or number the rooms. Preference is to provide numbers to each separate room and a list with the room names to be shown in the side of the layout drawings. Design engineers often change room names or enter incorrect names in their emails, drawings, and documents, leading to confusions and misinterpretations. It is better to communicate with the room ID numbers rather than names that tend to be inconsistent. The room numbering will also help the engineer to computerize the equipment schematics and cable lists.

4.4.2 Cable Floor Openings

The cable entry into the electrical cabinets and transformer termination boxes must be performed with proper gland connectors on the bottom (or top) termination plates. The termination plates must be made of nonmagnetic materials like aluminum, brass, bakelite, lexan, or nonmagnetic stainless steel. Magnetic materials surrounding single phase power cables must not be allowed, since it will cause severe induction and circulating currents on the termination plates. If use of steel cannot be avoided, cut a small slot to interrupt a magnetic circuit and break the flow of induced currents.

A serious construction issue is the floor cable openings for the cables exiting the electrical cabinets, either from the top or bottom. In case of cable bottom entry, cable openings are needed to be built on the concrete floors to match those in the equipment placed on the aforementioned floor. In this case, cable trays are installed under the floor.

Unfortunately, often the concrete floors are required to be designed and poured months ahead of the completion of the electrical design and even before receiving the suppliers' equipment drawings. Last minute design changes by the mechanical department are highly likely and unavoidable. A large tank may be added on the floor causing the electrical switchgear to be pushed around and away from the predesigned openings in concrete.

Seemingly, the easiest job becomes the most difficult. Very often the result is a mismatch of the floor openings against those on the equipment. The floor and re-bars will need to be re-cut later on during the equipment installation, thus weakening the finished floor. When it comes to the floor cable openings, Mr. Murphy (the most famous failure specialist) [2] and his associates are residents in your plant and hard at work to make your life less enjoyable. With the cable openings in the wrong place, you would have been better off to have done nothing. Now you may be faced with closing the openings and drilling new ones that may not fit.

Alternatively, the specifications could have been written to have the cables exit the electrical cabinets from the top and directly into the top mounted cable trays. In this case, concrete cable openings are not required. There are many benefits favoring this widely used option, including the following:

(1) Simplified floor plan design.
(2) Simplified cable installation and termination as well as accommodation of the changes, as the cables are visible at both ends. With the top cable entry, during commissioning, one can exchange hand signals with his partner at the other end; OK! or NOK!, for fast wiring checks.
(3) Simplified cable testing as you see them at both ends, instead of having someone else on the floor below shouting against the plant noise to tell you which cable goes where.
(4) Simplified fire detection and protection.

Therefore, avoid problems, where you do not need them.

4.4.3 Process Plant Enclosures

In the electrical rooms, floor cable openings can be designed to allow cable entry from top or bottom. However, in the process plant, the panels must be elevated, wall or column mounted due to a possibility of corrosive chemical spillages in the area. The cables to these elevated panels and cabinets must enter from bottom. Place the equipment on short 10 cm (4″) concrete pedestals. For the wall mounted panels, make sure that the mounting is not directly on the wall but on vertically mounted properly sized Unistrut channels to provide some clearance from the walls (see Figure 4.7).

4.4.4 NFPA Concerns for the Electrical Rooms

Due to the large concentration of cables, electrical rooms shall be designed to meet the NFPA guidelines, as follows:

- The electrical rooms inside the plants must be built with two-hour fire rated block walls.
- The cables must use nonflammable outer jackets.

Figure 4.7 Unistrut channels.

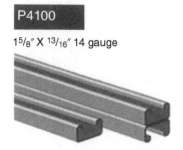

P4100

$1^5/_8″ \times 1^3/_{16}″$ 14 gauge

- The electrical rooms must have two doors on opposite sides, both built to two-hour fire rating.
- Cable openings, top and bottom, must be sealed with gland on nonmagnetic plates in the equipment cabinets and provided with approved electrical insulation to block air passage and heat transfer between the rooms.
- Upon a fire signal, the ventilation must stop and the fire dampers and duct vents must close.
- The electrical rooms require proper fire detection, annunciation, and suppression systems.

4.5 Transformer Vault Design

4.5.1 NFPA Fire Guidelines

Large oil immersed transformers will be located outdoors within concrete vaults adjacent to the buildings (Figure 4.8). The transformer primaries will be connected to the overhead lines or cables, while the transformer secondary bushings will be fed to the adjacent indoor electrical rooms by bus ducts or cable buses.

Based on the NFPA 850 [3] transformation separation guidelines, oil-type transformers may be allowed to be placed near the electrical rooms, providing

- The wall of the building facing the transformers is two-hour fire rated.
- The wall of the building facing the oil-filled transformers does not have cable openings, windows, or doors.

Figure 4.8 Transformers, fire barrier walls.

- A fire barrier wall is installed between the transformers to the height of 30 cm above the oil conservator tank or bushings, whichever is higher, and 60 cm width on both sides of cooling radiators.
- If the electrical room or other part of the building cannot be fire rated, the oil transformers will be located a minimum 8.5 m (25 ft) from the building for transformers having 1890–18900 l (500–5000 gal) of oil. Transformers with oil in excess of >18 900 l (>5000 gal) will be located a minimum 16 m (50 ft) away from the building.

4.5.2   Oil Containment

Proven engineering practices and design criteria are applied for the oil containment basins under the transformers or other large electrical equipment containing oil, as follows:

- Capacity of the overall containment for multiple transformer units will be based on a single unit, to contain 110% of the oil content of the largest unit. Two transformer coincidental failures are considered unlikely.
- The major rainfall coincidental with the oil leakage is to be considered unlikely. Regular daily average rainfall is to be taken into account.
- Rain storm water will be gravity drained continuously to leave the basin 10% filled with water. Water is to be drained close to the bottom, while any oil or effluent at the top of the basin, separate from the regular storm drain.

In the tropics, a light roof construction may be added to limit the basins to be overfilled by rain.

- The BC Hydro Engineering Standard [4] and IEEE Standard 980 Guide [5] stipulates the limit as 2500 l as the minimum threshold for a containment. Oil containment is recommended for transformers having more than 4000 l (1000 gal).
- Oil catchment area shall be designed to contain potential spill trajectory. It shall be extended, unless bounded by the walls, by 2 m from the transformer tank and radiators.
- The basins of multiple transformers will be interconnected and will be sloped to drain to a sump. It will be sized based on the assumption of a single fire outbreak for accessing and removal of the liquids inside the basins.
- In case of a fire outbreak and appropriate alarm given, the oil discharge valve will automatically close.
- The concrete curb around the containment will be not <150 mm (6 in.) high.

- The oil containment basin will be filled with prewashed gravel. Sometimes, if there is a space limitation, the gravel will be placed over the steel mesh covering the containment basin.
- If a deluge water spray system is part of the fire protection, the containment will allow for 90-minute water discharge.
- Water/oil separation may be required, depending on the local regulations.

4.6 Plant Control Rooms

The plant control room will be in a central location and very likely close to the main electrical room to reduce the cabling between the two rooms. The control room will be designed as a special area for the plant operators (Figure 4.9). From this room, the operators may also have a clear view over the mechanical processes. The operators will monitor the plant operation via the operating consoles and control system monitors.

The DCS I/O termination equipment may be located in an adjacent room together with the communication equipment. All these rooms are typically well sealed and protected by gas flooding systems or preaction sprinkler system in particular the cable spreading areas, which are located in a special tiled sub (raised) floor, made suitable for laying communication and instrument cables.

The tiled subfloor (Figure 4.10) is usually built 0.6–1 m high made of modular strong structural metallic members and tiles. The fire gas flooding will work in

Figure 4.9 Plant control room.

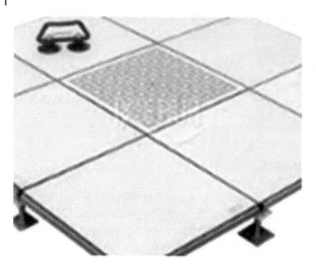

Figure 4.10 CR tile floor.

conjunction with the fire detection and notifying systems. The detectors, heat, smoke, and ionization type will be placed in the subfloor over the cables as well as in the ceiling of the control room. A Vesda dust aspiration system is also well accepted in these applications.

The fire detection together with Vesda systems are normally arranged to operate the gas flooding system upon a cross-zone activation of two different types of smoke or fire detectors (see Chapter 22).

References

1 Campbell, G.S. and Bristow, K.L. *The Effect of Soil Thermal Resistivity (rho) on Underground Power Cable Installations.* Decagon Devices www.decagon .com.
2 Murphy Laws and Corollaries. https://www.angelo.edu/faculty/kboudrea/ cheap/cheap3_murphy.htm.
3 NFPA (2005). Transformer Separation NFPA Guide 850–11-2005.
4 BC Hydro (n.d.) BC Hydro Engineering Standard ES 21-B0630 for Oil Containment.
5 IEEE (1994). IEEE Guide Std.980 for Containment of Oil Spills.

5

System Grounding

CHAPTER MENU
5.1 Methods of Grounding, 121
5.1.1 Ungrounded Systems, 122
5.1.2 Neutral Point Solid Grounding, 123
5.1.3 Neutral Point Resistance Grounding, 125
5.1.4 Reactance and Resonance Neutral Grounding Systems, 126
5.1.5 Summary, 126
5.2 Specific Applications, 127
5.2.1 Generator Grounding, 127
5.2.2 Transformer Grounding, 130
5.2.3 Grounding Transformers, 132
References, 135

5.1 Methods of Grounding

The various methods of system grounding have evolved over the years. This chapter evaluates each approach and explains why certain methods are regularly applied by the power systems engineers for the specific engineering applications. Proper selection of transformer winding configurations is essential in the design of power systems for utilities and industrial facilities with respect to the system grounding.

Throughout my working experience, it was evident that the field of grounding is the most ignored and lacking subject of electrical engineering. Fortunately, as a young engineer, I worked with a well-known and award winning electrical engineer Mr. Andy Sturton of Shawinigan Engineering Co. who gave me insights into this important subject and guided me in making proper engineering decisions.

What is system grounding, and what is equipment grounding? It really is confusing, in particular to those of different engineering backgrounds. So, what is it?

Practical Power Plant Engineering: A Guide for Early Career Engineers, First Edition. Zark Bedalov.
© 2020 John Wiley & Sons, Inc. Published 2020 by John Wiley & Sons, Inc.

System grounding is a connection to ground from live (when energized) conductors of a piece of electrical equipment, such as transformer, generator, capacitor bank, etc. Usually, the ground connection is made to the neutral point in case of a three phase system. This connection may be solid or through an intentional resistance. The grounding serves the purpose of stabilizing the system in cases of line to ground faults and provide means for applying selective relay protection for the operating equipment.

Equipment grounding is a connection to ground from the equipment enclosures, motor frames, footings, cable trays, cable armor, conduits, building structure and other metal parts, etc. This equipment is considered not live but may become "live" in case of a ground fault on the system or by live conductor touching the enclosure. The equipment connection to ground will minimize the potential rise as well as the step/touch potentials on the equipment (see Chapter 6).

Neutrals on smaller motors, if available, generally go ungrounded irrespective of the method of the neutral grounding of the incoming transformers.

A number of methods of the system grounding were practiced earlier, but like everything else, the power industry has standardized on several distinctive procedures. The selection of the methods of system grounding impacts the operation of the electrical equipment and systems differently as follows:

- Magnitude of the ground fault current and resulting damage to the point of fault,
- Transient overvoltages,
- Application of protective relaying, and
- Selection of lightning protection of the equipment.

5.1.1 Ungrounded Systems

This approach still has its day in some industrial applications in the low voltage (LV) systems, in particular mining operations, but none in the power industry. It is also used in some manufacturing plants to maintain the plant production during a single ground fault.

In case of a phase to ground fault, the transformer neutral point will be displaced and floating depending on the system impedances, while the healthy phases will be subjected to phase to phase and even higher voltages, up to six times of the line to line voltage in case of arcing faults. An ungrounded system does not provide an intentional return path to a ground fault. The insulation on the other two phases and in the equipment is overstressed to a full line to line voltage. However, the work can go on until a second ground fault is developed, thus creating a line to line fault, which will trip the system. In some situations, the faults may be of transitional nature and be self-cleared.

The ungrounded systems usually come with special equipment to detect and indicate the faulty phase. Having realized that there is a ground fault on the equipment and on which phase it is, one can delay the repair work until a convenient time for repairs. If a double fault occurs, the operation is stopped and the problem must be addressed immediately.

The line phases of the system are subjected to capacitive charges between the phases and ground and between the phases. Capacitances between phases are "delta" connected and are of no concern. The capacitances to ground do matter as they provide a return path to ground for the ground faults anywhere on the system. Therefore, the ground fault current is virtually all capacitive: $I_g = 3E_0/Z_0$. Z_0 is virtually all Z_c. We can say that an ungrounded system is a capacitance grounded system. As noted earlier, if the fault is intermittent (arcing), the voltage on the healthy phases may be significantly higher than ph–ph voltage due to the restrikes in the capacitive circuit.

Deficiencies of this method of grounding are as follows:

- The fault self-clearing feature for this method of grounding applies to smaller systems only.
- Ground fault relaying protection is difficult and insensitive.
- Arcing faults and its transient overvoltages are a serious concern.
- Lightning protection comes with the highest possible arrester ratings.
- Transformers operating as ungrounded must be built to the highest levels of basic insulation level (BIL) and insulation.

5.1.2   Neutral Point Solid Grounding

This method of grounding is provided with no intentional resistance between the neutral point and ground. It is applied virtually without exception on high voltage (HV) and extra high-voltage (EHV) transformers on their primary HV windings of the Υ/Δ transformers. It is often used on the LV systems, 600 V and lower.

In case of a phase to ground fault, the transformer neutral point is fixed while the healthy phases remain at the same phase to ground voltage without oscillations. Generally, the ground fault current for this method of grounding is equal or higher than the three phase fault current.

As shown in Figure 5.1 [1], *solid* neutral grounding is possible in three ways:

- Generator neutral grounding. This is no longer used or recommended due to the high destructive ground fault currents, which may flow in case of a ground fault.
- Transformer neutral grounding. This is common in HV and LV systems, but not on medium voltage (MV) systems.
- Application of a grounding Zig–Zag (Zg) transformer. The grounding transformer is of single Zg winding and connected to the bus bars. The

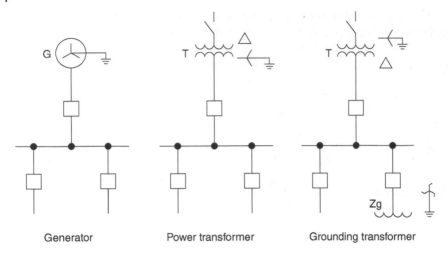

Generator   Power transformer   Grounding transformer

Figure 5.1 Neutral point solid grounding.

neutral of the winding is solidly grounded or through a resistor. The use of Zg transformers for grounding is more appropriate for the resistance grounded systems. If solidly grounding is required, the Zg transformer would have to be of large KVA capacity in line with the magnitude of the expected ground fault current.

Lately, the name of this method of grounding was changed from "solid" to "effective grounding" to better define the grounding for larger power systems. While transformer may be solidly grounded, the transformer capacity may be too small in comparison with the overall system to be effective in stabilizing the overvoltages on the system during an occurrence of ground fault. This mostly impacts the systems grounded with grounding transformers.

Let us hear the new definition for Effectively Grounded system: "*A system or a portion of a system is effectively grounded when for all points on the system or specified portion thereof the ratio of zero sequence resistance to positive sequence reactance is not greater than 1 and the ratio of zero sequence reactance to positive sequence reactance is not greater than 3 for any condition of operation and for any amount of generator capacity.*"

$$R_o/X_1 \leq 1 \text{ and } X_o/X_1 \leq 3 \tag{5.1}$$

Another acceptable criterion for defining the effective grounding is the line to ground fault current comparison against a three-phase fault on the same part of the system: $I_g = 3I_o > 0.6$ times the three phase fault current.

Benefits of this method of grounding far outweigh the disadvantages. Transformers effectively grounded use lower BIL designs on the primary winding of the transformer, thus reducing the cost. However, serious damage to the transformer windings is possible due to high fault currents. Therefore, the relaying protection must be set to be effective and clear the fault very quickly.

5.1.3 Neutral Point Resistance Grounding

It is used on MV systems 34.5 kV and lower to restrict the ground fault current to a reasonable value of 50–100 A, while permitting sufficient fault current flow to allow a meaningful relay protective scheme to be established, Figure 5.2 [1]. In case of a phase to ground fault, the transformer neutral point will be fully displaced while the healthy phases will be subjected to phase to phase voltage.

The resistor must be selected to ensure a sufficient resistive current flow, greater than the system charging capacitive current, which if not restricted may cause severe damage to the equipment. If the neutral fault current of 80 A is desired, the required R_n resistor can be calculated as follows:

$$I_n = \frac{V}{\sqrt{3}\, R_n} \quad \text{for instance:} \quad R_n = \frac{13\,800}{\sqrt{3} \times 80} = 100\,\Omega$$

where

V = line to line voltage
I_n = neutral current
R_n = neutral resistor

The line to ground fault current is limited by the neutral resistor. Less than 5 A would be considered a high resistance grounding, which approaches the

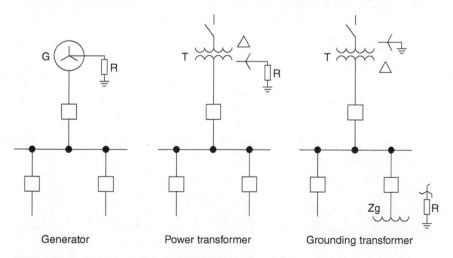

Generator   Power transformer   Grounding transformer

Figure 5.2 Neutral point resistance grounding.

situation similar to that of an ungrounded system. The ground fault current must be equal or greater than the capacitive current of the system to dissipate it and reduce the voltage rise due to the system capacitances. Generally, if the fault current $I_n > 10$ A through the neutral, in practical terms, one does not have to be concerned with the capacitive charging currents. Therefore, this method of grounding eases the issues of arcing faults. The grounding resistors dissipate a lot of energy and heat. They are made of stainless steel and rated to no more than 10 seconds for the fault to be cleared by the protective relaying.

5.1.4 Reactance and Resonance Neutral Grounding Systems

This method may be useful for some specific applications. I have not encountered a case for this method of grounding in the industrial and power plants.

5.1.5 Summary

The practical applications of the various methods of neutral grounding can be summarized in Table 5.1.

In the last 30 years, the industry has established that the *grounded systems* bring substantial benefits over the *ungrounded systems*, and for that reason, we will focus on the former for our projects.

The benefits of a grounding system are overwhelming, from safety, maintenance, equipment and cabling cost, relay protection, and reliability. The grounding systems with their equivalent diagrams are shown in Figure 5.3 [1].

Note: In Chapter 10, the resistance grounded systems are called ungrounded systems. This is because only the "effective" grounding offers less stringent requirements on the selection of the lightning arresters.

Table 5.1 Grounding.

	Ungrounded	Resistance	Effective
Equipment BIL and cost	Highest	Higher	Lowest
Continuity of operation	Possible	No	No
Maintenance	Difficult	Simple	Simple
Power cable insulation (%)	>133	133	100
Relay protection	Difficult	OK	OK
Arcing grounds	Yes	No	No
Double line faults	Yes	No	No
Lightning level of protection	Highest	Highest	Lowest
Radio interference	High	Minor	Lowest

Figure 5.3 Method of grounding and equivalent circuits.

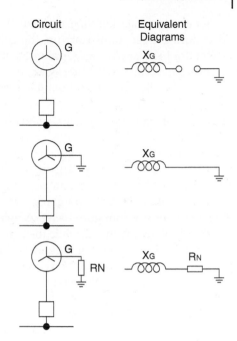

5.2 Specific Applications

Let us look at the specific applications of the system grounding.

5.2.1 Generator Grounding

There have been many methods of generator grounding through the history. At one time, the generators in smaller power plants were all grounded through a single neutral grounding resistor (NGR). Each generator neutral point was connected through a disconnect switch to a common NGR with a circuit breaker rated to the generator voltage.

Disadvantages of that concept were numerous, including:

- An additional neutral full voltage rated switchgear was needed.
- The intensity of a ground fault for a commonly grounded plant depended on the number of units on line.
- Since it was not possible to pin point the fault location, tripping of the entire plant might result.

An alternative to the grouped generator grounding setup was to equip the generator switchgear with a Zg grounding transformer and have all the generator neutrals left ungrounded. A disadvantage of that approach was that during a generator start up, the generator was ungrounded prior to the generator breaker closing. A ground fault on the isolated generator may cause a severe overvoltage if there is an arcing ground fault on the generator, as discussed earlier for the ungrounded systems.

5.2.1.1 Generator Unit System Grounding

In the last 50 years, the industry has evolved and standardized on a unit grounding with each generator having its own individual grounding resistor as shown in Figure 5.4 [1]. The high resistance method of grounding became a norm for the generators, in particular for the generators operating and connected directly to the outgoing transformers as shown. The output transformer is mostly configured as Y/Δ, whereby the Y side is connected to the grid and solidly grounded, while the generator is connected to the transformer Δ winding. This method of grounding does not allow the ground fault current to circulate among the units as was the case with the arrangement having generator neutrals connected directly to a common bus.

Grounding of the Δ system is accomplished by having the generator neutral grounded through a small single-phase distribution transformer with a low resistance secondary resistor (NGR). The primary of the distribution

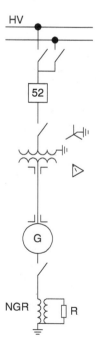

Figure 5.4 Unit G grounding.

transformer is connected directly to the generator neutral through a disconnect switch. This is considered a high resistance grounding as the ground fault current is restricted to a nondestructive value of 5–10 A. On large machines, the faults are rarely of three phase nature. They usually start as ground faults, which may develop into phase to phase faults. So, phase to ground faults must be cleared quickly. In some applications, instead of a distribution transformer/resistor combination, a high resistance resistor is applied, directly connected to the generator neutral and rated for the full voltage of the generator.

For this application, the distribution transformer is typically overload short time rated for five minutes. Hopefully, the unit is tripped within one second on its backup protection. The following criteria must be met while selecting the generator grounding resistor:

(1) The resistive current must match or exceed the system capacitive current: $R_o/X_{co} \leq 1$ and $X_o/X_1 \leq 3$ to restrict transient overvoltages.

X_{co} is determined by paralleling all the capacitances in the generator Delta system including the generator step up transformer (GSUT) transformer secondary winding.

(2) The NGR must dissipate all the capacitive current during a ground fault. In the calculations, it is better to err on the side of higher current than allow less current flow through the NGR.

Let us see how this works out in an example for a 3 ph, 13.8 kV, 60 Hz, 150 MW generator unit.

The *capacitance per phase to ground* for a total generator system inclusive of the main transformer secondary is $C_{1s} = 0.85\,\mu F$. This includes the generator, bus duct, surge pack, potential transformers (PTs), unit transformer, excitation transformer, step up transformer Δ winding, etc.

The phase to ground capacitances of all the elements mentioned previously are in parallel and are added up as follows.

➤ *The capacitance for all three phases is:* $C_{3s} = 3 \times C_{1s} = 3 \times 0.85 = 2.55\,\mu F,$
➤ *Capacitive reactance:* $X_{co} = \dfrac{10^6}{2 \times \pi \times 60 \times C_{3s}} = \dfrac{10^6}{2 \times \pi \times 60 \times 2.55} = 1044\,\Omega$
➤ *Select voltage of distribution Transformer:* 13.8 kV to 240 V
➤ *System capacitive current for three phases:* $I_c = \dfrac{13\,800}{\sqrt{3} \times 1044} = 7.66$ A to equal the resistive current.
➤ *Establish criteria:* $R_{pri} \leq X_{co} = 1044\,\Omega$ resistive current > capacitive current
➤ *Transformer KVA:* $T = 7.66 \times \dfrac{13.8}{\sqrt{3}} = 61$ KVA
➤ *Overload rating factor K for five minutes:* 2.8
➤ *Transformer five-minute rating:* $T_{5m} = \dfrac{61}{2.8} = 22.8$ KVA → select: 25 KVA, 1 ph, 13.8 kV to 240 V

➤ *Primary resistance*: $R_{\text{pri}} = 1044\,\Omega$

➤ *Actual secondary resistance*: $R_{\text{sec}} = R_{\text{pri}} \times \frac{240^2}{13\,800^2} = 0.315\,\Omega$

The generator protection for the ground faults is taken as an overvoltage across the resistor and fed as a (64 N) input to the generator multifunction protective relay. During a ground fault, the voltage on the generator neutral is raised to the full line to line voltage. In case a HV grounding resistor is used instead, the fault signal is taken to the generator multifunction relay from a current transformer (CT) connected in the path of the HV neutral resistor.

Smaller diesel engine generators up to 500 kW used for standby services normally go with ungrounded or solidly ground neutrals to make it compatible with the grounding system employed for the LV or MV system where the generator is connected to.

5.2.2 Transformer Grounding

5.2.2.1 Y/Δ or Δ/Y Transformers

If solid or resistance grounding is required, Y/Δ or Δ/Y transformers are used. The ground fault currents flow in the Y winding, while the Δ winding is essential in stabilizing the grounded neutral point by creating a cancelling effect by a flow of opposing (reflective) currents in the windings.

However, the Δ winding breaks the flow of ground fault current from the grounded Y side of the transformer to the LV side. Therefore, a Δ transformer winding segments the power system into several subsystems for the flow of ground currents. This is considered more of a benefit than a deficiency as different type of grounding of the MV Δ winding is often desired.

The transformer neutral fault current I_n passing through the system reactance is calculated as follows:

$$\text{Solid neutral grounding: } I_n = \frac{3\,E}{X_1 + X_2 + X_0},$$

$$E = \text{phase to neutral voltage} \tag{5.2}$$

$$\text{Resistance neutral, } R_n \text{ grounding: } I_n = \frac{3E}{3R_n + X_1 + X_2 + X_0}$$

$$= \frac{E}{R_n} \quad \begin{array}{l} \text{phase sequence reactances} \\ \text{are ignored} \end{array} \tag{5.3}$$

5.2.2.2 Y/Y Transformers

This type of transformer is generally not used in the industry for power distribution. The main reason is its inability to provide effective grounding even if their neutrals are grounded. Let us explain:

(1) Unlike the Δ winding where the third harmonics are shorted within the delta, the Y–Y connection allows a flow of third harmonics into the

external circuit. In balanced conditions, these harmonics are in phase with the transformer magnetizing current. Their three phase sum at the neutral of the star neutral point is not zero, and hence, it will distort the flux wave. This has an effect of producing a voltage having harmonics in each winding of the transformer.

(2) *Y/Y transformer*: Primary grounded. The zero sequence current I_{10} in the primary can flow, providing I_{20} in secondary winding can flow in opposing direction. Conditions for the creation of the opposing current are there, but unlike in a Δ circuit, the flow is severely restricted by the impedances of the Y external circuit. Since the opposing I_{20} is restricted, the I_{10} is limited to the magnetizing current with virtually infinite impedance. Therefore, the grounding path is blocked.

(3) *Y/Y transformer*: Secondary grounded. Similar situation as aforementioned. The same conclusions in reverse order.

(4) *Y/Y transformer*, both windings grounded. The zero sequence current I_{10} or I_{20} can flow, providing the reflective zero sequence current on the opposing side can find a closed circuit at some point along the connected circuit. That is inconclusive. Unlike in the Δ windings, the zero sequence currents flowing in the neutral wire are additive and do not cancel out, thus causing interference to telephone lines located along the same route.

5.2.2.3 Y/Y Transformers with Delta Tertiaries

Yes of course, a Y/Y transformer can be used in the industry, providing it comes with a third winding called Δ tertiary. It is called tertiary since it is an addition to the primary and secondary windings. The Δ winding offers the Y/Y transformer the grounding capability on either winding, and it is an ideal solution for power systems. The Δ winding is rated at about 35% of the nominal transformer capacity, capable to carry the full ground fault current on the solidly grounded neutral on the primary or secondary winding. Usually, the transformer secondary side is resistance grounded and has a lower ground fault current, though. Furthermore, the Δ winding also circulates the third harmonic component of the excitation current, which in fact mitigates the harmonic disturbance situation on the system.

The tertiary winding serves as a ground fault administrator for either Y winding; primary or secondary. The tertiary winding can even be loaded to provide some station service to the HV substation.

Suppose we have a 230 to −13.8 kV, Y/Y transformer with a 4.16 kV Δ tertiary, the primary would typically be solidly grounded, while the secondary 13.8 kV winding would be resistance grounded to limit the fault current to about 50–100 A. The delta tertiary can feed into a 4.16 kV service switchboard.

5.2.2.4 Autotransformers

This is basically a two winding transformer with both HV sides on the same winding, and a separate delta winding is provided for the circulation of the third

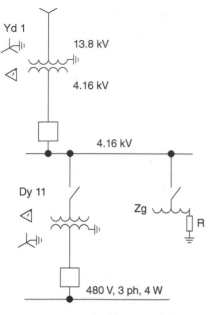

Figure 5.5 Grounding Zg transformer.

Yd 1
13.8 kV
4.16 kV

4.16 kV

Dy 11
Zg R

480 V, 3 ph, 4 W

harmonic. It has similar attributes as a YY transformer with a delta tertiary. It is mainly applied for connecting two system voltages such as 230–500 kV and is usually of single phase construction due to their large size. The delta winding may or may not be brought out. If brought out, it could be used as a station service supply or for an application of line capacitors (Figure 5.5).

5.2.3 Grounding Transformers

5.2.3.1 Zg or Y/Δ

Sometimes, we are faced with a situation where the power system at a certain voltage is delta. How do you ground a Δ system? [2] Well, there is a way. Suppose, your power system includes a 13.8/4.16 kV main transformer in a Y/Δ configuration for the reasons we have explained earlier. Further down, you would like to have LV as a 480 V solidly grounded system. For that you will need 4.16 kV to 480 V transformers in a Δ/Y configuration. The purpose of that would be to provide a possibility to ground the secondary winding and create a four wire system, which would be used for single phase loads, like lighting and other services.

So, now you have an ungrounded 4.16 kV delta system and with it all the problems of an ungrounded (unprotected) system. What to do?

Ground it! Use a grounding transformer, of course. There are two types of grounding transformers: (i) Zg in Figure 5.6 or (ii) Y/Δ transformer in Figure 5.7. Either one will do. In both cases, the primary of the grounding transformer would be 13.8 kV with a neutral point.

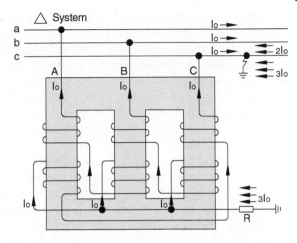

Figure 5.6 Zg grounding transformer.

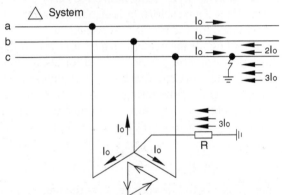

Figure 5.7 Fault currents in Y/Δ transformer.

The difference is that the Y/Δ is a two winding transformer, while the Zg is a single winding transformer. Both offer a neutral point, which can be grounded solidly or through a resistor, as you wish.

We mentioned earlier that MV levels are typically resistance grounded to about 50–100 A. Therefore, the grounding transformer will be resistance grounded. A current transformer will be installed on the neutral lead with a protective relay to serve as the grounding protection for the previously ungrounded Δ system.

The Zg grounding transformer, as shown in Figure 5.6, has one 13.8 kV winding. The Zg transformer in fact has two half windings for each transformer iron limb. The two part windings on each leg come from two different phases and are connected in opposing directions. The zero sequence currents in each phase flow in opposite directions, resulting that the magnetic flux in each core limb oppose and neutralize. The fault current is allowed to flow freely, effectively the same as in a Δ winding of a Y/Δ transformer.

As shown in Figure 5.7, the Y/Δ transformer primary side is grounded. In case of a fault to ground on the 13.8 kV, three zero sequence currents $(3 \times I_0)$ are created. All three currents are in phase flowing to the point of fault. Equally on the delta side, the zero sequence currents are flowing around the delta winding in opposite direction to cancel the flux creation by the I_0 currents in the primary. Without the flux, the V_0 zero sequence voltage is not created thus the neutral point is firmly pinned to its star point. Therefore, any Y/Δ, or Δ/Y transformer can serve as a grounding transformer as well as a service transformer.

Zg transformer terminates the harmonics of the power system. Furthermore, Zg grounding transformer can also serve as a service transformer if an extra winding, typically a Y winding, is added as its secondary winding. This secondary winding option is sometimes used in the industry. However, if the fault current is limited by a neutral resistor to a value of, for instance, $I_n = 90$ A only, the transformer kVA rating is sized to carry the ground fault current as a minimum. If the Zg transformer is also to be used for service, the kVA rating must be increased for the required service load.

Grounding transformers used for grounding purposes are typically short-time operating devices, rated for about one to five minutes for the full (ground) fault current. This by some estimates would be 20% of the continuous rating. The short time rating is based, as we noted earlier, on the time for the backup protection to clear the ground fault. The grounding transformers fitted with neutral resistors limit the flow of fault current I_f to:

$$I_f = \frac{V}{\sqrt{3}R_n} \tag{5.4}$$

where V is line to line voltage, R_n is the neutral resistor in Ohms.

The grounding transformers are permanently connected to the system, but loaded only during the system ground faults. The kVA rating of the grounding transformer, for instance for a 13.8 kV, neutral current $I_n = 90$ A is calculated as follows:

The loading current in neutral is: $I_n = 3 \times I_0 = 3 \times 30$ A $= 90$ A,

Grounding transformer, single phase kVA rating is

$$KVA = \frac{\sqrt{3}\ I_n\ V}{3\ K} = \frac{I_n\ V}{\sqrt{3}\ K}$$

$$= \frac{90 \times 13.8}{\sqrt{3} \times 2.8} = 256\ kVA \rightarrow \text{select: } 300\ kVA$$

$K = 2.8$, short time derating factor for five minutes.

Typical derating factor, K:

1 minute	4.7	30 minutes	1.8
5 minutes	2.8	60 minutes	1.6

The impedance to zero sequence currents of a effectively grounded transformer is slightly less than equal to the ohmic leakage impedance between the primary and secondary windings: $Z_0 = 0.9Z_1$. For the Z_g solidly grounded grounding transformer Z_0 can be assumed to be equal: $Z_0 = Z_{pg}$, impedance from primary to ground (see Chapter 16 for ground fault protection).

References

1 Donald Beeman General Electric Co. (1955). *Industrial Power Systems Handbook*. McGraw Hill Book Co.
2 Westinghouse Electric Corp (1964). *Electrical Transmission and Distribution Reference Book*. Westinghouse.

6

Site and Equipment Grounding

CHAPTER MENU

6.1 Requirements, 137
    6.1.1 Grounding Grid Design, 138
    6.1.2 Soil Conditions, 139
    6.1.3 Test Measurements, 140
6.2 Ground Potential Rise and Step and Touch Potential, 142
    6.2.1 Ground Potential Rise (GPR), 142
    6.2.2 Step Potential, 144
    6.2.3 Touch Potential, 145
    6.2.4 Reducing Step and Touch Potential Hazards, 145
    6.2.5 Human Tolerance, 146
6.3 Computer Study Report, 147
    6.3.1 The Study, 147
6.4 Below Ground Equipment Grounding, 148
    6.4.1 Connections to Rebar, 149
    6.4.2 Foundation Grounding (Ufer), 150
6.5 Above Ground Equipment Grounding, 150
6.6 Telecommunications in HV Substations, 152
6.7 Fence Grounding, 152
6.8 Plant Control System Grounding, 154
6.9 Overhead Line Grounding, 154
6.10 Remote Site Grounding, 154
6.11 Effect of Overhead Ground Wires and Neutral Conductors, 155
References, 155

6.1 Requirements

The grounding system must provide reliable and safe means of equipment grounding in order to

- Provide a low resistance path for the dissipation of ground fault currents, lightning, and switching surges into the earth without exceeding any operating and equipment limits.

Practical Power Plant Engineering: A Guide for Early Career Engineers, First Edition. Zark Bedalov.
© 2020 John Wiley & Sons, Inc. Published 2020 by John Wiley & Sons, Inc.

- Minimize the ground potential rise (GPR) between the equipment and the soil within safe limits throughout the site for the protection of the operating personnel and equipment.
- Provide a reasonably uniform potential rise in all parts of the plant to minimize the step and touch potentials throughout the plant for the protection of the plant personnel.
- Properly channel the ground fault and unbalanced currents to reduce electrical noise in the communication and data cables.

The equipment grounding covers the plant site grounding grid, the metallic structures, grounding connections below and above ground level, rebar connections, and interface with the lightning protection for all the plant areas, indoors and outdoors. It encompasses the system neutral ground connections, grounding of the plant transformers, motors, main electrical and mechanical equipment rooms, enclosures, conduits, cable trays, cable armor and shielding, plant fence and plant instrumentation.

In a plant, the equipment grounding can be divided in two parts, as discussed later in this chapter.

- *Below ground grounding*: This is the grounding hardware, cables, and steel bars that are buried or embedded in concrete walls and foundations, including the main substation. This work is executed by the civil contractor in the early phase of the plant construction, based on the electrical design.
- *Above ground grounding:* This is the grounding hardware and ground conductor connections to the below ground grounding grid. It includes the connections to the plant metal structures and electrical and mechanical equipment. This work is conducted by the electrical contractor during the installation phase of the project.

6.1.1 Grounding Grid Design

Prior to designing the plant grounding, the site soil tests must be conducted in all the areas of the plant, the electrical switching areas in particular to determine the soil resistivity (Ω-m), soil wetness and the acidity (pH factor). Soil resistivity ranges from 10 to 10 000 Ω-m. The higher the soil resistivity, the more grounding hardware will be needed to be able to establish satisfactory and safe grounding conditions.

Conceptual analysis of a grid system usually starts with inspection of the substation layout plan, showing all major equipment and metallic structures. The following points may serve as guidelines for starting a typical grounding grid design [1]:

(1) A continuous conductor loop shall surround the perimeter to enclose as much area as practical. This measure helps to avoid high current

concentration and, hence, high gradients in the grid area. Enclosing more area also reduces the resistance of the grounding grid.

(2) Within the grounding loop, conductors are typically laid in parallel lines and, where practical, along the structures or rows of equipment to provide for short ground connections.

(3) A typical grid system for a substation may include bare copper conductors buried below grade, spaced 3–7 m (10–20 ft) apart, in a grid pattern. At cross-connections, the conductors are securely bonded together. Ground rods may be placed at the grid corners and at junction points along the perimeter. Ground rods are also installed at major equipment, especially near surge arresters. In multilayered or high resistivity soils, it is useful to use longer rods to reach wetter layers.

(4) The grid system is extended over the entire substation and beyond the fence line. Multiple ground leads or large-sized conductors are used where high concentrations of current may occur, such as at a neutral-to-ground connection of generators, capacitor banks, or transformers.

(5) The ratio of the sides of the grid meshes usually is from 1 : 1 to 1 : 3, unless a precise computer analysis warrants more extreme values. Frequent cross-connections have a relatively small effect on lowering the resistance of a grid. Their primary role is to assure adequate control of the surface potentials. The cross-connections secure multiple paths for the fault current, minimizing the voltage drop in the grid itself, and provide a certain measure of redundancy in case of a conductor failure. Equal division of currents between multiple ground leads at cross-connections or similar junction points should not be assumed.

(6) The National Electrical Code (NEC) states that the resistance to ground shall not exceed 25 Ω. This is an upper limit and a guideline. Much lower resistance is required. The lower the ground resistance, the safer it is for personnel and equipment. Desired values in transmission substations should not exceed 1 Ω. In distribution substations, the maximum recommended resistance is 5 Ω or even 1 Ω. In most cases, the buried grid system of any substation will provide the desired resistance. In light industrial or in telecommunication central offices, 5 Ω is often the accepted value. For lightning protection, the arrestors should be coupled with a maximum ground resistance of 1 Ω.

6.1.2 Soil Conditions

An understanding of the soil resistivity and how it varies with depth in the soil is necessary to be able to design the grounding system in an electrical substation or for lightning arresters. Soil resistivity is a measure of how much the soil resists the flow of electricity, in this case the fault current.

Soil resistance is measured between two opposing surfaces of a 1 m^3 of homogeneous soil material. The SI unit of resistivity is the ohm-meter (Ω-m); the ohm-centimeter (Ω-cm) is often used.

$$1 \ \Omega\text{-m} = 100 \ \Omega\text{-cm}$$

Sometimes the conductivity, the reciprocal of the resistivity is quoted instead. Refer to IEEE Std. 80 [1] for further clarifications.

Salt content, wetness, and soil temperature in that order are the key factors in determining the soil conductivity. Rainwater has much smaller effect than sea water, 50-fold.

Moisture content changes seasonally. It varies according to the nature of the sub layers of earth, and the depth of the permanent water table. Since soil and water are generally more stable at deeper strata, it is recommended that the ground rods be placed as deep as possible into the earth, at the water table if feasible. Also, ground rods should be installed where there is a stable temperature, i.e. below the frost line. For a grounding system to be effective, it should be designed to withstand the worst possible conditions.

Soil resistivity readings of up to 250 Ω-m are considered good conducting soil. High soil acidity (salty soil) may dictate the copper conductors to be tinned to avoid excessive corrosion.

For instance, the following soil resistivity in Ω-m is expected:

• Wet organic and surface soil	10
• Surface soil	10
• Clay	50–100
• Moist soil	100
• Dry soil, sand, gravel	500–1000
• Bedrock and limestone	10 000

6.1.3 Test Measurements

6.1.3.1 Soil Resistivity Measurements

The soil test measurements [2, 3] are performed by specialized contractors having proper test measuring equipment. Estimates based on soil classification yield only a rough approximation of the resistivity. The tests should be made at a number of places within the site. Typically, there are several layers, each having a different resistivity. Lateral changes also occur, but in comparison to the vertical ones, these changes usually are more gradual. The number of readings taken should be greater where the variations are large. The tests must be extended to the depths of 3 m (9 ft) or more to determine the conductive layers, which may be the most beneficial in distributing the fault currents.

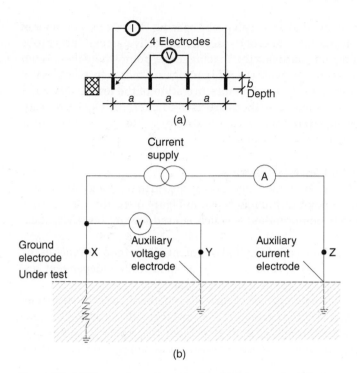

Figure 6.1 (a) Wenner four-pin method and (b) three-point fall of potential method.

The Wenner four-pin method, as shown in Figure 6.1a, is the most popular method in use. There are a number of reasons for its popularity. The four-pin method obtains the soil resistivity data for deeper layers without driving the test pins to those layers. No heavy equipment is needed to perform the test.

In brief, four probes are driven into the earth along a straight line, at equal distances apart, driven to a depth b. Using this method, if b is small compared to a, as is the case of probes penetrating the ground only for a short depth, the apparent soil resistivity value is

$$\rho_E = 2\pi a R_W \tag{6.1}$$

where

ρ_E = measured apparent soil resistivity (Ω-m)
a = electrode spacing (m)
b = depth of electrodes (m)
R_W = Wenner resistance measured as "V/I" (Ω)

This measurement need to include the temperature and moisture content of the soil at the time of the measurement.

The calculated resistivity is, to a first approximation, the average resistivity of the soil from the surface to a depth equal to the probe spacing a. Usually, the resistivity will be found to vary with probe spacing. Therefore, a number of readings are taken using different probe spacings, typically starting with a probe spacing of 0.5 or 1 m and increasing the spacing to the point where there is not enough space to measure. The wide spaced readings are more representative of the resistivity of deeper soil layers. The interpretation of the test results is not obvious, and usually, computer methods are used to interpret a series of readings into an equivalent layered soil resistivity structure. The depth of electrode shall typically not exceed the value of $a/20$.

6.1.3.2 Grounding Grid Test Measurements

The most commonly used method of measuring the earth resistance of an earth electrode is the three-point technique shown in Figure 6.1b. This method was derived from the four-point method – which is used for soil resistivity measurements.

The three-point method, called the "fall of potential" method, comprises the Earth Electrode to be measured and two other electrically independent test electrodes, usually labeled Y (potential) and Z (current).

The test Y and Z electrodes must be electrically independent of the electrodes to be measured. An alternating current (I) is passed through the outer electrode Z, and the voltage is measured, by means of an inner electrode P, at some intermediary point between them. The earth resistance is simply calculated using Ohm's law:

$$R_g = V/I \qquad\qquad (6.2)$$

Position of the auxiliary electrodes Y and Z on measurements is important. The goal in measuring the resistance to ground is to place the auxiliary current electrode Z far enough from the ground electrode under test so that the auxiliary potential electrode Y will be outside of the effective resistance areas of both the grid electrode and the auxiliary current electrode. The best way to find out if the auxiliary potential rod Y is outside the effective resistance areas is to move it between X and Z and to take a reading at each location. If the auxiliary potential rod Y is in an effective resistance area, the readings taken will vary noticeably in value. Under these conditions, no exact value for the resistance to ground may be determined.

6.2 Ground Potential Rise and Step and Touch Potential

6.2.1 Ground Potential Rise (GPR)

The GPR is a product of the fault current flowing into the grounding system times the resistance to the remote reference ground of the site grounding system. The remote ground or often called reference ground is assumed to be at the zero potential.

$$\text{GPR} = I_f \times R_g \; (\text{V}) \tag{6.3}$$

where

I_f = ground fault current (A)
R_g = ground resistance (Ω)

Under normal conditions, the grounded electrical equipment operates at near zero ground potential. That is, the potential of a grounded neutral conductor is nearly identical to the potential of **remote earth**. During a ground fault, the portion of fault current that is conducted by a grounding grid into the earth causes the rise of the grid potential with respect to remote earth.

Reducing resistance to ground of the site is often the best way to reduce the negative effects of any ground potential rise event, where practical.

For example, if the fault current at a high-voltage tower is 5000 A and the resistance to ground of the grounding system is 10 Ω, the GPR will be 50 000 V. If we reduce the resistance to ground down to 1 Ω, the fault current may increase to 11 000 A as a result, while the GPR becomes 11 000 V. The gradient of the potential rise over the step distance is the critical issue. Certainly, 11 kV GPR comes with much lower gradient than a 50 kV GPR/m.

Grounding grid is a system of horizontal interconnected bare conductors buried in the earth, providing a common ground for electrical devices or metallic structures, usually in one specific location. Grids buried horizontally near the earth's surface are effective in controlling the surface potential gradients. A typical grid usually is supplemented by a number of ground rods to lower its resistance with respect to remote earth.

Typically grounding grid loops (Figure 6.2) in electrical substations use bare copper conductor size of #4/0 AWG to 500 kcmil (120–500 mm²) medium drawn, depending on the fault current to be dissipated. Copper conductors, in addition to their high conductivity, have the advantage of being resistant

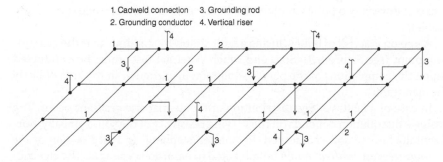

Figure 6.2 Part of plant grounding grid (mesh).

to most underground corrosion because copper is cathodic with respect to most other metals that are likely to be buried in the vicinity. The loop is buried in direct contact with earth, 0.9 m (3 ft) from the perimeter of the object, structure or building, buried 0.5 m (18 in.) below grade. The purpose of the loop is to minimize the voltage between the object and the earth surface where a person might be standing while touching the object, i.e. to minimize the touch potentials.

It is important that all the metallic objects in a GPR environment are bonded to the ground system to eliminate any difference in potentials. It is also important that the resistivity of soil as a function of depth be considered in computed touch and step voltages and in determining at what depth to place the ground conductors and grounding rods. For example, in a soil with a dry, high resistivity surface layer, conductors in this layer will be ineffective; a layer beneath the one with lower resistivity would be a better location for grounding conductors. On the other hand, long grounding rods or deep wells may be ineffective if high resistivity layer stretches further down. Permafrost layers are also high resistance grounds.

To combat the GPR, the Cu grounding mesh must be tighter (smaller mesh squares). Also, placing the horizontal grounding loop conductors closer to the surface may result in reduction in touch potentials. This is not necessarily always the case, as the conductors close to the surface are likely to be in drier soil, with a higher resistivity, thus reducing the effectiveness of these conductors. While touch potentials immediately over the loop may be reduced, touch potentials at a short distance away may actually increase, due to the decreased zone of influence of these conductors.

6.2.2 Step Potential

The step voltage is the in surface potential between the feet of a person standing near an energized grounded object. It is given by the voltage rise distribution curve, between two points at different distances from the electrode (grounding cable).

A person is at risk of injury during a fault simply by standing near the grounding point (see Figure 6.3). Step and touch potential tests must be conducted upon the completions of the plant installation in particular in the area of likely human traffic.

In case of a fault, voltage distribution and potential rise gradient occur. The voltage distribution depends on the varying resistivity of the soil. Typically, horizontally layered soil is assumed. Personnel "stepping" in the *direction of the voltage steepest gradient* could be subjected to hazardous voltages. The electricity will flow if a difference in potential exists between the two legs of a person.

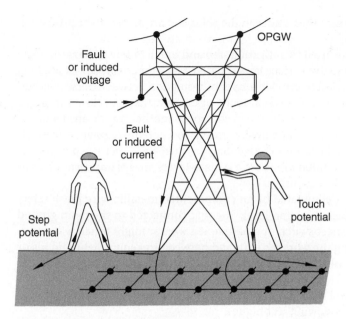

Figure 6.3 Step and touch potential.

The shorter the steps, the lower the potential. Calculations must be performed to determine the GPR and compare the results to the step voltages expected to occur between the buried conductors (electrodes) at the site. Suppose for a minute that a substation is completely covered by a grounded metallic plate. There would not be step potentials anywhere as the distribution would be one large equipotential plate without gradients. Unfortunately, that is not a practical solution.

6.2.3 Touch Potential

This is the voltage difference between an object energized (during a lightning or a fault) and the hand of a person standing and touching at a distance from the object. The closer the person stands to the object, the lower the voltage difference between the two points. The touch potential could be nearly the full voltage across a grounded object if that object is grounded at a point remote from the place where the person is in contact with it.

6.2.4 Reducing Step and Touch Potential Hazards

One of the simplest methods of reducing step and touch potential hazards is to wear electric hazard gloves and shoes. Dry, properly rated electric hazard shoes have millions of ohms of resistance in the soles and are an excellent protection for personnel safety.

Another technique used in mitigating step and touch potential hazards is the addition of more resistive surface layers. Often a layer of crushed rock is added to a tower or substation to provide a layer of insulation between personnel and the earth. This layer reduces the amount of current that can flow through a given person and into the earth. Weed control is another important factor, as plants become conductive and more hazardous to a person. Asphalt is an excellent alternative. It is far more resistive than crushed rock, and weed growth is not a problem. An addition of resistive surface layers always improves personnel safety during a GPR event.

The hazard is best eliminated by an equipotential metallic wire mesh safety mat installed just below ground level. The mat connected to the main ground grid and any disconnect switch or equipment a worker might touch will equalize the voltage along the worker's path and between the equipment and his or her feet. With the voltage difference (potential) thus essentially eliminated, the hazard to personnel is virtually eliminated as well.

Mitigating step and touch potential hazards is usually accomplished through one or more of the following techniques:

- Reduction in the resistance to ground of the grounding system,
- Proper placement of ground conductors,
- Addition of resistive surface layers, crushed stone, and asphalt, and
- Installing equipotential ground metallic mesh mats close to the equipment.

6.2.5 Human Tolerance

What are the tolerable limits for the step and touch voltages? According to the IEEE Std. 80 [1], the tolerance depends on the relay clearing time, the voltage, and frequency. Tests and experience show that the chance of severe injury or death is greatly reduced if the duration of a current flow through the body is brief. Furthermore, humans are vulnerable to the effects of electric current at frequencies of 50 or 60 Hz. Currents of approximately 0.1 A can be lethal. Research indicates that the human body can tolerate a slightly higher 25 Hz current and approximately five times higher direct current. It is not until current magnitudes in the range of 60–100 mA are reached that ventricular fibrillation or stoppage of heart occurs.

The probability of lethal exposure to electric shock is greatly reduced by fast fault clearing time, in contrast to situations in which fault currents could persist for several minutes or possibly hours.

6.3   Computer Study Report

6.3.1   The Study

The computer simulation report (Figure 6.4) must be prepared based on the soil resistivity, expected ground fault current/clearing time, and the grounding grid layout to determine a preliminary calculated overall ground resistance, GPR contours, and step touch profiles in all areas based on the IEEE Std. 80 [1]. The study must address the following principal issues:

(1) Determine the grounding grid to render the lowest possible ground resistance based on the soil conditions at site, to minimize the GPR.
(2) Provide step and touch potential profiles and tabulations along the site in all directions.
(3) Determine the amount of ground conductor locations and rods to safely dissipate the given projected fault current in kiloampere.
(4) Recommend the grid configuration and mesh density in all plant locations to minimize the potential gradients for touch and step potentials.

The report must present the details of the amount of conductors and grounding rods to be installed, the grid mesh and size pattern, where to install them, all based on the soil's capability to dissipate the fault current considered for the project. The report must give pointers on where to enhance the grounding by adding more conductors and/or grounding rods.

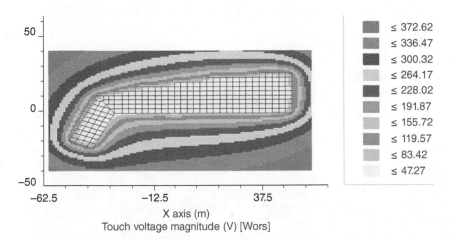

Figure 6.4 GPR equipotential contours.

The fault current considered for the project must be reasonable in kiloampere and duration (seconds) based on the plant system grounding design.

Following the field tests, additional rods may be added at specific locations to insure the compliance with the standards with respect to the gradients of potential rise and step and touch voltages.

The plant grounding grid will be connected to the line overhead ground wire (OHGW) at the line takeoff structures for connection to the utility over the transmission line. The effect of the OHGW connections to the utility is not taken into calculations of the overall site resistance and ground fault distribution.

6.4 Below Ground Equipment Grounding

A grounding grid made up of bare stranded, medium drawn 4/0 AWG (120 mm^2) copper conductors or alternatively steel flat bars is laid below the foundations under the buildings and in a loop around the facility buildings in all the areas of the plant including the outdoor transformers and fence. The grounding mesh density will depend on the ground fault current, which is usually given for the project.

The grounding conductors and connections (Figure 6.5) should be able to withstand the ground fault current for the duration of the fault, without being damaged by thermal, thermomechanical, or electromechanical stresses. The overall grid resistance to the remote reference ground should hopefully not exceed 1 Ω.

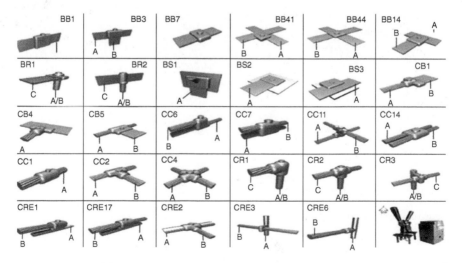

Figure 6.5 Cadweld underground connections.

The grounding conductors are connected to the foundation rebar and the steel piling driven under the building foundations to enhance dissipation of the fault current into ground (below the frost line if applicable) to achieve an acceptable equipotential ground plane within the plant thus creating a low touch and step gradient environment.

6.4.1 Connections to Rebar

For the interconnection of rebar on rebar (reinforcing steel) in power houses, it is not necessary to bond every reinforcing bar, but to every other one [4]. A few large diameter bars should be made continuous and used to bring out ground conductors. The other bars will become part of the ground electrode system through the many contact points and tie wires that are used to hold the reinforcing bar cages together during construction. The regular reinforcing steel is bare and cannot be welded. Weldable reinforcing steel should be interspersed with the regular steel so that bonds can be made to the complete steel cage. The weldable bars should be as long as possible and run parallel to major structural bars. They should be attached to the major bars with tie wrap wires at a maximum of 0.5 m (20 in.) spacing. The weldable bars should form an independent continuous cage within the nonweldable bars.

The continuous bars can be shaped to match the profile of the bottom of the structure. The layout should be such that cross-connections and vertical riser conductors discussed line up with the steel columns on floor above. When the steel columns are bonded to the rebar in this way, the requirements for lightning conductors are also met, and it is not necessary to install separate lightning conductors.

The underground joints between the grounding conductors are performed by exothermic welds (Cadweld, see Figure 6.6). The below ground grid is brought to the surface at a number of strategically located pigtail connections and wall plates to the building steel columns and structures and by exothermic welds. The buried exothermic connections, copper to steel piles, are covered with special protective paste.

On the outside, the grid and the top of the grounding rods are buried 50 cm (18 in.) below the final grade by earth fill. The grounding grid is typically

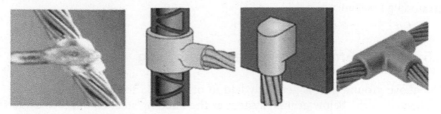

Figure 6.6 Cadweld connections.

Figure 6.7 Rebar connection.

anchored by 5/8 in. diameter copper clad steel rods, 3 m (16 ft) long, and a number of test wells. Ground rods can be extended and placed deeper if the soil at higher depths is found to be wetter and more conductive.

The area of the transformers and operator switching shall be provided with a separate grounding grid and operating mats, which are connected to the overall plant grounding grid.

6.4.2 Foundation Grounding (Ufer)

The use of foundation (Ufer) grounds has significantly increased in recent years, mainly due to the high cost of Cu conductors. Ufer grounds utilize the concrete foundation of a structure plus building steel and reinforcing bars as additional grounding electrodes. Even if the anchor bolts are not directly connected to the reinforcing bars, their close proximity and the semiconductive nature of concrete will provide an electrical path (see Figure 6.7).

The concrete itself has a relatively low resistivity when buried. The range of resistivity values given in the IEEE Guide is 30–90 Ω-m.

Proper design of Ufer grounds provides for connections between all steel members in the foundation and one or more metallic paths to an external ground rod or main ground grid, thus spreading the ground fault current to a larger area for dissipation of fault currents and reduction of the overall grounding resistance.

6.5 Above Ground Equipment Grounding

The above ground grid is generally laid in or attached to the cable trays and connected to the below ground system at the building steel columns and the

Figure 6.8 Mechanical crimp connections.

preinstalled grounding plates or pigtails. The connecting hardware is shown in Figure 6.8.

A layout drawing is generally not necessary for this engineering activity. Instead, the grounding requirements and instructions are described in the installation specifications.

The "Above Ground Grounding" includes the grounding connections to cable trays, motors, motor control centers (MCCs), unit substations, transformer neutrals, and metallic structure to the grounding grid as follows:

Cable trays: The bare ground conductor 120 mm^2 (4/0 AWG) is laid on the out-side of one of the trays in each bank of trays. Connections to the trays and all the electrical and mechanical equipment are by mechanical compression type crimp connections. The ground cable is connected to all the trays in the stack and to the main grid at intervals not more than 10 m (30 ft) apart.

Frames and enclosures: This electrical equipment is grounded by mechanical connectors. Connections to the equipment is by 20 mm^2 (#2 AWG) conduc-tors or larger. Large electrical equipment like generators, motors, transform-ers, switchgear, MCCs, and unit substations are grounded at two corners with 120 mm^2 (#4/0 AWG) grounding conductors.

Road lighting poles: The poles in the plant are grounded at every pole by a 3 m long grounding rod. The ground connection is made to the pole base if the poles are made of metallic material. If the poles are made of wood, the con-nection is made to the pole grounding cable, which connects all the pole metallic hardware.

Cable grounding: The cable sheath and armor (usually aluminum) of three-core and single-core power cables are commonly grounded at one end only at the source end to avoid circulating currents in the armor and sheath (see Chapter 12 for more details).

Motors: Motors up to 100 kW are grounded through the ground conductor in the power cable inside the motor cable box. Motors 100 kW and larger are grounded by means of the ground conductor in the power cable as well as by an external grounding cable to the motor base (Figure 6.9). Grounding conductors are contained together with phase wires in the same cable and brought to all 480 V electrical equipment.

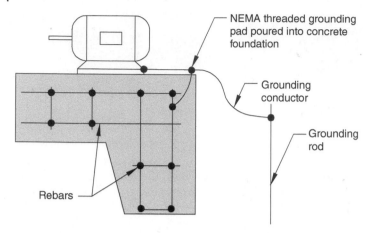

Figure 6.9 Motor base grounding.

6.6 Telecommunications in HV Substations

When telecommunication lines are needed at a high-voltage substation site, special precautions are required to protect the switching stations from unwanted voltages. Running any copper wire into a substation or tower is going to expose the other end of the wire to hazardous voltages and transfer the potential rise. Certain precautions are required.

Industry standards regarding these precautions and protective requirements are covered in IEEE Standard 487 and IEEE Standard 1590 [5, 6]. These standards require that a ground potential rise study be conducted so that the 300 V peak line can be properly calculated.

To ensure proper site grounding and telecommunication tower grounding, telecommunication standards require that fiber-optic cables be used instead of copper wires within the 300-V peak line. A copper-to-fiber conversion box must be located outside the GPR event area at a distance in excess of the 300-V peak or 212 V r.m.s. line. This means that based on the calculation results, copper wire from the telecommunications company may not come closer than the 300-volt peak distance to prevent any unwanted voltages from entering the phone companies' telecommunications network.

The most noted suggestion is a newer standard, IEEE Standard 1590-2003, that lists a 150-m (~500 ft) mark as a default distance of Cu–fiber conversion; if a GPR study has not been conducted at a given location.

6.7 Fence Grounding

In power and large industrial plants, substation perimeter grounding is part of the overall grounding system. Fence must be properly grounded as per

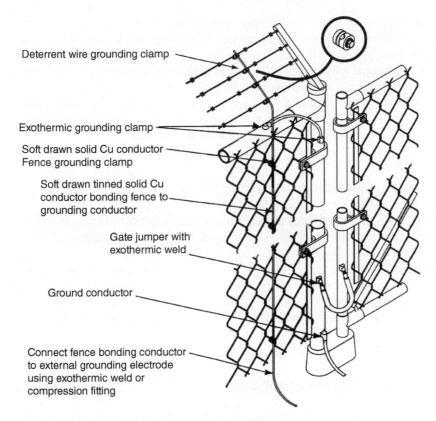

Deterrent wire grounding clamp

Exothermic grounding clamp

Soft drawn solid Cu conductor
Fence grounding clamp

Soft drawn tinned solid Cu
conductor bonding fence to
grounding conductor

Gate jumper with
exothermic weld

Ground conductor

Connect fence bonding conductor
to external grounding electrode
using exothermic weld or
compression fitting

Figure 6.10 Fence grounding.

Figure 6.10 for the protection of the operator or onlookers and livestock in the area for both; step and touch potential. In power and large industrial plants, substation perimeter grounding is part of the overall grounding system.

Touch and step potentials on both sides of the fence must be within the acceptable limit for a fault current of essentially the same maximum value as for the substation, if connected to the main substation grid. Actually, the fence is the most dangerous area of a substation, as the GPR gradient may be the highest at this location due to larger separation of the grounding conductors buried under the ground. A single 120 mm^2 (#4/0 AWG) grounding conductor is buried 50 cm (18 in.) deep and 0.9 m (3 ft) on the inside and outside of the fence.

Flexible braid ground straps are installed across all hinge points and gates. Connection of the fence to the buried conductor is at 30 m (100 ft) intervals, maximum.

Additional grounding protection and hardware is provided in the area of the fence where overhead lines crosses over, due to the induction of the overhead lines to the fences at times of ground faults.

The lines crossing a substation fence have shorter spans and are dead-ended at one or both ends. Hence, the danger of a line falling on a fence is usually not of great concern.

Also, additional grounding is provided in the fence entrance area, inside and outside the fence as an equipotential plane.

An area 2 m (6 ft) wide and 0.15 m (6 in.) deep along the fence, inside and outside, is covered with crushed stones to maintain 1000 Ω-m surface resistivity and have acceptable touch and step potentials throughout.

6.8   Plant Control System Grounding

The grounding system employed for the plant distributed control system (DCS) and instrumentation is made to suit and based on the recommendations by the supplier of the control equipment.

The instrumentation grounding is tied to the electrical grounding system. Electromagnetic induction (EMI) low-level signals in instrumentation and communication circuits must be restricted by proper separation. Proper grounding of protective shields will be practiced in order to minimize the effects of EMI.

6.9   Overhead Line Grounding

The medium voltage overhead lines hardware and poles are grounded at each pole to achieve a grounding of <25 Ω/pole. This may be achieved by adding grounding rods and counterpoise wires spread around the pole, or both. Wood distribution poles are encircled with grounding wire around its butt.

6.10   Remote Site Grounding

Suppose our plant site is spread around with some small facilities located several kilometers away from the central plant. Naturally, each remote site like main substation, camp, warehouse, tailing would has its own grounding system. Now, do we want to interconnect the grounding grid of these remote sites to the main plant site? The interconnections would supposedly enlarge the overall grounding grid, reduce the overall ground resistance, and reduce the potential rise.

Is it worth the additional cost of interconnection? The main substation in our case contains a mixture of voltages; LV, MV, and HV. It will have a large grounding grid on its own and may be a source of large fault currents during ground faults. The plant camp facility will also have its own small grounding grid and the voltages not greater than the low voltage. Similarly, for the remote warehouses, tailing, etc.

By interconnecting the grounding grids of these small facilities to the main grounding grid, there is a transfer of potential from the main site to the remote sites for every ground fault at the main grid. There may be some justification for interconnecting the main substation, but definitely not the camp, warehouses, and other small facilities. According to a BC Hydro Standard ES44Z1010 [7], there is no economic benefit of interconnecting the remote sites located more than 250 m from the main plant.

6.11 Effect of Overhead Ground Wires and Neutral Conductors

Where transmission line OHGWs or neutral conductors are connected to the substation ground, a substantial portion of the ground fault current is diverted away from the substation ground grid. Even if this beneficial situation exists, the designers often decline to take it into consideration in the design.

Connecting the substation ground to OHGWs or neutral conductors, or both, and through them to transmission line structures or distribution poles, will usually have the overall effect of increasing the GPR at tower bases, while lessening it at the substation. This is because each of the nearby towers will share in each voltage rise of the substation ground mat, instead of being affected only by a local insulation failure or flashover at one of the towers. Conversely, when a tower fault does occur, the effect of the connected substation ground system should decrease the magnitude of gradients near the tower bases.

References

1 IEEE Std. 80-2013: Guide for Grounding Safety in Substations.
2 AEMC (2019) Why Measure Soil Resistivity?
3 AEMC (2019) Understanding Ground Resistivity Testing.
4 Ground-it.com Consulting Ltd. (2014) Waneta Generating Station – Grounding Project Study.
5 IEEE Standard 487 and IEEE Standard 1590. Protection of Telecommunications.
6 IEEE Standard 1590-2003 Telecommunications in Substations.
7 BC Hydro Standard ES 44Z1010 (2011).

7

Plant Lighting

CHAPTER MENU

7.1 The Big Picture, 157
7.2 Lighting Design Criteria, 158
    7.2.1 Lighting Application, 160
7.3 Definitions, 162
7.4 Illumination Level, 164
    7.4.1 Candlepower Distribution Curves, 165
7.5 Outdoor Building and Road Lighting, 167
7.6 Lighting Hardware, 168
    7.6.1 Transformers for Lighting, 168
    7.6.2 Cables and Wiring, 169
    7.6.3 Lighting Fixtures, 169
7.7 Lamps Inside the Fixtures, 172
    7.7.1 Color Rendering Index (CRI), 172
    7.7.2 Fluorescent, 174
    7.7.3 Metal Halide (MH), 175
    7.7.4 Low-Pressure Sodium (LPS), 176
    7.7.5 High-Pressure Sodium (HPS), 176
    7.7.6 LEDs, 177
References, 178

7.1 The Big Picture

Lighting? That cannot be that tough, right? Everybody knows to change a light bulb or replace a fuse. I had a friend at our university who graduated as an engineer in lighting. I wondered, but never asked him, why he went to all that trouble of attending University to specialize in lighting?

I have met many good designers who were doing lighting, but also those who struggled and had absolutely not a clue how to do it. Well, they knew how to design lighting for a room but had a problem with the plant. They were just "not seeing the forest for the trees." It was not their skill that was missing. They were just not trained to look at "the big picture."

Practical Power Plant Engineering: A Guide for Early Career Engineers, First Edition. Zark Bedalov.
© 2020 John Wiley & Sons, Inc. Published 2020 by John Wiley & Sons, Inc.

Well, it turns out that there is a lot to a good lighting design. Lighting is a skill and a specialty when it comes to special effects in architecture design. We will not go into that type of artistic specialty, but concentrate on the lighting practiced in industrial and power plants.

First of all, plant lighting must be designed with a view of the whole plant, rather than designing it room by room and hope to arrive somehow at a compact and uniform plant design. The latter path would take you a long way around and be subject to endless changes and updates. Plant lighting is a power distribution system all by itself. It has its own one line diagrams, transformers, lighting panels, and distribution subpanels. It also has its own switching logic. Plant lighting design criteria must be established with principles and constraints that will be valid throughout the plant, rather than a single room.

Plant lighting systems must provide, in support of the plant production and operation, reliable uniform and substantial illumination in all areas of the plant to enable the plant operating staff to

- Safely operate and maintain plant facilities.
- Safely evacuate the plant in case of emergencies.

We will immediately make a decision that the lighting will be powered from the plant low voltage (LV) sources at 600 V (Canada), 480 V (USA and South America), and 400 V (Europe and Asia). The power will be sourced mainly from the LV motor control center (MCC)s and load-centers located at the various plant buildings. It will not be a single lighting system but a segmented plant system with individual panels and subpanels grouped around the load centers and MCCs. Having decided on the overall power distribution, we move on to establish the design criteria to define the various equipment choices and method of operation.

7.2   Lighting Design Criteria

The design criteria for plant lighting are the design basis and approach for our design. This allows us to start designing and selecting the equipment on a consistent basis. We should not be continually reinventing the criteria as we go along with our design.

(1) Transformers and panels

Lighting transformers	30 kVA, 480–480 V, 3 ph, Dy1, neutral solidly grounded
Outlets transformers	15 kVA, 480–120 V, 3 ph, Dy1, neutral solidly grounded
Lighting panel type	Surface mounted, 225 A mains, 3 ph, 480 V, 30–42 positions for 1p, 2p, 3p circuits

Subpanel type	Surface mounted, 3 ph, 480 V, 100 A mains, 12 positions for 1p, 2p, or 3p circuits
Panel main breakers	Provide ampere rating to match the mains ampere rating. (Mains are lighting panel busbars)
Branch breakers	Mixture of 15–100 A, 1p, 2p, 3p, molded case type, 10 kA s.c. interrupting
Branch breaker loading	80% maximum of its ampere rating, (e.g. ≤12 A for a 15 A breaker).

(2) Fixtures

Hi-bay	For rooms height > 6 m (>20 ft), 250–400 W, 480 V, high pressure sodium (HPS) or LED.
Low-bay	For rooms height < 6 m (<20 ft) height, 32 W, 277 V, compact fluorescent, or LED.
Stairways	Lighting circuit laid out from top to bottom and not part of individual floors.
Exit light	LED, 120 or 277 V, rechargeable.
Road lighting	HPS or LED, 250 W, 480 V, on 8–9 m poles. Operated on photo cell, pole span 40– 50 m.
Area lighting	HPS or LED, 250 W, mounted on buildings and structures.
Emergency light	Twin head, wall mount, 120 or 277 V, rechargeable.
Outdoor door lights	LED, 30 W, 120 or 277 V, controlled by photo cell.
Control room	Fluorescent, with dimming switches.

Right at the start, decide how the whole plant will be furnished, lit, and operated (see Table 7.1).

(3) Switching

Plant	Mostly, directly from panels and subpanels.
Offices	By room switches.

(4) Outlets

Plant	15 A, duplex, 120 V, metallic NEMA 4X enclosure with flip cover.
Offices	15 A, duplex, 120 V, hard plastic NEMA 1 enclosure.
Outdoors	15 A, duplex, 120 V, metallic NEMA 4 enclosure with flip cover.

(5) Cabling

Plant	Wiring in rigid conduits #12 AWG (2.5 mm^2) wires.
Offices	Wiring in electrical metallic tubing (EMT) conduits or armored cable (BX) cables in suspended ceilings.
Wet areas	PVC jacketed armored cables with water-tight connectors.

Table 7.1 Fixtures and switching by area (actual project).

Area		Fixture type	Exit	Switching			Power	
				LTG panel	Local	Light cell	Normal	Essential
1	Machine hall	LED + FL	Yes	LP			Yes	Yes
2	Offices	FL	Yes		Local		Yes	Yes
3	Diesel room	FL	Yes		Local		Yes	Yes
4	Switchgear room	FL	Yes		Local		Yes	Yes
5	Cable spread area	FL	Yes	LP			Yes	Yes
6	Toilets	FL			Local		Yes	
7	Control room	FL	Yes		Local		Yes	Yes
8	Battery room	Special			Local		Yes	Yes
9	Battery charger room	FL			Local		Yes	Yes
10	Transformer vault	FL		LP			Yes	
11	Staircase	FL		LP				Yes
12	Service bay	LED + FL	Yes	LP			Yes	Yes
13	Cooling water sump/ drainage pit room	FL	Yes	LP			Yes	Yes
14	Door lights	Wall mount		LP		Yes	Yes	
15	Outdoor road lights	LED pole mount		LP		Yes	Yes	

7.2.1 Lighting Application

The plant lighting system is installed for the following functions:

- Normal lighting (indoors and outdoors)
- Essential lighting
- Emergency and exit lighting
- Outlets

(1) *Normal lighting* is generally the main lighting for the plant. It covers 70–80% of all the lighting fixtures and is fed from the Normal power system. Most of the outdoor lighting and most of the outlet receptacles are also connected to the Normal power system.

(2) *Essential lighting* usually covers 20–30% of the overall lighting. Essential lighting is fed from the same source as the Normal lighting, but its lighting panels will remain energized if the "Normal" power comes down. The essential distribution panel (EDP) has a double feed through an automatic transfer switch (ATS) shown in Figure 7.1. It is switchable between Normal

Figure 7.1 ATS.

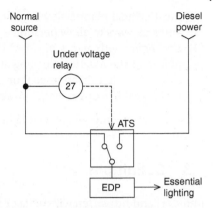

and Essential/power sources. Upon a loss of Normal power, a plant standby diesel engine generator energizes and feeds the Essential panel through the ATS. The essential power is expected to be on within 20 seconds following a loss of Normal power. Upon restoration of Normal power, retransfer back to the Normal power can be set for either manual or automatic switchover. For instance, in a large plant hall, there may be three rows of hi-bay lighting fixtures. The middle row would be fed from an essential lighting panel. In offices, one lamp will be connected to the essential power source and will not be switchable by the room switch.

Remote facilities, like camps for the plant operators will have their own small diesel engine generator to provide the essential power for lighting and cooking. In this facility, Essential power may assume 50% of lighting.

(3) *Emergency lighting* consists of incandescent or LED double-headed, industrial-type emergency power packs, rechargeable, and powered from the outlet circuits. The units are designed to be lit up on its integral battery immediately upon a loss of normal power. The battery is rechargeable from the essential power. The units (Figure 7.2) are placed in the areas where the

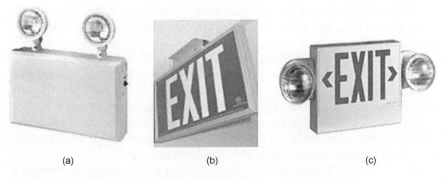

(a)   (b)   (c)

Figure 7.2 (a) Battery power pack, (b) exit light, and (c) exit light with lamps.

most critical plant switching and operation is performed and in hallways and stairways to allow personnel to exit the buildings.

(4) *Exit lights* with sign EXIT, SALIDA, KELUAR, SORTIE, AUSGANG are placed at the locations leading and pointing to exits. Illuminated exit fixtures are either incandescent or LED and are fed from an integral battery, rechargeable from essential power at 120 or 277 V.

Combination emergency–exit lights may be used for saving space and installation costs.

7.3 Definitions

Refer [1] and https://en.wikipedia.org/wiki/Luminance.

(1) *Luminance* is the density of luminous intensity in a given direction. Luminance falls within a given solid 90°. It is measured in candela per meter square (cd/m^2).

Luminance remains the same regardless of the distance from the light source. Luminance is apparent brightness; how bright an object appears to the human eye.

Since luminance is what we see, then light sources which we look at also have luminance. The sun's and moon's luminance give us a good idea of the huge range of brightness which the human eye can handle.

Luminance of the sun: $1\,600\,000\,000\,cd/m^2$. Luminance of the moon: $2500\,cd/m^2$

(2) Luminous intensity is a measure of the wavelength-weighted power emitted by a light source in a particular direction per unit solid angle. The SI unit of luminous intensity is Candela (cd).

(3) *Candela* (cd) is the base measurement for describing **luminous intensity**. It tells you how bright the light source is and showing how far away from an object you can be and still be able to see it. Any light source eventually becomes too dim to see the further away you are from the source. This is different from lumens (a total light output) because it is the value of light intensity from any point in a single direction from the light source.

(4) *Lumen (lm)* is the SI unit of luminous flux, equal to the amount of light emitted per second within a unit solid angle (90°) from a uniform source of one candela. It is a measure of the total quantity of visible light emitted by a source.

(5) *Illuminance* is the density of photons which fall within a given surface area. It is measured in **lux** or **foot-candle (fc)**. Illuminance can be measured with a lux meter. For a given light source, the closer to a light source the illuminated area is, the higher the illuminance value.

(6) *Lux* is unit of illuminance (lm/m^2). **Lux (lx)** is the amount of light on a surface per unit area. A single lux is equal to one lumen per square meter.

If the lamp displays its brightness as a measurement of lux, it usually lists a distance from the bulb since any change in distance of the bulb changes the lux level. As an example, if you place a 100 lm bulb in a flood light that shines on only one square meter of surface, then that surface will be lit at 100 lx. However, if you back the flood light away to shine on four square meters, the surface is now lit with 25 lx.

(7) *Foot-candle* is similar to Lux as a measurement of light intensity and is defined as the lumen illuminance on one-square foot surface (lm/ft^2) from a uniform source of light.

Lux is Lumens per square meter. Divide it with 11 for the Foot Candles if you prefer that.

Conversion: Lux (lx), (lm/m^2) $\rightarrow$ Foot $-$ candle (fc), (lm/ft^2);

$$E_{v\,(fc)} = E_{v\,(lx)} \times 0.0929 = E_{v\,(lx)}/10.76391$$

(8) *Luminous efficacy* is applicable to lighting fixtures. It is a measure of how well a light source produces visible light. It is the ratio of luminous flux to power, measured in **lumens per watt** in International System of Units (SI). Not all wavelengths of light are equally visible, or equally effective at stimulating human vision, due to the spectral sensitivity of the human eye; radiation in the infrared and ultraviolet parts of the spectrum is useless for illumination. The luminous efficacy of a source is the product of how well it converts energy to electromagnetic radiation and how well the emitted radiation is detected by the human eye.

(9) *Spacing criteria* is the spacing-to-mounting height ratio assigned to each fixture and an aid to laying out a pattern of fixtures. Multiply the ratio by the mounting height to get the maximum fixture spacing that will still provide a reasonably even illumination. The mounting height is the distance from the fixture to the **surface** you are lighting. The "End," "Diagonal," and "Across" designations refer to the orientation of the fixture and correspond to the "0°," "45°," and "90°" horizontal planes shown in the candlepower tabulation. The ratio numbers, "1.3," "1.4," and "1.6" are ratios respectively of the allowable spacing between fixtures in the particular orientation to the vertical distance between the work plane and the fixture mounting height.

The **surface** may be the floor or a desk top (typically 0.75 cm [30 in.] above the floor). The lighting designer must determine the **surface elevation** for their calculations.

(10) *Glare* is a very **harsh, bright**, or **dazzling light**. Lighting fixture/diffuser choices and their positioning minimize glare in the working environment. Light shining onto the object is what counts. Light shining into your eyes is waste of light and money. By eliminating glare, one can see more clearly with fewer fixtures installed.

7.4 Illumination Level

The level of illuminance provided here is based on the recommendations by the Illuminating Engineers Society (IES) for the various areas as listed in Table 7.2.

Here are a few more interesting places and their average lux illumination values requirements:

- Retail   400–500
- Classrooms   300–700
- Hospitals   100–200
- Operating rooms   50 000–60 000
- Factory assembly   200–500

The illuminance level (lux/m^2) is calculated (manual approach) by the amount of lux generated over the area covered as follows:

$$I = N \times L_f \times K_m \times \frac{K_r}{\text{Area}} \ (\text{lm/m}^2) \tag{7.1}$$

where:

N = number of fixtures
L_f = lumens per fixture
K_m = maintenance factor (dirt)
K_r = reflectance factor

Table 7.2 Illumination level for various areas.

Plant building/area	Illumination level (lx/m^2)
Building entries and stairways	200
Computer room	500
Control room	500–1000, dimmed
Laboratories and office	500
Machine shops, warehouses, storage	300
Main roads and parking lots	20
Outdoor pump areas	100
Process areas	50
Electrical rooms, switchgear, MCC, and battery area	300
Tank farms, stairs, and gauging area	100
Transformer areas and switchyards	100
Walkways and platforms	30
Offices	300

Now, that we have established design criteria, know the sources of power, have the required illumination levels, and have defined the lighting into functional subsystems, we are ready to start designing the plant lighting.

The aforementioned manual approach of calculating lighting distribution Eq. (7.1) has now been replaced with software programs which better present fixture light distribution patterns and fixture mounting heights. The software generates complete contour patterns for the whole area and indication of the darker spots that will have to be dealt with (see Figure 7.3). Suppliers like Hubbell, Phillips, GE, etc. have lux distribution patterns imbedded for their fixtures in their software for the illumination calculations.

Given a specific room configuration, including dimensions, defined surface and wall reflections, and a desired lumen count, a computer program will provide the lighting distribution for a chosen lamp and determine the number of fixtures required for an optimal average lighting illumination in accordance with our desired illumination levels.

7.4.1 Candlepower Distribution Curves

The photometric diagrams [2] tell if most of the flux intensity (the lumens, the "flow of light") goes upward, downward, or sideways. In the example provided in Figure 7.4a, the light flows in a downward direction all around making

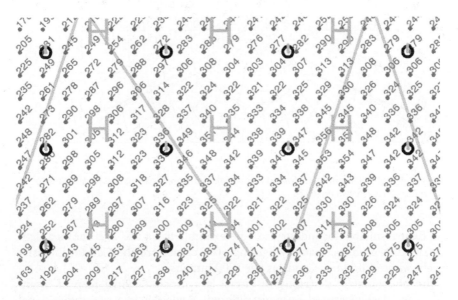

Figure 7.3 Illuminance level (an average illumination of 300 lx was required).

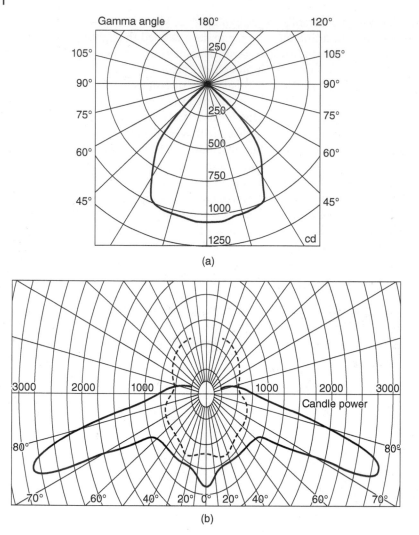

Figure 7.4 Light distribution (a) for indoors and (b) for street lighting downward. Source: Owen Ransen is referenced in [2].

the fixture suitable for indoor use. The example in Figure 7.4b shows a light distribution used for roadway lighting. The circles are the Cd/klm intensity levels, where Cd/klm = Candelas per 1000 lumens.

The **photometric distribution curve** is a cross-sectional map of intensity (candelas) measured at many different angles. It is a two-dimensional

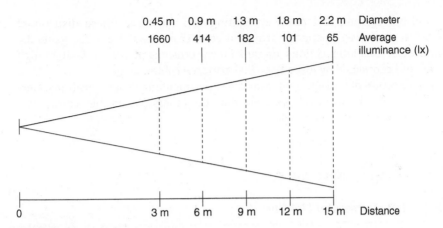

Figure 7.5 Photometric cone diagram.

representation and therefore shows data for one plane only. If the distribution of the unit is symmetric, the curve in one plane is sufficient for all calculations. If the distribution is asymmetric, such as for street lighting and fluorescent units, three or more planes are required. In general, incandescent and high intensity discharge (HID) reflector units are well defined by a single vertical plane of photometry. Fluorescent luminaires require a minimum of one plane along the lamp axis, one across the lamp axis, and one at a 45° angle. The greater the departure from symmetry, the more the planes needed for accurate calculations.

The luminous intensity data presented in photometric diagram allows lighting designers to observe both the total light output and the angular (polar) spread of the light output.

The angular information is also presented as polar diagrams and cone diagrams. The illuminance cone diagram in Figure 7.5 shows mean illuminance available within an area for various mounting heights (height above working plane). The diagram shows the light spread diameter and average Lux illuminance at various heights down from the fixture. This presentation is preferred because it shows how much of illuminance is available at the working plane.

7.5 Outdoor Building and Road Lighting

This lighting comprises wall or stanchion mounted 150 W HPS fixtures placed on the buildings and structures over the entrances and at other strategic locations to facilitate easy access to the building.

PVC jacketed armored cables are used for the outdoor application.

Outdoor lighting and roadway lighting will use three phase distributed systems, photo cell operated through a contactor. The fixtures are generally installed at road intersections, on tank farms, crushing plants, on 8–10 m high poles and connected by directly buried armored cables.

The selection of the light bulb must be experimented in some locations. Certain color of light attracts bugs more than the others. When they come, they come in thousands to die inside.

7.6 Lighting Hardware

7.6.1 Transformers for Lighting

Plant lighting transformers are fed at 600 or 480 or 400 V as applicable, shown in Figure 7.6. For our case, we will assume our plant's LV is 480 V. Our lighting transformers will be 3 ph, 480–480 V, Dy1, isolation transformers with solidly grounded secondary neutrals that will allow us the use of 480 V between the phases and 277 V between a phase and neutral.

Similarly, lighting transformers can be chosen as 3 ph, 480–120/240 V. These small transformers are typically wall bracket mounted, ventilated, or epoxy encapsulated, insulated for 150 °C temperature rise, and equipped with (4) 2.5% taps, high and low.

There are several important benefits and reasons worth mentioning here regarding the isolation 480–480 V lighting transformers chosen previously:

- Neutral solidly grounded for four wire system.
- Less impact on lighting due to large motor starting or plant short circuits.
- Reduced flicker.

Figure 7.6 Lighting transformers.

7.6.2 Cables and Wiring

In plants, incoming cables feeding the panels and transformers will be sized to suit the load and application, based on the lighting panel mains rating. The panel bus bars one wire or three wires are often called mains. They are usually rated 100 and 225 A. Therefore, the incoming cables will be of the same ampacity.

Cabling for the fixture distribution (branch circuits) will be wiring in conduits, 2.5 mm^2 (#12 AWG). In wet areas, indoors, and outdoors, PVC jacketed armored cables are used with water-tight connectors.

7.6.3 Lighting Fixtures

Indoor lighting includes the fixtures designed for high- and low-bay applications. Fixture choice depends upon the area configuration and height. The fixtures selected generally will conform to the environment of the area with respect to operator activity, ceiling height, glare, humidity, and dirt accumulation. Fixtures are selected for specific light distribution to suit the application, either circular for the indoors or elongated to suit the roadways or outdoor wall mounting (see Figures 7.7 and 7.8). The choice of fixture diffuser provides for comfortable lighting dedicated to each specific environment with minimum glare.

Figure 7.7 Fluorescent fixture with glare diffuser.

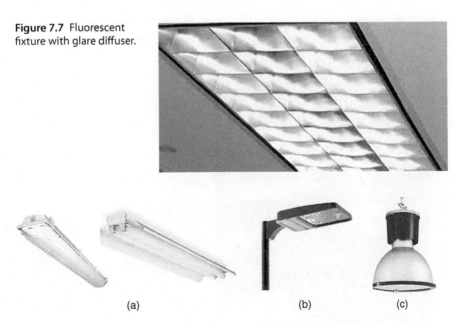

(a)	(b)	(c)

Figure 7.8 (a) Industrial fluorescent, (b) roadway LED, and (c) hi-bay HPS.

LED lighting fixtures are becoming more favored lately in the high- and low-bay areas and roads for their better efficiency (lm/W) and considerably longer life of up to 50 000 hours compared to 1000 for incandescent, 8000 for the fluorescent, and 20 000 hours for HPS lamps.

The purchase price difference between the HPS and LED fixtures is 1 : 3 (year 2017) in favor of HPS. However, the relamping costs often tilt the final decision on the type of fixtures used.

Relamping fluorescent fixtures in the offices is not a major expense, but to relamp them in the tunnels and on roadways may be a major maintenance cost concern.

Control room lighting uses fluorescent fixtures with special ballasts that allow light level control with dimming switches.

7.6.3.1 Lighting panels

Lighting panels and subpanels (Figure 7.9) will be three phase, four wire with solid neutral allowing use of three pole breakers for feeding subpanels and one, two, and three pole breakers for lighting circuits and outlets. The main lighting panels will have main incoming breakers.

Separate panel boards are provided for the following:

- Lighting (Normal and Emergency),
- Lighting, Essential,

Figure 7.9 Lighting panel.

- Convenience duplex receptacles and exit lights,
- Process power and control devices, including instrumentation.

How many lighting fixtures are you allowed to place on a single 15 A, 277 V circuit? Up to 80% of the breaker rating, i.e. 12 A. With the new efficient lighting fixtures, one can string a large number of units. However, the number of fixtures is also dependent on the area to be covered. It is unlikely that you wish to stretch a lighting circuit into two separate areas.

7.6.3.2 Outlets (Receptacles)

In the control and administrative rooms, there will be 4–6 duplex-type receptacles per office (Figure 7.10). Duplex receptacles of 120 V (220 V) come with a pin ground. They can be used indoors and, with appropriate cover plates, outdoors. In the plant, outlets are installed to generally cover the plant areas with a reach of 15 m. In control and administrative rooms, four to six receptacles are placed at 3 m intervals. No more than six duplex receptacles are usually connected to a single circuit. An outlet load is assumed to be 200 W, per each unit, total 1200 W, loaded or not.

The units intended for outdoors and in parking areas are of water tight design with flip covers. In Northern areas, special-type receptacles are used to provide for car heating, often with built-in cycle timers to reduce the overall connected load.

Duplex receptacles for process computers and programmable logic controller/distributed control system (PLC/DCS) power supplies are fed from uninterruptable power supplies (UPS), 1 ph panels, connected to an isolated ground system and provided with orange faceplates. Office receptacles are not connected to the UPS system.

7.6.3.3 Welding Receptacles

Welding receptacles (Figure 7.11) will be three phase and ground, 460 V, 60 A, with integral isolating switch interlocked to the outlet cover. Welding receptacles will be located throughout the project site, such that any area where maintenance welding may be carried out will be within 120 m of a welding receptacle. No more than four welding receptacles are connected to a single feeder circuit. It is assumed that a plant may have one or two welding machines to be used in various parts of the plant.

Figure 7.10 Receptacles (outlets).

Figure 7.11 Welding outlet.

7.7 Lamps Inside the Fixtures

7.7.1 Color Rendering Index (CRI)

Have you driven some brightly lit roads and noted that you cannot tell the color of the cars on the road? Yes, of course. So, why is that?

You are not alone. Once on a large outdoor parking lot, I could not find my car. I parked my dark red car during the day and returned to retrieve it at night under the lights. There were no dark red cars anywhere to be seen. Reason? The parking lot was lit with low pressure sodium (LPS) lamps.

This confusion has to do with the choice of lamp and its color rendering index (CRI) and light color temperature as shown in Figures 7.12 and 7.13 and illustrated by Refs. [3, 4]. The lamps may look the same, but they differ significantly. There are two systems of measurement commonly used to describe the color properties of a light source: CRI suggests how an object illuminated by that light will appear in relation to its appearance under other common light sources, and "color temperature" expresses in Kelvin the color appearance of the light itself.

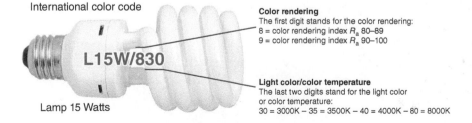

International color code

Color rendering
The first digit stands for the color rendering:
8 = color rendering index R_a 80–89
9 = color rendering index R_a 90–100

L15W/830

Light color/color temperature
The last two digits stand for the light color or color temperature:
30 = 3000K – 35 = 3500K – 40 = 4000K – 80 = 8000K

Lamp 15 Watts

Figure 7.12 Lamp CRI color rendering code.

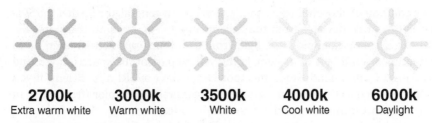

2700k
Extra warm white

3000k
Warm white

3500k
White

4000k
Cool white

6000k
Daylight

Figure 7.13 Light color temperature.

Both yardsticks are valuable in evaluating and specifying light sources and are shown in the fixture.

Some are "cool," while others are "warm," depending on its spectral power distribution (SPD) in the human visible range. Animals may see it differently, though.

Although most lamps emit "white" light, this varies from a cozy "warm white" to a cold or rather "cool white," according to the "color temperature" of the lamp. Color temperature is denoted by a numerical figure followed by the letter "K". The color temperature of a lamp gives us an idea of its light color. Colors and light sources from the blue end of the spectrum are referred to as cool (high temperature), and those toward the red/orange/yellow side of the spectrum are described as warm (low temperature) and more likely to have them in our living rooms. See Figure 7.14 for the color performance of different light bulbs. Incandescent lamps have an exceptional CRI, above 95. The light colors or color temperatures of fluorescent lamps are determined by the composition of the phosphor coating on the inside of the tubes.

Figure 7.14 Light spectra.
Source: [5].

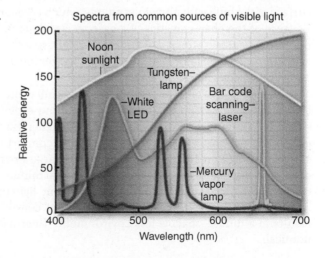

An addition of phosphor widens the spectrum. In general, the further the SPD of a light source deviates from the "full spectrum" of daylight, the worse the CRI. An extreme example is the light from a sodium-vapor street lamp (LPS) that emits around 90% of its light in the yellow part of the spectrum (around 589.3 nm). Consequently, sodium vapor lamps have a CRI of 0. Briefly, if you want to see reds, choose a light source that generates red color (frequency) in its spectrum. Or mix it up under your lamp shade.

Cool-white fluorescent lamps have a CRI of 62, but fluorescent lamps containing rare-earth phosphors are available with CRI values of 80 and above. Mercury-vapor lamps are poor performers with a CRI of 45, halogen lamps work well with a CRI of 90 or better, while compact fluorescent lights (CFLs) measure up to 80, which is acceptable.

LED lamps also capture different spectrum depending on the material compositions in them. Those with phosphorus content cover a wider range. The phosphor absorbs some of the blue light from the LED and then re-emits it across a broad range of wavelengths comprising some green and red, and a lot of yellow.

Natural sun light is classified as having a CRI of 100, the best possible. Look at its spectrum distribution over all wavelengths on the chart. Lower CRI values indicate that some colors may appear unnatural. The high value for incandescent (tungsten) bulbs means that all the colors are rendered well, with the exception of darker blues.

7.7.2 Fluorescent

A fluorescent lamp produces light by activating selected phosphors on the inner surface of the bulb with ultraviolet energy, which is generated by a mercury arc. Because of the characteristics of a gaseous arc, ballast is needed to start and operate fluorescent lamps. The fluorescent light source, compared to incandescent lamps, include improved efficacy and longer life. Efficacies for fluorescent lamps range anywhere from 50 to 100 lm/W.

There may not be much difference between cool white (CW) and warm white (WW) lamps in SPD charts (Figure 7.15), but our eyes see a substantial difference. Their low surface brightness and heat generation make them ideal for offices and schools where thermal and visual comfort is important. The disadvantages of fluorescent lamps include their large size for the amount of light produced, maximum brightness can take several minutes to occur, require special, expensive ballasts if dimming is required, and light output is reduced at low ambient temperatures. This makes indoor light control more difficult, which results in a diffuse, shadow less environment. Their use outdoors is not economical.

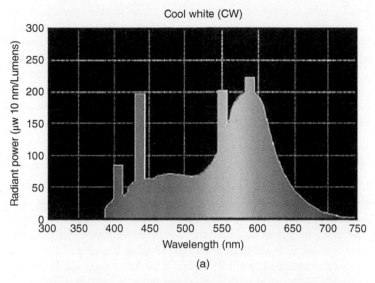

(a)

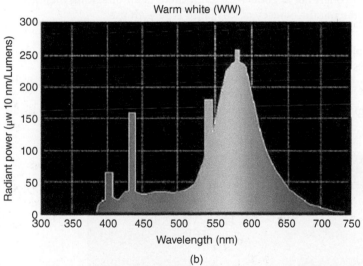

(b)

Figure 7.15 Fluorescent, (a) cool and (b) warm.

7.7.3 Metal Halide (MH)

Metal halide (MH) lamps are similar in construction to mercury-vapor lamps with the addition of various other metallic elements in the arc tube. The major benefit of this design is an increase in efficacy to 60–100 lm/W and an improvement in color rendition to the degree that this source is suitable for commercial areas. Light control of a metal halide lamp is also more precise than that of a deluxe mercury-vapor lamp since light emanates from the small arc tube, not the total outer bulb of a coated lamp. A disadvantage of the metal halide lamp is its shorter life (7500–20 000 hours) as compared to LEDs and HPS. Starting time of the metal halide lamp is approximately four to seven minutes depending on the ambient temperatures. Restriking after a voltage dip has extinguished the lamp, however, can take substantially longer, depending on the time required for the lamp to cool.

7.7.4 Low-Pressure Sodium (LPS)

LPS (Figure 7.16) offers the highest initial efficacy of all lamps on the market today, ranging from 100 to 180 lm/W.

However, because all the output is in the yellow portion of the visible spectrum, it produces extremely poor and unattractive color rendition (CRI). Control of this source is more difficult than with HID sources because of the large size of the arc tube. The average life of LPS lamps is 18 000 hours. While lumen maintenance through life is good with LPS, there is an offsetting increase in lamp watts, reducing the efficacy of this lamp type with use.

7.7.5 High-Pressure Sodium (HPS)

In the 1970s, as increasing energy costs placed more emphasis on lighting efficiency, HPS lamps (developed in the 1960s) gained widespread usage. With efficacies ranging from 80 to 140 lm/W, these lamps provide about seven times

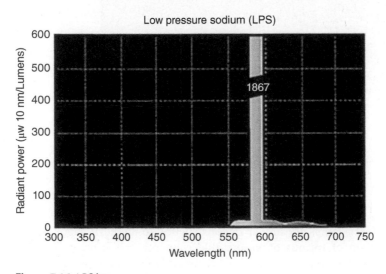

Figure 7.16 LPS lamp.

as much light per watt as incandescent and about twice as much as some mercury vapor or fluorescent lamps. The efficacy of this source is not its only advantage. An HPS lamp also offers long life (24 000+ hours) and the best lumen maintenance characteristics of all HID sources. The major objection to the use of HPS is its yellowish color and low color rendition. It is useful mainly for warehouse and outdoor applications.

7.7.6 LEDs

It appears LEDs [6] are the light sources of the future due to their light efficiency in lumens per watt, longevity as well as their capability to be color engineered to suit the market. LEDs produce light by the recombination of charge carried (electrons and holes) in the depletion region of the p–n junction of diode. During recombination, the electron–hole pair drops into a lower energy state by emitting a photon as a source of light.

In the process of recombination, the emitted photon has a specific energy, and therefore wavelength and "color," determined by the band gap of the material making up the LED, as seen in Figure 7.17. The larger the band gap, the shorter the wavelength of the emitted light.

That process can be controlled for the intensity as well as the color rendering.

Light has additive properties [7, 8] (see Figure 7.18). Red and green LED beams pointed at a white surface will overlap, with the resulting color perceived by humans as yellow.

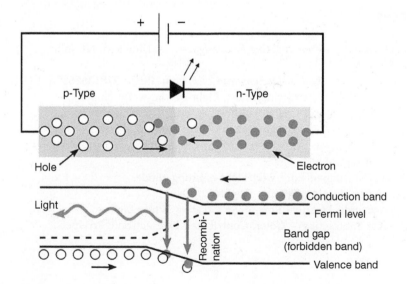

Figure 7.17 LED technology.

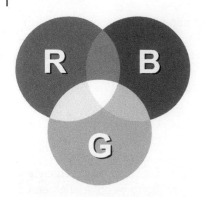

Figure 7.18 LED beams.

Likewise, equal intensities of red, green, and blue light from LEDs will produce white light. By varying those respective intensities, by means of integrated circuits, lighting engineers can produce LED-based luminaries – or flexible lighting solutions – that generate light similar to what consumers like in an old-fashioned incandescent light bulb.

Integrated circuits can vary the intensity of individual LEDs in an RGB array to produce a specified quality or temperature of white light. In fact, the color temperature of white light emitted from an LED-based system could also change based on the levels or colors to ambient light to create a desired solution for home, office, street, and factory floor. An external light measuring sensor may provide input to the integrated circuit to adapt.

References

1 Gordon, G. (2003). *Interior Lighting for Designers*, 4e. Hoboken, NJ: John Wiley & Sons, Inc.
2 Ransen, O. (2017). *Candelas, Lumens and Lux*. Lulu. ISBN: 9781365688621.
3 www.osram.de, Colour Rendering Index: Light-can-be-white-en.pdf.
4 Light Color Temperature? Online publication: https://www.osram.com/media/resource/hires/333565/light-can-be-white-en.pdf.
5 Sources of Visible Light. https://micro.magnet.fsu.edu/primer/lightandcolor/lightsourcesintro.html.
6 LEDs. https://en.wikipedia.org/wiki/Light-emitting_diode.
7 Digi-Key: Steven Keeping, Whiter, Brighter LEDs.
8 Digi-Key: Armando Emanuel Roggio, Sr. Creating White Light by Adding – Not Subtracting – Color, Contributed By Electronic Products.

8

DC System, UPS

```
CHAPTER MENU

8.1  Project Requirements, 179
8.2  DC Battery and Chargers, 180
       8.2.1  Battery Ampere-hour (Ah) Capacity, 181
       8.2.2  Battery Float/Boost Charge, 184
       8.2.3  Battery Types, 184
       8.2.4  Lead–Acid Battery Room Requirements, 185
8.3  Battery Chargers, 186
8.4  Ratings, 186
8.5  Uninterruptible Power Supply (UPS), 188
References, 190
```

8.1 Project Requirements

DC power maintains readiness of the electrical switchgear and the plant control system during the plant operation and plant outages. DC power is one of the first systems to be commissioned as it facilitates testing of the power distribution system, which in turn allows the mechanical equipment to be powered and tested.

In large plants, DC system and its equipment is needed in several places. In the case of this industrial project (see Chapter 1), it is needed in the main substations and the process plant. The plant DC system must produce the following voltage supplies:

- $125\,V_{dc}$ for switchgear breakers and disconnect switches in the main substation.
- $125\,V_{dc}$ in the process plant for the plant switchgear breakers.
- $125\,V_{dc}$ for the critical emergency lube systems.
- $24\,V_{dc}$ for field transmitters and distributed control system (DCS) I/O's at the process plant. $24\,V_{dc}$ will be obtained directly from $125/24\,V$ dc/dc converters.

Practical Power Plant Engineering: A Guide for Early Career Engineers, First Edition. Zark Bedalov.
© 2020 John Wiley & Sons, Inc. Published 2020 by John Wiley & Sons, Inc.

- 120 V_{ac} uninterruptible power supplies (UPS) for DCS at several locations of the process plant.

In the IEC countries, the equivalent DC voltage is 110 V_{dc}.

Evidently, these low-voltage sources cannot be transmitted from one building to the other over large distances. The individual power sources of this kind are to be limited to within 200 m.

8.2 DC Battery and Chargers

We, the electrical engineers will determine the ampere-hour capacity (Ah) of the station batteries and the DC output ratings of the chargers. The battery must carry full load without chargers for the duration of an outage, which for this plant we determine to be six hours.

Normally, the main plant battery will be fed by two chargers simultaneously feeding the battery and the plant operating load via a DC board. Each charger must be capable to boost the battery charge while at the same time caries its full DC load.

Battery bank consists of lead–acid cells, which are to be sized in accordance with IEEE 485 [1], by assessing the expected load profile, future growth, aging, temperature correction factor, and final discharge voltage. The storage battery stores electrical energy as chemical energy. This chemical energy is then converted into electrical energy when an electrical load is applied to its terminals, known as the battery discharge process. On the other hand, conversion of electrical energy into chemical energy by applying external electrical source is battery charging as shown in Figure 8.1.

The main active materials required to form a lead–acid battery are as follows:

(1) Lead peroxide (PbO_2) for (+) anode plate.
(2) Sponge lead (Pb) for (−) cathode plate.
(3) Sulfuric acid (H_2SO_4), diluted with distilled water; 3 : 1 in favor of water.

During the **discharge** (http://hyperphysics.phy-astr.gsu.edu/hbase/electric/leadacid.html), electrical current flows from anode (+) toward cathode (−) and the external load, as part of the following chemical process:

- Both plates get covered with $PbSO_4$.
- Specific gravity of sulfuric acid solution falls due to formation of water during reaction at PbO_2 plate.
- During this process, voltage difference between plates decreases.

During the **charge** (http://hyperphysics.phy-astr.gsu.edu/hbase/electric/leadacid.html), the chemical process is reversed, and the current flows from (−) cathode to (+) anode.

Figure 8.1 Battery (a) discharge and (b) charge process.

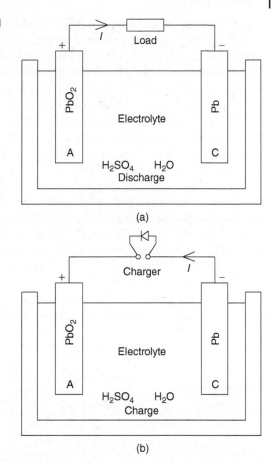

(a)

(b)

- Lead sulfate on anode converts into lead peroxide.
- Lead sulfate of cathode is converted to pure lead.
- Voltage of cell increases.
- Specific gravity of electrolyte increases.

The specific gravity of the electrolyte at 77 F (25 °C) fully charged should read 1.255 ± 0.005. To optimize battery life, it is recommended that the battery be disconnected from the load when the end voltage 1.70 V is reached. It is the point at which 100% of the usable capacity of the battery has been consumed and continuation of the discharge become useless and damaging to the battery.

8.2.1 Battery Ampere-hour (Ah) Capacity

The Ah capacity of a battery cell is determined by the size of the plates and the number of plates within the jar. The plates of the same polarity are paralleled together inside the cell generating the cell voltage of 2.1 V/cell, generally noted as 2 V. A typical battery will need 60 cells (2 V jars) interconnected in series to make a circuit in which the (−) terminal of one cell is connected to the (+) terminal of the next, for a total voltage of all cells of the plant storage battery of 125 V_{dc}, 55 cells for 110 V_{dc} in the IEC countries.

The electrical engineer must determine the overall Ah capacity of the battery, based on the requirements for the plant operation and plant restoration during and following an outage. The battery Ah capacity is measured by the discharge current at eight hours down to 1.75 V/end cell voltage. In Europe and Asia, it is at 10 hours/1.70 V end cell voltage.

The discharge table Figure 8.2 presented by Great Northern Battery (GNB) Systems [2], one of the battery suppliers, shows the capacity discharge rate at various time durations down to the 1.75 V end cell voltage. This battery capacity is rated at eight hours. The model MCX-11 in the table shows a capacity of $8 \text{ h} \times 54 \text{ A} = 432 \text{ Ah}$, which would be appropriate for our requirements. The discharge rate at 2 hours is 143 A, 8 hours is 54 A, and at 20 hours perhaps (guessing) in the order of 25 A.

MCX-11 means that each jar comprises 11 plates; 6 negative and 5 positive plates all connected in parallel (see Figure 8.3).

Cell	Hours							Minutes		
type	8	5	4	3	2	1.5	1	30	15	1
End voltage − 1.75										
2-MCX-5	22	30	36	44	57	68	85	117	157	234
2-MCX-7	32	46	54	66	86	102	128	176	231	339
MCX-9	43	61	72	88	114	136	170	234	304	448
MCX-11	54	76	90	110	143	170	213	293	380	550
MCX-13	65	91	108	132	171	204	255	351	457	654
MCX-15	76	106	126	154	200	238	298	410	535	756

Figure 8.2 Typical battery ampere-hour capacity and hourly discharge. Source: Courtesy of Great Northern Battery (GNB) Systems [2].

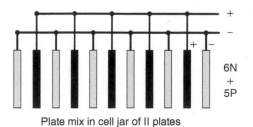

+
−
+ −

6N
+
5P

Plate mix in cell jar of II plates

Figure 8.3 Internal plate connections.

While in operation, the battery is continually being float charged by chargers, while at the same time also supplying the plant DC load (see Figure 8.7).

In "our" industrial plant, the DC running load is not high as there are no major users like DC oil lube pumps and other mechanical drives. The battery load consists of switching on and off the main breakers, disconnect switches, as well as continually feeding panel indicating lights and electronic hardware of protective relays and the plant control equipment.

Nowadays, most of the indicating lights on the electronic hardware are miniature low-consuming LED lights. Furthermore, the DC system feeds the plant UPS, its biggest load, which provides power to the plant sensitive computer control systems.

Figure 8.4 shows the battery life expectancy based on the ambient temperature. The battery room has to be well ventilated and kept at lower temperature for longer battery life.

Following a power outage, the standby generator will be energized and supply power to the chargers to insure the batteries are fully charged at all times. Therefore, the real outage time of the plant DC system is relatively short.

Assuming the normal load in the main substation is about 15 A continuous, a battery capacity of 432 Ah, MCX-11 would be sufficient. A charger rated at 40 Adc would be adequate for boost/float charging the battery and supplying the active load. Two chargers are required to meet our criteria of covering for a single contingency failure. Both chargers would be operating in parallel with the battery.

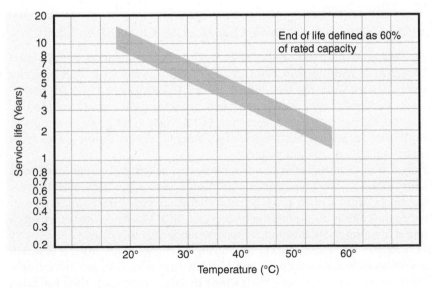

Figure 8.4 Battery life and temperature dependency.

Due to the remoteness of the process plant, an additional battery/charger sets would be required at the process plant for the plant control system, supply to UPS and some switchgear duty. In this instance, the best choice would be to furnish identical equipment as specified for the main substation. An additional charger would not be required as one extra charger can serve as a replacement for both applications.

8.2.2 Battery Float/Boost Charge

Float, often called trickle charging, is charging at a reduced voltage during normal operation. The charger ensures the battery is always in the charged condition and is therefore considered "floating." As the battery gets charged, its charging current reduces gradually. If the battery gets drained, the charger will again increase the charging voltage and the process continues.

Boost charge can be switched on, to enable a quick charge of depleted batteries. For instance, a 2.1 V lead acid battery which has been discharged will initially be boost charged with a charging voltage of around 2.35–2.4 V. However, as the battery voltage rises, the charger switches over to the float charge mode at a voltage of 2.25 V.

Follow factory proposed instructions regarding charging, filling with distilled water, measuring specific gravity, replenishing with electrolyte, and boost/floating the battery.

8.2.3 Battery Types

The most common battery types are lead–acid, Ni–Cd, and lately lithium ion.

Large lead–acid batteries are typically used by power plants. Recently, on a 2×125 MW hydro power plant project, the following lead–acid batteries were installed with capacities of two sets of 163 A chargers/1112 Ah batteries for the control system and then again two sets of 466 A chargers/2320 Ah batteries for two separate independent 110 V DC systems in the power plant. Both batteries at each location were operating in parallel. Power plants insist on having fully redundant systems.

Newer lead–acid batteries are of *sealed* type, called **valve-regulated lead acid (VRLA)**, which need considerably less maintenance. However, utilities like the traditional open-type batteries and like to follow the regular maintenance procedures to extend the battery life. VRLA batteries are more often found in industrial plants and are recommended for this plant.

Ni–Cd battery has a terminal voltage during discharge of around 1.2 V, which decreases little until nearly the end of discharge cycle. They are more costly. Their significant advantage is their ability to deliver practically their full rated capacity at high discharge rates to the end of the discharge cycle. They are

Figure 8.5 Battery rack.

generally used in the telecommunication industries due to their "cleaner" environmental aspects and their portability.

Lithium ion: Today, this is the fastest growing and most promising battery chemistry, in particular due to their 3.6 V/cell voltage. Lithium ion (Li–Ion) batteries have already taken a large part of the Ni–Cd market. Presently, Li–Ion batteries are becoming a strong contender to serve as energy storage for the wind and solar power generation to extend their daily producing cycle.

Battery racks and cables: The racks (Figure 8.5) are constructed of steel, coated with a material that resists the corrosive effects of spilled electrolyte and designed to meet the seismic rating specified for the area. Intercell connectors are lead-plated copper for lead–acid batteries. Stainless steel bolts, washers, and nuts are provided for terminations.

8.2.4 Lead–Acid Battery Room Requirements

The battery cell containers and covers are moulded of a material that is resistant to heat, shock, and chemical attack. Covers are equipped with explosion-resistant, flame-arresting vent caps. Explosive gases are likely developing during charging of battery. The battery room must meet the following requirements:

- Good ventilation is essential inside the battery room. Smoking should be strictly prohibited inside the battery room.
- Two redundant exhaust fans, preferably fed from two separate electrical circuits, of suitable size should be fitted in the battery room to keep the atmosphere free from gases.
- The temperature inside the battery room should be maintained in the range of 10–20 °C.
- The walls, ceilings, doors, window frames, fans, metal parts, and other apparatus in the battery room should be painted with anti-acid coating at regular intervals.
- The electrical wiring inside the room should be in metallic conduits. Lighting fixtures should be flameproof.

- All the switching elements including electrical fuses and plug sockets should be installed outside the battery room.
- The floor of the room should be well finished preferably by using ceramic tiles to allow for spray washing.

8.3 Battery Chargers

The charger as in Figure 8.6 is of solid state construction, filtered output, constant voltage type. Chargers are capable of parallel operation on the same DC bus. Input and output power circuits are provided with thermal/magnetic-type circuit breakers.

The charger provides voltage regulation of $\pm0.5\%$ of the DC float voltage and $\pm1\%$ of the equalization voltage under any load condition, for $\pm10\%$ input voltage and $\pm5\%$ input frequency variation. Each charger must have a capability of optimizing the charging voltage to minimize gassing of the battery, as well as to tweak the voltage when approaching the full charge.

8.4 Ratings

Two (2) batteries

• Type	Lead acid, VRLA
• Nominal voltage	$125\,V_{dc}$ $(110\,V_{dc})$
• Ampere-hour capacity	400 Ah approx.
• Final cell voltage	1.75 V
• Life expectancy	15 years

Two (2) chargers

• Input/output voltage	$480\,V_{ac}$, 3 ph/$125\,V_{dc}$
• Output capacity	40 Adc
• Current limit	125%
• Ripple (maximum)	2% when connected to battery
• Voltage regulation	$\pm10\%$ for load change 0–100%
• Enclosure	NEMA 1, Free standing

As shown in the DC system one line diagram Figure 8.7, courtesy of SNC-Lavalin, the chargers are fed from two independent sources and then

Figure 8.6 Charger and UPS.

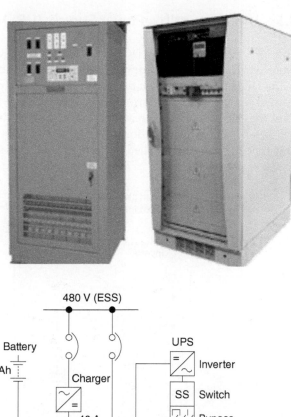

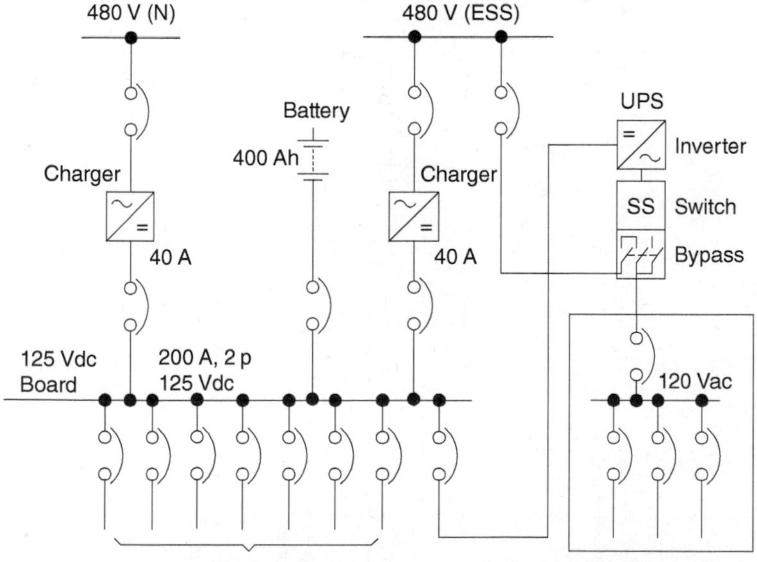

Figure 8.7 DC system connection diagram.

connected together with the battery to a DC distribution board with incoming DC breakers and several two pole DC branch breakers to feed individual users, like switchgear breakers, UPS, etc. all located in the charger room. The battery has its own breaker and is located in its own battery room.

8.5 Uninterruptible Power Supply (UPS)

The UPS in principle uses DC power input and inverts it to AC power of required voltage and frequency to feed the plant critical load.

The UPS internal battery if provided is continuously being charged from the normal source, and if the normal AC source fails, the battery keeps on feeding until the charging power is restored. In this case as shown in the diagram, the UPS is fed from the plant station battery which has considerably larger capacity. This type of operation is called a continuous UPS as in Figure 8.8, courtesy of SNC-Lavalin.

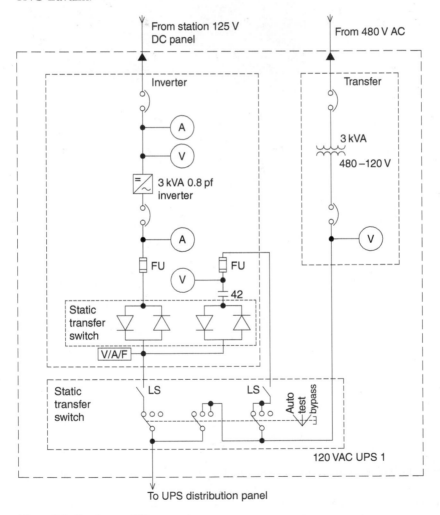

Figure 8.8 Continuous UPS.

DC power battery source is held as the primary source as it is considered to be more reliable than a regular AC source, which may be subject to voltage spikes, variations, and failures. For larger UPS units, the external plant station battery provides virtually unlimited source of power for feeding the UPS load even after the charging power has failed.

For this, a relatively large industrial plant, UPS must supply uninterruptible AC power to the DCS, programmable logic controllers (PLCs), certain instrumentation and control devices, fire and gas detection, security and alarm system as well as the plant administrative computers.

The best arrangement for our proposed plant would be to have a single large UPS with a large station battery support and a standby diesel generator feeding the essential bus to feed the UPS bypass feed. This centralized UPS unit with a diesel engine generator (DG) support would feed the other locations where UPS power is required. Since the distribution at $240\,V_{ac}$ is not economically feasible at the working distances of this plant, we must plan on having a main UPS at the process plant and a number of independent smaller UPS units at the peripheral smaller facilities. Therefore, two types of UPS will typically be used:

(1) Inverter with static transfer switch connected to the main station battery and chargers. This arrangement is used in power plants and major industrial plants. In this case, it will be in the process plant.

(2) Complete UPS equipment with its own integral 30 minute battery and chargers. The peripheral UPSs will be of all-inclusive type with integral Ni–Cd or Li–Ion batteries.

The main UPS has the design features as listed in the following:

- *Capacity*: 3 kVA, 0.8 power factor.
- *DC input*: $125\,V_{dc}$ from station battery.
- *Alternate input*: $480\,V_{ac}$, 3 ph, 3 w, 60 Hz with bypass switch.
- *Output*: $240/120\,V_{ac}$, 1 ph, 3 W, isolated solidly grounded output.
- *Maintenance bypass*: $480\,V_{ac}$, 3 ph, 3 W, 60 Hz.
- *Overloading capacity*: 125% – 10 minutes, 150% – 10 seconds.
- *Static transfer switch overloading*: 200% for 10 minutes.
- Input filter harmonic and input isolation transformer.
- *Total harmonic distortion (THD)*: 3%.
- Incoming and output circuit breaker.
- Diagnostic panel and trouble alarm for remote indication, with optical Ethernet communication interface.

The UPS will feed power to a standard $240/120\,V_{ac}$, one phase distribution panel with a main breaker and branch one pole circuit breakers.

The peripheral UPS unit will have the following design features:

- *Capacity*: 0.5 kVA.
- *Input*: $240\,V_{ac}$, 1 ph, 3 w, 60 Hz.

- *Output*: 120 V_{ac}, 1 ph, 2 w.
- Static transfer switch.
- Manual bypass switch.
- Integral 30 minute DC battery with AC/DC charger feed.
- Input filter harmonic and input isolation transformer.
- Incoming and output circuit breaker and distribution panel.
- Diagnostic panel and trouble alarm for remote indication, with optical Ethernet communication interface.

References

1 IEEE 485 – 2010 Recommended Practice for Sizing Lead–Acid Batteries.
2 Great Northern Battery (GNB) Systems. Product Catalog Information.

9

Plant Power Distribution

CHAPTER MENU

9.1 Plant Overhead Distribution, 191
   9.1.1 Introduction, 191
   9.1.2 Line Construction Elements, 193
9.2 Types of Distribution, 195
   9.2.1 Line Design, 196
9.3 Structure (Pole) Types, 197
9.4 Overhead vs. Underground, 198
9.5 Clearances, 201
   9.5.1 Line to Roads, 201
   9.5.2 Phase Clearances, 201
9.6 Line Voltage Drop Calculations, 202
9.7 Power Loss Calculations, 203
9.8 Line Conductor Sag and Tension, 204
9.9 Aerial Bundled Cable (ABC) Distribution, 205
9.10 Line and Cable Charging Current, 206
References, 207

9.1 Plant Overhead Distribution

9.1.1 Introduction

Each major industrial and power plant will require a spread-out power distribution system to interconnect the large number of plant loads to the main switchyard and switchgear distribution boards, as shown in Figure 2.7. Overhead power distribution is classified in the electrical power industry by the voltage levels, as listed in the following. For the voltages used throughout the world (see Chapter 11).

(1) *Low voltage*: Less than 1000 V; used for connections between residential or small commercial customers and the utility and within industrial facilities.
(2) *Distribution, medium voltage*: Between 1000 V (1 kV) and to about 66 kV; used for distribution in urban and rural areas and industrial facilities.

Practical Power Plant Engineering: A Guide for Early Career Engineers, First Edition. Zark Bedalov.
© 2020 John Wiley & Sons, Inc. Published 2020 by John Wiley & Sons, Inc.

(3) *Subtransmission*: 66–115 kV,
(4) *Transmission, HV*: Over 230 kV, up to about 800 kV; used for long distances.

The distribution lines (Figure 9.1) follow the most direct routes or along the roads to facilitate ease of construction and maintenance. Most of the countries use wood poles 8–10 m high with wood cross arms for their distribution lines. Galvanized steel and concrete poles are also common. In tropics, concrete poles are the rule. Wood poles do not last long as they are on the menu of ants and termites.

9.1.1.1 Standards for Design of Distribution Lines in USA

Any utility company, whether it is investor owned, cooperative, municipal, or telecommunications, will probably have its own way of designing distribution lines. While there are differences from company to company, the design methods will likely be similar because they are built to the same standards. There are two typical standards for distribution pole design – National Electrical Safety Code (NESC) and the American National Standards Institute (ANSI 05.1). Furthermore, there are additional guidelines for Electric Cooperatives imposed by the Rural Utilities Service.

Here are some definitions and distribution line terminology:

Span	Horizontal distance between supporting poles.
Basic (ruling) span	The span length adopted for sag/tension calculations.
Sag	The vertical distance, under any system of conductor loading, between the conductor and a straight line joining adjacent supporting points, measured at mid-span.
OPGW	Optical ground wire.
ACSR	Aluminum conductor steel reinforced conductor.
Section support	A support in a straight run of line on which the conductors are erected on either side of the cross arm on tension insulators.

Figure 9.1 Multicircuit overhead line.

9.1.2 Line Construction Elements

Structures: Structures for overhead lines (OHLs) take a variety of shapes depending on the type of line. Each structure must be designed for the loads imposed on it by conductors, wind, and snow accumulation, where applicable. Structures may be as simple as wood poles directly set in earth, carrying one or more cross-arm beams to support conductors, or "armless" construction with conductors supported on insulators attached to the sides of the poles. Tubular steel poles are typically used in urban areas. High-voltage lines are most often carried on lattice-type steel towers or pylons.

Guy wires: Structures may be considerably strengthened by the use of guy wires to resist some of the forces due to the conductors pulling from different angles.

Insulators: Insulators must support the conductors and withstand both the normal operating voltage and surges due to switching and lightning. Insulators are broadly classified as either pin-type, which support the conductor above the structure, or suspension type, where the conductor hangs below the cross arms. Insulators are usually made of wet-process porcelain or toughened glass, with increasing use of glass-reinforced polymer materials.

Conductors: Aluminum conductors reinforced with steel wire (known as ACSR) in Figure 9.2 are primarily used for medium and high voltage lines and may also be used for overhead services to individual customers. Aluminum conductors have better mechanical strength/weight than copper as well as being considerably less costly. Some copper cables are still used, especially at lower voltages and for grounding.

Bundled conductors: Bundled conductors are not used on distribution systems. They are needed for larger current capability at higher voltages starting at 220 kV with two subconductors. Bundled conductors consist of several subconductor 2–3–4 cables supported by nonconducting spacers. The bundled conductor arrangement simulates a larger conductor, which lowers the gradient of the electrical field across the conductors and reduces the corona effect. The following are the benefits of bundled conductors:

- Larger line power transfer.

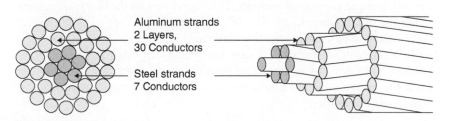

Aluminum strands
2 Layers,
30 Conductors

Steel strands
7 Conductors

Figure 9.2 ACSR conductor.

- Reduced corona effect and power loss, having appearance of larger conductor resulting in reduced voltage gradient on the surface of conductor for corona inception.
- Higher voltage required for inception of corona.
- Less telephone interference.
- Reduced line reactance.
- Lower surge impedance, resulting also to higher energy transfer.

Rain has the greatest impact on the corona effect and power loss. Twice as much of loss is expected during a rainy day in comparison with fair weather conditions, from 6 to 12 kW/km.

Ground cable: Medium-voltage distribution lines may have the grounded conductor strung below the phase conductors to provide some measure of protection against tall vehicles or equipment touching the energized line.

Sag: The sag of the conductor (vertical distance between the highest and lowest point of the curve) varies depending on the temperature. A minimum overhead clearance to roads and terrain must be maintained for safety. Shorter spans may usually be needed for road crossings.

Pole grounding: Overhead power lines are often equipped with a ground wire (shield wire or overhead earth wire) on the top of the structures. A ground conductor is a conductor that is usually grounded (earthed) at the top of the supporting structure and brought down to the pole grounding rod, to minimize the likelihood of direct lightning strikes to the phase conductors. Steel poles generally do not need grounding down conductor. The ground wire is also a parallel path with the earth for fault currents in earthed neutral circuits. Pole grounding resistance of $5\,\Omega$ is considered a good grounding condition. All the metallic pole hardware is grounded to a ground wire, which is then led down to the pole grounding rods. Depending on the soil conditions, the grounding rods will be installed at every pole in the dry areas and every several poles in the moist, low soil resistive areas.

Pole foundations: Typical tangent and guyed wood or concrete poles have setting depths based on a rule of thumb, 10% of the total pole length plus 0.6 m. That is, if the pole is 13 m long, approximately 2 m is buried while 11 m is above ground.

In the soils that are subjected to permafrost, the poles are often not buried but stood up and supported by rocks. This also allows for the line realignment to suit the plant changes.

Lattice steel poles use concrete foundations.

Right of way: A ROW provides a safe space on both sides at ground level between the lines and surrounding structures and vegetation. It provides space for inspections and access to towers and other line components, if maintenance is needed.

9.2 Types of Distribution

The choice of power distribution for a plant [1] includes the following:

Radial feeders: It goes straight from the main switchgear to the load. This arrangement lacks reliability. Lower reliability may be allowed to the circuits that can sustain lengthy outages, but certainly not to the circuits essential for the plant production (Figure 9.3).

Parallel feeders: Same as radial, but the switchboard has a tie breaker at LV level and can take the load to a healthy transformer source.

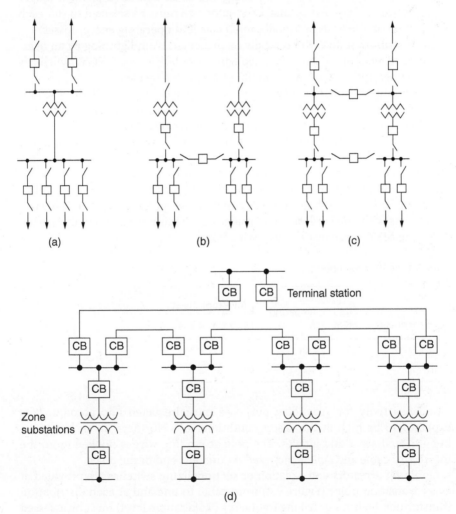

Figure 9.3 (a–d) Types of plant distribution.

Additional reliability and flexibility is accomplished by an additional tie breaker at the MV level Figure 9.3c.

Ring main or loop: As shown in the aforementioned diagram, each load center can be reached from two sources. Fairly reliable, but requires a lot of switching and managing.

9.2.1 Line Design

High-capacity conductors are used not only to maintain low voltage drops but also for their tensile strength for pulling and to be able to implement longer spans between poles. An optimal conductor size must be selected to suit each application on the basis of overall capital cost and operating energy losses.

The distribution lines will typically be protected from lightning by an overhead shield wire on the top. The shield wire, popularly called OPGW, will likely contain fiber optic, single mode, multi 24 core cable inside it.

Depending on the MVA capacity and distance to be transferred, the distribution voltage can be selected to be anywhere between 4.16 and 34.5 kV. For our 40 MVA plant as described in Chapter 1, we determined that the primary distribution from the main substation will be at 13.8 kV and the secondary within the plant buildings at 4.16 kV. The OHLs will run as double circuits to allow for connection of the double ended substations or to provide a provision for reconnection of single ended substations to different circuits. The lines will be designed for heavy loading, consisting of concrete poles, and galvanized steel cross-arms with line hardware and insulators. The line conductor will have the following MVA carrying capacity (Table 9.1:

Table 9.1 ACSR line conductors.

Line conductor	Size MCM	Ampacity A	MVA capacity at	
			13.8 kV	4.16 kV
Hawk	477	660	15.7	4.73
Partridge	267	460	10.9	3.3

Soil resistivity for grounding purposes must be taken into account. It is expected to be high in the desert conditions and Northern areas, but lower in cultivated areas and jungles. The pole grounding wire is hooked up to the messenger cable and a plate mounted on the butt end of the poles.

Lightning arrestors and isolation or sectionalizing switches are provided at every transition point (Figure 9.4) from cable to line and at each distribution transformer. Lightning striking frequency (isokeraunic level) must be assessed for the insulation coordination.

Figure 9.4 Circuit transition point.

Sectionalizing switches at the branch and loop circuits are provided to allow for isolation of faulted sections to minimize disruption to the rest of the system and to facilitate more rapid maintenance work on the lines. OHLs will sometimes have to convert to cables to pass through tight spaces within the plant. For the cables 5 kV and higher voltages, termination must be with stress cones to make sure the electrical field in the cable shield is smoothly and evenly expanded at the point of connection.

9.3 Structure (Pole) Types

Tangent structure: Tangent structures are the most commonly used pole structures on relatively straight portions of the lines. Because the conductors are in a relatively straight line, tangent structures are designed only to handle small line angles (changes in direction) of 0°–3°. Tangent structures are usually characterized by suspension (vertical) insulators, which support and insulate the conductors. They transfer the wind and weight loads to the structures.

Angle structure: Angle structures are used where transmission line conductors change direction. These types of structures are designed to withstand the forces placed on them by the change in direction of tension forces. Angle structures are guyed. They may be (i) similar to tangent structures, using suspension insulators to attach the conductors and transfer wind, weight, and line angle loads to the structure or (ii) similar to strain or dead-end structures, using insulators in series with the conductors to bring the wind, weight, and line angle loads directly to the structure.

Dead-end structure: A dead-end structure is typically used where transmission line conductors turn at a wide angle or end. Compared to tangent structures, a dead-end structure is designed to be stronger and often it is formed as an **H-Frame** structure. Typically, insulators on a dead-end structure are in series with the conductors (horizontal) to bring the total strain directly to the structure. A dead-end structure is designed to resist the full unbalanced tension that would occur if all conductors were suddenly removed from one face of the structure.

Tap-off structure: This structure serves for feeding the loads. It may be a termination or guyed tangent structure with tension insulators and equipped with arrestors and a three-pole switch for cable connection to the loads.

9.4 Overhead vs. Underground

The industrial projects may employ site distribution by underground cables or OHLs. Let us briefly review the cost difference and other technical considerations. Initial costs of OHL construction are considerably lower than for underground cabling. Cost of trenching, duct banks, and manholes is a labor intensive work. On the other hand, OHLs are exposed to environmental impacts such as storms, lightning, wind-blown debris, and traffic impact (on poles), which make them less reliable than underground distribution. Failures on underground cables demand longer outages for repairs.

From a technical point of view, conductor phase spacing on OHLs are larger, resulting in higher overall system inductance than on cables. This means OHLs are subjected to comparatively larger voltage drops than underground cables of equal current-carrying capacity. However, the overhead wires are much less costly, thus larger conductors can be used for less voltage drops in the conductor resistance.

OHLs offer more flexibility to add new loads along the route.

Underground cables have comparatively higher line capacitance, and under light load conditions can affect system power factor. This light-load performance can limit total underground feeder length. On the other hand, cables can supply heavy load, especially when it is inductive in nature, over longer distances than OHLs.

Bundled conductors are not used on distribution lines. Instead multiple circuits are installed as needed.

Let us open the ABB TD Blue Book [2] and make a comparison for a 15 kV OHL using 300 kcml ACSR conductors spaced at (4 ft) 1.2 m, against a 15 kV cable installation, using single Cu conductor cables of the same capacity, spaced

Table 9.2 Line characteristics.

	15 kV cable	15 kV O/H line
Self-inductive reactance, X_a (Ω/mi)	0.470	0.458
Spaced inductive reactance (X_d)	0	0.1682
Subtotal	*0.470*	*0.6262*
Self-capacitive reactance, (X_{ca})	0.1080	0.1057
Spaced capacitive reactance, (X_{cd})	0	0.0411
Subtotal	*0.1080*	*0.1468*
Overall: inductive – capacitive, (Ω/mi)	0.362	0.417
Comparison, % on a base of 100	100%	130%

at 30 cm apart. We will obtain data for the inductive and capacitive reactances in ohms per mile from the TD book. The two reactances namely inductive and capacitive will be deducted from each other for a total value of the line (cable) reactances at 60 Hz. The values are given in ohms per mile. Evidently, there are advantages and disadvantages to either option. The overall cost, maintenance, flexibility of installation, environmental conditions, and safety are the factors that would have to be evaluated to arrive to the most optimal solution for each project (Table 9.2).

The chart in Figure 9.5 demonstrates the difference in line characteristics, based on the conductor sizes.

It has been said that HV lines over 220 kV are predominantly capacitive. How do we explain that? The OHL characteristics say it differently. The line capacitance decreases with the phase conductor spacing (higher voltage). On the other hand, line inductance increases with larger phase conductor spacing.

The aforementioned conflicting statement is clarified by looking at the line loading in Figure 9.6, as applicable to long lines.

The statement is true for HV long light loaded lines (night operations). The line capacitive reactance (shunt capacitance) (V^2/X_c) per phase is immediately energized upon applying the voltage at any load.

This charging load is relatively constant and not dependable on the load. On the other hand, the inductive reactive load $I^2 \times X_1$ per phase is dependable on the load current flowing on the lines as shown on the chart. Initially at line light load, the line is capacitive, but at some increased load, the line becomes predominantly inductive. The longer the line, the greater the positive effect of the charging current on the receiving end voltage and a drop in line loss.

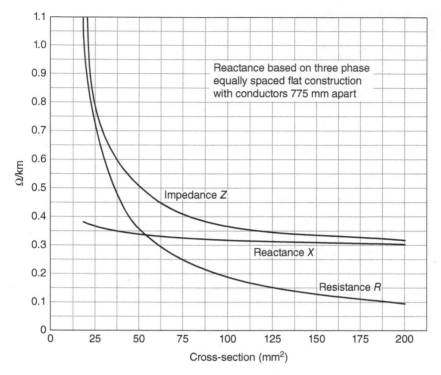

Figure 9.5 Relationship between resistance, reactance, and impedance for conductors of differing cross-sectional areas.

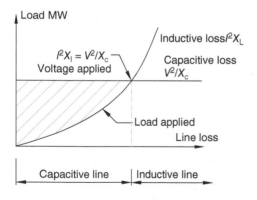

Figure 9.6 Line loading.

9.5 Clearances

9.5.1 Line to Roads

The line conductor clearances to roads and structures must be based on the utility standards and local site regulations. Here is for example a practice used on one of the mining projects in South America (Table 9.3);

Based on Table 9.3, the minimum calculated clearances for road crossings, for the category C voltages, are as follows:

- 36 kV   6.72 m
- 220 kV   7.82 m

It is recommended to establish a clearance of 10 m for the roads where big mining trucks pass. The OHLs must use larger poles at the crossings and shorter spans. For crossing the outdoor conveyors, it is recommended to maintain 7 m clearance between the bottom conductor and the equipment on ground.

9.5.2 Phase Clearances

Phase spacing depends on the conductor sag and the line voltage. The spacing will depend on the local circumstances. Table 9.4 shows the clearances that can be used for orientation purposes:

Table 9.3 Clearance in meters, vertical measurement.

Category	A	B	C
1. Less transmittable roads and grounds	5	5.50	5 + 0.006/kV
2. Principal roads	5	6	6.5 + 0.006/kV
3. Road crossings	5.5	6	6.5 + 0.006/kV

Category A: up to 1000 V, category B: up to 25 kV, and category C: over 25 kV.

Table 9.4 Phase clearances.

Voltage (kV)	Spacing (m)
Up to 15	1.2
Up to 36	2.0
66	2.5
132	3.5

9.6 Line Voltage Drop Calculations

Three cases of voltage drop were presented on spreadsheets as follows:

(1) Figure 9.7: The whole load is at the end of line.
(2) Figure 9.8: Load is unequally distributed along the line with varied loads and distances.
(3) Figure 9.9: Load is equal and uniformly distributed along the feeder. Typical for road lighting circuit.

The easiest method to calculate the voltage drop from the sending end to the receiving end is if the load is located at the end of the line. Since the voltage drop is highly influenced by the power factor of the load, the voltage drop calculations must be done at various loads and power factors to determine the worst case loading. Use the following spreadsheets listed with built-in formulae to calculate any condition on the line or feeder.

If the load is not equally distributed along the line, the voltage drops must be done on a segment by segment basis and then totalize the voltage drops for each segment. Equally distributed load may be the case of road lighting.

Project power distribution							
Line voltage drop (load at end of line)					zb: Sep 25. 2016		
Load at 33 kV		VD < 5%			------- Line --------		
MW	pf	MVA	Ppu	Qpu	Rpu	Xpu	VDpu
10.00	0.80	12.50	0.60	0.45	0.048	0.028	0.0414
10.00	0.85	11.76	0.60	0.37	0.048	0.028	0.0392
10.00	0.90	11.11	0.60	0.29	0.048	0.028	0.0369
10.00	1.00	10.00	0.60	0.00	0.048	0.028	0.0288
12.50	0.80	15.63	0.75	0.56	0.048	0.028	0.0517
12.50	0.85	14.71	0.75	0.46	0.048	0.028	0.0489
12.50	0.90	13.89	0.75	0.36	0.048	0.028	0.0461
12.50	1.00	12.50	0.75	0.00	0.048	0.028	0.0360
15.00	0.80	18.75	0.90	0.68	0.048	0.028	0.0621
15.00	0.85	17.65	0.90	0.56	0.048	0.028	0.0587
15.00	0.90	16.67	0.90	0.44	0.048	0.028	0.0554
15.00	1.00	15.00	0.90	0.00	0.048	0.028	0.0432
V = 33 kV				Distribution: 185 mm², AL, ABC cable in free air			
MVAb	16.67			Line Length: 15 km + 5% for Sag			
Rpu or Xpu = Ohms * MVAb/ kV*kV				Ohms = 15*R *1.05			
VDpu = Ppu*Rpu + Qpu*Xpu (J = F*H + G*I)				Ohms = 15*X*1.05			
Conductor Resistance taken at 75C: 0.2 Ω/km				0.164 Ω/km @20C, 0.211 @90C			
Conductor inductance X = 0.118 Ω/km							

Figure 9.7 Voltage drop for line load at end of line.

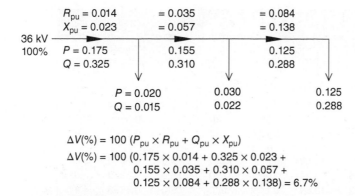

$$\Delta V(\%) = 100\ (P_{pu} \times R_{pu} + Q_{pu} \times X_{pu})$$
$$\Delta V(\%) = 100\ (0.175 \times 0.014 + 0.325 \times 0.023 +$$
$$0.155 \times 0.035 + 0.310 \times 0.057 +$$
$$0.125 \times 0.084 + 0.288 \times 0.138) = 6.7\%$$

Figure 9.8 Voltage drop for not equally distributed load.

Project: Voltage drop (for different line and lengths) Equally distributed load (road lighting)								zb Dec. 23/2013
Cu Conductor:	25 mm²							
Voltage (V)	415	415	415	415	415	415	415	415
Line distance (m)	1200	1200	1200	1200	1200	1500	1500	1500
Pole span (m)	50	45	40	35	30	30	80	100
Actual span with sag	60	54	48	42	36	36	96	120
No of spans (sp)	24	26.667	30	34.286	40	50	18.75	15
Luminaire (VA)	300	300	300	300	300	300	300	300
Luminaire (A)	0.418	0.418	0.418	0.418	0.418	0.418	0.418	0.418
Total (A) I1	10.032	11.147	12.54	14.331	16.72	20.9	7.8375	6.27
Cable size, mm²	25	25	25	25	25	25	25	25
Cable type	3c + n	3c + n	3c + n	3c + n	3c + n	3c + n	3c + n	3c + n
Cable material	Cu	Cu	Cu	Cu	Cu	Cu	Cu	Cu
Z (Ω/km)	0.817	0.817	0.817	0.817	0.817	0.817	0.817	0.817
Z1 (Ω/span)	0.049	0.0441	0.0392	0.0343	0.0294	0.0294	0.0784	0.09804
dV (V)	10.634	11.769	13.187	15.01	17.441	27.118	10.502	8.5076
dV (%)	2.5625	2.8359	3.1775	3.6168	4.2025	6.5344	2.5305	2.05002

Cu Conductor: $R = 0.814\ \Omega/km$, $X = 0.075\ \Omega/km$		Luminaire: 250 W + Balast
dV Voltage Drop at the last Luminaire		Connection: L-N, alternating phases.
$dV = 1.73*I1*Z1*0.5*sp*(sp + 1)$		
$dV(\%) = dV*100/V$	Allowable: <3%.	
Actual span is calculated as (1.2 * span) for the sag.		

Figure 9.9 Voltage drop for equally distributed load.

9.7 Power Loss Calculations

Power loss must be addressed for various loads and line lengths as presented in the calculations (Figure 9.10). The calculation is an actual case for a 33 kV

Project Power Transmission LINE LOSS CALCULATIONS			@ 33kV pf varied			zb: Sep 25. 2016		
---------- Load ----------			------- Line --------			Line Loss		
MW	pf	MVA	kA	Ohm/km	Line Km	MW loss	%
			I	R1	L		
10.00	0.80	12.50	0.219	0.20	15.0	0.4530	4.53
10.00	0.85	11.76	0.206	0.20	15.0	0.4013	4.01
10.00	0.90	11.11	0.195	0.20	15.0	0.3580	3.58
10.00	1.00	10.00	0.175	0.20	15.0	0.2899	2.90
12.50	0.80	15.63	0.274	0.20	15.0	0.7079	5.66
12.50	0.85	14.71	0.258	0.20	15.0	0.6270	5.02
12.50	0.90	13.89	0.243	0.20	15.0	0.5593	4.47
12.50	1.00	12.50	0.219	0.20	15.0	0.4530	3.62
15.00	0.80	18.75	0.328	0.20	15.0	1.0193	6.80
15.00	0.85	17.65	0.309	0.20	15.0	0.9029	6.02
15.00	0.90	16.67	0.292	0.20	15.0	0.8054	5.37
15.00	1.00	15.00	0.263	0.20	15.0	0.6524	4.35

33 kV		185 mm2, AL, ABC cable in free air
Line loss calculated for varied MW load and pf.		Lloss = 3 I*I *R1*1.05*L
Line conductor: ABC triplex cable, 185 mm2, Aluminum. In free air.		
Line conductor extended by 5% due to line sagging (factor 1.05).		
Conductor resistance taken at 75C: 0.2 Ω/km		0.164 Ω/km @20C, 0.211 @90C
Conductor inductance 0.118 Ω/km		

Figure 9.10 Line loss calculations.

line carrying maximum of 15 MVA. The line conductor used was the maximum cable available for the aerial bundled cable (ABC). The line loss is excessive at the maximum load and low power factor. It is essential to have the line operated at the power factor as close to 1.0 as possible.

Suppose that instead of 33 kV, it were possible to run the line at a 45 kV voltage. The line power losses would be significantly lower by a factor of $(33/45)^2 = 0.53$; therefore at 47% lower power loss.

9.8 Line Conductor Sag and Tension

The allowable conductor sag should be kept to a reasonable minimum in order to reduce the conductor material required, avoid extra high poles to give sufficient clearance above ground level, and maintain low tension in the conductors and pole supports.

The conductor tension is governed by conductor weight, current, wind, ice loading, and temperature variations for winter/summer conditions. It is

Table 9.5 Selected conductors.

Conductor	Weight (kg/m)	Tb (kg)	SF (>2)	Span (m)	Sag (m)
Hawk: ACSR, 477 MCM, 26/7 Strand	2.16	8931	3	200	3.63
Partridge: ACSR, 267MCM, 26/7 Strand	1.21	5175	3	100	0.87

a standard practice to keep conductor tension less than 50% of its breaking tensile strength i.e. minimum safety factor of 2.

In the calculations for the conductors Hawk and Partridge erected over a relatively flat terrain, we will use a factor of safety of 3 to allow for bird and rain loading. The span will be assumed to be 200 m for Hawk and 100 m for Partridge. The sag calculations are made based on the following equation:

$$\text{Sag} = \frac{W \times \text{SF} \times \text{sp}^2}{8\,T_b}\ (\text{m}) \tag{9.1}$$

where

W = conductor weight in kg/m
SF = safety factor >2
sp = span in meters
T_b = conductor breaking tension in kg

The equation stands in N (Newton) or kg without change (Table 9.5).

9.9 Aerial Bundled Cable (ABC) Distribution

This type of aerial distribution is popular in South Asia, India, Australia Middle East, Southern Europe, and other countries. Aluminum three phase cables with a messenger wire are bundled together and placed on the poles, as shown in Figures 9.11 and 9.12. Messenger wire takes most of the tension during the installation as well as the operation. The messenger wire is hanged on the hooks, while the conductors hang off the messenger wire.

Cables at 11 and 33 kV distributions in sizes of up to 250 mm^2 are used throughout. Typically, a pole can carry several cables. In most cases, the LV distribution cables for connecting to the houses run along the roads.

For tapping the line, simply terminate the line on an H frame structure to a fused line disconnect switch with arresters and feed it to a pole mounted unit substation. The cable comprises three circular compacted stranded aluminum conductors, cross-linked polyethilene (XLPE) insulated, Cu tape screen, and outer PVC jacket, all bundled around galvanized steel wire. Standard B2627, rated for 90 °C.

Figure 9.11 ABC cable installation.

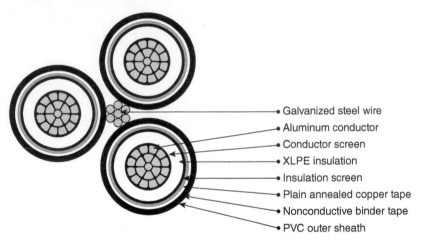

- Galvanized steel wire
- Aluminum conductor
- Conductor screen
- XLPE insulation
- Insulation screen
- Plain annealed copper tape
- Nonconductive binder tape
- PVC outer sheath

Figure 9.12 ABC cable composition.

Other than the low cost and easy installation, the fact is that the ABC cables tend to be more capacitive than inductive, causing the voltage drops to be lower. The poles are typically made of concrete as is common in most of the tropical areas. Due to heavy pole loading, pole spans tend to be short. Obviously, this design approach could not be recommended for Canada and northern parts of USA and Europe due to the snow and ice loading.

9.10   Line and Cable Charging Current

In transmission lines or cables, air in OHLs or insulation in cables acts as a dielectric medium between the conductors. When voltage is applied across the sending end, charging current associated with line (cable) shunt capacitance starts flowing between the conductors and ground. The magnitude of the

Table 9.6 Shunt capacitance and charging currents for selected circuits.

	Length (mi)	Shunt C_1 at 60 Hz (μF)	Shunt C_0 at 60 Hz (μF)	Charging current (A)
345 kV cable	5	1.9341	1.0468	145.2
345 kV OHL	100	2.1363	0.9892	160.48
765 kV OHL	150	2.9604	1.9889	492.9

charging current depends on the voltage, frequency, and the shunt capacitance. The capacitance is determined by the conductor size, the spacing (between conductors and between a conductor and ground), and the total length of the line.

For cables, the permittivity ε of the dielectric around the conductors is generally two to four times the dielectric constant ε_0 for the air. In addition, the spacing between each cable conductor and ground is much smaller than that of an OHL, so the cable will exhibit much higher capacitance than an OHL of the same length.

The charging current I_{ch} in A is calculated for a three phase circuit as follows:

$$I_{ch} = -j\,2\pi f\,l\,C\,V_n\,(A) \tag{9.2}$$

where j is imaginary unit, f is frequency 50 or 60 Hz, l is line length in km, C is capacitance to ground in F/km, and V_n is the circuit voltage, phase to ground.

Tabulation (*American Power* [3]) in Table 9.6 presents the shunt capacitance and charging currents of three example circuits: 345 kV cable, 345 kV OHL line, and 765 kV OHL.

The charging current are calculated per nominal voltage of the line.

From the tabulated examples, the charging current of a 5-mi cable is equivalent to that of a 100-mi OHL with the same voltage level. Table also shows that charging current is quite significant for long-distance extra high-voltage (EHV) lines.

References

1 Chrisholm Institute. *Electrical Transmission and Distribution Reference Book*, Last edition: 1964. Westinghouse Electric Corp.
2 ABB – T & D, Former Westinghouse Blue Book, Westinghouse Co.
3 Yiyan Xue, Dale Finney and Bin Le Schweitzer (2013). Charging current in long lines and high-voltage cables – protection application considerations, *Presented at the 67th Annual Georgia Tech Protective Relaying Conference Atlanta, Georgia May 8–10*.

10

Insulation Coordination, Lightning Protection

CHAPTER MENU

10.1 Economic Design, 209
10.2 Overvoltages, 210
10.3 Lightning Wave Phenomena and Propagation, 213
10.4 Equipment Testing, 215
    10.4.1 Switching Surge, 215
    10.4.2 Lightning Impulse Test, 216
    10.4.3 Chopped Wave Insulation Level, 216
10.5 Shielding, 217
    10.5.1 Transmission Lines, 217
    10.5.2 Substations (Switchyards), 217
10.6 Equipment Withstand Capability, 219
    10.6.1 Standard BIL Levels, 220
    10.6.2 Insulation Coordination, 221
    10.6.3 Arrester Charts, 221
    10.6.4 Arrester Energy Capability, 224
10.7 Arrester Selection, 225
    10.7.1 Arrester Classification, 226
    10.7.2 Method of System Grounding, 226
    10.7.3 MCOV(IEEE) $= U_c$ or U_k(IEC), 227
    10.7.4 Arrester Selection Steps, 227
10.8 Motor Surge Protection, 230
10.9 Building Lightning Protection, 231
    10.9.1 Material Classifications, 231
    10.9.2 Lightning Protection for Special Structures, 236
References, 237

10.1 Economic Design

Insulation coordination is an essential part of the design of electrical power systems in order to protect the electrical equipment located either indoors or outdoors from the various sources of overvoltages. Insulation coordination is

Practical Power Plant Engineering: A Guide for Early Career Engineers, First Edition. Zark Bedalov.
© 2020 John Wiley & Sons, Inc. Published 2020 by John Wiley & Sons, Inc.

Figure 10.1 Arresters in HV substation.

a process of determining adequate insulation levels to protect power system components. It is the selection of an insulation structure with the lightning arresters that will withstand voltage stresses to which the system or equipment will be subjected to.

For the economic reasons, the electrical insulation of the equipment cannot be designed to sustain all the overvoltages, which may confront and affect equipment operation.

Lightning arresters (Figure 10.1) are applied to limit the overvoltage to the levels allowed to reach the equipment. Similarly, high voltage (HV) lines are not designed to withstand the highest possible overvoltages. The high overvoltages are simply allowed to strike over the towers and led to ground.

10.2 Overvoltages

Overvoltages greatly endanger the insulation of electric equipment. System studies must be undertaken to determine the extent and magnitudes of the overvoltages, as these have a major economic impact in the selection of insulation of the system apparatus, particularly for the voltages above 10 kV.

The overvoltages may differ in characteristics such as amplitude, duration, waveform and frequency, etc. The following are the major sources of overvoltages:

(1) *Physical contact with higher voltage systems*: If the conductors of two different systems come in conduct, both systems will assume the voltage of the higher voltage.
(2) *Arcing intermittent short circuits*: Overvoltages can occur in ungrounded or high resistance systems by arcing ground fault conditions due to intermittent shorting contact. Line to ground overvoltages in the low voltage

(LV) systems, 400, 480, 600 V in excess of 1200 V were also observed. Left unchecked, this condition can rapidly break down the insulation of motors in the plant.

(3) *Continuous overvoltages in ungrounded systems*: Ungrounded systems are not uncommon in some industries. In these plants, the equipment is designed to tolerate a single ground fault, but interrupt the plant operation following another ground fault on a different phase. In this situation, the plant equipment is subjected to severe line to ground overvoltages whereby the phase to ground voltage increases to line to line voltage for an extended period of time.

(4) *Switching surges*: System-generated voltage surges appear in HV electrical equipment when abrupt changes occur in operating conditions. Switching surges can be produced by the repeated igniting and extinguishing of electric arcs, breaker, or capacitor switching in transmission line circuits. Also, by disconnecting unloaded lines or by the arcing ground faults on a three-phase system with an insulated neutral point. Voltage surges in lines are limited by good disconnecting capabilities of switches and by resistive losses. The crest height of the surge will depend on the time of contact parting against the current transition through zero.

(5) *System-generated voltage surges*: They usually do not present a danger to the insulation of electrical equipment that is operated at 230 kV or less. For equipment operated at or above 300 kV, it becomes necessary to limit system-generated voltage surges. Switching surges tend to have a slower front than the lightning strikes and are less damaging.

The breakers in the HV systems are equipped with bypass resistors, causing a higher power factor of the short circuit, which may bring the timing of the current and voltage passage to zero closer together thus lowering the voltage across the contacts and overvoltages during contact parting.

(6) *Lightning strikes*: In a direct strike, all the lightning current passes into the ground through the struck object. The duration of voltage surge caused by a direct lightning stroke is of the order of several tens of microseconds. The insulation of EHV electric equipment is not capable of withstanding direct lightning strikes. Lightning strikes (Figure 10.2) are overvoltages associated with lightning discharges either directly into the current-carrying parts of electric equipment (direct-strike surges) or into the ground adjacent to the equipment (induced surges). The world isokeraunic map is shown in Figure 10.3. Induced surges arise in the wires of power lines as a result of an abrupt change in the electromagnetic field near the ground at the time of a lightning strike. The amplitudes of the induced surges usually do not exceed 400–500 kV.

Lightning causes waves of short duration, typically rising rapidly (fast front) in 2–10 μs to crest and then decaying slowly and reaching 50% of the crest value within 20–150 μs.

Figure 10.2 Lightning strike.

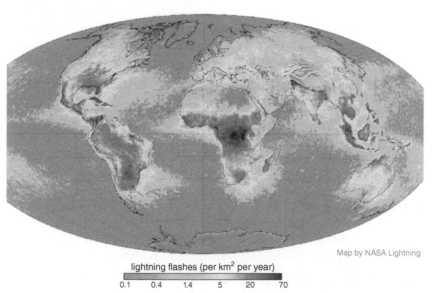

Map by NASA Lightning

lightning flashes (per km² per year)

0.1   0.4   1.4   5   20   70

Figure 10.3 World isokeraunic map by NASA.

Strike currents of up to 200 kA are possible. Direct hits to the electrical equipment are the most severe to the equipment insulation and must be avoided by proper substation design and shielding.

10.3 Lightning Wave Phenomena and Propagation

Lightning is a sudden electrostatic discharge that occurs between electrically charged regions of a cloud (called intracloud lightning or IC), between two clouds (CC lightning), or between a cloud and ground (CG lightning). The charged regions in the atmosphere temporarily equalize themselves through this discharge referred to as a strike if it hits an object on the ground and a *flash* if it occurs within a cloud. Cloud-to-ground is the most common type of lightning. It is usually negative in polarity.

Unlike the far more common "negative" lightning, positive lightning originates from the positively charged top of the clouds rather than the lower portion of the storm. Positive lightning typically makes up <5% of all lightning strikes. Because of the much greater distance to ground, the positively charged region can develop considerably larger levels of charge and voltages than the negative charge regions in the lower part of the cloud. Positive lightning bolts are considerably hotter and longer than negative lightning. They can develop 6–10 times the amount of charge and voltage of a negative bolt and the discharge current may last 10 times longer.

On Earth, the lightning frequency is approximately 40–50 times a second or nearly 1.4 billion flashes/yr, and the average duration is 0.2 seconds [1] made up from a number of much shorter flashes (strokes) of around 30 μs.

Most likely, the overvoltages will reach the equipment by traveling over a line or cable. The *voltage* surge travels in both directions from the point of impact with nearly the velocity of light, along the line to equalize the potential at all points of the line. The speed of the wave is $\frac{1}{\sqrt{LC}}$ in km/s, where L is the line inductance in H/km of line and C is the line capacitance in F/km of line. This works out to be the velocity of light if the line resistances are neglected. Or, about 90% of speed of light if resistance is included according to Professor J.R. Lucas [2].

The wave shape of these voltage surges is similar to that of the current in the lightning discharge. The discharge current splits itself equally on contact with the phase conductor. Using a typical value for the line surge impedance, say $Z_0 = 300\,\Omega$ and an average lightning current of $I = 20\,kA$, the voltage waves on the line would have a crest value of

$$E = \tfrac{1}{2}\,Z_0\,I = \tfrac{1}{2} \times 300 \times 20\ \text{kA} = 3000\ \text{kV}.$$

The *current* wave is of the same shape as the *voltage* wave. It travels inside the conductor along the voltage wave and is proportional to the voltage from the conductor to ground. The current in ampere is directly proportional to the voltage divided with the line surge impedance Z_0.

Transmission lines and cables have distributed inductance and capacitance as their inherent property. When the line is charged, the capacitance component feeds reactive power to the line, while the inductance component absorbs the reactive power. Now, if we assume a balance of the two reactive powers, we arrive at the following equation for surge impedance (characteristic impedance), which is an ohmic value independent from the line (cable) length and its voltage:

$$\text{Capacitive VAR} = \text{Inductive VAR} \rightarrow \frac{V^2}{X_C} = I^2 X_L \rightarrow Z_0 = \sqrt{L/C} \quad (10.1)$$

$Z_0 = \sqrt{L/C}$ in Ω, typically for a HV line
$Z_0 = 300–500$ and $40–80\,\Omega$ for HV cables.

A transmission line of finite length (lossless) that is terminated at one end with impedance equal to the surge impedance Z_0 appears to the source like an infinitely long transmission line and produces no reflections.

The surge impedance loading (SIL) or natural loading is the power loading at which reactive power is neither produced nor absorbed as follows:

$$P_{sil} = \frac{V_r}{Z_0} = \frac{V_r}{\sqrt{L/C}} \quad (W), \text{ based on the receiving end voltage } V_r. \quad (10.2)$$

The velocity of wave propagation $\frac{1}{\sqrt{LC}}$ in km/s differs for the traveling media. The velocity in cables is about half of the speed of that in the lines.

A wave amplitude changes when the traveling wave reaches a junction point of line with cable or the equipment of differing Z_0 surge impedance. The original wave V_i splits into two parts: transmitted wave V_2, which continues ahead onto the cable, while the reflected wave V_1 travels back over the original path of Z_1. The resulting increase may cause a flashover on the line and a discharge on the arrestors. The magnitudes of the two waves, neglecting the resistances, at the change point, are as follows:

Z_0 transition: From line, $Z_1 \rightarrow$ To cable, Z_2

Transmitted wave: $\quad V_2 = 2\,V_i\,\dfrac{Z_2}{(Z_1 + Z_2)}$ $\hfill (10.3)$

Current wave: $\quad I_2 = \dfrac{V_2}{Z_2}$ $\hfill (10.4)$

Reflected wave: $\quad V_1 = V_i\,\dfrac{(Z_2 - Z_1)}{(Z_1 + Z_2)}$ $\hfill (10.5)$

Current wave: $\quad I_1 = \dfrac{V_1}{Z_1}$ $\hfill (10.6)$

➤ If $Z_2 = Z_1 \rightarrow$ There is no reflection. V_i remains unchanged.
➤ If $Z_2 =$ infinite, (open circuit) $\rightarrow V_1 = 2V_i$, $V_2 = 0$
➤ If $Z_2 = 0$, (short circuited) $\rightarrow V_1 = 0$, $V_2 = 0$. The V_i wave reflects as negative to bring V_i to zero.
➤ If $Z_2 > Z_1 \rightarrow$ the incident (original) wave V_i reflects positively at the junction. V_2 increases into the Z_2. V_1 decreases.
➤ If $Z_2 < Z_1 \rightarrow$ the incident (original) wave V_i reflects negatively at the junction. V_2 decreases from V_i into the Z_2. V_1 increases.
➤ If $Z_2 =$ infinite (open circuit) $\rightarrow$ the incident (original) wave V_i reflects back at the junction. V_2 decreases to zero, while V_1 doubles up.

Lightning waves gradually decay as they travel over the lines due to the line resistances.

10.4 Equipment Testing

10.4.1 Switching Surge

The switching impulse test is conducted to confirm the withstand capability of the transformer's insulation against excessive voltages that may occur during system switching. The insulation between windings and between winding and earth and withstand between different terminals are checked using the negative wave shapes as shown in Figure 10.4.

(1) Negative voltage wave shape
(2) Current wave shape.

According to IEC 60076-4 standard, front: $T_1 \geq 100\,\mu s$, 90% value: $T_d \geq 200\,\mu s$, full: T_2: $\geq 500\,\mu s$.

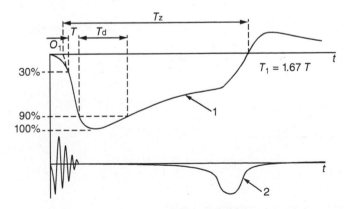

Figure 10.4 Test voltage and current waveshapes.

Sudden collapses of the voltage during the test show deformation of the insulation in the transformer.

10.4.2 Lightning Impulse Test

Basic impulse insulation level (BIL). As per American Standard IEEE C62.41, the impulse wave shape is 1.2/50 μs of negative polarity and 8/20 current waveform. In IEC standard, the wave shape is defined as 1.5 × 40 μs (see Figure 10.5).

This representation of the wave has a special significance. It is a unidirectional wave with a steep rise to its peak value from 0 in 1.2 μs and then falling to 50% of peak value in 50 μs, produced by an impulse generator. It has been established as a standard test vehicle, though it does not resemble to actual lightning strikes.

By conducting this test, the transformer insulation is judged on the same basis throughout the world.

For a three phase transformer, impulse test is carried out on all three phases in succession (Figure 10.6). The voltage is applied on each of the line terminal in succession, while keeping the other terminals earthed. The current and voltage wave shapes are recorded on the oscilloscope, and a distortion in the wave shape is the criteria for failure. Repeated tests at site are conducted at 80% of the standard test.

10.4.3 Chopped Wave Insulation Level

This is determined by using impulse waves that are of the same shape as that of the impulse waveform, with the exception that the wave is chopped after 3 μs. Generally, it is assumed that the chopped wave level is 1.15 times the BIL level for oil filled equipment such as transformers. However, for dry type equipment, it is assumed that the chopped wave level is equal to the BIL level.

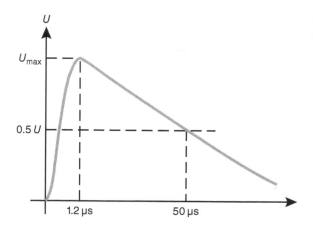

Figure 10.5 Impulse wave shape.

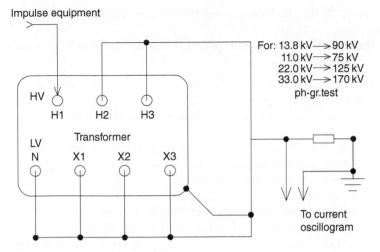

Figure 10.6 Impulse test connection.

10.5 Shielding

10.5.1 Transmission Lines

Clearly, for the higher line voltages, the higher the insulation clearances to ground are required, the higher are the overvoltages that can be sustained on the line, which results in the tower dimensions becoming economically intolerable. Therefore, more protection is implemented in terms of lightning arresters of adequate ratings to maintain the line design within reasonable limits. Furthermore, the lines are shielded with one or two overhead shielding wires to channel the lightning strikes to flashover to ground rather than to the line conductors.

Therefore, transmission line towers are designed for surge withstand voltage to be set just below the critical flashover rating having 50% flashover probability. All the tower insulation dimensions are selected that applied surges result in an acceptable flashover rate. That is, the tower design allows a chance that 50% of lightning strikes will flash over to ground.

10.5.2 Substations (Switchyards)

The switchyard shielding coverage [3] must be extensive to minimize a probability of direct lightning hitting the electrical equipment. The protective shielding over the transformers within switchyards is extremely important to reduce a chance of lightning strikes hitting directly onto the equipment bushings. The transformer bushings often have step higher BIL ratings than the transformer itself. For instance, 1050 kV for bushings and 900 kV for the 230 kV transformer HV winding.

There are three methods of overvoltage protection for the switchyards and substations:

(1) *Switchyard shielding* with high protective masts and/or shielding wires. The switchyard shielding is a good engineering practice. The design requirement may follow the ground flash density number in the area (see Figure 10.7).

Utilities typically have their area charted for flash density, which generally corresponds to the isokeraunic levels of the region, i.e. the number of lightning days a year as shown in the chart for the United States. This forms a base for defining the density zones as follows:

Zone 1 – Area with a ground flash density <0.5 flashes/km², year

Zone 2 – Area with a ground flash density >0.5 flashes/km², year.

The switchyard shielding design for overvoltage protection should provide a coverage to force the lightning strikes to hit the masts or shielding wires strung between the shielding masts or gantry structures, rather than the equipment.

Substations in **Zone 2** area shall be shielded with overhead shield wires and/or lightning masts. The fixed angle per IEEE 998 is used with 45° angle for heights to 20 m and a 30° angle for heights >20 m. In my work with the switchyards, I have always used a 30° angle for the whole switchyard.

(2) *Rod or spark gaps*: These devices are easy and cheap to install and are usually installed in parallel with insulators between the live equipment

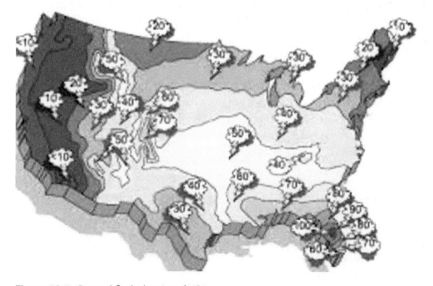

Figure 10.7 Ground flash density of USA.

terminal and earth. When the gaps operate, they cause a short circuit fault, which may cause protection to operate and isolate the equipment. A sudden reduction in the voltage during operation causes high stresses on the transformer interturn insulation.

(3) *Lightning (surge) arresters*: Modern surge arresters are of the gapless zinc oxide (ZnO) type. Previously, silicon carbide (SiC) arresters were used. The arresters have an inverse nonlinear resistance characteristic, thus offering lower resistance to ground for higher overvoltages.

The leakage mostly capacitive currents passing through a ZnO arrester within the operating power frequency voltages are so small (0.1–1 mA) that the arrestor behaves almost like an insulator. However, during the overvoltages, massive discharge currents are let through leaving low voltages across its terminals to protect the insulation of the apparatus from the effects of overvoltage.

The arrester must be located as close to the transformer bushings as feasible to limit the wave reflections along the lead connections.

10.6 Equipment Withstand Capability

For a reliable operation of the equipment, it is necessary to implement protective measures, like adding protective arresters. Figure 10.8 taken from Siemens Literature displays the arrester protective coverage and tolerance over a power system, depending on the duration of overvoltages.

The chart by Volker Hinrichsen represents the magnitude of voltages and overvoltages in an HV system vs. duration of their appearance $\left(1 \text{ pu} = \sqrt{2}\frac{U_s}{\sqrt{3}}\right)$, a peak value of an L_L voltage). As shown, there are four categories of overvoltages with respect to the duration and steepness of the front as follows:

Type of overvoltages	Duration	Strike front
a. Continuous	Continuous	—
b. Temporary	Seconds	—
c. Switching	Milliseconds	Slow
d. Lightning	Microseconds	Fast

The withstand capability is called basic insulation level or simply BIL. The withstand capability against switching surges is called SIL.

In the first two categories on the aforementioned chart, the overvoltages exceed the withstand ratings of the equipment. To enhance the equipment withstand capability in this region, the equipment is protected by lightning arresters. The arresters are generally connected between phases and ground.

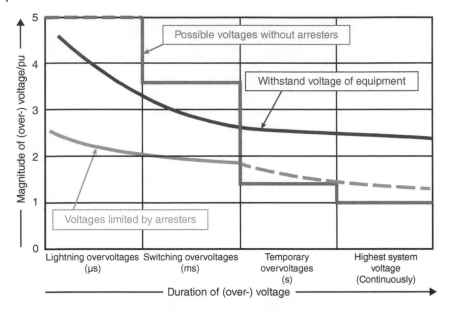

Figure 10.8 Equipment withstand capability. Source: Courtesy of Siemens [4].

10.6.1 Standard BIL Levels

The most common test to verify the withstand of equipment is the lightning impulse test comprising an application of a full waveform voltage surge of a specified crest value to the equipment insulation. Complete tabulations of the BIL levels for the applicable voltages are available in the Standards. Listed here is a short list of the usual voltages (Table 10.1).

Table 10.1 Standard BIL levels.

Reference (kV)	BIL (kV)
5	60
15	95/110 indoors/out
46	250
115	550
138	650
230	900/1050
345	1550

10.6.2 Insulation Coordination

Insulation coordination is defined in ANSI C92.1-1982 [5] as "the process of correlating the insulation strengths of electrical equipment with expected overvoltages and with the characteristics of surge-protective devices." Degree of coordination is measured by the protective ratio (PR). The fundamental definition of PR is insulation withstand level voltage at protected equipment.

Needless to say, the voltage withstand capacity of all equipment in an electrical substation or on an electrical power transmission system must be determined based on its operating system voltage. The operating voltage level of surge protecting devices must be lower than the minimum voltage withstand level BIL of the equipment.

The insulation provided on any piece of apparatus and particularly the transformers constitutes quite an appreciable part of the cost. The international standardizing bodies have had in mind to fix the basic insulation level or BIL as low as is commensurate with safety.

Of course, the transformers are rated for two BIL levels. For instance, a 66 kV/11 kV transformer would be rated for BIL 450/75 kV. Also, higher system voltages often have several BIL levels for the transformer HV windings, depending on the method of grounding of the transformer neutral.

If the primary star winding is rated 230 kV and it is solidly (effectively) grounded with zero impedance, the voltage rise during a lightning strike will be lower, and it can merit a lower BIL 900 kV and consequently a lower transformer cost. However, if the primary winding is grounded through a resistor, or if the primary winding is Delta, the BIL of the primary winding would be 1050 kV. That is why the primaries of the HV transformers are always star and solidly grounded with no intentional impedance to ground. HV bushings are typically rated for the highest applicable BIL since they are the first in line facing the incoming lightning strikes.

There is another wrinkle to be noted for the HV transformers. Often it is specified that the neutral point must be rated to the full BIL. There is no necessity for an effectively grounded transformer having minimal resistance to the grounding electrodes. Tappered insulation to 10% at the neutral point would suffice.

10.6.3 Arrester Charts

Lightning arresters operate as valves to discharge (sparkover) the overvoltage strikes to ground, while a smaller cutoff portion of the overvoltage continues to the equipment.

Discharge voltage $(V_{ref}) = (I_d)$ discharge current × arrester resistance at the applied overvoltage.

A varistor is a variable resistor in ZnO arresters with its resistance controlled by the voltage stress impressed across the device. The varistor and arrester voltage–current characteristic curve (*V–I* curve) shows how the resistance changes as a function of the applied voltage. This characteristic is relied upon to understand and predict the performance of an arrester.

Here are some of the important points pointed along the discharge curve, Figure 10.10 for a pole mounted 15.3 kV lighting arrester in Figure 10.9:

V–I characteristic curve: Varistors are bidirectional devices. The curve shows only one polarity of the conduction curve of a varistor.

Figure 10.9 MV pole arrester.

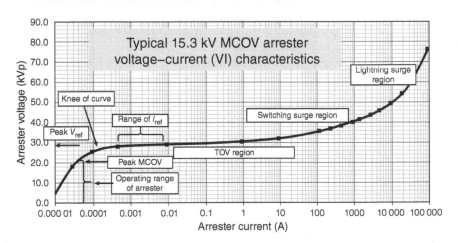

Figure 10.10 MV arrester performance chart. Source: [8].

Knee of the curve: This is a term that roughly describes the voltage stress region where the conduction path through the varistor rapidly changes from the operating or leakage region to the conduction (protection) region. On a log-linear scale, it appears as a knee on the curve.

V_{ref}: The protective level of arrester is defined as its residual voltage (V cutoff or V_d discharge voltage) at a nominal discharge current I_d. The arrester can withstand higher discharge currents but it will cause even higher residual voltage across the equipment terminals, depending on its $V–I$ characteristic. At some point, the arrestor may be subjected to an excessive overvoltage/discharge current over the limit of its energy capability.

Duty-cycle voltage rating (IEEE): The designated maximum permissible voltage between its terminals at which an arrester is designed to perform its duty cycle test.

U_r-*rated voltage of an arrester (IEC)*: Maximum permissible r.m.s. Value of power frequency voltage between its terminals at which it is designed to operate correctly under *temporary overvoltage (TOV)* conditions as established in the operating duty tests.

MCOV requirements: The equipment BIL value defines the highest discharge voltage the arrester must maintain with a proper margin to the equipment withstand BIL level. The higher the BIL, the higher are the discharge currents expected and the higher will be the arrester cutoff voltage generated by the discharge current, but hopefully well below the equipment BIL rating, as shown in Figure 10.11. Incorrectly selected arresters (too high a MCOV voltage) will sustain the arrester life but will pass higher protective (discharge) voltages to the equipment.

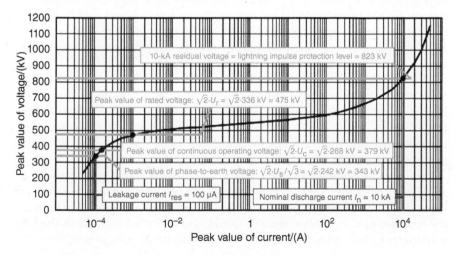

Figure 10.11 *V–I* chart of a ZnO arrester for 420 kV system with earthed neutral. BIL: 1425 kV, SIL: 1050 kV EHV arrester chart. Source: Courtesy of Siemens [4].

The MCOV (U_c) is most likely determined to be just over the system voltage during a line to ground fault. In case of a solidly grounded transformer neutral, that voltage is the nominal line to ground voltage ($V_{line}/\sqrt{3}$). Ungrounded and resistance grounded transformers will demand arresters with higher MCOV ratings. The ungrounded classification includes resistance grounded systems, both high and low resistance connected to the transformer neutrals. In these cases, during the line to ground faults, the healthy phases will reach V_{line} voltage.

ANSI standards propose the following minimum margins of protection for the equipment protected by lightning arresters:

Switching surge withstand	>1.15
Lightning surge withstand	>1.20
Chopped wave withstand	>1.25

Transformer voltage tap regulation must also be included in the process of the selection of lightning arresters. Typically, for an effective grounded system, one would consider the 110% × line to ground voltage as the minimum MCOV or U_c (IEC) voltage for this application.

10.6.4 Arrester Energy Capability

Arresters dissipate switching surges and lightning strikes by absorbing thermal energy. The higher the strike (kA), the less time is allowed of excessive current passage based on factor (I^2t). Given the discharge current, the duration of the current flow is responsible for heating up the arrester resistor discs with limited capacity to dissipate the heat. If, for instance, an arrester is designed for 10 kA discharge current but it is forced to pass much higher currents for longer durations, it is likely the arrester would be destroyed.

Also, switching surges may be occurring in bunches. The amount of energy is related to the prospective switching surge magnitude, its wave shape, line length, prospective strike current, discharge voltage, and the number of operations (single, multiple events). The selected arrester should have an energy capability greater than the energy associated with the expected switching surges on the system.

Arresters are defined by its energy class in kJ/kVMCOV according to standard ANSI/IEEE C62.11. This energy is not a fixed value, but instead depends on the arrester's protective level. The higher the discharge voltage, the less energy the arrester absorbs during the line discharge, since the line will discharge less intensely when the arrester discharge voltage is higher.

Energy class	A	B	C	D	E	F	G	H	J	K	L	M	N
Energy rating (two shot) kJ/kVMCOV	3.0	4.5	6.0	7.5	9.0	11	13	15	18	21	24	27	30

These energy ratings assume that switching surges occur in systems having surge impedances of several hundred ohms, which is typical for overhead transmission circuits. In circuits having low surge impedance involving cables or shunt capacitors, the energy capability of metal oxide arresters may be derated because the discharge currents can exceed the values stated for the lines.

Similarly, if the arrester is subjected to higher voltages than MCOV for an extended period of time, it will be damaged. Should the arrester fail, by creating an arc, it must be replaced.

The discs must have sufficient capacity to contain the heat. The diameter of the resistor discs is the decisive parameter for the energy level of the arresters, ranging from 30 mm diameter to >100 mm diameter for the higher voltages.

A 10 kA rating is not a limitation on particular arrester. A 10 kA arrester can withstand much higher lightning current waves satisfactorily without sustaining damage.

10.7 Arrester Selection

It is important to properly select arresters to suit the application. Present time engineering employs computer study; Electromagnetic TransientProgram (EMTP), which takes into account arrester performance for possible lightning strikes, analyzing the incident and reflection surge voltages entering the environment of drastically different surge impedances. Refer to a study performed for an actual project [7]. Special situation may arise worth looking into is the interface transition point of an medium voltage (MV) bus duct (IPB) with a generator breaker and the need to install arresters ahead of the generator. An open generator may be a major reflection point worth evaluating for the overvoltages coming down the IPB.

The selection greatly depends on the following important factors:

- Classification, depending on the discharge current (kA) expected in the area,
- Maximum continuous voltage (MCOV). Greatly dependent on the system grounding.
- Equipment BIL withstand requirements.
- Arrester placement and shielding within the area of overvoltages.

For system voltages >300 kV, switching surges are often more dominant over the lightning as a criteria for protecting the transmission lines, HV transformers, and switchyard equipment.

10.7.1 Arrester Classification

Nowadays, all the arresters are of metal oxide (MO) type, requiring no external spark gaps. The resistors are made of metal oxide disks, the most popular being Zn Oxide having an extremely inverse nonlinear resistance characteristic. At nominal voltages, the arrester appears as an open circuit. The resistance is high allowing a minute leakage current through the arresters. At the extreme overvoltages, the resistance approaches zero to allow a large discharge current from 10 to 200 kA to flow to the ground for a brief period of time.

The arrester classification in accordance with their principal characteristics and field of application is mostly dependent on the likely level of discharge current as follows:

(1) *Distribution type*, for lower voltages up to 20 kV, 20 kA discharge for pole mounted transformers, located on the common crossarms and transformers up to 1000 kVA.
(2) *Line type*, for up to 138 kV for medium size transformers and discharge currents to 80 kA. Often used for the protection of cables and dry-type transformers, which have lower BIL ratings.
(3) *Station type*, for HV systems and the highest possible discharge currents and switching surge applications. In some instances, the current may be excessive to require paralleling the arresters to conduct the discharge current.

10.7.2 Method of System Grounding

Typical arresters selected for transformer protection, as proposed by a North American utility, are listed below as a guide. Note the difference in protected coverage between the grounded and ungrounded systems. For higher voltages, the systems are exclusively effectively grounded, i.e. a star winding for HV winding with the neutral solidly grounded with zero impedance. The arrester must be grounded to the transformer body and to the grounding grid with as short a lead as possible.

The solidly grounded equipment merits 80% arrester rating for the insulation protection. The ungrounded systems essentially have a floating neutral points, thus for any strike to ground the windings are stressed for line to line over voltages, compared to the solidly grounded systems where the neutral point is fixed

Table 10.2 Arrester ratings for grounded/ungrounded systems.

Transfer voltage (kV)	Transfer BIL (kV peak)	Arrester voltage (kV) Grounded	Arrester voltage (kV) Ungrounded	MCOV (kV)	
				Grounded	Ungrounded
345	1050	264	NA	212	NA
230	900, 1050	172, 190	228, 240	144	NA
138	550	108, 120	132, 144	98	118
69	350	54, 60	66, 72	48	57
34.5	200	27, 30	36, 39	24.4	29
13.8	110	10, 12	15, 18	10.2	12.7

in the middle and to ground. Read more in Chapter 5 on the star/delta transformers used for solidly grounded systems (Table 10.2).

10.7.3 MCOV(IEEE) = U_c or U_k(IEC)

It is the maximum continuous operating r.m.s. (root mean square) voltage that can be continuously applied to the arrester terminals, line to ground [8], inclusive of the voltage regulation that may be applicable to the operating system.

Table 10.3 indicates the minimum MCOV and U_c voltages in $kV_{r.m.s.}$ that can be applied to the arresters selected to protect the equipment operating on the system voltages IEEE and IEC respectively, and applicable to the effectively grounded systems.

10.7.4 Arrester Selection Steps

The following are the arrester selection steps [8]:

(1) V_s: System nominal voltage
(2) $V_{MCOV} \geq 1.05 \times V_s/\sqrt{3}$: For solidly grounded systems.
   $V_{MCOV} \leq V_s$: For solidly grounded systems.
   $V_{MCOV} \geq V_s$: For ungrounded and resistance grounded systems.
(3) $V_r \geq 1.25 \times V_{MCOV}$ rated voltage V_r. For protective margin consider the TOV in duration of 10 seconds. The most common source of TOV is voltage rise on unfaulted phases during a line-to-ground fault. This is an important factor for the ungrounded systems.

Table 10.3 IEEE MCOV and IEC (U_k) minimum ratings [6, 9].

Nominal line to line voltage (kV$_{r.m.s.}$)	Maximun line to line voltage (kV$_{r.m.s.}$)	Maximum line to ground voltage (kV$_{r.m.s.}$)	Minimum MCOV (kV$_{r.m.s.}$)	Nominal line to line voltage (kV$_{r.m.s.}$)	Typical maximum line to line voltage (kV$_{r.m.s.}$)	Maximum line to ground voltage (kV$_{r.m.s.}$)	Minimum U_c (kV$_{r.m.s.}$)
Typical IEEE system voltages				Typical IEC system voltages			
2.40	2.52	1.46	1.46	3.3	3.7	2.1	2.1
4.16	4.37	2.52	2.52	6.6	7.3	4.2	4.2
4.80	5.04	2.91	2.91	10.0	11.5	6.6	6.6
6.90	7.25	4.19	4.19	11.0	12.0	6.9	6.9
8.32	8.74	5.05	5.05	16.4	18.0	10.4	10.4
12.0	12.6	7.28	7.28	22.0	24.0	13.9	13.9
12.5	13.1	7.57	7.57	33.0	36.3	21.0	21.0
13.2	13.9	8.01	8.01	47.0	52	30.1	30.1
13.8	14.5	8.38	8.38	66.0	72	41.6	41.6
20.8	21.8	12.6	12.6	91.0	100	57.8	57.8
22.9	24.0	13.9	13.9	110.0	123	71.1	71.1
23.0	24.2	14.0	14.0	132.0	145	83.8	83.8
24.9	26.2	15.1	15.1	155.0	170	98.3	98.3
27.6	29.0	16.8	16.8	220.0	245	142	142
34.5	36.2	20.9	20.9	275.0	300	173	173
46.0	48.3	27.9	27.9	330.0	362	209	209
69.0	72.5	41.9	41.9	400.0	420	243	243
115.0	121	69.8	69.8				
138.0	145	83.8	83.8				
161.0	169	98	97.7				
230.0	242	140	140				
345.0	362	209	209				
500.0	525	303	303				
765.0	800	462	462				

Source: Courtesy of ArresterWorks [8].

(4) Determine the arrester nominal discharge current Id as recommended by the standards to classify the arresters for the nominal voltage levels:

$V_s \leq 36\,kV \rightarrow 2500\,A, \rightarrow V_s \leq 72.5\,kV \rightarrow 5000\,A$

$V_s \leq 360\,kV \rightarrow 10\,000\,A, \rightarrow V_s \leq 756\,kV \rightarrow 20\,000\,A$

Table 10.4 Line shielding.

Line shielding coverage	Chances of stn. arrester discharge(%)
0.5 mi (0.8 km)	50
1.0 mi (1.6 km)	35
1.5 mi (2.4 km)	25

The discharge current is expected to be higher for higher line voltage due to higher insulation levels and clearances, as well as due to longer line lengths. Switchyards must be shielded for the protection from lightning discharges. Furthermore, the lines tied to the switchyards must also be protected for at least 2 km out of the switchyards. Lower voltage ≤ 35 kV lines entering stations sometimes are not shielded, which increases the likelihood of the station arresters being hit with an increased number of strikes and of high magnitudes up to 20 kA in particular if the lines pass through the areas of high lightning ground flash density. The chances of the substation arresters being discharged when the lines are shielded are estimated as follows (see Table 10.4).

The following probabilities of discharge current were derived based on statistical field investigations:

99% of time <10 kA.

95% of time <5 kA.

90% of time <3 kA.

70% of time <1 kA.

(5) Calculate the protective margin. The discharge voltage, often also referred to as residual voltage V_{ref}, is the voltage that appears between the line and arrester ground terminals during the passage of the nominal discharge current. As noted earlier, this voltage is an indication of the protection margin. It has to be safely below the BIL rating of the protected equipment.

Due to the nonlinearity of the ZnO arresters, the impact of the discharge current on the discharge voltage is small, as shown in the $V-I$ chart of Figure 10.10. For instance, for a 15 kV arrester, an increase of discharge current from 1 to 3 kA (300%) results in the discharge voltage increase from 40 to 44 kV (11%).

(6) *Housing*: Unless, the application has some seismic test requirements, the housing of the arrester can be either polymer or porcelain. Porcelain insulator is sturdier and preferred. The selection of housing should consider the following factors:

Pollution, creepage distance, height of installation, and energy discharge requirements (diameter, number of units in parallel).

Example

	66 kV	110 kV
Voltage system	**66 kV**	**110 kV**
Maximum voltage (V_m)	72.5 kV	123 kV
System voltage ($V_s = V_m$)	72.5 kV	123 kV
Method of grounding	**Effective**	**Resistance**
BIL	325 kV	550 kV
Nominal discharge current[a]	10 kA	10 kA
$V_{MCOV} (U_c) = 1.05 \times V_s / \sqrt{3}$	44 kV	N.A.
$V_{MCOV} (U_c) = V_s$ (for ungrounded systems)	N.A.	123 kV
Rated voltage, $V_r = 1.25 \times V_{MCOV}$	55 kV	153 (rounded)
Lightning impulse protective characteristic for 10 kA	168 kV	370 kV (discharge voltage)
Protection margin BIL/for 10 kA wave	1.93 OK! (325/168)	1.5 OK! (550/370)

a) The standard waveform for the discharge current is taken as 8/20 μs.

10.8 Motor Surge Protection

This protection comprising a set of capacitors and distribution arresters (surge pack) is often used for large MV motors starting at 1500 kW and higher.

The insulation withstand of the motor is relatively low in comparison to the other equipment at the same voltage level. Generally, it is expected that the LV and MV motors are not connected directly to the overhead lines, but to the switchboards fed from transformers. If the transformers are provided with proper lightning protection, there is general consensus that 575 V motors and lower do not need additional overvoltage protection.

Surge protection is readily provided for the plant generators, but rarely for standby generators.

The test requirement is $2V + 1$ kV for motors. For example, for 4000 V motors, this is

Motor test voltage $= 2 \times 4.16 + 1 = 9.32$ kV.

The MV systems are generally resistance grounded to allow 50–100 A ground fault current. In view of that, arresters must be rated as an application for ungrounded systems. A 4 kV motor would require 4.5 kV MCOV arrester rating. The capacitors in the surge packs are employed to reduce the steepness of the switching strikes, which may cause a rise of voltage on the motor windings and neutral.

The capacitor rating would also be specified as line to line voltage of the ungrounded systems. The capacitance typically used is 0.5 μF for the 6.6 kV or less and 0.25 μF per pole for 15 kV systems.

10.9   Building Lightning Protection

10.9.1   Material Classifications

The building lightning protection is designed in compliance with the UL 96A "Installation Requirement for Lightning Protection Systems" and the NFPA 780 [10, 11].

Structure lightning protection is subject to the site inspector's approval. An approval will be given if proper materials classified for lightning protection are used.

Air terminals and associated hardware and materials are installed on all the buildings and structures needing lightning protection, depending on the type of building, height, type of construction, and the structural materials used. The lightning protection system together with the ground electrode system forms a common low impedance path for the lightning strikes to effectively dissipate the lightning energy into the ground. Copper or aluminum materials can be chosen based on the metallic materials of the roof. All the plant metallic structures are bonded to the plant grounding system.

Copper lightning protection materials shall not be installed on aluminum roofing, aluminum siding, or other aluminum surfaces.

Class I and Class II materials (Figure 10.12 and Table 10.5) are used throughout the plant, depending on the structure height in accordance with the NFPA Standards 96 and 980 as follows:

- *Class I materials*: Lightning conductors, air terminals, ground terminals, and associated fittings are required for the protection of structures up to 23 m (75 ft) in height.
- *Class II materials*: Lightning conductors, air terminals, ground terminals, and associated fittings are required for the protection of structures >23 m (75 ft) in height.

Figure 10.12 Lightning aerial base.

Table 10.5 NFPA Std. 780 Material Classifications for Lightning Systems[10,11]

Table 4.1.1.1(A) Minimum Class I Material Requirements

Type of Conductor	Parameter	Copper	
		SI	U.S.
Air terminal, solid	Diameter	9.5 mm	$\frac{3}{8}$ in.
Air terminal, tubular	Diameter	15.9 mm	$\frac{5}{8}$ in.
	Wall thickness	0.8 mm	0.033 in.
Main conductor, cable	Size each strand		17 AWG
	Weight per length	278 g/m	187 lb/1000 ft
	Cross section area	29 mm^2	57 400 cir. mils
Bonding conductor, cable (solid or stranded)	Size each strand		17 AWG
	Cross section area		26 240 cir. mils
Bonding conductor, solid strip	Thickness	1.30 mm	0.051 in.
	Width	12.7 mm	$\frac{1}{2}$ in.
Main conductor, solid strip	Thickness	1.30 mm	0.051 in.
	Cross section area	29 mm^2	57 400 cir. mils

Table 4.1.1.1(B) Minimum Class II Material Requirements

Type of Conductor	Parameter	Copper	
		SI	U.S.
Air terminal, solid	Diameter	12.7 mm	$\frac{1}{2}$ in.
Main conductor, cable	Size each strand		15 AWG
	Weight per length	558 g/m	375 lb/1000 ft
	Cross section area	58 mm^2	1 15 000 cir. mils
Bonding conductor, cable (solid or stranded)	Size each strand		17 AWG
	Cross section area		26 240 cir. mils
Bonding conductor, solid strip	Thickness	1.30 mm	0.051 in.
	Width	12.7 mm	$\frac{1}{2}$ in.

Copyright NFPA

Ordinary structure height classifications are defined as:

- ≤23 m (75 ft) shall be protected with Class I materials as shown in NFPA Table 4.1.1.1(A).
- >23 m (75 ft) shall be protected with Class II materials as shown in NFPA Table 4.1.1.1(B).

Conductors: Lightning conductors do not fall into the basic wire size categories used in other installations. Typically, lightning conductors are more stranded and more flexible.

Class I main conductors: Stranded copper conductors at least 57.4 kcml and aluminum conductors at least 98.6 kcml. For secondary or bonding conductors, the minimum size shall be 26.2 kcml for copper and 41.1 kcml for aluminum. Table 10.5, clarify the material classifications used for installation of lightning protection systems.

Class II main conductors: Not <115 kcml for stranded copper conductor and 192 kcml for aluminum conductors.

For secondary or bonding conductors, use the minimum size 26.4 kcmil for copper and 41.1 kcmil for aluminum. Copper materials must be of the grade required for commercial electrical work of 95% conductivity. Copper alloy materials if used must be as resistant to corrosion as copper.

Main conductors must interconnect all the roof air terminals and form two or more paths from each air terminal downward for connections with ground terminals (Table 10.5).

Figure 10.13 Air terminal height. Source: Courtesy of NFPA Standard 780.

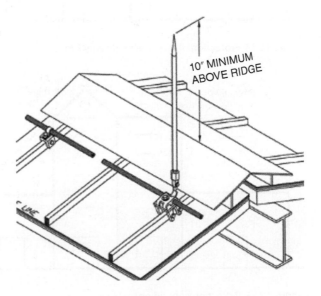

10″ MINIMUM ABOVE RIDGE

Air terminal height: It must be ≥254 mm (10 in) up to 1000 mm (40 in.), above the object or area it is to protect (Figure 10.13).

Placement of air terminals [10, 11]: Air terminal devices are not required for those parts of a structure located within the zone of protection as shown in Figures 10.14 and 10.15. The zone of protection forms a cone whose apex is located at the highest point of the air terminal, with walls forming a 45° or 63° angle from the vertical, depending on the building height.

Structures shorter than 7.6 m (25 ft) above earth shall be considered to protect lower portions of structure located within a 2–1 (63°) of protection.

For structures exceeding 15 m (50 ft), the protection zone over the lower sections will be under the 1–1 cone (45°) of zone of protection.

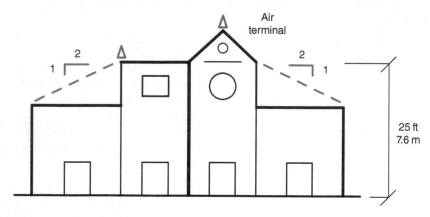

Figure 10.14 Protection of low-rise buildings ≤7.6 m.

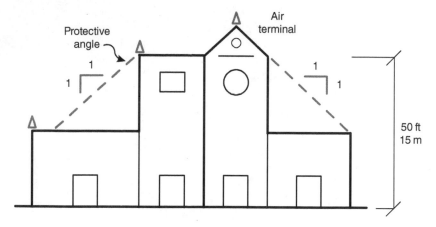

Figure 10.15 Protection of buildings >15 m.

Pitched roofs are defined as roofs having a span of 12 m (40 ft) or less and a pitch 1/8 or greater and roofs having a span of more than 12 m (40 ft) and a pitch ¼ or greater. All other roofs shall be considered gently sloping and are to be treated as flat.

Separation of air terminals [10, 11]: Air terminals shall be placed at or within 0.6 m (2 ft) of ridge ends on pitched roofs or at edges and outside corners of flat or gently sloping roofs at intervals not exceeding 6 m (20 ft).

Flat or gently sloping roofs that exceed 15 m (50 ft) in width or length shall have air terminals located at intervals not to exceed 15 m (50 ft) on the flat or gently sloping areas or such area can also be protected using taller air terminals that enlarge zones of protection. Along the roof edges, the maximum spacings shall not exceed 7.6 m (25 ft).

There shall be no more than 45 m (150 ft) maximum length of cross run conductor without connection of the cross run conductor of the main perimeter or down conductor.

Down conductors: Provide at least two down conductors on any kind of structure. Structures exceeding 76 m (250 ft) in perimeter shall have a down conductor for every 30 m (100 ft) of perimeter or fraction thereof. The total number of down conductors on structures having flat or gently sloping roofs shall be such that the average distance between all down conductors does not exceed 30 m (100 ft).

Each down conductor shall terminate at a ground terminal dedicated to the lightning protection system.

Cable and cable connectors for Class II shall be continuous from air terminal to ground and interconnected with the balance of the system.

Ground rods: Ground terminals shall be copper-clad steel, solid copper, hot-dipped galvanized steel, or stainless steel. It shall be not <12.7 mm (½ in.) in diameter and 2.4 m (8 ft) long. Rods shall be free of paint or other nonconductive coatings and driven vertically not <3 m (10 ft) into the earth. Electrical system and telecommunication grounding electrodes shall not be used in lieu of lighting ground electrodes. This provision does not prohibit the bonding together of grounding electrodes of different systems.

Common grounding: All the grounding in or on a structure shall be interconnected to provide a common ground potential. This interconnection shall include lightning protection, electric grounding, telephone, and antenna system grounds, as well as underground metallic piping systems. Underground metallic piping systems may include water service, well casings located within 7.6 m (25 ft) of the structure, gas piping, underground conduits, underground liquefied petroleum gas piping systems, and so on. Buried installations preferably use Cadweld connections (Figure 10.16).

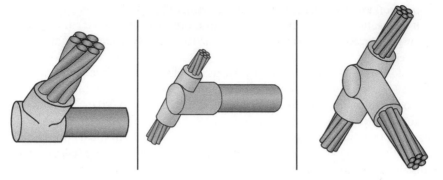

Figure 10.16 Cadweld underground connections.

10.9.2 Lightning Protection for Special Structures

Metallic structures: Metallic structures exposed to direct lightning flashes and having a structural metal thickness of 5 mm (3/16 in.) or greater are considered conductive and shall require only structure connections to the lightning and grounding protection system. It shall not need air terminals.

Masts: Electrically continuous metal masts shall require only bonding to ground terminals.

Metal towers and tanks: Metal towers and tanks constructed so as to receive a stroke of lightning without damage shall require only bonding to ground terminals.

Concrete tanks and silos: Lightning protection systems for concrete (including prestressed concrete) tanks containing flammable vapors, flammable gases, and liquids that produce flammable vapors and for concrete silos containing materials susceptible to dust explosions shall be provided with either external lightning conductors or with conductors embedded in concrete.

To be considered properly protected against lightning, the stacks, chimney, condenser, and boilers are required to be properly grounded on at least two places to the grounding system, at 18 m (60 ft) apart or less or every second structural column. The connections are made to cleaned spots on the structural steel members with appropriate bonding plates or exothermic connections, having contact surface area not less than 50 cm² (8 in.²).

Steel prefabricated buildings: Air terminals of appropriate material class are installed on the laminated metal roof elevations (if the metal thickness is <5 mm (3/16 in.) and if falling outside the zone of protection afforded by the balance of the lightning protection system. The air terminals are bonded directly to the steel framework by short cables. The connections are made by bonding plates or exothermic connections on cleaned areas of the steel framework. The building steelwork is considered and used as the primary lightning conductor.

Ground connections are made at every second steel column at the foundation level by exothermic process around the building or structure and tied directly to the closest building grounding loop or grounding rod.

References

1 World Lightning Map by NASA, 2006: https://earthobservatory.nasa.gov/IOTD/view.php?id=6679.
2 Lucas, J.R. Lecture course EE4020 - insulation co-ordination. In: *Wave Propagation*, Chapter 4. University of Moratuwa.
3 IEEE 998-2012 - IEEE Guide for Direct Lightning Stroke Shielding of Substations.
4 Hinrichsen, V. and Siemens (2012). *Metal Oxide Surge Arrester in High-voltage Power Systems: Fundamentals*. Siemens.
5 ANSI C92.1-1982 Insulation Coordination, 1982.
6 IEEE Std. C62.22-1991 IEEE Guide for the Application of MO Arresters.
7 Fung, E. (2017). *Surge and Lightning Protection, Waneta Expansion Limited Partnership*. SNC-Lavalin.
8 Woodworth, J.J. (2009). *Part 1 of Arrester Selection Guide*. Arrester Works.
9 IEC Standards 6099-1,2,4,5 Insulation Coordination and Selection of Lightning Arresters.
10 UL 96A Installation Requirement for Lightning Protection Systems.
11 NFPA 780. Standard for the Installation of Lightning Protection Systems.

11

Voltage and Phasing Standards

CHAPTER MENU

11.1 Supply and Utilization Voltages, 239
11.2 System Phase Sequence, 243
    11.2.1 Motors, 243
    11.2.2 Generation, 244
    11.2.3 Phase Sequence Convention CCW and CW, 245
    11.2.4 Phase Sequence Blunders, 247
    11.2.5 Conclusions, 249
11.3 World Plugs/Sockets, 250
References, 251

11.1 Supply and Utilization Voltages

There is nothing that confuses nonelectrical engineers and managers on your project team more than reading the "confusing" voltage numbers used in the specifications, drawings, and reports.

Well, there is one even more confusing item for them and that is the reactive power, discussed in-depth in Chapter 13.

Typical low voltage (LV) and medium voltage (MV) in the world are shown in Table 11.1.

Mechanical engineers keep on asking me: "Is it 15 kV, 14.4 kV, 13.8 kV, or is it 13.2 kV? Make up your mind, please." You may explain it to your friend mechanical engineer, but don't get surprised if he pops the same question again, two months later. If he is confused, that's to be expected. But you should not be. Let us clarify.

Yes, there exist a variety of voltages. The aforementioned list is not even complete. Simply explained, at every voltage level, there is *supply* and *utilization* or receiving voltages that are available.

The voltages on the rows listed above belong to the same system group. The higher voltages in a group are typically the supply or sending voltages assigned

Practical Power Plant Engineering: A Guide for Early Career Engineers, First Edition. Zark Bedalov.
© 2020 John Wiley & Sons, Inc. Published 2020 by John Wiley & Sons, Inc.

Table 11.1 World's typical voltages.

North America	Europe, Asia
36, 33, 27.6 kV	34.5 kV
15, 13.8, **14.4**, **13.2** kV	11, **12**, 10.5 kV
5, 4.16, 4 kV	**3.3**, 3, 6.6, 7.2 kV
600, 575, **480**, 460 V	**400**, 380 V
120, 115 V	231, 220 V

Bold are nominal values.

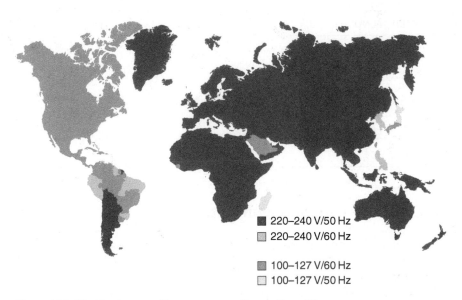

Figure 11.1 World voltage and frequency map. Source: From [1].

to the supply equipment, while those with lower numbers are the receiving or utilization voltages.

The world is not only divided on the voltages. In addition, the world operates on different frequency; 50 and 60 Hz as seen in Figure 11.1. The ANSI C84.1-1970 standard [2] defines the voltages and voltage ranges for the 60 Hz operating regions. The term "nominal system voltage" designates not a single voltage but a range of voltages at any point on the system.

The voltages may vary and still provide satisfactory operation of the equipment on the system. The voltage range for all the nominal system voltages differ in the utilization and distribution range for two critical points on the distribution: the point of delivery by utility and the point of connection to utilization equipment. For transmission voltages, the maximum voltages are specified because the voltages are unregulated, and only a maximum is given to establish the design insulation level for the line and associated apparatus.

Suppose (in Canada) a transformer feeds a motor control center (MCC). On the transformer, let the secondary (supply) voltage be 600 V. At the MCC, the (receiving) voltage be anywhere between 575 and 600 V. The same transformer is fed from MV switchgear, which has a rating of 13.8 kV. The transformer primary windings may be rated either 13.8 or 13.2 kV, as its (receiving) primary voltage. Then, using 13.8 kV is suitable if the transformer is expected to be lightly loaded. If, on the other hand, the transformer is expected to be heavily loaded or it is remote from the switchgear, then it is specified the transformer has 13.2 kV as its primary voltage. By lowering the voltage, voltage will be boosted on its secondary side for the transformer load and have a satisfactory voltage at the MCC.

Motors invariably use 440, 460 V in the 480 V group or 550, 575 V in the 600 V group. Or in Europe, 380 V in the 400 V group. MV motors are 4000 V, used on the 5 kV systems.

Overhead lines are usually called 5 or 15 kV lines, though the voltages are 4.16 and 13.8 kV.

The aforementioned concept allows us to accommodate for the voltage drops along the power distribution flow, thus avoiding the use of tap setting on transformer tap changers or to minimize their use. This has an effect of the MCC receiving approximately proper voltage by canceling the voltage drops in the cables feeding the MCC. It makes sense. Why should you buy a 13.8 kV to 480 V transformer, when immediately you must set it to its −2.5% tap setting on the 13.8 kV side, to make it right on the secondary side?

Here, it is in a picture form on the Canadian industrial scene:

→13.2 kV—[T1]—4.16 kV → 4.0 kV—[T2]—600 V → 575 V—[MCC] →
575 V Motors

Mechanical engineers should relate voltages to the pressures on their systems. The pressures have supply pressure, receiving pressure, and even pressure drops, just like the voltages.

IEC Standard 60038 – 1997 [3]

This international standard lists the present usage of voltages and attempts to cement the present practice of loose voltage selections. Let us start with definitions. For the alternating voltages, the voltages stated are r.m.s. values.

Nominal system voltage: Voltage by which a system is designated.
Highest and lowest voltages of a system (excluding transient or abnormal conditions): The highest value of voltage which occurs under normal operating conditions at any time and at any point in the system.

Lowest voltage of a system: The lowest value of voltage which occurs under normal operating conditions at any time and at any point on the system. It excludes voltage transients, such as those due to system switching, and temporary voltage variations.

Supply voltage: The phase-to-phase or phase-to-neutral voltage at the supply terminals.

Supply voltage range: The voltage range at the supply terminals.

Utilization voltage: The phase-to-phase or phase-to-neutral voltage at the outlets or at the terminals of operating equipment.

Utilization voltage range: The voltage range at the outlets or at the terminals of equipment.

Rated voltage (of equipment): The voltage assigned generally by a manufacturer, for a specified operating condition of a component, device, or equipment.

Highest voltage for equipment: Highest voltage for which the equipment is specified regarding the insulation. The highest voltage for equipment is the maximum value of the "highest system voltage" for which the equipment may be used.

Here is a tabulation of recommended voltages <1000 V (Table 11.2).

Concerning the supply voltage range, under normal service conditions, it is recommended that the voltage at the supply terminals should not differ from the nominal voltage of the system by more than ±10%.

For the utilization voltage, in addition to the voltage variations at the supply terminals, voltage drops may occur within the consumer's installations. For low-voltage installations, this voltage drop is limited to 4%, therefore, the utilization voltage range is +10% to −14% (Tables 11.3 and 11.4).

Table 11.2 Nominal low voltages.

50 Hz	60 Hz
3 ph. 3 W and 4 W	
—	120/208 V
—	240 V
—	120/240 V
231/400	277/480 V
400/690	480 V
—	347/600 V

Table 11.3 Equipment voltages, three phase, AC > 1 kV < 35 kV.

	50 Hz			60 Hz	
Nominal	Lowest	Highest		Nominal	Highest
3.3	3	3.6		4.16	4.4
6.6	6	7.2		—	—
11	10	12		—	—
—	—	—		13.8	14.52
22	20	24		—	—
33	—	36		34.5	36.6

Table 11.4 Transmission voltages (kV).

50 Hz	60 Hz
66 (69)	72.5
110 (115)	123
132 (138)	145
220 (230)	245

11.2 System Phase Sequence

Phase sequence or phase rotation is one of the most confusing concepts in the electrical engineering. It is often confused with motor rotation. For instance, one would say: "If it doesn't turn the way you want it, just rewire two phases and the motor will turn correctly." That applies well for most of the motors, but not for the large generators. The issue of a proper phase sequence, clockwise (CW) or counterclockwise (CCW), has a different impact on the electrical equipment as follows:

(1) Phase sequence of the source equipment (generation).
(2) Phase sequence of utilization equipment (motors).

11.2.1 Motors

Typically, the pumps and fans are designed for a specific direction of rotation. The mechanical rotation is usually marked on the driven equipment. Before starting the pump, the pump is uncoupled from the motor and the motor direction of rotation is checked. If the rotation does not match that of the related fan or pump, the phase sequence is changed by reversing two phases. This can be done at the motor terminal box or at the source; or preferably at MCC or switchgear. One does not even pay attention if the required phase sequence is CW or CCW. Just make it right for the driven equipment.

The problem here is that one day you may take that motor–pump set and move it to another location in the same plant thinking that it will turn in the same direction, which may not be true. This is generally due to an inconsistent approach of phasing of the plant MCCs, which feed plant motors. Well, before one reconnects it at a different place make sure you uncouple the motor and go through the bumping procedure again.

The phase sequence is of no concern for three phase feeding of lighting panels and similar electrical boards, which do not feed rotating equipment.

11.2.2   Generation

This issue applicable to generation is more difficult to understand and more costly to correct if incorrectly built or installed. It concerns larger (generation) equipment and integration of a generating plant into a power grid to operate in synchronism. A generator wired to incorrect phase sequence cannot be electrically synchronized with the rest of the grid, to which it belongs to.

Unlike the pumps, the electrical world is not visible until measured and referenced against something that is known. A pump or generator cannot be forced to rotate differently than it was originally built to rotate or to pump a liquid in a predefined direction. Unfortunately, a phase-change modification on large generators is often not possible, or it is extremely costly and subject to delays to the synchronizing. If a large generator is wired for an incorrect phase sequence, the only solution may be to rephase two outgoing phases.

A 100 MW, 13.8 kV, 0.85 pf, generator produces almost 5 kA per phase. That particular generator requires huge conductors, which are not easy to rephase. The large phase bus ducts are cut to measure to suit the plant layout arrangement and cannot be moved. It may be possible for small generators but certainly not for the big ones. The simplest way is to reverse the phases on the high voltage side of the step up transformer before the HV substation of the generating station, if feasible. In that case, the generator is let (electrically) turn CW, while the phases beyond the step up transformer are switched to provide a plant output in a CCW phase sequence to be able to synchronize with the grid. That will not be the first time it was done. That remedial approach has been employed often.

The generator is still built to a wrong phase sequence convention, but its output is now correct. In fact, if you change the phasing, often you may have to change the wiring of current transformers (CTs) and potential transformers (PTs) as well as the protective relay settings. The service motors within the power plant must rotate in the same direction when powered from the generator or from the external source. So changing the phasing on the generator must be followed by changing the phasing on the plant service output.

11.2.3   Phase Sequence Convention CCW and CW

Let us go back and discuss the accepted convention of the phase sequence. The phase sequence phasor diagrams by convention are drawn with vector A pointing up while vector B is on the right following vector A, and vector C on the left following vector B, all in CCW direction. Furthermore, the leading power factor is established as negative and lagging as positive. The North American grid phase sequence is CCW as the accepted convention as A ← B ← C, as shown in Figure 11.2.

The "Rotation" on the aforementioned graph indicates the stator AC phases or electrical vectors and not the rotor rotation.

A professor who commented on this chapter noted that it was wrong to use the term "phase rotation" as is the case in North America and noted in Figure 11.2. The rotation should be reserved for the rotors, while the vectors should use the term phase sequence.

The European and Asian grid phase sequence is CW as A → B → C. The arrows are not the rotor rotation, but the resulting phase sequence, one after the other, is shown in brackets (see Figure 11.2).

Phase sequence is the order in which the voltage waveforms of a polyphase AC source reach their respective peaks. For a three-phase system, there are only two possible phase sequences:

CCW: A ← B ← C and CW: A → B → C.

The CCW is assumed to be a standard sequence for the discussions as shown in Figure 11.3.

If the generator stator phases are wired and laid in sequence ABC and they are subjected to CW rotor rotation (looking from the generator to the turbine),

Figure 11.2 CCW phase sequence.

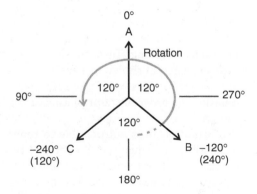

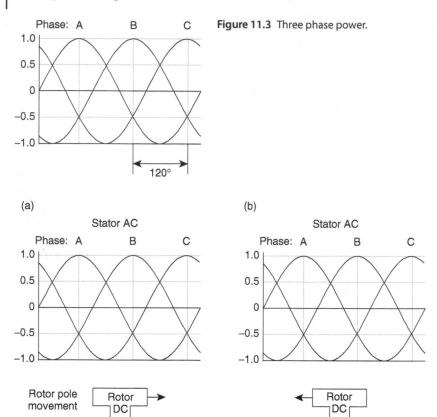

Figure 11.3 Three phase power.

Figure 11.4 (a, b) Phase sequence based on rotor rotation.

the phase sequence will be CCW. If the rotor turns the other way, the generator phase sequence will be CW.

Let us look at it another way. The stator phase bars are stationary, laid out in ABC formation. The rotor turns CW across the bars. It crosses A, followed by B and then C as shown in Figure 11.4. That causes the CCW phase sequence in the stator winding. Turn the rotor in opposite direction; it causes the CW phase sequence, as shown in a and b.

To change to a desired phase rotation, a change must be done either by (i) rephasing the order of stator phase bars or (ii) by changing the rotor physical rotation.

Unfortunately, it is not possible to rephase the stator bars at site. It is also not possible to change the direction of rotation of the turbine runner either. Therefore, in order to change the generator phase sequence, the only way is to swap two of three cable phases at the generator output.

The fact is there are operating CCW islands within the CW territories and opposite. I am aware of CW islands in the CCW territory, whereby the transmission lines from the CW islands were transposed to match the interconnection with the CCW grid.

11.2.4 Phase Sequence Blunders

One would think that the manufacturers of the generators would pay more attention to this issue and produce the generators suitable for the plant location but that is not the case. Major and costly errors are often being made.

In the power plants around the world, I have encountered problems in this area, I would say in at least 30% of the plants. In several cases, the problem surfaced only after the installation reached the point of synchronizing, which is virtually at the end of the commissioning. In the cases of smaller generators, say <50 MW, it was possible to rewire two generator phases and turn the sequence from CW to CCW or opposite, with some expense, but without considerable consequences.

The following examples from my personal experience will highlight the importance of proper design with respect to the system phase sequence and put some light on the prevailing confusion in the power industry:

(1) In Minnesota, a Japanese supplier of a 55 MW generator was warned that, based on the drawings, the generator appeared it had been wired to a CW phasing convention. The supplier answered that the generator was built to CCW convention as it should be. We checked with an independent engineer, and he was of the opinion that the generator was indeed of CW type. We cautioned the supplier again, but we were reassured again that everything was correct. The unit was finally installed and energized. The phase sequence was incorrect: →ABC and not ←ABC, as required. The supplier rephased the generator terminations at some cost and made the CW generator to produce a CCW output.

(2) An oil fired 150 MW plant was built in 1953 by a British consultant in Canada. It was built to a CW (British) convention, which was unknown to us. The plant was later integrated into the CCW North American (NA) grid, likely connected through a 230 kV rephased line.

Our company, 50 years later, was asked to add 6 × 35 MW generators to the same plant. The old units were planned to be scrapped.

We set the protective relays to CCW phase sequence as the plant was running on the North American CCW grid. The owner also "confirmed to me" that the old plant is a CCW fully integrated grid operating plant. A week later, during the commissioning of the first unit, the unit was energized, but it tripped after a few seconds. The tripping was happening without a bang or any other trace indication. Having tried everything to resolve this issue within a whole week without a success, we changed the relays settings to the CW convention. The unit started and stayed on! How do you like that? Then, we decided to check the phase rotation in the old plant. It was CW! How about that?

This was before the synchronizing. Then I asked the generator suppliers to tell me what the generator was built to? They told me it was CW. Now, everything became clear! Then they informed us that years back they checked and proved the existing plant was turning as a CW plant and built the generators to CW convention. It was evident that after 50 years, there was no one in the plant who knew that bit of critical information. The plant was operating as a CW island within the CCW grid. The Japanese contractor thought it was important to check the plant phase rotation before they build the generators. In this case, there were neither delays nor rework on the units, except for the resetting of the relays.

(3) Recently, in an Asian country in which we were erecting two 125 MW units, we were aware the grid was operating in CW convention. On the phasing diagram of the new Hydro project, we noted the symbols indicated a CCW convention. The CW convention was confirmed with the local utility as the plant phase sequence and that information was passed to the supplier of the generators. They immediately changed the phasing diagram on the drawing to suit and promised to comply and have the generator built to CW convention. Or, if late and not possible, they would transpose the 300 kV output cables in the tunnel during the installation.

During the cable installation a year later, they ignored our warnings to transpose the cables. Later, they admitted that they were unwilling to transpose the cables as they were not sure, which way the generator was wired, to CW or CCW. Imagine that? Finally, both units were installed, and the first generator energized. Unfortunately, the generator phase sequence was not compatible with the grid.

We faced a huge problem. It was suggested to rephase two phases on the 13.8 kV generator terminals ahead of the 6500 A isolated phase bus. They declined this approach as costly, due to the tightness for space. And the CTs, PTs, and excitation would have to be reterminated too. The biggest problem was that this change would also affect the rotation of all the motors in the plant. They estimated that the suggested work would require a massive amount of rewiring and software changes and terminations.

To rephase the 300 kV cables at the 300 kV switchyard was also out of question as one of the 300 kV cable phases was going to be short, and the termination was already filled with special tar that should not be disturbed. Costly!

Finally, it was decided to rephase two 300 kV cable phases at the bushings of the two main output transformers as shown on the simplified diagram in Figure 11.5 for unit generator G1. Same change was done for G2. The rework took several weeks by a number of specialists at a huge cost and embarrassment. The Unit Auxiliary transformers (UATs) were also reterminated to maintain the plant motor rotation unchanged. The generator protective relays were left set on CCW because the phasing change was executed outside of the differential protection zone.

Figure 11.5 Changes due to phase sequence error. UAT = unit auxiliary transformer, SST = station service transformer.

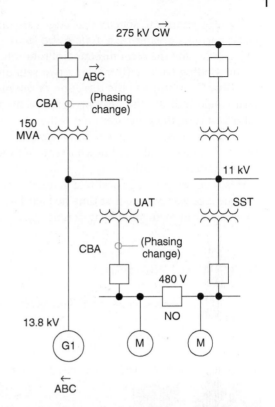

11.2.5 Conclusions

Vector phasing sequence is a confusing issue to all engineers. In some cases, engineers are unsure until the time the unit is energized. By that time, it may be too late to make a change, in particular if the generators are >50 MW.

So, what is the reason for such major blunders by the big international man-ufacturers of the generators? In my opinion, it is to do with the lack of coor-dination between the turbine and generator suppliers. These two major pieces of the equipment are likely being assembled in two different and mostly unre-lated places. Turbine is made to have a certain direction of physical rotation. The generator stator on the other hand was assembled with the phase bars laid in either one or the other physical order: ABC or ACB. Either bar laying order can generate a CCW or CW phase sequence that will depend on the turbine physical rotation when it is coupled with the generator. Once the turbine starts turning, the generator rotor will become CCW or CW. The phase sequence is now determined, and it cannot be changed without a major intervention.

In my opinion, it is absolutely unacceptable to have a turbine generator installed and still not know the actual phase sequence until the unit is energized.

My suggestion is to specify that large generators, in particular those having phase currents greater than 1000 A, be factory marked to indicate the phase laying order and the rotor direction of rotation. This could be identified on the stator winding bars by (i) marking two sets of bars painted **Red A – Yellow B – Blue C** and (ii) add the direction of the mechanical rotation of the rotor. That would indicate the peaking order of the phase sinusoidal waves, as the rotor is cutting the generator phases (bars). This would not necessarily resolve the error in phase sequence if it has already happened, but at least the actual phase sequence would be known ahead of time so other corrective measures can be undertaken earlier.

This suggestion was passed to a generator specialist. The specialist told me that my idea was not the first time he heard it. It was considered, but unfortunately never implemented by the company.

11.3 World Plugs/Sockets

The world is greatly divided on the plugs and sockets as shown in Figures 11.6 and 11.7.

The most common ones from the top row are:

Type A, B in North America, C in Europe, D in Australia, G in England.

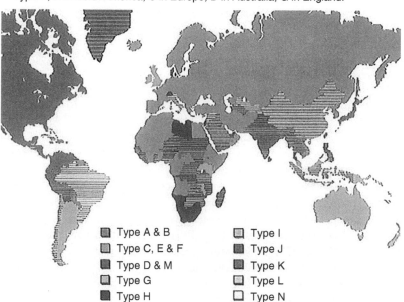

□ Type A & B   □ Type I
□ Type C, E & F   ■ Type J
■ Type D & M   ■ Type K
□ Type G   □ Type L
■ Type H   □ Type N

Figure 11.6 World plug map by type. Source: From [1].

Figure 11.7 World plug and socket types.

References

1 www.wikitravel.org – World map of Voltages and Frequency.
2 ANSI C84.1-2016 (2006). *Electric Power Systems Equipment – Voltage Ratings (60 Hz)*. New York, NY: American National Standards Institute.
3 IEC Standard 60038 (1997). *Standard Voltages*. International Electro-technical Commission.

12

Cables and Supporting Equipment

CHAPTER MENU

12.1 Cables, 254
    12.1.1 Cable Shielding, 254
    12.1.2 Conductor Insulation, 256
    12.1.3 Armoring and Jackets, 256
    12.1.4 Current Ratings, 257
    12.1.5 Single or Three Core Power Cables, 260
12.2 Power Cables, 261
    12.2.1 Power Cable Selection, 262
    12.2.2 Sample MV Cable Specification, 264
    12.2.3 LV Power Cables, 264
    12.2.4 EHV Cables, 265
    12.2.5 Power Cable Terminations, 267
    12.2.6 Cable Jacket and Wire Color Coding, 268
12.3 Control and Instrumentation Cables, 268
    12.3.1 Typical Characteristics, 269
12.4 Specialty Cables, 270
    12.4.1 Ethernet Cables, 270
    12.4.2 DeviceNet Cables, 270
    12.4.3 Fiber Optic Cables, 271
    12.4.4 Thermocouples, 272
    12.4.5 FieldBus Cables, 272
12.5 Cable Trays, 273
    12.5.1 Tray Materials, Support Span, and Loading, 273
    12.5.2 Cable Fill and Classification, 273
    12.5.3 Cable Minimum Bending Radius, 278
12.6 Conduits and Accessories, 278
12.7 Bus Duct or Cable Bus Systems, 279
    12.7.1 Bus Duct, 279
    12.7.2 Cable Bus, 280
    12.7.3 Typical Cable Bus Data Sheet, 282
References, 283

Practical Power Plant Engineering: A Guide for Early Career Engineers, First Edition. Zark Bedalov.
© 2020 John Wiley & Sons, Inc. Published 2020 by John Wiley & Sons, Inc.

12.1 Cables

Cables and raceways are installed in all the plant areas, indoors, and outdoors at all voltage levels. The plant cable and raceway systems are of design and capability to carry electrical power and data throughout the plant under all the plant operating conditions. Power cables are laid out with spacing and configuration to minimize the effects of overheating, cable induction, voltage drops, and to insure balanced loading on all phases (see Figure 12.1).

12.1.1 Cable Shielding

Dielectric field [1]: In all the electrical cables, irrespective of their voltage ratings, a dielectric field is present when the conductor is energized. The dielectric field is created by **electrostatic flux lines** and **equipotential lines**, as shown in Figure 12.2 between the conductor and electrical ground.

The density of the flux lines is dependent on the magnitude of the potential difference between the conductor and ground. The electrostatic flux lines are crowded in the insulation closest to the ground. For cables of <1 kV, the dielectric field is less of a concern. The density of the equipotential lines represents a voltage differential in the cable insulation. For a given voltage differential, the lines are closer together nearer the conductor. In the shielded cables, the shield represents the ground and the flux lines are radial toward the shield.

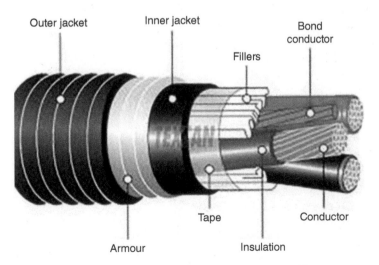

Figure 12.1 3c+g (Three conductors + ground) LV power cable.

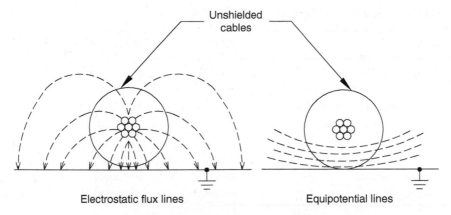

Figure 12.2 Cable electrical field. Source: courtesy of SouthWire.

Figure 12.3 Cables with armor and jacket.

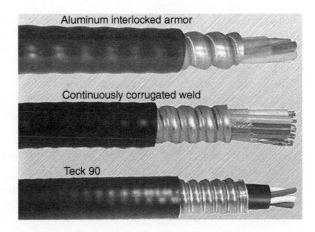

Conductor shield (screen): It serves to minimize the distortion of the field lines. The conductor is furnished with a semiconducting tape to smooth out the conductor surface and reduce the effect of the field lines.

Insulation shield (screen): An energized cable is effectively a capacitor made up by the conductor and ground as the plates and the insulation between them. This shield comprises a semiconducting tape applied directly over the insulation and a metallic Cu tape or wires applied over the Semicon tape. This shield must be capable to continuously carry the capacitive leakage current to the ground connection usually located at the end of cable. In addition, the screen must be capable of carrying the full ground fault current for about 0.5 seconds, which is typically defined as the clearing time of the relay back up protection (see Figure 12.3).

Table 12.1 Insulation characteristics.

	XLPE	EPR
Dielectric constant	5	2.5
Continuous operation, rated to	90 °C	90 °C
Continuous operation, possible to	105 °C	105 °C
Intermittent operation to	130 °C	130 °C
Emergency operation to	140 °C	140 °C
Low temperature operation	Good	Good
Corona resistance	Good	Not noted
Treeing effect	Retardant	Not noted

12.1.2 Conductor Insulation

Thermoplastic or thermoset materials are generally used as the cable insulation. The insulation thickness depends on the cable rated voltage. Thermoplastic materials tend to lose their form upon subsequent heating more than thermosetting materials.

The most commonly used materials are cross-linked polyethylene (XLPE) and ethylene–propylene rubber (EPR). Both are thermoset type of materials.

These two compounds offer the following characteristics (see Table 12.1):

Insulation level classification: In North America, power cables are built to distinctive levels; 100% and 133%. Naturally, 100% insulated cables are less costly than 133% cables.

100%: Cables in this category are generally intended for the effectively grounded systems. It shall be permitted to be used where the system is equipped with relay protection, which will rapidly clear the fault condition (ground fault), and in any case within one minute. This cable insulation may be allowed to be used on other systems, provided that the above clearing requirements are met in completely de-energizing the faulted section.

133%: Cables in this category shall be permitted to be applied in situations where the clearing time requirements mentioned previously cannot be met, and yet there is adequate assurance that the faulty section will be de-energized in time not exceeding one hour.

12.1.3 Armoring and Jackets

Metallic armoring (aluminum or corrugated steel) is primarily used to protect the cables mechanically and to add strength to the cable for pulling. Industrial and a majority of installations within power plants use solely armored cables.

The armor for the single conductor power cables must be of nonmagnetic materials, usually aluminum (see Figures 12.3 and 12.6).

Additional advantage of the continuous armored cables is for being impervious to water. They are also gas and vapor tight. These cables are also explosion proof and classified by NEC for use in hazardous locations. Applications include, as a traditional alternative to conduit installations, for direct burial, concrete encasing, open tray or placing on rigid cable supports.

The jackets are the overall nonmetallic insulation and protection for the cables, generally made of nonflammable and vermin proof materials, PVC, and similar. The jacket material also must be capable of withstanding solar radiation. In the 1980s in Saudi Arabia, this author has seen cable jackets cracked within a year after being left exposed to the sun. In Malaysia in 2016, the sun exposed cable jackets did not show any deterioration. I suppose, the suppliers have improved on the jacket endurance against the sun radiation.

12.1.4 Current Ratings

Power cable continuous current ratings [2, 3] based on its cross-sectional area must be selected for carrying plant loads without exceeding their insulation temperature limits. In most applications, 90 °C is used as the criterion in the selection tables. Low voltage (LV) cable conductors are selected from National Electrical Code (NEC) (or Canadian Standards Association, CSA or International Electro-technical Commission, IEC Tables). Medium voltage (MV) and high voltage (HV) cable conductors are selected from Insulated Cable Engineers Association (ICEA) conductor ampacity tables. The cable ampacities must be derated based on their method of installation. For the ampacity ratings, open cable tray is considered as installation in air. Cables installed buried or encased in concrete must be derated from those in air.

12.1.4.1 Designations

In North America, it is prevalent to identify cable sizes by the American Wire Gage (AWG) for sizes up to #4/0 and circular mills (kcmil) above. For instance #12 AWG, 250 kcmil, respectively. This may sound confusing. But, there is more. Please remember that for instance #12 AWG cable is larger in size than #14 AWG cable. On the other hand, cables increase in size with the numbers from #1/0 to 4/0 AWG, as well as the cables of 250–1000 kcmil (circular mills). The AWG sizes range from #26 (smallest) to #4/0 AWG (largest). The AWG cables are prefixed with "#".

In the rest of the world and in Canada, cables are listed in mm^2.

1 kcmil = 1000 cmil, 1 mil = 0.001 in. Circular mill area (cmil) of a solid conductor is defined as an area of a circle of 1 mil in diameter. kcmil is often called thousand Circular Mills (MCM).

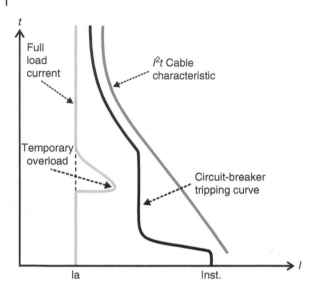

Figure 12.4 Cable protection by circuit breaker (CB).

12.1.4.2 Sizing of Power Cables

A lot goes into the consideration in selecting a cable for a particular situation. Typically one starts by selecting the conductor with an appropriate ampacity allowing for all the derating factors and allowing for a 15–25% overload e.g. $1.25 \times I_n$ for feeders or motors. Engineers should study the NEC code and appropriate guidelines with respect to the cable tray fill and spacing to determine the proper derating factors for the conductor sizing.

Furthermore, the cable conductor size must always be larger than the thermal setting of the upstream circuit breaker or fuse. Observe the conductor curve in chart Figure 12.4. Plant construction inspectors are all familiar with this requirement and will check the cable sizing with respect to the breaker settings and fuse ratings.

This is to insure the circuit breaker always trips or the fuse always blows, ahead of burning the cable.

12.1.4.3 Voltage Drops

The allowable voltage drops in feeders and branch circuits must be <3% for each branch, and <5% total. During motor starting, <15% maximum voltage drop is allowed.

12.1.4.4 Derating Factors

Site altitude, ambient temperature and cable grouping within cable trays, cable bus, and conduits must all be included in the derating factor calculations. Derating factors will be considered based on the local and IEC or NEC standards and codes. Derating factors (always <1) are multiplied together to arrive at the overall derating factor and then with the corresponding ampacity value.

12.1.4.5 Short Circuit Rating

Cable short circuit rating is based on using the total operating time for protective relays and breaker, or the fuses, to clear the maximum short circuit fault condition at the bus. 250 °C thermal rating for short circuit is considered a standard for cables (see Figure 12.5).

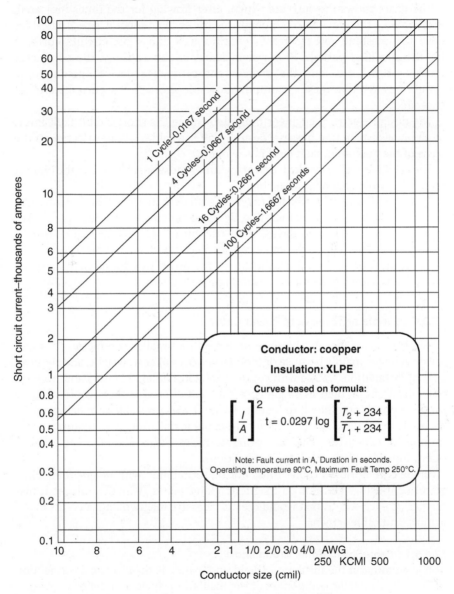

Figure 12.5 Cable short circuit thermal withstand. Source: From [2].

The chart (Figure 12.5) shows the short-circuit withstand capability of copper cables from #10 AWG to 1000 kcmil, with 90 °C XLPE insulation for rated current and 250 °C for short circuit conditions, based on ICEA formulae [4].

Similar charts are also available for aluminum conductors and different types of insulation.

The chart shows the currents which, after flowing for the times indicated, will produce the maximum temperatures for each conductor size. The cables have to withstand rated currents continuously and short circuits currents for the specified periods. The system short circuit capacity, the conductor cross-sectional area and the overcurrent protective device opening time should be such that these maximum allowable short-circuit currents are not exceeded.

The design electrical engineer must determine the short circuit fault levels at all plant distribution busses. Naturally, all the cables will be selected on the basis of the maximum fault level in the plant.

The calculations for the device clearing times must be set based on the highest fault level, which happens at the bus.

The cable size must be selected to have a sufficient thermal value to withstand the let through fault current allowed by the protective device fuse or breaker based on the coordinated device time-current ($T–C$) characteristic. The I^2t thermal characteristic of the cable can be placed on the relay protection $T–C$ chart. It must be seen on the right side of the protective device characteristic. $I = I_{r.m.s.}$, as shown on the $T–C$ graph Figure 12.4 earlier.

12.1.5 Single or Three Core Power Cables

Three conductor (core) cables are preferred for smaller conductor sizes and lower voltages mainly because of their lower overall cost, weight, and reduced cost of installation. Cables with conductors 250 kcmil (125 mm^2) and over are more likely to be built as single core cables. Otherwise, they would become too heavy and too difficult to handle. Large three core cables are delivered on the reels in smaller lengths. Three core cable installations will require a large number of expensive joints and generate more waste. Should a cable fail in operation, very likely it will fail on the joints.

Single core armored cables must not be terminated with *magnetic* metallic gland connectors and not on magnetic gland plates (window current transformer [CT] effect), as that would create excessive magnetizing currents in the armors and cable screens.

12.1.5.1 Cable Shield/Armor Grounding

The grounding of the insulation shield and armor is the electrical connection between the metallic component of the insulation shield and the plant ground system. The grounding makes the shield to be effectively at the ground potential thus resulting in no distortion of the flux and equipotential lines. The proper cable design and manufacture of shielding, if properly grounded, results in electrostatic flux lines being spaced symmetrically and perpendicularly to the equipotential lines. Providing proper shielding and grounding of the shield is essential to avoid tracking and arcing discharges, which may be present in nonshielded cables if used for higher voltages.

Furthermore, circulating currents along and within the cable armor must not be allowed to flow. This causes power losses and unnecessary heating of the cable. Three conductor cables with a single common armor do not allow circulating currents to flow in their armor as the magnetic induction is canceled by the balanced three conductor magnetic field. On the other hand, if single core cable armors are grounded at both ends, it causes circulating currents to flow in their armor. If the armor or shield resistance is low, the magnitude of the return current flowing may approach the magnitude of the current in the main conductors. For that reason, cables are grounded at one end only, preferably at the source end. In the case of single core cables, the cables armors and shields are grounded at the grounded end. The other side is left isolated. The armor is grounded by the cable glands. This grounding method consisting of grounding at one end only will eliminate circulating currents but will cause a potential rise at the ungrounded end. The potential rise is a function of separation spacing between the phases (the larger the spacing, the larger the impact) and is proportional to the cable length. IEEE Standard 525 recommends limiting the ground potential to 25 V.

To keep the potential rise low at the ungrounded end, it is preferred to tie rap the three single core cables in a trefoil formation rather than placing them horizontally on the tray with appropriate spacing between phases as shown in Figure 12.16. This minimizes the distance between the phases and makes for a symmetrical interaction. For extra-long cables (>500 m), one can prevent the excessive voltage rise at the open end, (e.g. >25 V), by using a shield/armor interrupter kit. As an alternative, it may be allowed to ground single conductor cables at both ends and split the armor at the middle and leave it ungrounded, but care must be taken to prevent an excessive voltage developed across the gap.

12.2 Power Cables

Power cables, single conductor or three conductor types (Figure 12.6), for industrial and power plants are generally XLPE or EPR insulated, armored or nonarmored (depending upon the application), copper (Cu)-stranded conductors, 90 °C rated. Some jurisdictions have changed cable temperature rating to 75 °C due to limitations of the switchgear to which the cables are connected.

Figure 12.6 MV power cables.

Manufacturing process of heating the conductor (strand wires) to elevated temperatures for specific periods called annealing is used to soften the conductor for better flexibility. Typically, medium hard drawn conductors are fabricated and preferred.

The cable insulation class for medium voltages (MV) 4.16 kV and higher will be 100% or 133% for effectively grounded and ungrounded systems, respectively. Medium voltages are generally resistance grounded, considered ungrounded for the cable selections.

12.2.1 Power Cable Selection

Let us summarize the selection of power cables for industrial applications.

(1) Select voltage class: 600 V, 1000 V, 5 kV, 15 kV.
(2) Select exclusively Cu conductors, to be stranded, annealed as medium hard drawn.
(3) Select 90 °C cable temperature rating. 75 °C rated cables are also used for special applications.
(4) For medium voltages over 1000 V, verify method of system grounding. If ungrounded or resistance grounded, select 133% insulation level. For effectively grounded systems, select 100% insulation level. In Asia, I have noted that this selection is not available. Perhaps, the insulation is rated for ungrounded cables only.
(5) For MV cables, specify the grounding method (typically; resistance grounding) and the maximum ground fault current to let the supplier determine the size of the concentric grounding conductor.

(6) Select for method of installation: Generally use *aluminum* armored cables for buried or tray installations. For short cables placed in conduits, unarmored cables can be used. As much as feasible, place all three phases within the same conduit if using metallic magnetic conduits.

(7) Make a selection for 1c or 3c cables, depending on the size of conductors. Cables with conductors $\geq 125\,mm^2$ (250 kcmil) use 1c cables. They are easier to install. On drawings, cables are marked 3-1c-250 kcmil or 1-3c-250 kcmil for single and three core cables, respectively.

(8) Use grounding conductors within the 3c cables. On MV cables, the grounding conductor may not be utilized. Select the grounding conductor to be equal to the phase conductor for up to $50\,mm^2$ and then the $50\,mm^2$ size for larger cables.

(9) Select general plant derating factor for altitude if installation is at >1000 m.

(10) Use standard NEC, CEC tables. Select cable derating factor for the type of installation; tray or buried. Cables placed on cable trays are considered as being in free air in the ampacity tables.

(11) Use standard NEC tables for cable tray fill derating factor. Attempt to arrange power cables on the trays for the fill derating factor = 1.

(12) Calculate load current. Use standard tables for selecting the cable conductor size by applying applicable continuous rating, multiply with all the applicable derating factors and add 15–25% temporary overloading margin.

(13) Estimate three phase fault current available at the switchboard, motor control center (MCC). For larger size cables, verify according to the chart Figure 12.5 the temperature tolerance for the fault and fault duration.

There may be several thousands of individual power cables in a large plant installation.

It seems, the aforementioned selection criteria require a large and tedious effort. It is not. Most of the requirements are of general nature based on pre-established design criteria applicable for the whole plant.

MV cables: Teck 90; 5–35 kV, 1 Cu conductor, 133% rated XLPE insulation, insulation shield extruded semiconducting, aluminum interlocked armor, outer PVC jacket, fire retardant, and sunlight resistant. Also, rated for hazardous applications.

LV cables: Teck 90; 1000 V, 1 and 3 Cu conductors, ground conductor Cu bare, XLPE insulation, aluminum interlocked armor, PVC jacket, fire retardant, sunlight resistant. Also, rated for hazardous applications.

12.2.2 Sample MV Cable Specification

Conductor size	*Specify* all sizes
One or three conductors	*Specify*
Voltage	*Specify:* 5–35 kV
Switchgear type	Metal-clad
Application	*Specify*: indoors or outdoors or buried
Single or three conductor	Single for >125 mm² , triple for <125 mm²
Insulation	XLPE
Insulation class	133% (for ungrounded systems)
	100% (for effectively grounded systems)
Shielding	Cu tape or concentric wires
Grounding conductor	*Not required*
Armor	Aluminum interlocked
Temperature	90 °C operating, 250 °C short circuit (if allowed by Switchgear)
Outer jacket	PVC, fire retardant, UV resistant
Quantity	*Specify* for each size.

12.2.3 LV Power Cables

480 and 600 V service use 1000 V class cables, armored or nonarmored (depending upon the application), PVC/XLPE insulated. In USA, for 480 V service, often 600 V class insulated cables are used.

The minimum conductor size for LV power cables in industrial installations is 2.5 mm² (#12 AWG) (see Table 12.2).

The sizing of the MV cables is usually determined to be close to 100% of the load (1 × In). The MV cables are properly spaced by at least one diameter on the trays to insure a 100% derating factor or preferably placed in trefoil formation.

Cables for 2 × 100% motors (pumps) can be laid next to each other on the tray. Since one of the pumps is always off, thus one of the cables counts as an empty space. I have not found this rule in the NEC code, but the inspectors have always given me a pass on this design concept.

Here is how an engineer has established derating rules for the multiconductor (MC) and single conductor (SC) cable tray installation;

$$\text{DF} = K_t \times K_g$$

For the ambient temperature of 35 °C → $K_t = 0.96$
For cable grouping → $K_g = 0.86$ (SC) = 0.64 for (MC)
Derating factor DF = $0.96 \times 0.86 = 0.83$ for (SC)
DF = $0.96 \times 0.64 = 0.61$ for (MC)

Table 12.2 Cable comparisons in USA and IEC used in Europe and Asia.

AWG American Wire Gauge #	mm²	IEC mm² Equivalent	AWG American Wire Gauge #	mm²	IEC mm² Equivalent
24	0.20	0.20	1/0	53.0	55.0
20	0.52	0.50	2/0	67.4	70.0
18	0.82	0.75	3/0	95.0	85.0
16	1.31	1.50	4/0	107	120
14	2.08	2.50	250 (MCM[a])	127	150
12	3.31	4.00	350 (MCM)	177	185
10	5.26	6.00	500 (MCM)	253	240
8	8.36	10.0	750 (MCM)	380	400
6	13.3	16.0	1000 (MCM)	507	500
2	33.6	35.0			
1	42.5	50.0			

a) MCM → Thousands of circular mills (kcmil).

12.2.4 EHV Cables

In power plants, generators are typically installed in buildings with the generator step-up transformers (GSUTs) located just outside the building walls. The generator output is taken to the transformers by one of two means. For generators up to 6000 A output, the GSUT connection may be a bus duct. For generators 6000 A and above (up to 20 kA), the GSUT connection will be isolated phase bus duct (IPB or isophase). Typical generator output voltages will be 11–24 kV. From the transformer, the power output is taken by a HV overhead transmission line.

When a hydro plant is built underground, (e.g. under a mountain), it is preferable to have the GSUTs located underground close to the generators. From the underground transformers, the only choice is by HV single conductor cables (Figure 12.7) rated for transmission voltages at 230–400 kV to the external switchyard several kilometers away. Separate cables are usually used for each generator. On an underground project in India, the 400 kV switchyard in form of a 400 kV gas insulated switchgear (GIS) was also placed indoors adjacent to the six 155 MW generators. The combined output of the plant was taken by two sets of 400 kV, 2 km long cables to the 400 kV transmission poles outside the tunnel.

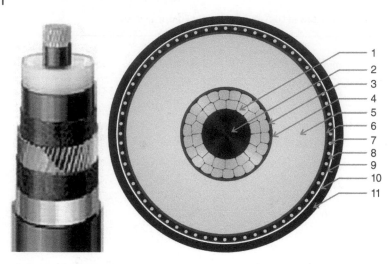

Figure 12.7 300 kV cable.

Here is typical fabrication data for a 300 kV single conductor cable, which you may find useful for your future projects:

1. Conductor, Cu stranded, 400 mm^2
2. Core, polyethylene
3. Bedding, fabric tape, semiconducting, 0.4 mm
4. Conductor screen, conductive, 0.8 mm
5. Insulation, XLPE, 23.5, 82.8 mm diameter
6. Insulation screen, semiconducting, 0.8 mm
7. Concentric neutral, 69 Cu wires, 1.27 mm^2
8. Bedding, Swelling tape semiconducting, 0.8 mm
9. Bedding, fabric tape semiconducting, 0.2 mm
10. Outer jacket, 4 mm, with rodent protective layer, 0.5 mm
11. Weight: 11.5 kg/m, 91.0 mm outer diameter.

12.2.4.1 HV Cable (300 kV) Terminations

Seven single cable runs ($2 \times 3 + 1$ spare) for two generators were laid and clamped on ground sleepers every 2 m, slightly snaking to allow for creep and expansion. The cable termination at the transformer end was immersed into the transformer oil in bushings and at the switchyard in a special compound. Cable sheath of each phase is grounded at both ends. At the transformer sealing end, the sheath is grounded through metal oxide surge voltage arrestors, while at the switchyard end, it is grounded directly to the switchyard grounding system.

12.2.4.2 HV Cable Data (300 kV)

Standard	IEC 62067-2011 power cables with extruded insulation for rated voltage over 150 kV.
Generator rated current	400 A at 275 kV, 50 Hz, Maximum 300 kV
Basic impulse level	1050 kV peak
Switching impulse level	850 kV peak
Conductor, Cu	400 mm^2
Rated current	936 A at 90 °C, 443 MVA
System grounding	Solid (effective)
Rated SC, one second for conductor	50 kA
Short circuit temperature	250 °C
DC resistance: at 20 °C/90 °C	0.047 Ω/km/0.0613 Ω/km
Cable length per phase	400 m
Delivered	7 runs total for two generators
Capacitance	0.137 µF/km
Charging current	6.9 A/km
Inductance/phase	0.72 mH/km
Losses for three phases, rated I^2R	162 kW/km
Layout separation between phases	300 mm (1 ft), snaking allowance for expansion.

12.2.5 Power Cable Terminations

As shown in Figure 12.8 a–d, each cable is terminated with an appropriate approved cable gland connector to terminate ground and isolate the armor for cable connections to the electrical cabinets. Cable glands are typically made of brass or aluminum. Steel glands are not used for single conductor power cables.

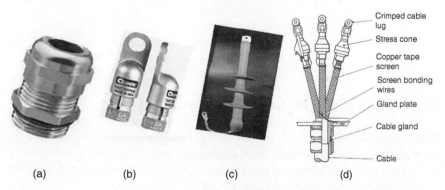

Figure 12.8 (a) Cable gland, (b) conductor lugs, (c) one phase stress cone, and (d) three phase stress cone.

Conductors of the power cables are terminated with cable lugs for larger conductor sizes and with crimp compression ring lugs for conductors # 8 AWG and less.

The shields of MV and HV power cables are terminated with appropriate stress relief, heat, or cold shrink terminations (stress cones) to allow for a smooth transition of the electrical field on the cable insulation and shielding.

12.2.5.1   Phisterer Connectors (CONNEX™)

This fully encapsulated connection system, with solid insulation, is becoming more common because of the increasing prevalence of the GIS.

12.2.6   Cable Jacket and Wire Color Coding

As used in North America

• 36 kV	Red
• 5 kV	Yellow
• 1000 V and lower	Black
• 300 V control	Black
• Instrument cables	Gray
• Communications	Blue
• DeviceNet	Gray

All individual wiring shall be color coded as follows:

• Phase A/B/C	Red/Black/Blue
• Ground	Green
• Neutral	White
• Live	Black
• DC+/−	Black/White

12.3   Control and Instrumentation Cables

Cables must be designed with appropriate shielding and twisting to be able to carry control signals and data, both analog and digital, without noise during the normal as well as emergency operating situations. The cables used to carry status of equipment contacts do not need shielding.

Figure 12.9 Spade lug.

Figure 12.10 Control cables.

Control and instrumentation cable conductors are typically #14 or 16 AWG (2.5, 1.5 mm²) (see Figure 12.10). Lately, 16 AWG (1.5 mm²) seems favored because it uses smaller spade-type compression lugs (Figure 12.9) used on control system equipment terminals.

Control cables are specified with standardized number of conductors; 2, 3, 5, 7, 10, 15, and 20 to suit the most desired and common applications.

Instrumentation cables are specified with standardized multiple, individually shielded, pairs of conductors: 1, 2, 4, 8, 12, and 24 pairs.

12.3.1 Typical Characteristics

XLPE insulation with sun – UV resistant and flame resistant chlorinated polyethylene (CPE) or PVC jacket, aluminum interlocked armor, stranded and twisted copper conductors, National Electrical Manufacturers Association (NEMA)/ICEA S-73 Table E1 or E2 [5] color coded or phase identity numbered. Rated voltage 300 or 600 V, recommended for operation in wet or dry locations. Armored control cables may be installed in open air, in ducts or conduits, in tray or trough, and are suitable for direct burial.

12.4   Specialty Cables

These cables are used for specific applications as follows.

12.4.1   Ethernet Cables

see Figure 12.11

• Type	CAT 5E 300 V
• Construction	PVC jacketed
• Pairs	4
• Conductors	# 23 AWG, Cu solid

12.4.2   DeviceNet Cables

Cables for MCC, programmable logic controller (PLC) communications and automation (see Figure 12.12):

• Rating	600 V, UL type CL2, bulk 600 V, rated
• Thick cables	Used for trunk lines
• Thin or flat cables	Used for droplines
• Construction	2 foil shielded pairs with 1c copper drain wire
• Power pair	2c tinned Cu, red/black
• Data pair	2c tinned Cu, white/blue
• Cable jacket	PVC, gray
• Connectors	Connectors provided at MCC buckets or external devices are male type

Cat5e UTP patch cables

Figure 12.11 Ethernet cable.

Figure 12.12 DeviceNet cable.

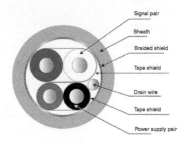

Cable specifications

Product code		DVN18		DVN24	
Type		Thick cable		Thin cable	
		Signal pair	Power pair	Signal pair	Power pair
Conductor size (AWG)		18	15	24	22
Insulator color		Blue/White	Red/Black	Blue/White	Red/Black
Conductor resistance (Ω/km) (20 °C)		22.6	11.8	91.9	57.4
Specific impedance (Ω) (1 MHz)		120 ±12	–	120 ±12	–
Maximum attenuation (db/100 m)	125 KHz	0.43	–	0.95	–
	500 KHz	0.82	–	1.64	–
	1 MHz	1.31	–	2.30	–
Outer diameter (mm)		Approx. 12		Approx 7.0	
Approx. weight (kg/km)		180		60	

※ The specifications of DVN18SF and DVN24SF are the same as above.
※ Standard sheath colour is gray, but can also be made in pale blue.

Cross sectional diagram

Signal pair
Sheath
Braided shield
Tape shield
Drain wire
Tape shield
Power supply pair

12.4.3 Fiber Optic Cables

For short distances, <200 m: see Figure 12.13.

• Brands	Corning, Altos, or AMP
• Type	50/125 multimode, 12 or 24 cores
• Armor	Metallic
• Fabrication	Gel-free, waterlocked

Figure 12.13 Fiber optic (FO) cables.

Indoor/outdoor

Applications:

• Outdoors	Hang on messenger wire, armored.
• Indoors	Within electric metallic tubing (EMT) conduit
• Max attenuation	<2.6/<0.6 dB/km for (850/1300 nm)
• Connector type	UniCam Pretium

For long distances, >200 m:

• Brands	Corning, Altos, or AMP
• Type	8.3/125 single mode, 6 or 12 cores
• Armor	Metallic
• Fabrication	Gel-free, waterlocked

Applications:

• Outdoors	Hang on messenger wire or armored
• Indoors	Within EMT conduit
• Performance	Serial gigaBit ethernet distance: 5 km/1 GB
• Max attenuation	<0.34/<0.22 dB/km for (850/1300 nm)
• Connector type	UniCam Pretium

12.4.4 Thermocouples

Individual pair shielded, 1, 4, 8 pairs, K type.

12.4.5 FieldBus Cables

Used for instruments and valves. The FF-844 standard specifies the following cable details: Fieldbus cable is a shielded single or multi-pair cable with #18 or #16 AWG tinned copper conductor, 23.5 Ω/km at 20 °C.

Each twisted pair is individually shielded with metalized polyester, 90% coverage drain wire <51 Ω/km. The shielded pair is to be twisted to a lay length of 10–22 twists/m. Note: twisted pair lay length is a factor in magnetic noise rejection. Characteristic impedance:

$$Z_0 = \sqrt{\frac{L}{C}} = 100 \pm 20\ \Omega \text{ at } 31.25\ \text{kHz}$$

12.5 Cable Trays

12.5.1 Tray Materials, Support Span, and Loading

Cable trays must be designed to safely carry, with 10% spare capacity, power and control cables without exceeding the deflection limits defined by the codes. Cable trays must also be designed to safely carry and provide separations for control cables to ensure the cable rating and noise-free environment.

Cable trays are made from various materials depending on the application requirements, including aluminum, galvanized steel (either mill galvanized or hot dip galvanized after fabrication [HDGAF]), stainless steel and fiberglass reinforced plastic (FRP). They are constructed in different types, including ladder, ventilated trough, channel, single rail, and wire mesh. The latter two are mostly used for control and instrumentation cables.

Available standardized tray widths – 150, 300, 450, 600, 750, and 900 mm (9, 12, 18, 24, 30, and 36 in.). Available standardized depths – 50, 100, 150, 200, and 250 mm (2, 4, 6, 8, and 10 in.) (see Figure 12.14).

For industrial plant applications, the most common cable tray is heavy duty ladder Tray, loading class NEMA 12B, 75 lb/ft (112 kg/m), for a 12 ft (3.6 m) support span. These cable trays are specified in standardized widths – 300, 600, and 900 mm (12, 24, and 36 in.) and depth – 100 mm (4 in.) (see Table 12.3).

Cable trays will be supported at 8–10 ft (2.5–3 m) intervals and also at both sides of elbows and risers.

12.5.2 Cable Fill and Classification

Cable fill [2, 3, 6, 7] into cable trays is in accordance with the local and NEC and IEC standards. Tray loading capacity, per foot of tray, is defined by NEMA standard VE-1. Assume a safety factor of 1.5.

The loading is based on the allowed deflection measured at the middle of the span. NEMA classification is based on the working load (the total weight of the cables, ice, wind and snow) and the support span (the distance between

Figure 12.14 Ladder tray.

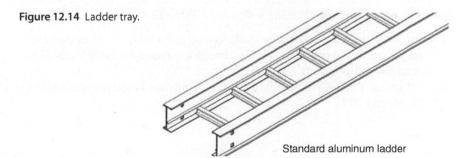

Standard aluminum ladder

Table 12.3 NEMA tray class.

NEMA tray	Support span		Load	
Class	m	ft	kg/m	lb/ft
5AA	1.5	5	37	25
5A	1.5	5	74	50
8AA	2.4	8	37	25
8A	2.4	8	74	50
10AA	3	10	37	25
10A	3	10	74	50
12B	3.7	12	112	75
20AA	6	20	37	25
20A	6	20	74	50
20C	6	20	112	75

Table 12.4 CSA load classification (maximum design load for maximum associated support spacing) [3].

Class	Design load		Design load spacing	
	kg/m	lb/ft	m	ft
A	37	25	3	10
C	97	65	3	10
D	179	120	3	10
D	67	45	6	20
E	299	200	3	10
E	112	75	6	20

supports). Table 12.3 summarizes the NEMA classes based on cable/working load. The equivalent CSA loading is shown in Table 12.4.

Working load: The cable load or the working load is the total weight of the cables to be placed in the tray. The NEMA classes are based on cable loads of 50, 75, and 100 lb/lineal foot.

Support spans: The NEMA standard support spans are based on maximum of 8, 12, 16, and 20 ft.

CSA classifies the cable trays as follows:

Class A: Light duty for control and instrumentation cables with supports up to 3 m.
Class C: Standard duty for maximum support span of 3 m.
Class D: Heavy duty for maximum 6 m support span.
Class E: Extra heavy duty.

12.5.2.1 Cable Tray Fill

Cables are laid in trays in accordance with NEC standards for ampacity derating and cable fill. According to National Electrical Code Article NEMA 392-9(b), cable tray fill will be 50% filled when using control or signal wiring. The sum of the cross-sectional areas of all the cables cannot exceed 50% of the tray's fill area. The cable tray sizing equals the width times the loading depth. Generally, cable tray sizing is correct when it contains a 50% fill of cables or wires.

Cable trays are laid in banks of two to six trays on top of each other, as in Figure 12.15.

Separation between cables is required for power cables (see the following illustrations). Here is the basic rule when dealing with NEC for power cables in Figure 12.16a,b.

Find the minimum ampacity that you need for your cables, make sure that they will be installed per Rule 392.22 and use Table 310.15(B)(16), for multiconductor or single conductor use 310.15x(x). Table 392.80 defines the ampacity for your specific installation. There are many variables and tables to use depending on how and what size of cable you are installing.

Figure 12.15 Cable tray installation.

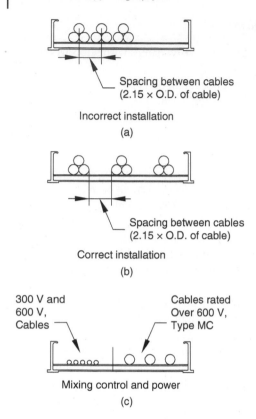

Figure 12.16 (a) Cable fill, (b) spacing, and (c) mixing.

Spacing between cables
(2.15 × O.D. of cable)

Incorrect installation

(a)

Spacing between cables
(2.15 × O.D. of cable)

Correct installation

(b)

300 V and
600 V,
Cables

Cables rated
Over 600 V,
Type MC

Mixing control and power

(c)

When it comes to the cable derating and fill, the Canadian Electrical Code (CEC) [3] and the IEC are different to the NEC [2], so one must apply the appropriate code to your installation. That is, the code applicable to the jurisdiction where the installation is being made.

Cable trays erected as banks of several trays allow for laying the cables of different voltages and/or applications. Separate cable trays are employed for medium voltage power, low voltage power, control and instrumentation cables. The vertical separation between trays in a bank is minimum 30 cm (12 in.). The highest voltage generally occupies the tray on the top. Preferred fill of single conductor cables is bunching them in triangle (trefoil) formation with separation between the triangles of $2.15 \times D$ (diameter of the single cable), or one diameter for the single cables installed flat on cable tray.

In limited applications, LV power and control cables can be placed in the same cable tray with an appropriate metallic divider as shown in Figure 12.16c. Based on NEC 39.80 [6] drawings, the cables with insulation rated 600 V or less can be installed with cables >600 V voltages, providing

- Cables over 600 V are armored cables (MC cables, or Teck cables).
- The two groups are separated and partitioned with a solid barrier of material compatible with cable tray.

Separation between power and control cables is important. In one case, a builder hung a 15 kV cable and a number of control cables together in air, hanging off a common messenger wire over a distance of 15 m. We experienced multiple daily malfunctions on a 15 kV breaker. A problem was finally assigned to the joint cable installation. When the cables were separated by about 1 m, the faults stopped.

In highly corrosive areas, cable trays and junction boxes (JBs) are made of either stainless steel or fiberglass (FRP). The panels will use NEMA 4X enclosures (where 4 means totally enclosed water-tight and the X signifies corrosion resistance).

Cable trays for the areas having dusty environment, like the fuel hall and ash hall are provided with solid covers. Cable trays within the fuel storage area in the fuel hall are laid on the side in order to avoid dirt accumulation. In both instances, cooling through convection is impaired, (i.e. free air rating no longer applies), and power cable ampacities must be derated.

Here is an extract Table 12.5 from a tabulation of the Canadian Code for the cable ampacity ratings for 90 °C, R90 cables, based on ambient of 30 °C[1] in free air or open cable trays; Cu: single and three conductors, Al: single and three conductors.

Table 12.5 Cable ampacity.

Allowed current (A)	Cu		Al	
Cable size (AWG)	1c	3c	1c	3c
12	25	20	20	15
10	40	30	30	25
8	70	45	45	30
4	135	85	105	65
2	180	120	140	95
4/0	385	235	300	185
250	425	265	330	215
500	660	395	515	330

Source: From [3].

---

1 Derating factor for ambient 40 °C is 0.90.

12.5.3 Cable Minimum Bending Radius

The industry accepted practices and NEC-2004, Section 300.34 [2]. Conductor bend radius for unshielded and shielded single and multiconductor cables, for voltages >1000 V are as follows:

Single or multi conductor cables without shielding	8 × Overall cable diameter
Single or multi conductor cables with shielding	12 × Overall cable diameter
Multi conductor cable with individual shielding	12 × diameter of single cable or 7 × overall cable
	Overall cable, whichever is greater
Interlocked armor or corrugated sheath (MC cables)	7 × Overall cable diameter

12.6 Conduits and Accessories

There are several types of metallic and nonmetallic conduits used in the industry that are detailed in the NEC, including rigid galvanic steel (RGS), electric metallic tubing (EMT), flexible conduit, and PVC conduit. Conduits may be used above ground, buried in cable trenches, manholes, or to connect to the electrical equipment enclosures.

The maximum number of insulated cables permitted in conduits can be found in the tables provided courtesy of General Cable "Cable Installation Manual for Power and Control Cables [8]," developed based on NEC Chapter tables [2]. Accordingly, it is permitted to have a conduit fill of 40% for three or more cables, 31% for two cables and 53% for single cable to avoid jamming while pulling.

In cold climates, another consideration for conduit installations to equipment in the yard is frost. Frost can cause metallic conduit to push or pull equipment cabinets. To minimize the upward movement of conduit into enclosures, use flexible conduit or expansion fittings. Also, a layer of sand under the conduits may help with this installation issue.

PVC conduits are susceptible to cracking when exposed to colder climates. NEC permits PVC conduit to be installed at a depth of 18 in. (45 cm). For circuits exceeding 600 V additional depth requirements are covered under Section 300.50. PVC conduits used for circuits from 22 to 40 kV are required to be laid at least 24 in. (60 cm) deep.

Rigid metallic conduits are more likely used in the aforementioned ground installations. Electro-metallic tubing (EMT) is allowed for buried installations due to their capability of bending. When equipped with proper fittings, it can be used as the equipment grounding conductor. PVC conduits and premade elbows are often used for the below-grade installations in the substations. The maximum size of PVC conduit allowed by NEC is 6 in. (15 cm). For concrete duct banks and buried cable installations (see Chapter 4).

Consideration must be given to local codes, frost depth, cable ampacity, cable pulling, and bending radius for the fill requirements. The bending radius for each trade size of conduit NEC covers in their Table 9.2.

Manholes are often sized to allow for conduit bending requirements and to allow for the cable pulling. The distance between manholes shall be less than 300 ft (100 m). It is important to consider the length of the conduit run and the number of bends in the conduit runs when determining the amount of fill that is acceptable and the number of manholes needed.

12.7 Bus Duct or Cable Bus Systems

12.7.1 Bus Duct

Nonsegregated phase bus duct is defined by NEC 368.2 as a grounded metal enclosure containing factory mounted bare or insulated conductors, which are usually Cu, Al bars, rods, or tubes (see Figure 12.17).

Therefore, a nonsegregated phase bus duct is an assembly of bare or insulated (PVC tape or epoxy) copper conductors with associated connections, joints,

Figure 12.17 Bus duct.

and insulating supports confined within a single metal enclosure (ventilated indoors and totally enclosed outdoors) without interphase barriers. Bus duct is designed for use on circuits from 480 V to 15 kV and capable of carrying rated current continuously without exceeding a conductor temperature rise of 65 °C above an outside ambient temperature of 40 °C.

A major problem with bus duct is that it is completely factory prefabricated and lacks any flexibility during installation. The prefabricated sections can be very heavy and difficult to manoeuver. Flexibility is what counts at the construction sites as the layouts are hardly ever built to a precision needed to accommodate this product.

Unfortunately, things change at site, so one must be able to adapt to the changes quickly and economically. Another major issue with bus duct is that outdoor installations must be in totally enclosed steel housings, which are subject to condensation (when de-energized), corrosion, and failures when moisture gets in.

12.7.2 Cable Bus

Cable bus [9] is shown in Figures 12.18 and 12.19. Cable bus looks a lot like a ventilated trough cable tray with a cover. Cable bus is typically defined as a metal enclosed bus system provided with all the necessary fittings, tap boxes, enclosure connectors, entrance fittings, insulated conductors, electrical connectors, terminating kits, and other accessories as required. These systems are

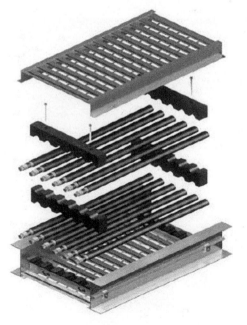

Figure 12.18 Cable bus assembly. Source: Courtesy of HP Husky [9].

Figure 12.19 Cable bus installation. Source: Courtesy of HP Husky [9].

suitable for indoor or outdoor use with the same ventilated enclosure. The cable bus is favored due to the more economical approach, flexible routing, more tolerance on the routing and terminations. Cable bus enclosures are most often fabricated from extrusions of aluminum alloy but are available in mill galvanized steel and stainless steel.

Cable bus enclosure fittings have the same radius options as cable tray and NEC standards, unless the minimum bending radius of the conductor requires a larger fitting radius. The enclosure will have a continuous current rating of not <1000 A (50 °C rise), and the resistance across the enclosure section splice may not exceed 50 μΩ. The enclosure is grounded at sufficient intervals for preventing a potential above ground on the bus enclosure during a fault. Voltage rise concerns are only a problem for extremely long cable bus runs, (i.e. >300 m).

Conductors are continuous length and are pulled in after the bus enclosure is in place. Electrical connectors, typically two-hole long barrel compression lugs, are used only at the termination of conductor runs. The conductors are arranged in a phasing pattern which minimizes interphase and intraphase imbalance, which is an advantage over the bus duct. Conductor temperature rise calculations and current balance calculations can be obtained from the manufacturer.

Table 12.6 Cable bus ampacity comparison.

Voltage (kV)	Conductor size (kcmil)	Cable bus (A)	Armored cable in tray (A)	3c cables in conduit in air
0.6	500	637	405	477
	750	805	500	598
	1000	960	585	689
5	500	695	425	473
	750	900	525	579
	1000	1061	590	659
15	500	685	470	481
	750	885	450	588
	1000	1040	650	677

Based on NEC and ICEA tables, for 90 °C cables in 40 °C ambient.
Source: Courtesy of MP Husky.

Conductor support blocks are designed in segments and maintain a minimum one conductor diameter in both the horizontal and vertical planes, as required for free air conductor rating. The construction and spacing typically provides short circuit bracing of up to 100 kA r.m.s. symmetrical. Cable bus is flexible and easily adjusted to suit the site contours and dimensions (refer to the ratings in Table 12.6).

The electrical engineer must provide the layout and desired ampacity and let the supplier do the design and selection of the cables, based on their calculations and practices.

The conductors are cut to size at site to the precise location of the equipment bushings and terminals, which is a huge benefit over the predesigned bus duct assembled with insulated Cu bars.

12.7.3 Typical Cable Bus Data Sheet

Voltage	480–69 kV
Ampacity	800–6000 A, 3 ph, 3 W (or 4 W)
Short circuit rating	100 000 A r.m.s.
Conductors	Insulated EPR or XLPE, Cu conductor, 133% rated insulation
Enclosure	Aluminum, unpainted, ventilated
Operating temperature	90 °C for cables
Stress cones	Yes, for 5 kV and higher
Cable lug type	Yes, two-hole long barrel compression.

Table 12.6 illustrates the greater current carrying capacity of cable bus as compared to alternate cabling methods. The increased capacity is due to free air ventilation, engineered individual cable separation, and balanced phasing arrangement.

References

1 Southwire (2005). *Power Cable Manual*, 4e. Southwire.
2 NEC (National Electrical Code), Ampacity Tables for Power Cables.
3 CEC (Canadian Electrical Code) (2017). Ampacity Tables for Power Cables.
4 ICEA (2018). Short Circuit Current Withstand for Cables.
5 Standard for Control, Thermocouple and Instrumentation Cables ANSI/ICEA S-73-532, (2014).
6 NEMA (National Electrical Manufacturers Association) (2017). Standard VE-1.
7 General Cable: Cable Installation Manual for Power and Control Cables, (2014).
8 Eaton – Cooper Industries: Cable Tray Manual (2015). MP Husky Cable Bus System, Brochure.
9 HP Husky for Cable Bus.

13

Power Factor Correction

CHAPTER MENU

13.1 Power Factor and Penalties for Low pf, 285
13.2 Leading and Lagging Power Factor, 286
13.3 pf Correction, 288
    13.3.1 Calculations, 288
    13.3.2 Capacitor Applications and Switching, 290
    13.3.3 Synchronous Motors, Condensers, and Generators, 292
13.4 Power Factor at Diesel Engine Generating Plant, 295
13.5 Voltage Improvement by Adding Capacitors, 295
    13.5.1 Other Relations, 297
13.6 Harmonic Issues with the Capacitors, 297
    13.6.1 Nonlinear Loads, 297
    13.6.2 How Does pf Correction Affect Harmonics?, 299
    13.6.3 Capacitor Fusing and Grounding, 300
13.7 Other Applications, 300
    13.7.1 Surge Packs, 300
    13.7.2 Series Capacitors, 300
    13.7.3 Reactors, 301
References, 302

13.1 Power Factor and Penalties for Low pf

A "power factor study" must be conducted with an effort to reduce ohmic losses in the plant cables, lines, and transformers caused by motors due to poor power factor performance and magnetizing currents. Appropriate actions must be implemented to maintain a proper pf at the point of interconnection (POI) with the utility and to reduce the import of reactive power into the plant. The pf correction equipment must be installed at the plant load centers where they are the most effective in reducing the flows of reactive power through the system.

The utility supplying the industrial load will insist on power factor of 0.95 or better. A poor power factor results in financial losses to the utility and the industrial plant in several ways:

Practical Power Plant Engineering: A Guide for Early Career Engineers, First Edition. Zark Bedalov.
© 2020 John Wiley & Sons, Inc. Published 2020 by John Wiley & Sons, Inc.

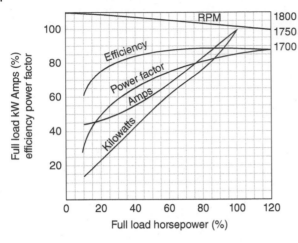

Figure 13.1 Motor partially loaded performance.

- Higher kWh tariff payment and additional capacity charges due to higher demand, typically if power factor is below 90%. Naturally, the utility is restricted in the sale of kWh if they are forced to deliver reactive power (kVARh).
- Financial penalties applied by the utility for causing restriction to their generating MW capacity.
- Additional kW and energy losses in the lines, cables, and equipment are a direct result of the increased currents flowing through the electrical equipment.

The following are some benefits of power factor correction:

- Compliance with the utility requirements.
- Savings in reduction of maximum demand.
- Lower line and plant losses, lower tariffs.
- Better voltage regulation.
- Releasing the power capacity in the plant transformers and distribution system.

Power factor is often called cos φ. Generally, low power factor is in large part caused by the plant motors, mainly due to *partial loading*. Motor ratings are selected to match the peak load but are usually operated at a considerably lower load and draw more reactive power. Instead of motoring at 0.8 pf, the motor operating at lower load may be drawing 0.5–0.6 pf according to the graph in Figure 13.1 [1].

13.2 Leading and Lagging Power Factor

Figure 13.2 and Table 13.1 clarify the difference in the kVAR flows for various equipment.

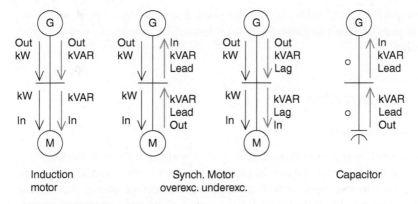

Figure 13.2 Reactive power flows [2].

Table 13.1 VAR power flow.

Type of load	At generator (source)			At load		
	kW	kVAR	pf[a]	kW	kVAR	pf[b]
Induction motor	Out	Out	Lag	In	In	Lag
Synchronous motor overexcited	Out	In	Lead	In	Out	Lead
Synchronous motor underexcited	Out	Out	Lag	In	In	Lag
Capacitor bank	0	In	Lead	0	Out	Lead

a) pf measured at source.
b) pf measured at load.

An induction motor always has a lagging power factor. It draws reactive and active power from the source. Power factor can be improved by employing synchronous motors used for larger mills, and by adding capacitor banks at the locations of concentrated plant motor loads as a source of reactive (magnetizing) power. Synchronous motor, by its field control, can be overexcited or underexcited. The synchronous motor is expected to perform as overexcited to furnish inductive power to the loads at the bus to which it is connected. A synchronous motor operating underexcited is counterproductive as it takes inductive power from the grid like a regular motor.

Capacitor bank installed at the right place will export reactive power to the users (motors). Capacitor bank actually imports capacitive power, which can be interpreted as exporting inductive power. In that sense, a capacitor is a generator of the inductive (magnetizing) power for motors.

Installing large capacitor banks at the POI with the utility is a less effective method of reducing power factor. Large switchable capacitors at the main substation centers do reduce the overall power factor at the POI, but the losses in the equipment and cables within the plant due to the reactive power flows are still present. The plant energy losses must still be imported and paid for to the utility.

13.3 pf Correction

13.3.1 Calculations

Ideally, each larger motor should be provided with an integral capacitor. This would not only improve the overall plant power factor, but it would reduce the power losses in the lines, as well as improve the voltage profile throughout the plant and provide better start for the motor. Without a corrective action, each motor is importing reactive power all the way from the utility. The best location for the capacitors is next to the motors just after the motor starter. It is usually economical to install one capacitor for each motor >25 kW motor and switch the capacitor and motor together. If the plant consists of many small motors, one can consider the motors as a group load and install one capacitor at a central point in the distribution system. Often, the best approach for plants with large and small motors is to use both types of capacitor installations. A typical motor power factor performance of 0.8 is shown in Figure 13.3.

There is no intent to correct power factor all the way to unity (1.0). The power factor of an installation can be improved from 0.7 to 0.8 or 0.9 by smaller capacitor kVAR steps. Improving power factor from 0.95 to unity takes a lot more capacitors. Table 13.2 illustrates what it takes to correct pf in steps of 5% for a motor of 100 kW.

The basic power factor correction formulae:

$$\text{pf} = \cos\,\varphi = \frac{\text{kW}}{\text{kVA}}, \tag{13.1}$$

$$\text{kVA} = \frac{\text{kW}}{\text{pf}}, \quad \text{kW} = \text{kVA} \times \text{pf} \tag{13.2}$$

$$\text{kVA} = \sqrt{\text{kW}^2 + \text{kVAR}^2}, \quad \text{kW} = \sqrt{\text{kVA}^2 - \text{kVAR}^2} \tag{13.3}$$

For instance: $P = 8\,\text{kW}, Q = 6\,\text{kVAR}, \quad S = \sqrt{P^2 + Q^2} = 10\,\text{kVA}$
$\cos\varphi = 8/10 = 0.8$ or 80%. $\varphi = \cos^{-1}(\text{kW/kVA}) = 36.86°$.
Also, $\tan\varphi = 6/8 = 0.75$ $\varphi = \tan^{-1}(\text{kVAR/kW}) = 36.86°$.

If the reactive power flow (VAR) flow at the load bus is balanced (In = Out), the particular bus operates at unity power factor, though the motors are likely working at lower power factors to satisfy their magnetizing needs.

Capacitors come in different sizes, thus there may be a case of overshooting or undershooting the correction for each application. There is no intent to correct the power factor exactly to unity (100%). The operating conditions for the

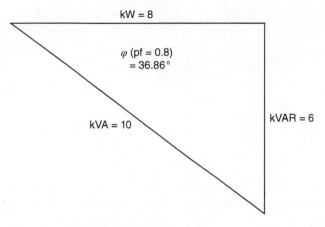

Figure 13.3 Typical motor loading.

Table 13.2 Correction needed for 100 kW motor.

pf	Angle (°)	tan φ	kVAR	Capacitor steps
0.65	49.4	1.17	116.8	—
0.7	45.5	1.017	101.7	15.2
0.75	41.4	0.882	88.2	13.5
0.8	36.8	0.748	74.8	13.4
0.85	31.78	0.619	61.9	12.9
0.9	25.84	0.484	48.4	13.5
0.95	18.19	0.328	32.8	15.6
1.0	0	0	0	32.8

motors are satisfactory if the overall power factor at the bus is at 95% lag to 95% lead. As the motor load changes, the power factor on the bus will also vary.

Example Let us work out the method of power factor improvement by capacitors added to the motor bus.

Refer to Figure 13.4. Suppose the 480 V, 3 ph bus is operating at pf = 0.77 (77%). The intent is to correct the power factor to 0.97 (97%) or better. The bus mixed kW load is $P = 100$ kW.

Initial pf = 0.77, cos $\varphi_1 = 0.77$	Power factor angle is $\varphi_1 = \cos^{-1} 0.77 = 39.64°$
Desired pf = 0.97, cos $\varphi_2 = 0.97$	Power factor angle is $\varphi_2 = \cos^{-1} 0.97 = 14.07°$

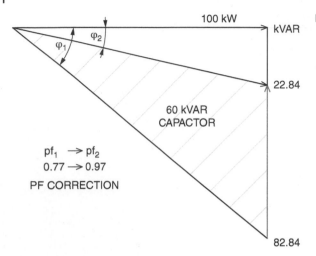

Figure 13.4 pf correction.

Initial kVAR at φ_1 = kW × tan φ_1 = 100 × tan 39.64° = 82.84 kVAR
Desired kVAR at φ_2 = kW × tan φ_2 = 100 × tan 14.07° = 25.06 kVAR
Capacitors to be added: 82.84 − 25.06 = 57.78 kVAR
Install 60 kVAR capacitor bank: 82.84 − 60 = 22.84 kVAR
Final power factor: pf = cos (tan^{-1} $\frac{22.84}{100}$) = 0.9749 for a power factor angle of 12.86°.

Calculate the corresponding current flows:

pf$_1$ = 0.77 = 39.64° → 100 kW, 129.8 kVA → 163.2 A
pf$_2$ = 0.975 = 12.86° → 100 kW, 102.6 kVA → 128.9 A

By adding a 60 kVAR capacitor, the motor current was reduced by 21% at 480 V, (128.9/163.5).

$$\tan \varphi = \frac{kVAr}{kW}, \quad \varphi = \tan^{-1}\frac{kVAr}{kW} \tag{13.4}$$

13.3.2 Capacitor Applications and Switching

Capacitors installed close to the loads of lagging power factor offer numerous benefits as follows:

- Reduces lagging component of the load current.
- Increases voltage level at the load and reduces the voltage drop.
- Reduces kW (I^2R) and kVAR (I^2X) losses due to the reduced current.
- Increases the power factor of the supply generators.
- Increases the loading capability of the generators by not carrying the kVARs for the loads.

By adding capacitors, it lowers the current carried in the supply circuit from the source as the capacitors supply most of the kVARs needed by the load, while the source supplies the kW component. Thus, the useful load can be increased to the motor to its temperature limit. Clearly, the capacitors increase the circuit capacity for additional loads or they relieve overloaded circuits.

One has to pay attention as not to overcompensate the group load if load decreases. As the load decreases, the capacitive current I_c remains the same or even greater as it is dependent solely on the voltage at the point of its connection. The voltage at a lighter load may increase. Thus, for the variable group loads, the best solution would be to shed the capacitor banks in relation to the magnitude of the load currents in the circuit.

Automatic regulation of kVAR flows can be added to control the load at the peak conditions, thus reducing the cost of power as well as the cost of power demand. The demand is not a peaking load, but an average load over a 15-minute period of peak load, as monitored by the utilities. Certainly the most beneficial location for the capacitors is as close as possible near the critical "dirty" loads of low power factors, like arc furnaces in metal smelting facilities, which at some instances in its operating cycle may draw power factor as low as 0.2, thus causing excessive currents, voltage drops, and lamp flickers. These loads must be supported by switching capacitors or preferably by more costly nonswitching synchronous condensers. Low-speed motors draw more magnetizing current, thus require larger capacitors, as shown in Table 13.3.

Capacitors on distribution lines are typically applied at the load locations to rid the line from reactive currents. Capacitors are usually not switchable, thus maximum amount of capacitors is limited to power factor correction near the daily minimum loads, so as not to bring the plant power factor to

Table 13.3 Maximum capacitor kVAR for use with 3 ph 60 Hz motors (NEC code).

Motor	1800		1200		900		720 rpm	
HP	kVAR	a)	kVAR	a)	kVAR	a)	kVAR	a)
10	4	11	4	12	5	17	5	28
20	5	10	5	11	7.5	15	10	24
30	7.5	10	7.5	10	10	13	12.5	21
50	12.5	8	12.5	9	15	12	20	17
75	17.5	8	17.5	8	20	11	27.5	15
100	22.5	8	22.5	8	25	10	35	14
150	35	8	35	8	37.5	9	47.5	13
200	42.5	8	42.5	8	45	9	60	12

a) % Reduction of line current.

the leading side and make the equipment to function counterproductively by raising voltage too high.

Large capacitor banks clustered in several switchable groups can be applied at larger substations. The number of groups will depend on the voltage steps caused by the capacitor switching on and off. We had several applications of several tens of MVAR capacitors switchable in large banks for electrical arc furnaces and draglines to maintain a reasonable constant operating voltage of the installation in an environment of highly variable power factor ranging from 0.8 down to 0.3 lagging in several seconds.

The continually changing voltage may seriously affect the operation of lighting in the plant installation, unless the power factor is properly coordinated. Modern high lumen lighting sources need time to reignite after a major voltage drop. In these cases, it is important to utilize a mix of lighting fixtures, some of them with quick acting ballasts to ride through the brief outages.

An important point to remember is that if the capacitor used with the motor is too large, self-excitation may cause a motor-damaging overvoltage when the motor and capacitor combination is disconnected from the line. In addition, high-transient torques capable of damaging the motor shaft or coupling can occur if the motor is reconnected to the line while still rotating and generating a voltage of self-excitation.

13.3.3 Synchronous Motors, Condensers, and Generators

All of these apparatus can serve to operate as reactive power generators. Unlike a capacitor bank, the amount of reactive power from a rotating apparatus can continuously be adjusted in both directions without switching; leading and lagging to suit the load power factor and the load magnitude. In comparison, this quality of regulation cannot be duplicated by capacitor banks. The synchronous motors in addition of doing its active work as a mill or a pump, when overexcited generate inductive power in addition to that needed for its own needs. The amount of the inductive kVARs will also depend on the load it is producing to drive a mill or a pump. If the load demand is low, it can produce more reactive power. The synchronous motors are rated at either pf = 1 or at 0.8 leading, thus serving as the generators of inductive power.

Synchronous condensers are typically used for supplying (exporting) reactive power and have no capability of providing mechanical load. They are generally made for and used by the utilities.

Synchronous condensers may also be referred to as dynamic power factor correction systems. These machines can prove very effective when advanced controls are utilized. They can be regulated to allow the system to meet a given power factor or to produce a specified amount of reactive power.

Synchronous generators are also used by the utilities to generate and export inductive power. Often, at night at the time of light load, the generators instead of shutting them down, the utilities keep them operating as synchronous condensers lightly overexcited to serve both as standby generating units and as well as spinning reserve in case of a failure of other generating units on the system.

Conveniently, a hydrogenerator can be designed and built for double duty as a generator and synchronous condenser. It can be switched from one duty to the other within one minute as required by the dispatch authority, which is directing the power usage in an area. To operate as a generator, water drives the prime mover, which may be a Francis turbine. To operate as a synchronous condenser, water is pushed and pumped out from the turbine runner allowing the turbine to spin freely in air instead of water at a minimum motoring power. The generator excitation system then generates the reactive power on demand.

A synchronous condenser is a more expensive apparatus in comparison with a shunt capacitor bank on the basis of MVAR capacity. However, it does a better job in all the areas of reactive power applications. Generation (leading) or consumption (lagging) of reactive power is achieved by regulating the excitation current by their automatic voltage regulators (AVRs) tuned to the operating conditions on the system. Depending on the application, the condensers are powered at 4.16, 11, 13.8 up to 20 kV, just like generators.

A synchronous condenser offers the following benefits:

- Provision of sufficient short-circuit capacity and inertia. Very important for the system stability!
- Steady-state and dynamic voltage control, and
- Reactive power control of dynamic loads in both leading and lagging conditions.

A synchronous condenser typically comprises a horizontal synchronous generator connected to the high-voltage transmission network via a step-up transformer. The unit is started up and stopped with a frequency-stepped up electric motor (pony induction motor connected with a clutch) or a starting frequency converter. When the unit reaches its operating synchronous speed, it synchronizes with the transmission network. Without active power delivery (or consumption), the machine can act like a capacitor or like a reactor, depending on the excitation field current, overexcited or underexcited, respectively.

The efficiency of long power transmission lines may be increased by placing synchronous condensers along the line to compensate lagging currents caused by the line inductance, thus allowing more real power to be transmitted through a fixed size line. The ability of synchronous condensers to absorb or produce reactive power on a transient basis stabilizes the power grid against transient fault conditions. An important benefit of a synchronous condenser is that it contributes to the overall short circuit capacity of the grid. This, in turn, improves the chances that equipment connected to the network will be able to "ride through" network fault conditions.

As with any synchronous motor/generator, the electrical dynamics are largely determined by the system reactances and by the nature of its excitation system. Low-transient reactances and comparably high rotor inertia ensure high-transient stability margins and excellent fault ride-through capability for rapidly fluctuating loads such as electric arc furnaces or mine hoists.

Large installations of synchronous condensers are also used in association with high voltage direct current (HVDC) converter stations to supply reactive power to the grid.

Synchronous machines have higher energy losses than static capacitor banks. Assume 2% compared to almost zero for capacitors. Capacitors can also be better distributed within the system and placed closer to the load.

The synchronous condensers connected to electrical grids are rated between 20 and 200 MVAR, and many of them are hydrogen cooled. There is no explosion hazard as long as the hydrogen concentration is maintained above 70%, typically held above 91%.

A synchronous condenser operates at nearly zero real power and does not result in any mechanical torque. As the machine passes from underexcited to overexcited, its stator current passes through its minimum as shown on the V curves (see Figure 13.5).

For the same mechanical load, the motor or generator armature current can be made to vary with the field excitation over a wide range and cause the power factor to vary accordingly. In between the under-excitation and over-excitation, the power factor is unity. The minimum armature current corresponds to the point of unity power factor (voltage and current in phase).

A synchronous condenser does not generate dangerous switching transients on the system. Also, the condensers are free of electrical harmonics.

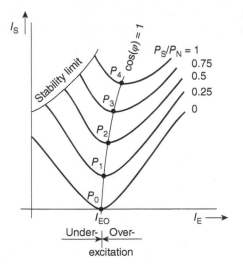

Figure 13.5 Generator excited at pf change.

Synchronous condensers are not susceptible to electrical resonances at harmonic frequencies, while the unwanted harmonics cause overheating and failures in capacitors.

The reactive power produced by a capacitor bank is in direct proportion to the square of its terminal voltage, and if the system voltage decreases, the capacitors produce less reactive power, when it is most needed, while if the system voltage increases, the capacitors produce more reactive power, which exacerbates the problem. In contrast, a synchronous condenser naturally supplies more reactive power to a low voltage and absorbs more reactive power at times of higher voltage. This reactive power improves voltage regulation in situations such as when starting large motors, or where power must travel long distances from where it is generated to where it is used, as is the case with power wheeling, i.e. transmission of electric power from one geographic region to another.

13.4 Power Factor at Diesel Engine Generating Plant

A diesel-generating plant serving a remote community or a remote mining load will supply all the active power needs as well as the reactive power needed by the plant transformers and motors. The generators will be excited for the combined load of the plant active and reactive power. This may not exactly be the most economic method of powering the plant as the excessive reactive power demand reduces the capacity of the plant generators to produce active power. More fuel is being burned than necessary.

The remote plants which use expensive fuel need to do a power factor study to determine how and where capacitors should be installed to reduce the fuel consumption and to release generator electrical capacity to produce more useful power.

13.5 Voltage Improvement by Adding Capacitors

The reactive output P_c of capacitors in kVAR is a function of the square of the line voltage V and directly with the supply frequency. The capacitance C is expressed in microFarads.

$$P_c = \frac{\sqrt{3} I V}{1000} \quad \text{or} \quad 2\pi f C V^2 \times 10^{-9} \text{ (kVAR)} \tag{13.5}$$

An improvement of the plant voltage profile stems from these relations:

$$\Delta V = RI_r \pm XI_x = RI \cos \varphi \pm XI \sin \varphi \text{ (V)} \tag{13.6}$$

where

R = the line resistance

X $\qquad$ = line reactance

I_r and I_x = the resistive and reactive components of the load current flowing in the circuit.

In this relation, $\pm$ can be + or − depending on lagging or leading power factor, respectively. Lagging power factor will cause larger voltage drop, while the leading power factor will raise the voltage on the system. In other words, the voltage will improve for the improved power factor, while the voltage will drop with a sagging power factor.

By decreasing the I_x, one can achieve substantial voltage improvement as the reactive and capacitive components are opposing each other. So the equation can be expanded to read:

$$\Delta V = RI_r + X(I_x - I_c)\,(V) \tag{13.7}$$

here, I_c is the charging current by the capacitor.

The same can be calculated on a pu basis by this relation: $\Delta V = PX + QR$ or expanded:

$$\Delta V = PR + (Q_x - Q_c)\,X \text{ (pu or V)} \tag{13.8}$$

Let us calculate a voltage drop and power loss for a power transfer of 30 MW over a 50 km long line, at 132 kV at two different power factors 0.8 and 0.93 at the sending end.

Load is $P = 30$ MW, $V = 132$ kV, 50 km line, $\text{pf}_1 = 0.8$, $\rightarrow$ $\text{pf}_2 = 0.93$.
Per unit base: MVAb = 30 MVA.
$P = 30$ MW $\qquad\qquad\qquad\qquad\qquad\qquad\qquad$ $\rightarrow 1.0$ pu

Calculate reactive power draw:
at pf = 0.8 $\qquad$ $Q_1 = 30 \times \tan (\cos^{-1} 0.8) = 22.5$ MVAR $\qquad$ $\rightarrow 0.75$ pu
at pf = 0.93 $\qquad$ $Q_2 = 30 \times \tan (\cos^{-1} 0.93) = 11.85$ MVAR $\qquad$ $\rightarrow 0.375$ pu

Let us assume the line conductor resistance is 11.7 Ω, reactance is 30.5 Ω. Line parameters expressed in per unit;

$$R_{\text{pu}} = R_\Omega \times \frac{\text{MVAb}}{\text{kV}^2} = 11.7 \times \frac{30}{132^2} = 0.0195 \text{ pu} \tag{13.9}$$

$$X_{\text{pu}} = X_\Omega \times \frac{\text{MVAb}}{\text{kV}^2} = 30.5 \times \frac{30}{132^2} = 0.0525 \text{ pu} \tag{13.10}$$

Voltage loss:
At pf = 0.8 $\quad$ $\Delta V = PR + QX = 1 \times 0.0195 + 0.75 \times 0.0525 = 0.0588$ pu $= 5.88\%$
At pf = 0.93 $\quad$ $\Delta V = PR + QX = 1 \times 0.0195 + 0.375 \times 0.0525 = 0.0392$
$\qquad\qquad\qquad\qquad\qquad\qquad\qquad\qquad\qquad\qquad\qquad$ pu $= 3.92\%$

Voltage improvement $\Delta V = 1.96\%$ for a pf change from 0.8% to 0.93 %

Power loss on the line : $\Delta P = 3\,I^2 R$ (W) $\tag{13.11}$

at pf $= 0.8$ MVA $= \sqrt{P^2 + Q^2} = \sqrt{30^2 + 22.5^2} = 37.5\,\text{MVA}; \quad I = 0.164\,\text{kA}$
at pf $= 0.93$ MVA $= \sqrt{P^2 + Q^2} = \sqrt{30^2 + 11.85^2} = 32.25\,\text{MVA}; I = 0.141\,\text{kA}$

Loss $= 3 \times 0.164^2 \times 11.7 = 0.944\,\text{MW}$ at 0.8 pf
Loss $= 3 \times 0.141^2 \times 11.7 = 0.697\,\text{MW}$ at 0.93 pf
   Reduction in power loss: $\Delta P = 0.247\,\text{MW}$.
   Annual cost of loss at \$70/MWh, assuming an 80% (0.8) average load.
\$/yr $= 0.8 \times 0.944 \times 8760\,\text{h} \times 70 = \$463\,084/\text{yr}$ at 0.8 pf
\$/yr $= 0.8 \times 0.697 \times 8760\,\text{h} \times 70 = \$341\,920/\text{yr}$ at 0.93 pf
   Annual saving on line losses due to an improvement in power factor: \$121 000/yr.
   That annual saved amount may pay for a lot of capacitors to improve the line operation.

13.5.1 Other Relations

Voltage rise ($\%V_R$) due to addition of capacitors:

$$\%V_R = \text{kVAR (cap)} \times \text{Zt\%}/\text{kVA}_{\text{tranformer}} \tag{13.12}$$

Power loss reduction ($\%L_R$) due to added capacitors:

$$\%L_R = 100 - 100 \left(\frac{\text{original pf}_1}{\text{improved pf}_2} \right)^2 \tag{13.13}$$

Capacitor kVAr output when operating at different voltage.

$$\text{kVAR}_2 = \text{kVAr}_1 \left(\frac{V_2}{V_1} \right)^2 \tag{13.14}$$

13.6 Harmonic Issues with the Capacitors

13.6.1 Nonlinear Loads

Installation of capacitors is not exactly a smooth sailing. Well, not any more. Addition of nonlinear and direct current (DC) apparatus into the plants has greatly changed waveforms of the plant currents and voltages, thus affecting the capacitors. Significant harmonics have been added to the waveforms due to the nonsinusoidal requirements for energizing the nonlinear loads.

   Earlier, harmonics were present in the magnetizing currents of the motors and transformers. Nonlinear loads like large variable frequency drives (VFD, in Chapter 15) for motors have added even more harmonics into the power system flows.

The utility will insist that the industrial plant meets the standard IEEE-519 [3] harmonic requirements at the POI to protect other customers connected to their system.

What are the Harmonics? IEEE 519: "Harmonic is a sinusoidal component of a periodic wave or quantity having a frequency that is an integral multiple of the fundamental frequency." Harmonic distortion is a specific type of sine wave power that is usually associated with an industrial plant's increased use of VFDs and other devices that use solid-state switching. The graph in Figure 13.6 shows a 180 Hz (3×60 Hz), the third order harmonic of the fundamental frequency combined with the fundamental frequency and creating a distorted wave.

The harmonic currents or voltages that are the dominant harmonic orders created by three-phase nonlinear loads are 5th, 7th, 11th, and 13th. The level of voltage or current harmonic distortion existing at any one point on a power system can be expressed in terms of the total harmonic distortion (THD) of the current or voltage waveform. The THD (for a voltage waveform) for a six-pulse system is given by the following formulae:

$$V_{\text{THD}} = \frac{\sqrt{(V_5^2 + V_7^2 + \cdots + V_n^2)}}{V_1} \tag{13.15}$$

In most of the cases it must be kept <5%. Where,

V_1 = fundamental voltage value. V_n ($n = 1, 2, 3, 4, 5,$ etc.) = harmonic voltage order. The even and triple harmonics are not included for the VFD operations of six and higher pulse rectifiers.

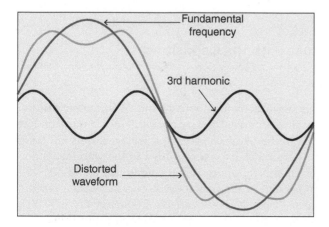

Figure 13.6 Harmonics.

Source: Courtesy of IEEE Standard 519-2012 [2]. This governing standard represents the most recent effort to establish a standard level of acceptable harmonic distortion levels on a power system.

More on the mitigation (see Chapter 15).

13.6.2 How Does pf Correction Affect Harmonics?

In an industrial plant containing power factor correction capacitors, harmonic currents and voltages can be magnified considerably due to the interaction of the capacitors with the service transformers. This is referred to as *harmonic resonance* or *parallel resonance*. Capacitors themselves do not cause harmonics, but only aggravate potential harmonic problems.

Harmonics are generally to be restricted by limiting the nonlinear load to 30% of the transformer capacity. However, with power factor capacitors installed in an industrial system resonating conditions may occur that can potentially limit the percentage of the nonlinear loads to about 15% of the transformer capacity.

The following equation can be used to determine if the resonant conditions may occur.

$$h_r = \sqrt{\frac{kVA_{sc}}{kVAR_c}} \tag{13.16}$$

where

h_r = resonant frequency as a multiple of the fundamental frequency $(=f_r/f_1)$

kVA_{sc} = short circuit kVA at the point of study

$kVAR_c$ = capacitor kVAR rating of the system voltage at the point of study.

There is a likely a chance of resonant condition if h_r is equal or close to a harmonic order fifth or seventh.

Large industrial plants must conduct a harmonic study to determine the presence of harmonics at all of its system buses. If the plant contains power factor correction capacitors, the current into the capacitors should be measured using a "true r.m.s." current meter. If this value is higher than the capacitor's rated current at the nominal system voltage (by >5% or so), then the presence of harmonic voltage distortion is likely.

The critical information on the ratio of the total nonlinear kVA to the service transformer kVA rating *must* be known before attempting to correct power factor in the presence of nonlinear loads. This ratio alone can often be used to determine whether harmonic filters are necessary to correct power factor or whether capacitors can be added without experiencing problems, as follows:

- If the plant's total three-phase nonlinear load (in kVA, 1 HP ~ 1 kVA) is more than 25% of the main transformer capacity, harmonic filters will almost always be required for power factor correction.
- If the plant's total three-phase nonlinear load is <15% of the main transformer capacity, capacitors can usually be applied without problems.
- If the plant's total nonlinear load is between 15% and 25% of the plant load, other factors should be considered.

13.6.3 Capacitor Fusing and Grounding

Since the large capacitor banks require parallel–series combinations of connected capacitors, it is essential to make sure each capacitor unit is individually fused. The fuses are typically provided with indication contacts to quickly identify the faulty units. When a fuse operates, it will cause a voltage rise across the rest of the units connected in parallel.

Capacitor banks can be connected in wye (star) ungrounded, wye grounded, or delta. The wye ungrounded system is preferred in the distribution systems used for VAR and voltage control. In this arrangement, the fault currents are lower, and there is no path for harmonic currents to ground. For the capacitors exposed to lightning, it is recommended that lightning arresters be provided for all wye ungrounded and delta-connected capacitor banks.

13.7 Other Applications

13.7.1 Surge Packs

Capacitors employed in parallel with arresters are called "surge packs." The surge pack neutral is solidly grounded. They are used as protection for generators, transformers, and large motors operating in high resistance grounded systems. The typical value taken by the power industry is $3 \times 0.25\,\mu F$, connected across the bus duct connecting the generator with transformer. The effect of the surge pack is to slope down the fronts of switching surges before they enter generator windings.

13.7.2 Series Capacitors

This application serves a distinctive purpose of increasing power transfers over transmission lines and to add to the stability of the overall power system. Let us look at this power transfer equation:

$$P = \frac{E_s\, E_r\, \sin\,\delta}{X_1} \tag{13.17}$$

or a compensated line with a series capacitor:

$$P = \frac{E_s\,E_r\,\sin\,\delta}{X_l - X_c} \tag{13.18}$$

where

E_s, E_r = sending and receiving voltages, respectively
X_l and X_c = line reactance and capacitance.
δ = power angle between the line end voltages. Briefly, line compensation increases active (MW) power transfer, improves voltage profile, reduces line power angle, and increases system stability.

Therefore, given the phase angle between the ends of the line, the power transfer depends on the line impedance. Placing a series capacitor into the line, the line impedance is reduced, and with it (MW) power transfer is considerably increased. This is achieved by installing a series fixed 85% and variable 15% capacitor bank. Power transfer is affected primarily by the power angle between the bus voltages and not by power factor. Series capacitors are not switchable.

Determine the kVA rating of a series capacitor bank.

Line power loss of capacitors, kW: $\qquad P_d = 3\,I^2\,X_c \tag{13.19}$

Line power, kW: $\qquad P_r = \sqrt{3}\,E_r\,I \tag{13.20}$

Power loss, for 30%compensated line: $\qquad \dfrac{P_d}{P_r} = 30\%\dfrac{\sqrt{3}\,I\,X_c}{E_r}\ (\%)$

$$\tag{13.21}$$

Therefore, a 30% rated capacitor also has a 30% of line voltage across its terminals. If this is a 138 kV line, 80 kV line to ground, the capacitor voltage rating would be $V_c = 30\% \times 80\,\text{kV} = 24\,\text{kV}$.

13.7.3 Reactors

Until the sixties, reactors were often used as fault-limiting reactors, applied between various switchgear buses as an added impedance to reduce the fault levels when the buses were tied together. The switchgear breakers have become more powerful with higher interrupting capabilities thus making the current limiting reactors not necessary.

Reactors have found their place and usage on long transmission lines, often applied together with capacitors or synchronous condensers to control the VAR flows to facilitate transmission of power with an effect of reducing the line losses.

They are typically dry-type coils with no magnetic core. This gives them constant inductance to the changing currents. This is not a case for the reactors with magnetic core. Saturated iron core at high fault current offers a reduced coil reactance, which is exactly opposite to that what is required.

References

1 Power factor for partially loaded motors. https://www.kele.com/templates/content.aspx?id=4636.
2 Beeman, D. (1955). *Industrial Power Systems Handbook*. McGraw Hill Book Co.
3 IEEE 519 – 2014: Recommended Practice and Requirements for Harmonic Control in Electrical Power Systems.

14

Motor Selection

CHAPTER MENU
14.1 Motor Selection, 303
14.2 Motor Characteristics, 304
14.3 NEMA Torque Classification (Design Code), 306
14.4 NEMA, IEC Frame Sizes, 307
14.5 NEMA Starter Sizes, 310
14.6 Motor Enclosures, 310
14.7 Large Motor Starting, 311
14.7.1 Induction Motors, Short Circuit Requirements, 311
14.8 Synchronous Motors, 315
14.9 Motor Service Factor, 315
14.10 Motor Starting Criteria, 316
14.11 Premium Efficiency Motors, 316
14.11.1 NEMA Premium™ Motors, the New Standard, 316
14.11.2 Motor Efficiencies by IEC, 317
14.11.3 Replacing a Serviceable Standard Efficiency Motor, 318
14.11.4 Premium Motor Inrush Current and Starting Issues, 319
References, 319

14.1 Motor Selection

Electrical motors are the main users of electrical power in any industrial plant. Motors drive mechanical equipment and systems: pumping, milling, ventilation, air supply, conveying, compressors, mills, heating, ventilating and air conditioning (HVAC), etc. Each of these services is a mechanical specialty by itself. The mechanical engineer is responsible to provide the specific requirements and environmental conditions to the equipment suppliers to determine the motor characteristics to perform the tasks required; kW (HP), torque curves A, B, C, D, motor starting requirements, variable speed (if applicable), enclosure type, and frame size. The motor usually comes to the plant site coupled together with its mechanical equipment.

Practical Power Plant Engineering: A Guide for Early Career Engineers, First Edition. Zark Bedalov.
© 2020 John Wiley & Sons, Inc. Published 2020 by John Wiley & Sons, Inc.

So, what is the role of an electrical engineer in the process of selecting a motor? Quite a bit, actually. Electrical engineer is responsible to provide a technical specification (TS) with a data sheet (DS) written for the motors up to 200 kW operated at low voltage (LV) as well as for motors over 200 kW fed at medium voltage (MV). Once the motor is selected, he/she must also select the power supply and the starter for the motor (see Chapter 24 for more details).

The motor data sheet is to be filled by the equipment supplier and returned completed with their tenders. In it, included is the relevant information of the project; detailing the site location and conditions, indoor or outdoor, voltage, frequency, phasing, anticondensation heaters and protection sensors. A space is allowed for the suppliers to fill in the applicable technical data associated with their offer. A decision may also have to be made to determine if the MV motors are to be of induction or synchronous type.

Upon receipt of tenders (supplier's offers), the project mechanical engineers will review the bid documents to verify if the motor/pumps meet the operating requirements for their desired services, while the electrical engineers will be responsible to make sure the electrical data matches the plant electrical standards. Motor efficiency together with the equipment energy consumptions will be compared and also evaluated over the life of the project. Once it has been decided to procure the equipment, the electrical engineer will allocate appropriate motor controllers of correct ratings in motor control centers (MCCs) or switchgear to power the motors. Should there be a need for a variable frequency drive (VFD); the electrical engineer will also write an appropriate specification to procure the equipment.

14.2 Motor Characteristics

Voltage change effect: Should the motor operating voltage vary, performance of the motor will be affected as follows (see Table 14.1):

Table 14.1 Supply voltage change.

V change	−10%	+10%
Starting torque	−19% ($-V^2$)	+21% ($+V^2$)
Slip	+20% to 30%	−15% to 20%
Load current	+1% to 7%	−5% to −15%
Temperature	+10% to 15%	+2% to 15%

Induction motors torque: The torque developed by asynchronous (induction) motors vary with the speed of the motor as it accelerates from full stop to the maximum operating speed according to the motor built in torque characteristic. The most common motor speed is 1800 rpm at 60 Hz (or 1500 rpm at 50 Hz) for a four pole motor. Lower 1200 (1000) rpm motors are often selected for higher torque requirements (see Figure 14.1).

Slip: The motor meets its load at its torque characteristic curve at a speed lower than its nominal synchronous speed. The difference in speed (1–5%) is slip. The slip will compensate for the variable load. The higher the load requirement, the higher the slips, according to the torque curve.

$$\text{Slip: } s = \frac{n_s - n}{n} \times 100\% \tag{14.1}$$

n_s = synchronous speed

n = operating speed.

Starting torque: It is the torque the motor develops when its starts at zero speed. When selecting a motor, high starting torque is important for the machines hard to start – as positive displacement pumps, cranes etc. A lower starting torque is selected for centrifugal fans or pumps where the starting load is low or close to zero.

Accelerating torque: This is the available motor torque based on the motor torque characteristic less than the load operating torque. It is the torque that will bring the motor to its full operating speed. Lack of it, may stall the motor or stretch its start-up time. The accelerating torque must be sufficient to allow the motor to reach its nominal speed for voltages down to 80%.

Break-down torque: The highest torque available before the torque decreases when the machine continues to accelerate to the nominal operating conditions.

Figure 14.1 Motor torque characteristics.

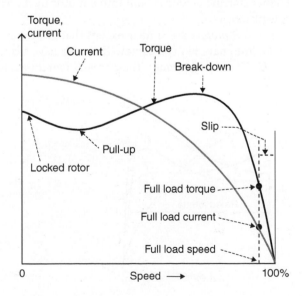

Full-load (rated) torque or braking torque: Motor produces the rated power P at full-load speed.

The full torque T is the torque when $T_{motor} = T_{load}$

$$T = 5252 \frac{P}{n} \text{ in imperial units; } T \text{ in lb} - \text{ft}, P \text{ in HP} \qquad (14.2)$$

$$T = 9550 \frac{P}{n} \text{ in metric units; } T \text{ in Nm}, P \text{ in kW} \qquad (14.3)$$

n = rated speed (rev/min, rpm)

14.3 NEMA Torque Classification (Design Code)

We may say that a motor is built for a specific torque, which is proportional to the designed V/Hz ratio.

The load torque is not always flat as it is often presented on the charts. Fans, for instance, are a centrifugal load, easy to start. They have a low starting torque, but once started, the load curve accelerates proportionally to the motor speed on cube (Fans → $T \lozenge n^3$). On the other hand, loaded conveyors have a high starting torque requirement, which gradually flattens down as the conveyor gains speed.

Motors are National Electrical Manufacturers Association (NEMA) classified as A, B, C, D, or E motors based on their torque characteristic for which they are designed (Figure 14.2). The classification mainly concerns the starting torque. Higher starting torque is built into a motor by increasing the resistance of the rotor bars.

A and B motors are similar except that A motors have higher starting currents. Both have high efficiency. They also have a large range of starting torque (70–275% of rated torque). Their breakdown torque is 65–190% and 175–300%,

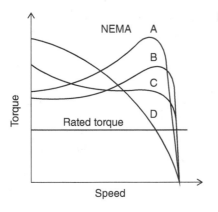

Figure 14.2 NEMA design code.

respectfully. A and B motors are commonly used for fans, rotary pumps, blowers, and machine tools, where large starting torques are not necessary and the motor does not need to support a large load. Slip 3–5%.

C motors are built for high locked (starting) rotor torque (200–285%) along with high pull-up (140–195%) and break down (190–225%). The locked rotor current averages at around 600–700%. C motors are mid-range when it comes to efficiency. They are best used in machines that require the motor to start under a load such as crushers, conveyors, compressors, centrifugal blowers, and reciprocating pumps. One can use a B motor for the same application, but it will need to be oversized by 20–30%.

D motors have similar torques to C motors, except that the break down torque can reach up to 275%. They also have a much higher slip, ranging from 5% to 13%. This creates a stronger torque but makes the motors inefficient. They are used for machinery with high peak loads such as elevators, hoists, forming machine tools, and punch presses.

E motors, not shown in the chart, are mid-range in locked rotor torque (75–190%), pull-up torque (60–140%), and break down torque (160–200%). They have the highest locked rotor current (800–1000%). Their slip is also somewhat smaller than the other designs (0.5–3%). These aspects give the E motors the highest efficiency out of the NEMA ratings. They can be used in similar applications to A and B motors like fans, pumps, and blowers with low starting torque.

As noted earlier, the approximate kW rating of the motors will be picked by mechanical engineer and recalculated by supplier. The torque characteristic will be picked by the equipment supplier. The rest is up to an electrical engineer. If it is more than 200 kW, it will go into a group of motors to be fed from 4.16 kV, to ensure the voltage drop during the start is not affecting the performance of the rest of the installation. Motors powered at 4.16 kV (5 kV system) are significantly more expensive. If feasible, it may be more economical to operate the motor at a lower voltage with a VFD, for up to 500 kW.

14.4 NEMA, IEC Frame Sizes

Motor manufacturers build motors to specific motor frame sizes (see Figure 14.3 for NEMA [1] and Figure 14.4 for IEC), popularly called T-frame motors. Each frame size determines the basic mounting/dimensions of the motors as well as the coordinates of the motor shaft. This information may not be of interest to electrical engineers, but it is of great importance to mechanical engineers to arrange for proper motor mounting and assembly to the driven equipment. Based on this information, the purchaser will know the basic dimensions and be able to immediately start defining the layout around the motor. The other motor dimensions of less concern to the layout

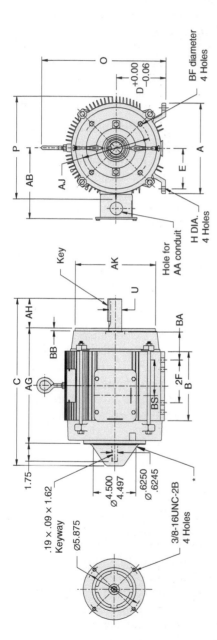

TENV • C Face • Rigid base • Dimensions (in.)

Frame size	D	E	2F	H	O maximum	P maximum	U	AA	AB maximum	AG maximum	AH	AJ	AK	BA	BB	BF	BS	Key
143TC	3.50	2.75	4.00	0.38	7.45	7.98	.875	1.09	7.04	9.25	2.12	5.875	4.50	2.62	0.13	3/8-16	2.81	0.19 × .19 × 1.38
145TC	3.50	2.75	5.00	0.35	7.45	7.98	.875	1.09	7.04	10.25	2.12	5.875	4.50	2.62	0.13	3/8-16	3.81	0.19 × .19 × 1.38
182TC	4.50	3.75	4.50	0.41	9.36	9.76	1.125	1.09	8.08	11.75	2.62	7.250	8.50	3.50	0.27	1/2-13	2.25	0.25 × .25 × 1.75
184TC	4.50	3.75	5.50	0.41	9.36	9.76	1.125	1.09	8.08	12.75	2.62	7.250	8.50	3.50	0.27	1/2-13	2.75	0.25 × .25 × 1.75
213TC	5.25	4.25	5.50	0.44	10.97	11.50	1.375	1.09	9.31	15.12	3.12	7.250	8.50	4.25	0.27	1/2-13	3.50	0.31 × .31 × 2.38
215TC	5.25	4.25	7.00	0.44	10.97	11.50	1.375	1.09	9.31	17.31	3.12	7.250	8.50	4.25	0.27	1/2-13	6.18	0.31 × .31 × 2.38
254TC	6.25	5.00	8.25	0.53	14.09	14.27	1.625	1.25	9.83	19.06	3.75	7.250	8.50	4.75	0.27	1/2-13	6.31	0.38 × .38 × 2.88
256TC	6.25	5.00	10.00	0.56	14.09	14.27	1.625	1.25	11.08	18.82	3.75	7.250	8.50	4.75	0.27	1/2-13	5.00	0.38 × .38 × 2.88
284TC	7.00	5.50	9.50	0.56	14.16	14.32	1.875	1.50	12.31	20.57	4.38	9.000	10.50	4.75	0.27	1/2-13	5.50	0.50 × .50 × 3.25

Figure 14.3 NEMA "Standard T Frame" dimensions. Source: [1].

Metric (IEC) frame dimensions (mm)

Frame	Mounting						Shaft						General			B5 flange						B14 face					
	A	B	C	H	AB	K	D	E	F	G	ED	DH	AC	AD	HD	M	N	P	S	T	LA	M	N	P	S	T	LA
D56	90	71	36	56	107	6	9	20	3	7.5	8	M3 × 8	–	–	–	100	80	120	7	2.5	7	65	50	80	M5	2.5	7
D63	100	80	40	63	122	7	11	23	4	8.5	10	M4 × 10	126	84	171	115	95	140	10	3.0	7	75	60	90	M5	2.5	7
D71	112	90	45	71	136	7	14	30	5	11.0	20	M5 × 12.5	141	94	191	130	110	160	10	3.5	7	85	70	105	M6	2.5	9
D80	125	100	50	80	154	10	19	40	6	15.5	25	M6 × 16	159	102	206	165	130	200	12	3.5	12	100	80	120	M6	3.0	9
D90S	140	100	56	90	172	10	24	50	8	20.0	32	M8 × 19	180	112	229	165	130	200	12	3.5	12	115	95	140	M8	3.0	9
D90L	140	125	56	90	172	10	24	50	8	20.0	32	M8 × 19	180	112	229	165	130	200	12	3.5	12	115	95	140	M8	3.0	9
DF100L	160	140	63	100	205	12	28	60	8	24.0	40	M10 × 22	205	130	270	215	180	250	15	4.0	11	130	110	160	M8	3.5	14
DF112M	190	140	70	112	230	12	28	60	8	24.0	40	M10 × 22	240	150	300	215	180	250	15	4.0	11	130	110	160	M8	3.5	11
DF132S	216	140	89	132	270	12	38	80	10	33.0	56	M12 × 28	275	180	345	265	230	300	15	4.0	12	165	130	200	M10	3.5	14
DF132M	216	178	89	132	270	12	38	80	10	33.0	56	M12 × 28	275	180	345	265	230	300	15	4.0	12	165	130	200	M10	3.5	14
DF160M	254	210	108	160	320	15	42	110	12	37.0	80	M16 × 36	330	210	420	300	250	350	19	5.0	13	215	180	250	M12	4.0	13
DF160L	254	254	108	160	320	15	42	110	12	37.0	80	M16 × 36	330	210	420	300	250	350	19	5.0	13	215	180	250	M12	4.0	13

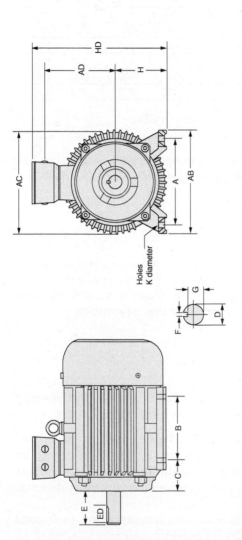

Holes
K diameter

Figure 14.4 IEC "Standard Frame" dimensions. Source: [1].

NEMA size	Continuous ampere rating	Maximum UL Horsepower single-phase 115 V	(two-pole) 230 V	Three-phase 208 V	240 V	480 V	600 V
00	9	1/3	1	1–1/2	1–1/2	2	2
0	18	1	2	2	3	5	5
1	27	2	3	7–1/2	7–1/2	10	10
2	45	3	7–1/2	10	15	25	25
3	90			25	30	50	50
4	135			40	50	100	100
5	270			75	100	200	200
6	540			150	200	400	400

Figure 14.5 Motor starter sizes. Source: courtesy of Eaton Corp. [2].

are not standardized and may be different from one supplier to the other. The tabulations NEMA or IEC given in the following are provided in part only.

14.5 NEMA Starter Sizes

The MCC starters (contactors) are standardized based on the contactor continuous ampere current. The Figure 14.5 provides the nonreversing starter size selection for three phase motors HP or kW. A similar table is available for single phase motors. The list was sourced from Eaton catalog information.

14.6 Motor Enclosures

The enclosures of electrical motors are standardized by NEMA. The following list is a list of the most common motor enclosures used in the industry.

Drip-proof: Ventilation openings in shield and/or frame prevent drops of liquid from falling into motor within up to 15° angle from vertical. It is designed for reasonably dry, clean, and well-ventilated (usually indoors) areas. Outdoors installation requires the motor to be protected with a cover that does not restrict the flow of air to the motor.

Totally enclosed fan cooled (TEFC): This is the most popular motor. An external fan is an integral part of the motor. The fan provides cooling by blowing air on the outside of the motor. It can be used in wet conditions also (Figure 14.6a).

Totally enclosed nonventilated (TENV): No ventilation openings, enclosed to prevent free exchange of air (not airtight). No external cooling fan. Relies on convection cooling. Suitable where the motor is exposed to dirt or dampness. Not suited in very moist humid or hazardous (explosive) air (Figure 14.6b).

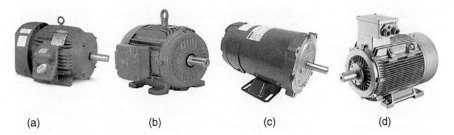

(a) (b) (c) (d)

Figure 14.6 The most common motor enclosures. (a) TEFC, (b) TENV, (c) explosion proof Cl 1, and (d) explosion proof Cl 2.

Explosion-proof motors: The explosion proof motor is a totally enclosed machine and is designed to withstand an explosion of specified gas or vapor inside the motor casing and prevent the ignition outside the motor by sparks, flashing, or explosion.

The motor ambient temperature must not exceed +40 °C. Motors are approved for the following explosion classes:

Class I (gases, vapors) (Figure 14.6c)

- *Group A*: Acetylene.
- *Group B*: Butadiene, ethylene oxide, hydrogen, and propylene oxide.
- *Group C*: Acetaldehyde, cyclopropane, diethyl ether, ethylene, and isoprene.
- *Group D*: Acetone, acrylonitrile, ammonia, benzene, butane, ethylene dichloride, gasoline, hexane, methane, methanol, naphtha, propane, propylene, styrene, toluene, vinyl acetate, vinyl chloride, and xylem.

Class II (combustible dusts) (Figure 14.6d)

- *Group E*: Aluminum, magnesium, and other metal dusts with similar characteristics.
- *Group F*: Carbon black, coke, or coal dust.
- *Group G*: Flour, starch, or grain dust.

14.7 Large Motor Starting

14.7.1 Induction Motors, Short Circuit Requirements

14.7.1.1 Case 1

Let me illustrate an actual case for large motor starting. We designed and built a gold mine plant up North of Canada. As usual in the Northern regions, the plant must have its own generating plant. We built a plant with five diesel generating units of 3 MW each. The Colomac mine was a low-grade gold mine. To make it profitable, it had to process 100 000 tons of ore daily, 24 hours a day. This was a huge milling operation.

Briefly, the ore was crushed, milled, grinded, and then leached and further processed to produce several ounces of gold every day. The equipment within the plant was mostly "used equipment," scrapped from other mines and reassembled at this plant. Consequently, the used equipment was not exactly rated and sized for this plant. Mostly, it was larger than necessary. The prominent pieces of equipment were one SAG mill of 7000 kW, similar to that shown in photo, and two synchronous motor Ball mills, 3000 kW each. The SAG mill was helped to start gradually with a liquid resistor connected on its rotor circuit.

The liquid resistor was a huge monster equipment using Blue Vitriol, normally employed in vineyards, as electrolyte. Depending on the outside temperature, the liquid was giving more or less resistance for the motor starting conditions.

Generally, it was needed three, sometimes four generators running to overcome its starting current to get it going on empty. In some situations, they would wake me at night if they were unable to start the 7 MW, 9.5 m diameter. SAG mill (Figure 14.7), when they had three generators only available to run the plant.

Figure 14.7 Sag mill.

I managed to get the plant going by starting one 3 MW Ball Mill first, followed by another Ball Mill and then I started the 7 MW SAG mill. Then, I would shut down the Ball Mills if they were not needed. The operators would look in owe, wondering how this was possible. What has got a Ball Mill to do with the SAG Mill?

The shift foremen asked me why I started the Ball mills ahead of starting the SAG mill. "That's sounds like silly to me, he might say."

"Silly perhaps, but I got it going, didn't I," I answered?

Well, here it is. A large motor needs a large source of short circuit MVA capacity to help it start. When the motor starts, at the first instant, the motor is virtually in a short circuit and drawing a huge inrush current, six to eight times the nominal current. This SAG mill was started by the rotor variable (liquid) resistor to develop a large starting torque.

The mill usually starts empty, and the load is added later while running. This time, the mill stopped suddenly fully loaded with 20 tons of frozen ore remaining inside its drum. This situation needed a much larger motor starting torque to overcome the load torque of the fully loaded mill. Because of this, the accelerating torque was considerably reduced and the starting current was "stretched in time."

The short-circuit current is basically a reactive power that is drawn from all the sources of short circuit (rotating motors and generators) to the point of short circuit. In this case, it is toward the motor which is starting as the short circuit is the point of zero voltage.

We had enough of short-circuit MVA to start the Ball Mill (BM), but not the Sag Mill. With the BMs running, we increased the system fault capacity to be able to start The Big Fellow. If the Ball Mills were not sufficient, I would have run the conveyors and other large motors to increase the available source of short-circuit power.

Not all the rotating machines contribute fault current. DC drives do not. Motors operated via VFDs do not. The motor inrush current lasts typically six to eight seconds as the motor gains its torque and speed. After that it starts drawing current based on the load it is pulling, pumping, conveying, etc.

If the motor does not have enough power at the start to develop its torque to exceed the torque of the load, the motor will stall and perhaps overheat and fail, unless the overload protection trips it out. Motor must have a torque characteristic greater than that of the load. The greater the torque difference in favor of the motor $T_{motor} - T_{load}$, the faster the motor gains its speed. This difference is called the accelerating torque, which drives the motor to the equilibrium with its load.

The motor inrush current six to eight times nominal is defined by its impedance, typically considered 15–17% on the motor MVA base. The inrush current at the instant of the motor start is not dependent on the mechanical load. It will typically be $I_s = (100/16) \times I_n = 6.25 \times I_n$. However, the duration of the inrush current six seconds or more will depend on the load and the accelerating torque, which brings the motor to its nominal speed. If the motor is too small and its torque characteristic is below the load characteristic, the motor will not be able to develop the torque needed and it will stall and fail. If the accelerating torque is just a bit larger than the load torque, the motor will accelerate, but it may take more time to reach its nominal speed and it may overheat.

14.7.1.2 Case 2

This is the case of an induction motor, 1000 kW, 6 kV, also a mill, starting fully loaded directly on line (DOL). The inrush current was seven times the nominal and causing a severe voltage dip on a weak power grid. The motor was specifically built for *this duty*, having a starting load torque requirement of 150% of nominal, as shown in Figure 14.8. We were forced to look into a softer start and to review several options (Figure 14.9).

Reduced voltage starters: Line autotransformers, line reactors, star/delta, soft start. All these starters operate on a principle of reduced voltage. Since the reduced voltage causes drop in torque proportionally by voltage square, these starters are not suitable for this application. These starters are inexpensive and suitable for the motors requiring low starting torque, like fans, etc.

Rotor resistor starters: This option would require a complete replacement of the motor to be designed for this starting duty. Furthermore, the mill owner

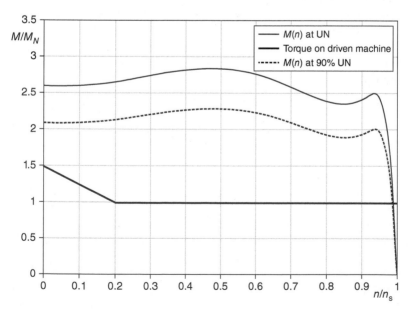

Figure 14.8 Motor torque characteristic.

Figure 14.9 Soft start comparison.

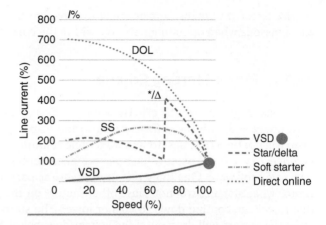

already had an experience with this type of motors and was not satisfied with the maintenance duty of the ring brushes.

Synchronous motors: These drives would also require a replacement of the induction motors as well. The motors would have to be provided with rotor cage equal to those of the original induction motors. That would not help with the inrush currents.

Variable frequency drives (VFDs): Figure 14.8 shows the difference in the performance between the options. VFD is a reasonable choice to soft start the motor on a high initial torque. When reaching the full speed, the VFD must use a synchronized bypass switchover to the main circuit breaker.

14.8 Synchronous Motors

Larger MV motors may be requested to be of synchronous type, rather than regular induction (asynchronous) type. They are more expensive, but they can be used to correct the plant power factor and reduce the penalty one must pay to the utility for having low power factor. The starting inrush current of the synchronous motors is three times nominal current, compared to six to eight times for an equivalent induction motor.

14.9 Motor Service Factor

There is another thing about the motors; it is the service factor (SF). Typically, the LV motors (400–600 V) are specified with a service factor of 1.15, while the MV motors are specified with the service factor of 1.0. Of course, you do not wish to throw away money on having extra reserve for larger motors. Motor service factor (SF) is the percentage of overloading the motor can handle for short periods when operating normally within the correct voltage tolerances.

14.10   Motor Starting Criteria

When starting a motor, the criteria will be to design a plant power system in such a way and with the sufficient transformer capacity to insure the voltage drop will not be >15% on start and >3% while running. It is the job of electrical engineer to verify the plant voltage performance for the large motors using system study software. One thing one must assure as part of the study is that the motor while starting will not cause other motors on the same bus to drop out due to voltage coming down to 80% or lower. The starters (contactors) operating the motors will drop out if the voltage goes below 80%. This may lead to further failures in the same area of the plant.

Most of the utilities limit the allowable voltage variation at the point of common coupling caused by a single motor start based on the governing standards. The voltage variation on the distribution system is determined by the impedance of the distribution system supply in relation to the impedance of the step-down transformer and secondary cabling to the motor. If the motor starting operation results in voltage sag that causes tripping of equipment within the facility, the following steps (if available) can be taken to improve the motor starting conditions:

(1) Keep large motors on a separate supply bus from other sensitive loads.
(2) Use soft starters to ramp up the starting sequence.
(3) Distribution transformer must be of sufficient kVA capacity to ease the motor start.
(4) Distribution transformer must be selected with lower impedance.

14.11   Premium Efficiency Motors

14.11.1   NEMA Premium™ Motors, the New Standard

The old-style "standard efficiency" motors remained popular because they generally cost less than the new models. Purchasing agents were not inclined to spend more money up front in order to save on energy costs later on. Because of the national energy implications of motor efficiency, the US Congress enacted the Energy Policy Act of 1992, to set minimum efficiency standards for certain classes of electric motors. All motors 1–200 HP sold in the USA after that date were required to have efficiency ratings equal to or better than those listed in NEMA MG-1 Table 12-11: Full-load efficiencies of EPAct efficient electric motors [3–5].

In June 2001, NEMA granted such "better-than-EPAct" motors special recognition by creating a designation called NEMA Premium™. Going a step beyond EPAct, NEMA Premium applies to single-speed, polyphase, 1–500 HP, 2-, 4-, and 6-pole (3600, 1800, and 1200 rpm) squirrel cage induction motors, NEMA Designs A or B, 600 V or less (5 kV or less for medium voltage motors) and rated for continuous operation.

It goes without saying that more-efficient motors consume less energy and reduce the electric bills over the operating time. Motors that operate intermittently may or may not save enough to justify replacement except in cases where utility rates are especially high. But, in evaluating motors that operate at a high duty cycle, or continuously, replacement with energy-efficient motors can usually result in very rapid payback, and save many times their initial cost.

The total annual energy consumption due to motor-driven equipment in the US industrial, commercial, residential, and transportation sectors was approximately 1431 billion kilowatt-hours (kWh) in 2006. This amounted to 38.4% of total US electrical energy use. Motor driven systems in the industrial sector consume approximately 632 billion kWh/year or 44% of all motor-driven system energy use. This industrial sector motor use equates to about 17% of the total US electrical energy use. Within the industrial sector, about 62.5% of the total electrical energy use is for motor-driven equipment [4].

It is estimated that the NEMA Premium motor program saves over 5.8 billion kWh of electricity a year.

What does "efficiency" mean? Like all electromechanical equipment, motors consume some "extra" energy in order to make the conversion of the electrical to mechanical energy. Efficiency is a measure of how much total energy a motor uses in relation to the rated power delivered to the shaft.

A motor's nameplate rating is based on its output, which is fixed for continuous operation at full load. The amount of input power needed to produce rated power will vary from motor to motor, with more-efficient motors requiring less input wattage than less-efficient models to produce the same output.

Electrical energy input is measured in Watts, while mechanical output is given in HP or kW. One HP is equivalent to 0.746 kW. There are several ways to express motor efficiency, but the basic concept and the numerical results are the same. For example:

$$\text{Efficiency\%} = \frac{\text{kW(Output)}}{\text{kW (input)}} 100 \quad \text{Efficiency\%} = 0.746 \frac{\text{HP (Output)}}{\text{kW (Input)}} 100$$

$$(14.4)$$

14.11.2 Motor Efficiencies by IEC

IEC 60034-30-1 [6]. Figure 14.10 defines four classes for all electric motors that are rated for sinusoidal voltage. The graph shows the correlation between

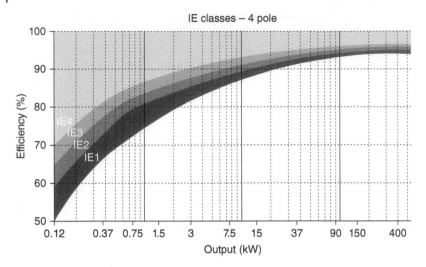

Figure 14.10 Motor efficiencies by IEC.

required efficiency and output for the four efficiency classes for 50 Hz motors:

IE1	Standard efficiency
IE2	High efficiency
IE3	Premium efficiency. It roughly corresponds to the NEMA Premium efficiency motors in the United States.
IE4	Super premium efficiency.

14.11.3 Replacing a Serviceable Standard Efficiency Motor

There is a software called **MotorMaster+** to evaluate the motor performance in terms of kWh and $/yr. Assume you have a serviceable standard-efficiency, 4 kW, 1800 rpm, 208–230/460-V, general-purpose, T frame, TEFC, NEMA Design B motor. Assume the motor operates 8000 hours (11 months) per year at 75% of full load and that power costs $0.075/kWh. Such motors have an average efficiency rating of 84% at full load. (Efficiency ratings for motors of this type at 75% loading range from 81% to 88.8%, averaging 84.06%).

Using the operating parameters and power cost given previously, the motor will consume 26 644 kWh of energy annually. The annual cost of operating this motor will be as follows:

$$\frac{4kW \times 0.75 \text{ (load factor)} \times 8000(h/yr) \times 0.075 \text{ \$/kWh}}{0.84 \text{ (efficiency)}} = \$2143.00$$

If, on the other hand, we upgrade to a NEMA Premium motor that has an efficiency of 90.5% at 75% of full load, annual energy and cost savings rise to

1989 kWh and $154, respectively, over the standard model. One such motor would cost approximately $302, and it would pay back its purchase price in about two years.

14.11.4   Premium Motor Inrush Current and Starting Issues

The motor inrush current for the older motors is expected to be between 600% and 700% of full load current (FLC). The expected motor subtransient impedance is $100/650 = 0.155\%$ or 15.5%. In that sense, we can say that the fault contribution by a motor can be calculated for a 35 kW motor as follows:

$$\frac{\text{kWmotor}}{Z''\text{m}} = \frac{35}{0.155} = 225\,\text{kVA} \tag{14.5}$$

Switching to the premium efficiency motors, it became known, that higher motor inrush currents and therefore lower impedances of the motor are expected. The motor impedances can now be expected to be 12.5%. In 1996, The *National Electrical Code (NEC)* was required to address this problem. The NEC now allows certain settings for the motor HCMP breaker magnetic elements to be raised from *currently 800% of FLA (Full Load Amps)* to exceed 800%.

The most significant problem associated with inrush currents is the resulting voltage sag within the plant. ANSI C50.41-2000 [7], *"American National Standard for Poly-phase Induction Motors for Power Generation Stations,"* states that motors must be able to start as long as the voltage is not <85% of the rated voltage, thus allowing 15% voltage drop during a large motor start.

References

1 Motor Dimensions http://www.leeson.com/Literature/pdf/1050/MotorDimensions.pdf.

2 Eaton, (2019). NEMA CN15 Contactors and Starters, Non-Reversing. Vol.5 V5-T2.

3 NEMA Premium energy efficiency motor program. https://www.nema.org/Technical/Pages/NEMA-Premium.aspx.

4 US Dept of Energy (2014). *Premium Efficiency Motor Selection Guide. Handbook for Industry*. United States Department of Energy.

5 NEMA MG-1 Table 12-11, pages 258–259 Full Load Efficiencies of EPAct Efficient Electric Motors.

6 IEC 60034-30-1-2014 Rotating electrical machines – Part 30-1: Efficiency classes of line operated AC motors (IE code).

7 ANSI C50.41-2000, (2012) American National Standard for Poly-phase Induction Motors for Power Generation Stations.

15

Variable Frequency Drives (VFDs) and Harmonics

CHAPTER MENU
15.1 Why Are Variable Frequency Drives (VFDs) Needed?, 321
15.1.1 Introduction, 321
15.1.2 Principles of Operation, 323
15.1.3 Power and Torque, 327
15.2 Vector VFDs for Low-Speed Operation, 328
15.3 VFDs: Variable or Constant Torque?, 329
15.4 Regenerative VFDs, 332
15.5 Motor and Cable Harmonics Issues, 334
15.6 How to Mitigate the Harmonics?, 335
15.6.1 What are Harmonics?, 335
15.6.2 Mitigation of Harmonics, 336
15.7 Harmonic Order Limits, 337
References, 339

15.1 Why Are Variable Frequency Drives (VFDs) Needed?

15.1.1 Introduction

Electric motor systems are responsible for more than 65% of power consumption in industry today. Optimizing motor control systems by upgrading to variable frequency drive VFDs can reduce energy consumption for some applications by as much as 70%. Combining energy efficiency tax incentives and utility rebates, returns on investment for VFD installations can be as little as six months.

Since 1985, motor control technology has changed, and a new method of speed control using variable frequency has been accepted for the industry automation. VFDs come in different sizes to match the motor kW ratings (see Figure 15.1).

Figure 15.1 VFD cabinets.

Before the advent of VFDs, a common method of varying motor speed was in steps by means of changing the resistance of the rotors of induction motors through external resistors. Another option was to use DC motors, which offered a linear speed control, but were expensive and inefficient. Induction motors with brushes and rotor resistors may still be found in some older installations. VFDs can be used for driving synchronous motors as well. Early VFD designs included current source VFDs and voltage source VFDs; nowadays, the industry has accepted the voltage source drives as a better choice for running both; the constant torque as well as the variable torque drives.

VFDs are used for driving pumps, fans, and occasionally conveyors to regulate and automatically vary the flow of liquid, air, and ore into tanks and stockpiles, in accordance with the set points communicated from the plant control system. The flows are directly controlled by transmitters from level, pressure, or temperature sensors, which act on the speed of the VFDs to insure the product does not overflow the tank or stockpile, or to regulate the temperature as dictated by the operating conditions (see Figure 15.2).

Nowadays, VFDs can automatically change motor speed from 0% to 120% in a continuous linear manner as demanded by the process control. This capability has resulted in considerable savings in plant layout design, efficiency, and sizing of mechanical equipment. For example, tank capacities have been significantly reduced to contain just enough content by automatically controlling the flows in and out of the tank as demanded by the downstream production. They no

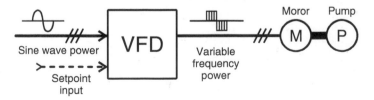

Figure 15.2 VFD: AC → → DC → AC conversion.

longer need to store several days' worth of material. Pumps no longer run at full capacity and overflow the tanks if demand changes. Similarly, plant fans modulate their flows of air to meet the ambient temperature needs more economically by varying their speed as opposed to the old way of adjusting the mechanical vanes while the motor runs at full speed.

Continuous modulation of motor speed to match demand contributes to significant savings in the energy required in plant production. For instance, if a fan is run at reduced speed instead of at full speed, the power P and energy savings $\$$ can be calculated proportionally to the power of 3, based on the following relations:

P_1/P_2, $\$_1/\$_2$ proportional to $(n_1/n_2)^3$, $(f_1/f_2)^3$

$P_{\text{at 80\% speed}} = \left(\frac{80}{100}\right)^3 \times P_n = 0.5\,P_{\text{nominal}}$

$\frac{T1}{T_2}$ proportional to $\left(\frac{V_1}{V_2}\right)^2$, provided the ratio of $\frac{V}{f}$ remains constant.

By running the motor at 80% of its nominal speed, the motor consumes 50% of its nominal power, as shown previously. Actually, the VFD does not change the motor speed but its operating synchronous speed. Thus, the motor does not operate at an inefficient partial load but at its new nominal speed. By having a motor running close to its synchronous speed, it also needs less reactive power.

During motor starting, VFDs can ramp up the voltage to facilitate a smooth start and remove severe motor inrush currents. This further mitigates voltage drops within the plant during large motor starts. The torque of the driven equipment can be considered to be proportional to voltage squared on power of 2.

On conveyors, belts, and gear drives, VFDs eliminate jerks on start-ups due to extended inrush currents, thus allowing high throughput and operation with less maintenance and fewer broken belts (see Figure 14.1).

15.1.2 Principles of Operation

A VFD is a type of motor controller that drives an electric motor by varying the frequency and voltage supplied to the electric motor. Other names for a VFD are "variable speed drive," "adjustable speed drive," and "adjustable frequency drive," all meaning the same thing.

VFDs come in many varieties, including sine-weighted, pulse width modulation (PWM), and vector controlled for low-speed applications. All the varieties come with variable voltage/frequency, for optimum speed control of any conventional squirrel cage induction or synchronous motor.

Generally, VFDs will include the following basic components:

- Incoming circuit breaker.
- Full-wave rectifier with metal oxide varistor (MOV) protection.

- Intermediate section consisting of DC link with capacitive filtering.
- Inverter (output section) to convert DC to variable frequency/voltage AC.
- Optical ethernet controls and communications.

How does a variable frequency drive work? (see Figure 15.3).

A VFD takes an input fixed frequency (60 or 50 Hz) AC voltage, converts it to DC voltage, and then inverts it back to a variable AC voltage and frequency (speed), as demanded by the load flow, pressure, and torque requirements. The transition follows this path:

$$\text{AC: } V_1, f_1 \rightarrow \text{DC: } V_{dc}, f = 0 \rightarrow \text{AC: } V_2, f_2$$

Let us go a bit slower. The first stage of a variable frequency AC drive, or VFD, is an AC/DC six diode converter, which generates six current "pulses" as each diode opens and closes. This is called a "six-pulse VFD," which is the standard configuration for modern VFDs. In the schematic representation, the 480 V_{ac} voltage is "r.m.s." or root-mean-squared. The voltage peaks on a 480 V_{ac} system are ±679 V. After conversion, the VFD DC bus has a DC voltage of between approximately 580 and 680 V with an AC ripple. By adding a capacitor into the DC bus circuit, called a DC link, the AC ripple on the DC bus goes away. The capacitor absorbs the AC ripple and delivers a smooth flat DC voltage to within less than a 3 V ripple. Thus, the voltage on the DC bus becomes "approximately" 650 V_{dc}.

The output section that converts the DC back to AC is referred to as an "inverter" to distinguish it from the input diode converter. By closing one of the

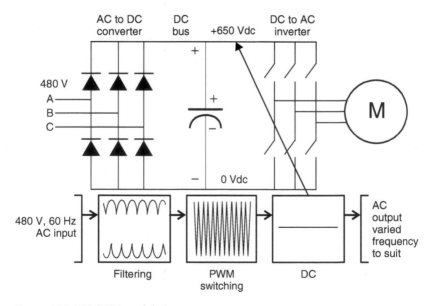

Figure 15.3 VFD PWM modulation.

top switches in the inverter, that phase of the motor is connected to the positive DC bus, and the voltage on that phase becomes positive. Then when one of the bottom switches in the inverter closes, that phase is connected to the negative DC bus and becomes negative. The switches are actually IGBT (insulated gate bipolar transistor) power transistors, which are biased to cut in and out at will, depending on the load demand. Thus, the controller makes any phase on the motor become positive or negative at will at any frequency desired, based on the speed reference issued by the control system and fed into the VFD.

VFDs do not produce a sinusoidal output (see Figure 15.4). They produce rectangular waveform, which motors can tolerate. To reduce the motor frequency to 30 Hz, for example, the controller simply switches the inverter output power transistors more slowly. By the switching process called PWM pulsing the output, any average voltage and frequency of the VFD can be achieved. The output switching is modulated at frequency of around 1–3 kHz. The graph shows the desired sine wave voltage and the actual produced +/− chopped voltage for motor.

Voltage-source inverter (VSI) drive topologies: In a VSI drive, the DC output of the diode-bridge converter stores energy in the **capacitor** bus to supply flat voltage input to the inverter. The vast majority of drives are VSI type with PWM voltage output (see Figure 15.5).

Current-source inverter (CSI) drive topologies: In a CSI drive, the DC output of the silicon controlled rectifier (SCR)-bridge converter stores energy in **series-reactor** connection to supply flat current input to the inverter.

Impact of voltage: There were a number of cases I have witnessed where the motor failed to start while connected in Y while starting on a Y-Δ starter, due to low torque proportionally reduced to (57%) of voltage on square. Fortunately, the motor was able to start directly on Δ, and the motor connection was permanently left bypassed to delta.

A VFD reduces the motor voltage in a different way. Starting torque is often called locked rotor torque. The key here for maintaining a strong starting torque is the ratio of V/Hz, which must be held constant for a particular motor and its design NEMA classification. That is why VFDs are able to soft start a motor by

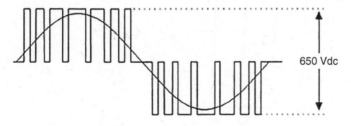

650 Vdc

Figure 15.4 VFD output voltage.

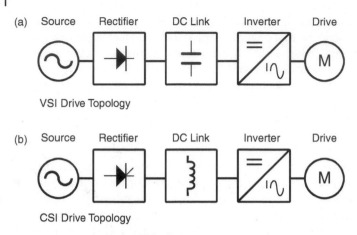

Figure 15.5 (a) VSI and (b) CSI topology.

matching the frequency to that of the voltage to maintain the V/Hz constant even at the lowest voltage. Typically for the 460 V motors, the ratio is 7.66 at 60 Hz, the same as for the IEC motors; 380 V/50 Hz = 7.66. As long as this ratio is held constant, the motor can develop constant rated torque at every point of the accelerating cycle [1].

Observe the chart in Figure 15.6. The starting torque is in fact higher at lower speeds than at higher speeds.

Here is the proportionality: Torque → Flux density within air gap → V/Hz.

The VFD typically starts an induction motor by beginning at low voltage and low frequency and increasing the voltage and frequency to the desired

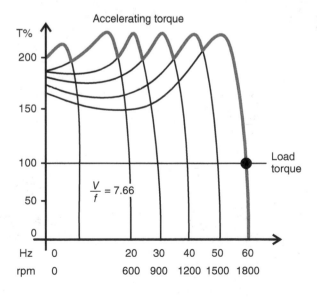

Figure 15.6 Motor start on VFD.

operating point. By starting the motor with low voltage and low frequency, the inrush current associated with across-the-line starting is completely eliminated. This contrasts the typical way of starting induction motors by applying full voltage (say 480 V at 60 Hz) immediately to the motor. With a VFD, motor is operating at the breakdown torque and synchronous speed as soon as it is started. The VFD accelerates the motor in a linear action, and the breakdown torque point moves along from left to right on the chart.

15.1.3 Power and Torque

When the speed of an AC motor is controlled by a VFD, power or torque will change depending on the change in frequency. The Figure 15.7 provides a graphical illustration of these changes. The x-axis is motor speed from 0 to 120 Hz. The y-axis is the percent of HP and torque. At 60 Hz (base motor speed), both HP and torque are at 100%. When the VFD reduces frequency and motor speed, it also reduces voltage to keep the volts/hertz ratio constant. Torque remains at 100%, but HP is reduced in direct proportion to the change in speed.

At 30 Hz, the motor's HP is just 50% of the 60 Hz HP. The reason this occurs is because the total torque produced per unit of time is also reduced by 50% due to fewer motor rotations. Use the HP and torque T equations to verify this relationship [2]. Power P and torque T change at different speed n (or frequency), as follows:

$$P = \frac{T \times n}{5252} \text{ (HP)} \tag{15.1}$$

$$T = \frac{HP \times 5252}{n} \text{ (lb ft)} \tag{15.2}$$

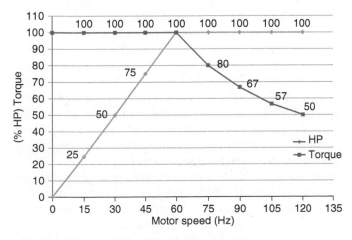

Figure 15.7 Power and torque–speed relationship.

The two equations show the relationship of HP and torque when a motor's speed changes. That is the reason a VFD is capable of starting a big motor seemingly without a struggle. Less speed demands less sweat and effort.

When a VFD increases frequency above 60 Hz, HP and torque do a complete flip flop. HP remains at 100%, and torque decreases as frequency increases. The torque reduction occurs because motor impedance increases with increasing frequency. Since a VFD cannot increase the voltage above its supply voltage, the current decreases as frequency increases, decreasing the available torque.

Based on that relationship, torque must double if power is to remain constant when speed is reduced by one half. To produce the same power at lower speed, motor has to do twice as much work per rotation, which requires twice as much torque. That is why the shaft and frame of a 900 rpm motor are usually larger than those of a 1800-rpm motor of the same HP. Theoretically, torque is reduced by the ratio of the base speed to the higher speed (60 Hz/90 Hz = 67%). However, continuous operation at lower speeds may cause motor overheating due to lower fan cooling. Instead of motors built with Class B insulation and 1.15 service factor, motors built with Class F insulation and 1.0 service factor are recommended for the VFD duty.

15.2   Vector VFDs for Low-Speed Operation

Vector VFDs are a technological advancement over regular VFDs, improving on the speed and torque requirements for the driven loads at lower speeds. Vector drives are typically used for hoisting and conveyor applications, where the operation may require greater than $10:1$ speed variation, such as below 6 Hz. The speed error at this frequency is generally due to changes in torque demand and motor slip at low speed. In most cases, a slight discrepancy in the torque is not critical as the process control will continually readjust the pump speed to maintain a specific level in the receiving tank or a flow.

A vector drive offers tighter speed regulation. A standard VFD is sometimes referred to as a "V/Hz" drive to differentiate it from a vector drive. A standard VFD maintains a certain V/Hz ratio to the motor at all times. As we noted earlier, a 460 V_{ac}, 60 Hz motor maintains a V/Hz ratio of 7.66 (460/60 = 7.66). When a V/Hz drive changes speed (frequency), it also changes the output voltage to keep the V/Hz ratio constant, such that at half speed, the motor will see the same V/Hz = 7.66 (230/30) output from the drive. This constant ratio is maintained in order to maintain a constant flux in the motor.

For most applications, standard VFDs will work well, as long as they are operating within the range of 6–60 Hz. Below 6 Hz, a motor operating on a V/Hz type VFD cannot generate much torque because at the lowest speeds, the V/Hz ratio required to achieve maximum torque is different than at higher speeds. Therefore, vector drives are designed to separately manipulate the voltage and frequency to produce an optimum V/Hz ratio for a maximum torque at any speed. In this way, vector drives can generate more starting torque and can provide full torque down to 1 Hz or less.

How is this done? Well, there are basically two types of vector drives: closed-loop with a shaft encoder and open-loop or sensorless.

Close loop with shaft encoder: Closed-loop vector drives require special motors with encoder feedback that provide motor shaft position to the VFD controller. The VFD drive uses this information to constantly alter the V/Hz ratio to produce a maximum torque. This type of VFD offers the best performance, but it is costly.

Sensorless: These vector drives work with standard motors. The performance is much better than that of the constant V/Hz drives. Instead of getting real-time info from an encoder, a sensorless vector drive estimates the actual motor speed along with other information that the drive measures when it is connected to the motor. It may not be the real field data, but it is still far better than having no feedback at all.

These days, most VFD suppliers offer sensorless vector drives for low-speed applications. They cost about the same as the standard V/Hz type drives.

V/Hz drives always maintain the V/Hz ratio to come up with a proper torque, whereas vector drives on the other hand come up with a flux density resulting from the vector sum of two separate motor current components to further modify the PWM pattern to maintain more precise control of the desired operating parameter, be that speed or torque.

The *first* current component is the magnetizing current needed to induce an electromagnetic field in the rotor. As the rotor is spinning, this magnetic field is rotated synchronously with the rotor to maintain a static orientation of the magnetic field relative to the rotor.

The *second* current is the current derived from the load torque. This current is generated in the stator to produce another electromagnetic field that is simultaneously induced in the rotor. The magnetizing current always lags (inductive) voltage by 90° and that of the torque producing current, which is always in phase with the voltage. As the intensity of the load current increases, the torque applied to the motor shaft increases. It is the vector sum of these two currents in the stator that is solved for by a vector VFD drive application.

15.3 VFDs: Variable or Constant Torque?

That is another question one must answer when selecting a VFD to suit a specific application. Many types of loads require reduced torque when driven at speeds less than the base speed of the load. Conversely, such loads may require increased torque when driven at speeds greater than the base speed of the load. These are classified as variable [3] torque loads (Figure 15.8). Many variable

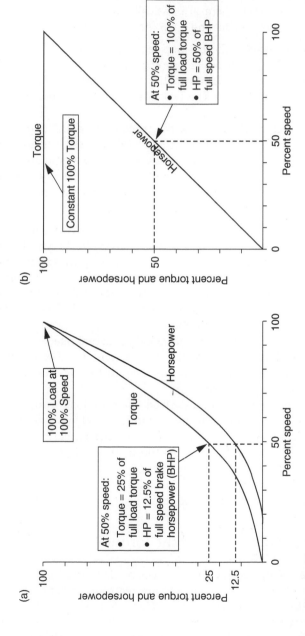

Figure 15.8 Variable/constant torque characteristics.

torque loads decrease with the square of the speed. This is the characteristic of centrifugal pumps and certain types of fans and blowers. Typically, as the speed decreases, the torque decreases with the square of the speed, and power decreases with the cube of the speed.

With constant torque loads, the torque loading is not a function of speed. Typical applications are traction drives, conveyors, positive displacement pumps, and hoists. Constant torque loads cause motors to draw relatively high current at low speeds when compared to variable torque applications. The same size drive may have a lower HP rating when used for constant torque applications.

Pumps can be grouped into two broad categories: **dynamic** (centrifugal) and **positive displacement.**

(1) *Dynamic* (centrifugal) pumps develop pressure and continuously impart energy to the liquid by a centrifugal force, dynamic lift, or momentum exchange. These pumps usually present a variable torque load.
(2) *Positive displacement* pumps discharge a given volume for each stroke or revolution of the pump.
    Energy is added in intermittent pulses. These pumps usually present, as an average, a **constant torque** load (Figure 15.9).

Air movers can be grouped into two broad categories: **variable torque fans and blowers** and **constant torque compressors.**

(1) *Variable torque fans and blowers* with centrifugal and axial designs develop static pressure and continuously pass on energy to the gas by a centrifugal force. These fans and blowers follow the fan *affinity laws*[1] and usually

---

1 *Wikipedia*: The **affinity laws** [4] for pumps/fans are used in hydraulics and/or HVAC to express the relationship between variables involved in pump or fan performance (such as head, volumetric flow rate, shaft speed) and power. They apply to pumps, fans, and hydraulic turbines. In these rotary implements, the affinity laws apply both to centrifugal and axial flows.

The laws are derived using the Buckingham π theorem. The affinity laws are useful as they allow prediction of the head discharge characteristic of a pump or fan from a known characteristic measured at a different speed or impeller diameter. The only requirement is that the two pumps or fans are dynamically similar, that is the ratios of the fluid forced are the same.

Law 1: With impeller diameter (D) held constant:

Law 1a: Q flow is proportional to shaft speed n:

$$Q_1/Q_2 = n_1/n_2$$

Law 1b: Head H is proportional to the square of shaft speed:

$$H_1/H_2 = (n_1/n_2)^2$$

Law 1c: Power is proportional to the cube of shaft speed:

$$P_1/P_2 = (n_1/n_2)^3$$

Law 2: Variable impeller diameter, while speed held constant: Replace n with D in the aforementioned equations.

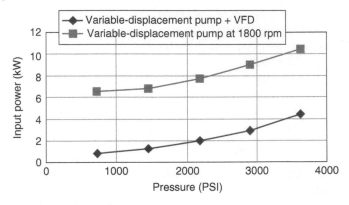

Figure 15.9 Displacement pumps.

present a variable torque load. The load varies drastically as a function of gas density (for example, hot air density vs. cold air density).

(2) *Constant torque compressors* with reciprocating and rotodynamic designs develop static pressure by passing on energy to the gas in intermittent pulses. These types of compressor loads do not follow the fan affinity laws (even when the compressor is centrifugal) and should be considered a constant torque load. As the speed changes, the load torque remains constant, and the horsepower changes linearly with speed.

15.4 Regenerative VFDs

A typical VFD controls the motor by supplying it with energy which then powers the load. However, occasionally the energy flow will reverse, that is, it starts flowing from the load through the motor and back to the VFD drive (see Figure 15.10).

This occurs if the load is giving up energy, such as when an elevator is lowering a load, or when a conveyer is transporting material downhill. Regeneration, as it is called, will also take place if a high inertia load is decelerated

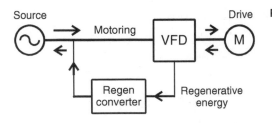

Figure 15.10 Regenerative VFDs.

(braked), the speed goes over its synchronous speed, and the motor enters into a generating mode.

With a variable frequency source, regenerative braking of induction motor can occur for speeds lower than synchronous speed of the grid source. The energy stored in the rotating mass flows back through the motor to the drive and then to the grid.

The continued operation of the drive will maintain a voltage on the motor to maintain the magnetic flux, but the phase direction of the currents will change, so energy – that is the current – will flow from the motor into the drive.

In a regular VFD, during the braking operation (downhill conveyor), current flows into the DC link and onto the DC link capacitor. Here, it charges the capacitor and the voltage rises. The current cannot go back to the supply as the front rectifiers are blocking it, so if regeneration continues, the voltage on the DC link capacitor will continue to rise, and the drive will trip on overvoltage. Regular VFD is incapable of operating in a regenerative mode. The aforementioned issues can be resolved in two ways:

(1) *Dynamic braking*: Add a braking resistor to the VFD to allow full load current to be returned to the drive and the power dissipated as heat in the resistor (Figure 15.11a). This solution may be adequate for low and medium size motors, and for control of braking that occurs occasionally.
Yes, but why throw out this energy?
(2) *Regeneration (Figure 15.11.b)*: If you are operating a container crane that is continuously lifting and lowering large containers or operating a downhill conveyor, it is wasteful to be burning all that energy in a resistor. The real solution is a fully **regenerative VFD** that will feed energy back into

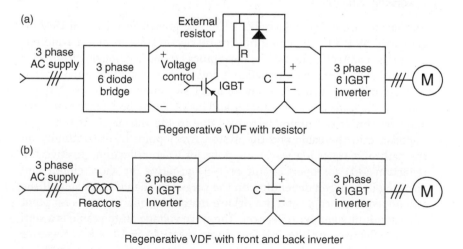

Regenerative VDF with resistor

Regenerative VDF with front and back inverter

Figure 15.11 Regenerative drives options.

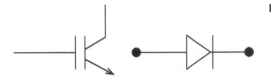

Figure 15.12 IGBT and diode.

the power supply to be used by other VFDs which are motoring. This is accomplished by replacing the front diode rectifier with a fully functioning inverter with IGBTs as shown in (Figure 15.12) to make the VFD conductive in both directions for motoring an regenerating.

The key to understanding regeneration is to recall the difference between a diode and an IGBT switching device. The diode allows current flow in only one direction and cannot be turned off. Current flows whenever the diode is forward biased. The IGBT, on the other hand, can pass current in either direction. In the "reverse" direction, it acts simply as an uncontrolled diode. In the "forward" direction, the current can be switched on and off. This is what makes the IGBT useful for both regeneration and motor control. IGBT is best suited for this application due to its highest switching speed compared to the all other electronic switching devices, 12–16 kHz, compared to 1–3 kHz for standard transistors [1].

15.5 Motor and Cable Harmonics Issues

There are two issues that affect the performance and longevity of the motors when working with VFDs:

(1) The harmonics as discussed in IEEE Transactions of March/April 1997 [5] generated by PWM switching may cause overheating in motor windings and in the connecting leads. The harmonics are more damaging to smaller motors due to their higher impedance and less slot insulation, as compared to larger motors. Typical rise time of IGBT switching wave is 0.4–0.6 μs.
(2) Motor cable leads between a motor and its VFD may be a cause of voltage amplification at the motor terminals due to the mismatch of the surge impedance in the cable and the motor. Every point of discontinuity on the path of a travelling wave is a point of wave reflection; positive full reflection (+1) for open circuit or (−1) negative for short circuit and anything in between depending on the surge impedance mismatch. If the two impedances (cable and motor) are matched (equal), there is no point of discontinuity and no reflection. The wave voltage on the cable is a sum of the reflective and the incident component: $V_c = V^+ + V^-$. Negative reflection (incident wave) is no concern to voltage amplification as most of the wave continues on in the same direction.

However, if the $Z_{cable} \ll Z_{motor}$, there is a considerable wave reflection and voltage amplification at the motor terminals, accentuated for smaller motors due to their large surge impedances. The maximum reflection expected is as follows:

$2 \times 600\,V \times 1.41 = 1692\,V_{peak}$, $-575\,V$ motors on the 600 V system (Canada)
$2 \times 480\,V \times 1.41 = 1354\,V_{peak}$ $-460\,V$ motors on the 480 V systems (USA)

15.6 How to Mitigate the Harmonics?

15.6.1 What are Harmonics?

Harmonics [6] in the power system are generated by nonlinear loads. Unlike the regular loads, these are the loads that change their impedances with the changes of the applied voltage. These include VFDs, computers, and other electronic loads. An example of a waveform of a nonlinear load can be seen in Figure 15.13.

Harmonics are site specific and dependant upon distribution transformer's short circuit capacity, its impedance, and maximum demand load of linear and nonlinear load.

The nonsinusoidal current contains harmonic currents that by interacting with the system impedances create voltage distortions at the point of common coupling (PCC) with the utility. The harmonics can affect both the distribution system equipment and the loads connected to it. The notches in the voltage wave (not shown in Figure 15.13) are a result of the switching action of the VFD rectifiers. The distortion is more present in the weak systems feeding the VFD, while the distortions are less significant on the stronger power sources.

The harmonic order is generated by the type of the rectifier bridge VFD front end. The most common bridge for low voltage, for three phase VFDs is a six

Figure 15.13 Nonlinear waveform.

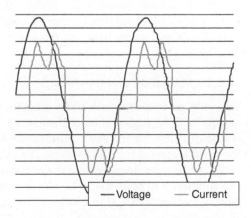

diode pulse bridge. Large VFD units use 12, 18, and 24 diode bridges to reduce the harmonics. The harmonic order is defined by the following formula:

$$h = np \pm 1 \qquad\qquad\qquad\qquad (15.3)$$

where

h = harmonic order

n = integer number 1, 2, 3, 4, 5, ...

p = number of pulses.

For instance, a 6 pulse VFD will ideally generate only the 5th, 7th, 11th, 13th, 17th, 19th, ..., while a 12 pulse bridge can generate the harmonic order starting from 11th and up from there. The order of even numbers is not created. The triplen (3, 6, 9, 12, ...) harmonics are also not created due the front end star/delta transformer winding configurations.

According to IEEE 519-1992, the total effect is of distortion in the current waveform at the PCC is measured by a ratio %THD. It is defined as a ratio of the sum of root mean square of the harmonic content up to 50th order to the maximum demand load current at the PCC, expressed as a percentage.

For instance, a total harmonic distortion (THD) of 30% at 100% load, for a 50% load results in a %THD of 15%

When a nonlinear load draws distorted current through the impedances from the supply to the load, it causes voltage drops and harmonics. The vector sum of all the individual drops results in total voltage distortion, the magnitude of which depends on the system impedance, available system fault current levels, and the levels of the harmonic currents at each harmonic frequency.

In a strong power supply system having a high fault current, the system impedances are low, while harmonic currents are high. On the other hand, in a weak system having high system impedances, voltage distortion is high, while the current draws are low.

The electrical equipment is greatly impacted due to harmonic distortion by increased currents, losses in iron and copper as well as eddy currents and skin effect at higher frequencies. Further damage can be evident in the bearings due to the bearing currents which can degrade the bearing lubrication.

15.6.2 Mitigation of Harmonics

Typically, motors used with VFDs must be specified for "inverter duty," which usually calls for higher class of insulation; 1600 V for 575 V motors (Canada) and slightly less for the 460 V motors (USA). A typical critical length of cable for *small motors* is from 20 to 45 m. The extra-long length is not in the critical range, as the wave decays due to the resistances over the cable length. The obvious solution is to place the VFDs close to the motors, rather than at the motor control centers (MCCs) in electrical rooms. That is feasible and practicable in particular for smaller motors. The VFDs can be ordered with reactances (chokes) at the end of the DC link to reduce voltage rises reaching the motors.

Mitigation can be included as an integral part of the nonlinear equipment (VFD) in form of a reactor (choke) or a passive filter. Or, as external equipment connected to the switchboard for multiple unit protection. The methods used in the industry are listed in the following.

1. *Reactance*: It is the simplest and effective protection available. It is added to the AC line side of the DC link of a VFD in particular for the six pulse drives. It serves better than the isolation transformer at the front end of the device. The standard ratings are 3%, 5%, and 7.5% of the VFD rating. A voltage is induced across the reactor in the opposite direction of the applied voltage, which opposes the rate of change of harmonic currents.

2. *Passive filters*: Also known as line harmonic filters (LHFs) are used to trap and eliminate the dominant lower order harmonics: 5, 7, 9, 11, and 13. They can be added in parallel to the individual drives or connected to the switchboard to deal with multiple drives. It includes an LC circuit, tuned to a specific harmonic frequency to be mitigated by offering low impedance sink for a particular range of frequencies. The filter is inherently capacitive below the tuned range and reactive above.

   The filters can also help with power factor correction. The filter capacitors are normally switched off during small loads to restrict pf going into the leading side.

3. *Active filters*: These devices are relatively complex and expensive. They employ an IGBT bridge and DC bus to filter and pass the fundamental harmonic and generate a compensating current to match those of the harmonic currents.

4. *Transformers with multiple secondary windings:* They are used as front ends for large drives. A combination of star/delta windings phase shifts the voltages by 30° to operate 12, 18, 24 diode pulse bridges to in effect totally eliminate the lower order harmonics depending on the choice of the bridge according to Eq. (15.3)

5. *Isolation transformers:* They offer voltage matching solution by stepping up or down the system voltage.

The effectiveness (I_h/I_1) of the aforementioned corrective actions is listed in Table 15.2 [7].

15.7 Harmonic Order Limits

With their many benefits, VFDs have grown rapidly in usage. This is particularly true in industries where their use in pumping, supplying air, and other applications has led to significant energy savings, improved process control, increased productivity, and higher reliability. In order to prevent the harmonics from negatively affecting the utility supplies, **IEEE 519 Standard** has been established as the "Recommended Practices and Requirements for Harmonic Control in Electrical Power Systems."

As the use of VFDs has increased, so has the negative impact of the harmonics they generate. This has resulted in more utilities enforcing compliance with the **IEEE Std. 519** Tables 15.1 and 15.2 [7, 8]. An unfortunate side effect of the VFD usage however is the introduction of harmonic distortion in the power systems. As a nonlinear load, a VFD draws current in a nonsinusoidal manner, rich in harmonic components. These harmonics flowing through the power system distort the supply voltage, overload electrical distribution equipment (such as transformers), and resonate with the service transformers among other issues (see Chapter 13).

IEEE 519-2012 is presently an industry standard that determines allowable limits for harmonic current distortion in percentage of the load current for individual order of harmonics. The standard defines the current limits based on the various levels of system short circuit over the load current I_{sc}/I_L. The short circuit ratio (I_{SC}/I_L) is the ratio of short circuit current (I_{SC}) at the PCC with the utility, to the customer's maximum load or demand current (I_L).

Table 15.1 Motor/cable wave reflection.

Motor (kW)	Z_{motor} (Ω)	Z_{cable} (Ω)	Positive reflection (%)
20	1500	80	90
40	750	70	83
75	375	50	76
150	188	40	65
300	94	30	52

Table 15.2 Harmonic correction.

Harmonic order (*h*)	5	7	11	13	17
6 pulse bridge without line reactor	80	58	18	10	7 (% of I_h/I_1)
6 pulse with 2–3% line reactor	40	15	5	4	4%
6 pulse with 5% line reactor	32	9	4	3	3%
6 pulse with passive harmonics filter	2.5	2.5	2	2	1.5
12 pulse bridge front end	3.7	1.2	6.9	3.2	0.3%
18 pulse bridge front end	0.6	0.8	0.5	0.4	3%

The higher the short circuit level in relation to the load current, the higher the allowable harmonic level. For instance for the harmonic order of $3 \leq 11$ (the most common harmonics) the limit at $<20\,I_{sc}/I_L$ is 4%, while at $50 < 100\,I_{sc}/I_L$ the limit is 10%.

Also, the standard establishes the harmonic limits on voltages from 1 to 161 kV for individual and THD. The higher the voltage, the lower is the THD allowed. For instance for voltages $V \leq 1.0\,kV$ the individual harmonics allowed are 5% and THD 8%. At the voltages higher than 161 kV the individual harmonics allowed are 1% and THD 1.5%.

This standard has been widely adopted, particularly in North America. The voltage and current harmonic limits presented in the standard were designed to be applied while taking the entire system into consideration, including all linear and nonlinear loading. The philosophy adopted by the utilities was to develop the limits for these indices to restrict harmonic current injection from individual customers so that they would not cause unacceptable voltage distortion levels when applied to normal power systems.

The justification for these limits states that "Computers and allied equipment, such as programmable controllers, frequently require ac sources that have limitations on harmonic voltage. Higher levels of harmonics result in erratic, sometimes subtle, malfunctions of the equipment that can, in some cases, have serious consequences. Instruments can be affected, giving erroneous data or otherwise performing unpredictably. Perhaps the most serious of these malfunctions are in medical instruments."

References

1 Polka, D. (2003). *Motors and Drives*. ISA.
2 Joe Evans, PhD. https://www.pumpsandsystems.com/topics/pumps/motor-horsepower-torque-versus-vfd-frequency.
3 Schneider. (1995) Adjustable Frequency Drives, SC100 Guide. Variable and Constant Torque.
4 Wikipedia – Affinity Laws https://en.wikipedia.org/wiki/Affinity_laws.
5 Bentley, J. and Link, P.J. (1997). Evaluation of motor power cables for PWM AC drives. *IEEE Transactions on Industry Applications* 33 (2).
6 Hoevenaars, T., LeDoux, K., and Colosino, M. (2003). Interpreting IEEE Std. 519 and meeting its harmonic limits in VFD applications. In: *IEEE Industry Applications Society 50th Annual Petroleum and Chemical Industry Conference*. Houston, TX: IEEE.
7 USA. Siemens, Whitepaper (2013). *Harmonics in Power Systems – Causes, Effects and Control*. Siemens Industry, Inc.
8 IEEE 519, (2012) *Recommended Practice and Requirements for Harmonic Control in Electric Power Systems*.

16

Relay Protection and Coordination

CHAPTER MENU

16.1 The Objective, 342
    16.1.1 Relay Operation, 342
16.2 IEEE Equipment and Device Designation, 343
16.3 CTs and PTs, 345
    16.3.1 Introduction, 345
    16.3.2 Polarity, 345
    16.3.3 Metering Accuracy Class, 347
    16.3.4 CT Relaying Accuracy Class, 349
16.4 Relay Protection, 351
    16.4.1 Multifunction Relays (MFR), 351
    16.4.2 Terminology, 353
16.5 Major Equipment Protection, 354
    16.5.1 Transformer and Generators, 354
    16.5.2 Motors, 356
    16.5.3 Transformers Current Reflections, 357
    16.5.4 Synchronizing and Synchrocheck Relays: What is the Difference?, 357
16.6 Relay Coordination, 359
16.7 Protection Function Elements, 360
    16.7.1 Overcurrent (50/51, 50/51N), 360
    16.7.2 Overcurrent Instantaneous (50, 50N), 364
16.8 Time Grading, 369
16.9 Time–Current Grading, 371
16.10 Reclosing, 373
    16.10.1 Breaker Duty Cycle and Interrupting Capability, 373
16.11 Load Shedding and Automatic Quick Start of Generators, 374
16.12 (86) Lockout and (94) Self-Reset Trip Relays, 375
    16.12.1 Lockout Trip Relays, 375
    16.12.2 Self-Reset Trip Relays, 375
    16.12.3 Trip Supervision Relay, 376
References, 376

Practical Power Plant Engineering: A Guide for Early Career Engineers, First Edition. Zark Bedalov.
© 2020 John Wiley & Sons, Inc. Published 2020 by John Wiley & Sons, Inc.

16.1   The Objective

The primary objective of the power systems is to maintain continuity of service to customers. Relay protection as part of the power system is essential in providing safety to personnel, equipment, integrity, and continuity of supply. It impacts all the areas of the power system such as power generation, transmission, distribution, and utilization. The protection system is expected to maintain the power system in a stable and operable state under all the operating conditions inclusive of overloads and system faults. The protection must act in a reliable, coordinated, and selective manner to insure minimum of damage to the equipment, the least of interruption to the power systems, and to allow a quick power restoration following abnormal conditions.

Relay protection and coordination is a specialized and ever changing activity. As an electrical engineer, you may not become a relay protection specialist, but you should acquire the basic knowledge to be able to understand and discuss the major issues of the system protection. To become a relay protection specialist, one requires dedication to this subject, tutoring, and a lot of field work testing of the equipment and the operating systems during the plant commissioning and operation.

Causes of abnormal conditions may be natural events, physical accidents, equipment failure, or human error. The natural events are lightning strikes, earthquake, snow storm, falling tree, or fire. Physical accidents can result from vehicle crashing into electricity distribution pole, animal or human coming into contact with live equipment, or digging into underground cables during construction. Human error may be an operator's careless action to energize equipment in incorrect sequence or closing a breaker to grounded equipment. Most of the faults, about 70% are phase to ground faults. The rest are multiphase faults.

Usual practice is to divide the whole system into protective zones. When a fault occurs in a given zone, only the circuit breakers within that zone are activated to isolate the faulty element without disturbing the rest of the system. The zones are partly overlapped to provide backup protection to the primary protection and increase protection reliability. An incorrect action may happen due to a failure of DC power source, relay, incorrect setting, breaker mechanism, or breakage of current transformer (CT)/potential transformer (PT) leads.

16.1.1   Relay Operation

Protective relays are fed from the instrument transformers CTs and PTs at 1 A (5 A) and 120 V, respectively. To properly condition the relay settings, additional inputs may be fed at 24 or $120 V_{dc}$ from other relays or circuit breakers. In case of a fault, when the protective relay reaches its pick-up value and then its trip setting, a signal is issued to the relay output contact. This contact is connected to the breaker $125 V_{dc}$ circuit with fast speed lockout trip relay (86), which activates the breaker trip mechanism. Once the breaker trips, a change of status is sent to the plant control system distributed control system (DCS) over a 24 V_{dc} supervisory circuit.

The following protective elements must be coordinated to work together in a selective way:

- Protective relays for voltages >600 V.
- LV breakers with integral trip devices for the equipment and feeders ≤600 V. Protective relays are rarely used here.
- LV and HV fuses.
- Equipment thermal overloads.

A protective relay is a device which detects an abnormal condition in the system and it acts on it. According to its algorithm and parameter settings, a protective relay will provide alarms or trips of the associated circuit breaker or several breakers. Furthermore, the new generation of multifunction relays (MFRs) allows an operator also to observe three phase situation of the protective elements in real time as the operating variables approach the pickup settings. The relays use the logic of breaker status inputs, measurement, determination and output as follows:

(1) Inputs to a relay include analogs of current or voltage applied to the protected equipment. Analog input sources include CTs, voltage transformers, resistance thermal device (RTD) sensors, and transducers. Furthermore, the plant control system and/or other relays provide (On/Off or Open/Closed) logic inputs of their status.
(2) The protective relay continually measures analog and digital input values and compares them against the parameter settings.
(3) The relay determines if the operating conditions are within normal safe values.
(4) If the conditions are different to those preset, due to an overload, an outage, or a fault, the relay will, depending upon its protection settings, output an alarm to the control system or trip the breakers to isolate the faulty equipment.
(5) The control system will initiate a restoration procedure to continue operating under the new operating conditions.

16.2 IEEE Equipment and Device Designation

In order to simplify the drawings and document descriptions, in 1928, someone came up with an idea of numbering the electrical relays, devices, and equipment as previously noted. The idea caught on and became popular throughout the world. Anywhere in the world, the drawings seem to be standardized on the IEEE numbering procedure [1]. Drawings are filled with these designations, and the protection logic became easy to understand.

The designations are regularly updated to keep up with new technology and are recorded in ANSI/IEEE Standard 37.2. It seems the numbers were not assigned in any particular sequence. The following list is not a complete list, but it does include the most popular devices. While the numerical symbols are standardized, device drawing symbols are not.

IEEE symbols for relays and transformers differ significantly to those of IEC. IEC has created their own figurative symbols to compete with the IEEE numbers, like $\gg$I for extremely overcurrent, or $<$V, for undervoltage, etc. The IEC symbols seem intuitive on paper but are difficult to communicate in discussion. Engineers like to talk in terms of the IEEE numbers, for instance 25, 51, 87, 21, 50BF, etc., and they instantly understand each other. As a young engineer you will too. One can be creative and add additional designations to them to be more precise of the equipment being protected. For instance:

52/B1, 52/B2 → circuit breaker B1, etc.

52a/B1, 52b/B1 → "a" contact of B1 circuit breaker, etc.

87T3, 87G1, 87U → differential protection for transformer 3, generator 1, and overall G + T

2 – Timer	56 – Field application relay
21 – Distance relay	57 – Short-circuiting or grounding device
24 – Volts per Hertz relay	59 – Overvoltage relay
25 – Synchronizing relay	60 – Voltage or current balance relay
25CH – Synchro-check device	64N – Neutral voltage relay
27 – Undervoltage relay	65 – Governor
30 – Annunciator relay	66 – Number of motor starts
32 – Directional or reverse power relay	67 – AC directional overcurrent relay
37 – Undercurrent/underpower	67G – Directional ground overcurrent relay
40 – Field (over/under excitation) relay	68 – Blocking or "out-of-step" relay
40 – Field circuit breaker	71 – Liquid level switch
46 – Current unbalance	72 – DC circuit breaker
48 – Incomplete sequence relay	76 – DC overcurrent relay
47 – Phase reversal	78 – Out of step protection
49 – Machine or transformer, thermal relay	79 – AC reclosing relay
50 – Instantaneous overcurrent (O/C) relay	80 – Flow switch
50V – Voltage restraint overcurrent	81 – Over frequency relay
50G – Instantaneous ground fault relay	85 – Communications, pilot-wire relay
50N – Instantaneous neutral fault relay	86 – Lockout trip relay
50BF – Breaker fail relay	87 – Differential protective relay
51 – inverse time overcurrent relay	89 – Line switch
51G – Inverse time O/C ground relay	91 – Voltage directional relay
51N – Inverse time O/C neutral relay	92 – Voltage and power directional relay
52 – AC circuit breaker	94 – Self-reset trip free relay
55 – Power factor relay	

16.3 CTs and PTs

16.3.1 Introduction

A CT (Figure 16.1) takes the primary current and reduces it typically to 1 or 5 A on the secondary winding, to be used with a watt-hour meter, ammeter, or protection relay. There are two types of CTs; wound type and window type. In a window type, the main conductor serves as the primary winding. A CT 400 : 5 A expresses the ratio of a CT. It means that when 400 A flows through the primary, 5 A will flow through the secondary winding.

The ratio of an instrument transformer is the relationship of its primary rating to its secondary rating. For example, a PT having a rating of 480 : 120 V will have a ratio of 4 : 1, and a CT having a rating of 400 : 5 A will have a ratio of 80 : 1.

The insulation class indicates the magnitude of voltage which an instrument transformer can safely withstand between its primary and secondary winding and between its primary or secondary winding and ground (core, case, or tank) without insulation breakdown. Industry standards have established insulation classes ranging from 600 V up through to the EHV. Industry recommendations are that the insulation class of a PT should be at least equal to the maximum line-to-line voltage existing on the system at the point of connection. For example, the insulation class of a PT used on a 7200Y/12 470 V system should be at least 15 kV even though the PT has a primary rating of 7200 V and is connected phase-to-ground. During a fault condition, the PT could be subjected to line-to-line voltage.

16.3.2 Polarity

Polarity markings on CTs and PTs and diagrams determine the correct coil wounding orientation. The polarities are extremely important for the proper functioning of the protection schemes. In a CT, the current flowing in the secondary winding is in a direction opposite to the current flowing in the primary winding, that is, the currents are 180° out of phase (see Figure 16.2). At any instant, when the current is flowing into one of the primary terminals, it will be flowing out of the corresponding secondary terminal.

All the instrument transformers, whether current or potential will have polarity marks associated with at least one primary terminal and one secondary

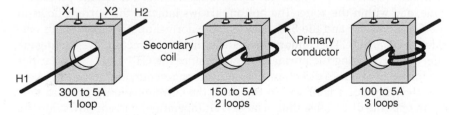

Figure 16.1 Window type CTs, ratios.

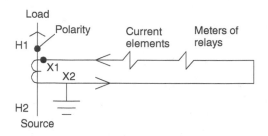

Figure 16.2 Polarity markings and current flows.

terminal on each phase. These markings usually appear as white dots or letter and number combinations. When number and letter combinations are used, IEEE refers to H1 as the primary terminal marking and to X1 for the secondary polarity mark.

The CT polarities must be clearly marked on the protection diagrams to define the flows of currents to suit the desired operating logic. Window CTs placed on the primary cables, with polarities pointing in an incorrect direction, will not activate the differential protection as required. Even worse case would be if one of six CTs is placed upside down. This author has seen this situation many times in the field, which has caused faulty commissioning situations. If all three phase CTs on one side are incorrectly oriented, the correction can be done by reversing the CT secondary leads, from X1–X2 to X2–X1. If one of the three CTs in a set is placed upside down, the primary connection must be disconnected and the offending CT reversed.

Let us see the significance of the CT polarity on the operation of a differential scheme. What happens when a differential scheme includes a service auxiliary transformer (Figure 16.3) connected and taking power out from within the transformer differential zone?

The transformer protective zone is defined by the CTs on both sides of the transformer. The issue here is that the CT1 sees up to 16% less current than CT2 (after the transformer ratio conversion) due to the power outflow to the station service, if the system is fully loaded. That is a mismatch which may cause the 87T to initiate a trip even during the normal operation. Mismatch = 7.5/47 = 16%, assuming generator MVA = 40/0.85 = 47 MVA, if CT3 was not used.

The upper arrows next to the CTs in Figure 16.3 show the directions of current flows in the CT primaries for three cases: normal, outside the protective zone, and within the zone. The bottom arrows indicate the current flows in the CT secondary and the 87T relay. For the normal operation and fault outside the zone, the CT secondary flows are in the same direction. Neglecting the effect of CT3 during the normal operation, since the CT1 and CT2 polarities are turned in opposite directions the currents cancel out, and there is no flow into the relay. For a fault within the zone, the currents add up because they are in opposite directions, thus, a large fault (operating) current flows into the relay.

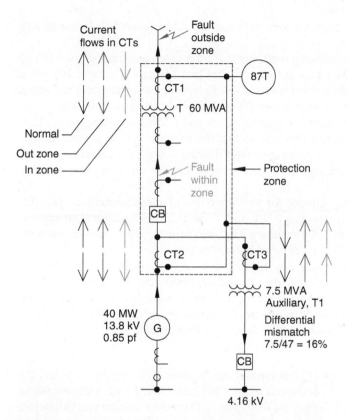

Figure 16.3 Transformer differential protection.

A differential relay is typically set to trip if there is a mismatch between two sides of 20–30%. Therefore, the mismatch 16% is insufficient to cause a trip during normal operation. CT3 connection may not be necessary for this arrangement. However, it is close to comfort. The sensitivity of the scheme is reduced. Our practice is to add the CT3 connection if the mismatch is >5%. The CT3 must be connected with the polarity positions in such a way to insure it subtracts its current from the current flow from CT2 to make the readings of CT1 to be equal to that of CT2 during the normal operation. CT3, if correctly wired, eliminates the mismatch.

16.3.3 Metering Accuracy Class

Different accuracy classes are applied to the instrument transformers; one for metering and the other for relaying as explained in the following.

16.3.3.1 Metering Accuracy
In a given instrument transformer, the metering error is a combination of the ratio and phase angle errors. The permissible error in a PT, for a given accuracy

class, must remain constant over the range of voltage from 10% below to 10% above rated voltage.

The permissible error in a metering CT, for a given accuracy class, has one value at 100% rated current and allows twice that amount of error at 10% rated current. Typically, 0.3% error is acceptable for revenue kilowatt-hour metering, while 0.6–1% error is assigned for CTs used for indicating instruments.

Metering CTs are built for high accuracy and a low saturation point. They are deliberately designed that way so that the measuring instruments are not damaged by high currents during a fault. During a fault, the measuring transformers get saturated, and the output stays within the range of the measuring instruments.

Metering accuracy classes for various types of measurement as per IEC 61869-1 Standard are 0.1, 0.2, 0.5, 1, and 3. The class designation is an approximate measure of the CT's accuracy. The ratio (primary to secondary current) error of a Class 1 CT is 1% at rated current or less.

Metering CTs are classified by IEEE into 0.1, 0.3, 0.5, and 1 categories. The values indicate the percentage error at the rated primary current. Thus, a 100 : 5 A current transformer with 0.3 accuracy will have a maximum error of 0.3 when a current of 100 A passes through the primary. A typical metering CT would be designated as 0.3B0.5, where B0.5 designates burden of 0.5 Ω.

16.3.3.2 Burden

The secondary load of a CT is termed the "burden." The burden includes the impedances of the CT, leads, and relays. This means a CT with a burden rating of B0.2 can tolerate an impedance of up to 0.2 Ω on the secondary circuit before its accuracy falls outside of its classification.

The leads resistance may be a substantial part of the burden if the CTs are located in remote substations. This problem can be reduced by using thicker cables and CTs with lower secondary currents (1 A), both of which will produce less voltage drop between the CTs and its operating devices. New digital numerical relays have significantly lower burden impedances compared to the old electromagnetic relays (see Table 16.1).

Table 16.1 Standard PT burdens.

Application	Designation	Impedance (Ω)	VA	pf
Measurements and relaying	W	1152	12.5	0.1
	X	576	25	0.7
	M	411	35	0.2
	Y	192	75	0.85
	Z	72	200	0.85

Source: ANSI.

16.3.3.3 0.15 Accuracy Instrument Transformers

New accuracy classes have been developed by IEEE C57.13.6 to accommodate the shift toward electronic relays and meters from the traditional induction electromechanical devices. Consequently, manufacturers have begun to improve the accuracy of instrument transformers to take advantage of the lower impedance (burden) of modern relays and meters.

The new high accuracy standard for these instrument transformers also have new burden ratings:

- E 0.4, (1.0 VA at 5 A, unity pf),
- E 0.2, (5.0 VA at 5 A, unity pf), and
- Low current test point of 5% vs. the traditional 10% rated current, are now required.
- 0.15% accuracy instrument transformers. CTs must maintain 0.15% accuracy from 5% rated current through rating factor (RF) at rated burden. No accuracy is guaranteed at levels below 5%.
- Voltage transformers have 0.15% accuracy from 90% to 110% of rated voltage.

16.3.4 CT Relaying Accuracy Class

It applies to CTs that operate relays for control and system protection. The CT must be able to withstand the high primary currents and transform them to lower values suitable for application of the relay with a reasonable accuracy. A typical relay accuracy classification, as noted below, may be C100 or T100. The letter "C" stands for calculated accuracy and the "T" stands for tested accuracy. Window and bar type CTs with fully distributed secondary winding on a low leakage flux core lend themselves to calculated values. Wound-type CTs that do not have fully distributed windings must be tested because the leakage reactance is not predictable. The number represents the secondary voltage that can be developed at the secondary terminals without saturation at 20 times the nominal current and nominal allowed burden.

The meaning of the relay classification 10C200 would be 10% accuracy inferred at 20× nominal primary current × secondary impedance. Thus, this CT would have an error of no larger than 10% at 20 times normal secondary current with a secondary burden of $2.0\,\Omega$ ($20 \times 5\,A \times 2\,\Omega = 200\,V$). The same CT with an actual relay burden of $1\,\Omega$ would operate at 100 V, which is in the region well below the saturation knee. Engineers often make a mistake by selecting the primary current of the relaying CTs based on the load current instead of the fault current. This set of CTs would quickly saturate and will not send a proper signal to the relay to coordinate with the other relays on the protective equipment.

Often, one can find the CT ratios and burdens incorrectly selected on switchgear or metering devices and relaying devices connected to the same set of CTs. Let us give an example:

Load current I_n: 150 A, 575 V motor control center (MCC) fault level I_f: 50 kA, CT ratio selected: 200 : 5 A, 10C50.

The CT exposure to maximum fault current: 50 000/200 = 250 >> 20 time the ratio. Depending on the actual burden, these CTs would likely saturate during a fault at about 4 kA and would not coordinate with the other relays. If the CTs are used for instantaneous trips, say set at $12 \times I_n = 1800$ A, the CTs may correctly act if the burden voltage rise is held to <50 V.

Here is an example of a relaying CT defined according to the IEC standards IEC 60044-1; 200 to 5 A ratio, 15 VA burden, Class 5P10. It means the maximum permissible limit of error (accuracy limit) is 5% at 10 times rated current (accuracy limit factor). CT will give allowable ratio error of 5% if fault current reaches 10 times the nominal value (I_n) and maximum burden. After that CT will start to saturate. P means the protection class CT. The assumed impedance for a 15 VA burden and 5 A secondary CT is calculated as shown in the following. Higher impedance would require higher VA burden.

$$VA = 5^2 \times Z; Z = \frac{VA}{5^2} = \frac{15\ VA}{25} = 0.6\ \Omega\ \text{protective circuit impedance.}$$

16.3.4.1 Continuous Thermal Rating Factor (RF)

This RF applies to CTs and is the number representing the amount by which the primary load current may be increased over the CT's nameplate rating without exceeding the CT's allowable temperature rise. It is a designation of the CT's overload capability. In the CT manufacturer's literature, a typical statement would be 2.0 at 30 °C ambient with RF 1.5 at 55 °C ambient. These statements mean that in a 30 °C ambient, the CT will safely carry on a continuous basis two times the nameplate rating and at 55 °C ambient 1.5 times the nameplate rating. Note that a fault current will exceed the nominal CT current many times over, before the equipment is tripped.

16.3.4.2 Multiratio CTs

CTs sometimes can be ordered with multiple ratio taps. For instance, CT in HV bushings can be built as 400/800/1200-5 A, C class: 10C400. This CT can operate at any of those taps. The burden classification is defined on the highest tap. In this case, it is 1200 A, 400 V, and prorated to the lower ratios.

16.3.4.3 Connections

Power metering with the new digital meters has become rather simple for connecting the instruments. It generally comprises 3 CTs + 3 PTs as the input and digital communication ports for remote indications. The meter with its software algorithm develops all the indications, including A,V, W, Wh, pf, Demand, etc., all in digital form. Nothing else is needed. This author likes to see an additional analog voltmeter on each switchgear and MCC incoming feeder to quickly tell the status of the operating assembly.

Table 16.2 Standard CT burdens.

Application	Designation	Impedance (Ω)	VA	pf	Volts on CT secondary (winding)
Measurements	B0.1	0.1	2.5	0.9	C10
	B0.2	0.2	5	0.9	C20
	B0.5	0.5	12.5	0.2	C50
	B0.9	0.9	22.5	0.9	
Relaying	B1	1	25	0.5	C100
	B2	2	50	0.5	C200
	B4	4	100	0.5	C400
	B8	8	200	0.5	C800

Source: ANSI relaying class.

PTs are generally connected on three phase PTs in a Y connection, phase to ground, or two PTs in V connection, phase to phase (see Table 16.2).

$$C100 = 20 \times 5\,A \times B = 20 \times 5 \times 1 = 100\,V, \text{ at 20 times nominal current.}$$

where, C100 is CT relaying class, B is burden in Ω.

Therefore, this CT can be loaded 20×5 A against 1 Ω to meet its class. If the burden is higher, the CT will fail to meet its accuracy class at that current.

16.3.4.4 Conclusion
(1) Select metering CTs for its desired accuracy class with ratios to suit the metering range required.
(2) *ANSI*: Select relaying CTs for its desired accuracy class with ratios to be within 1/20 of the expected fault current to avoid saturation. Minimize the burden to extend the operating range before saturation.
(3) *IEC*: Same as (2), but consider that the accuracy class before saturation for IEC designations is 1/10 of the fault current.

16.4 Relay Protection

16.4.1 Multifunction Relays (MFR)

In this book, we will present relay protection on the basis of protecting specific pieces of electrical equipment, such as generators, transformers, motors, switchgear, MCCs, etc. It must be noted that each MFR will include the protective elements specific for the type of equipment being protected. For example, all the equipment will require overcurrent protection, but only generators would include reverse power protection, etc.

The old electromechanical protective relays did not offer much in a way of protection. Each specific function required a set of separate relays. For instance, the overcurrent protection was simple. One would need a separate relay on each phase and required to set the tap and select the time dial for each overcurrent protective curve. That was all that was needed to be done. With the newer MFRs, things are considerably more complex as there are more functions to be set and configured to make a better fit to the equipment protected. To be able to set an MFR, one will need intuitive instruction manuals and software to guide you through the settings and configuration process. Having done it once, one will become familiar with the process of the relay configuration and set point management for all the relays of the same family. Each supplier has a different approach to the method of settings, but having done it once, it becomes a bit simpler to understand and implement it.

On projects throughout the world, this author has had an opportunity to use and set the new generation of relays by all the major suppliers, including GE Multilin, Alstom, Siemens, ABB, Schweitzer, Basler, and Beckwith. The new relays have changed the world of protection significantly for better. These relays offer multitude of settings previously unavailable for the operating logic. Unfortunately, having thousands of settings to be set and configured on each relay, one encounters problems and realizes that not everything is perfect. Some suppliers do not provide appropriate manuals. In some cases, one has to be on the phone with suppliers for hours getting help to understand the setting procedures, which should be explained in the manuals.

In other instances, software/firmware problems were uncovered, for which one has to wait on the next relay official update to be able to make a correct setting. In one case, the settings of the relays partly defaulted back to the factory settings after the relay was disconnected from DC supply. Imagine if you set everything on 5 A CT secondary, and then by some freak accident, DC supply is lost and your relay defaults back to 1 A. That is a serious issue. The worst part is that the relay does not even tell you about this change. It takes a while to detect this situation. By that time you find out about it from your angry client, you may be 3000 mi away on another project and the client is in panic as he cannot restart the generator and does not know why? Naturally, the relay trips all the breakers as soon as the plant is energized.

Finally, when you figure out that there was a relay software problem, the supplier tells you that they were aware of it and were working to resolve the issue for the next revision.

The MFRs must work properly. Possible damage from incorrect settings is enormous.

The settings and operating logic information must be well sequenced in the instruction manuals and written in a way that is intuitive and easy to understand.

The manuals and software must be broken down into proper protection function segments with setting steps, appropriate defaults, and supporting engineering notes to help you design the protection system. Of all the relays I have used, GE Multilin [2] relays are the easiest to read, understand, set, and manage. So, to simplify the presentation in this book, we will use the GE Multilin explanations as a guide. This choice will make it easier for the student to focus upon the technical application data while moving from one protective relaying application to the next. For example, the user interface of the modern digital relays, within a single family of relays, all follow the same sequence of steps. Once the students learn the fundamentals of protective relaying, they will be better equipped to learn their operation, setting up, and use of different brands of protective relays and decide for themselves which relay brand, software, manuals, and technical assistance best suits their needs.

16.4.2 Terminology

Pick up	Setting value (A, V, MW, %, etc.) at which the relay is activated and waiting for the timing element to initiate the trip action.
Reset	Relay activation, manual or automatic, by which the relay is returned to its original state.
Tap setting	Value of the setting, usually current I_n, is considered as 100% on the T–C charts, or 1 Tap. Value of 2.50 I_n is read as 2.5 Tap. Charts are read in Taps.
Lockout	It refers to the trip relay (86), which must be reset by human action in order to reset the trip relay.
Time–current	The relay operating curve as shown in chart to the right.
Time dial	Selection of an operating curve on the time–current family of curves by selecting 1–10 or any fraction in between (Figure 16.4).
Trip supervisory	The auxiliary relay or system which supervises the trip readiness of the breaker trip circuit.
Interlock	This is a contact from the first circuit inserted into the second circuit to condition the operation of the second circuit.
REF protection	Restricted earth fault protection (50/51REF) is provided in electrical power transformer for sensing internal earth fault of the transformer. This specific CT connection between the phase and neutral CT discriminate the internal from external faults.
Permissive	One of many interlocks needed to be "valid or closed" in order to allow the operating sequence to proceed.
Setting	The values of V, A, MW, %, m, seconds, etc., which form the relay activation criteria.
Back up	Secondary group of protection typically set with additional time delay behind the primary protection.
Restraint coil	The principal coil in the differential relay which measures two separate values and it is activated when the difference meets the set point.
Transfer trip	Activation of an additional breaker to trip over a communication line.
Transfer blocking	Some breaker trips are blocked when a certain action has occurred.

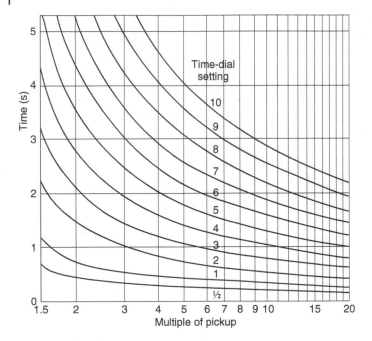

Figure 16.4 Time dial settings.

16.5 Major Equipment Protection

Functional protection diagrams for the major electrical equipment such as generators, motors, transformers, busses, lines, feeders, etc. show the functions and operations available to the programmers and the operators. These supervising features or protective elements are similar to all the relays within the same family of the same supplier. The functional diagrams are available in the suppliers' manuals. In addition, the manuals include typical equipment programmable tripping and control templates used for equipment monitoring and control.

16.5.1 Transformer and Generators

Refer to the IEEE equipment and device designation noted earlier in this chapter. Select the phase CTs for the fault conditions to maintain the relay accuracy and metering CTs accommodate for the load current.

A functional protection diagram is presented in Figure 16.5. The key element in this diagram is the differential element 87T, connected to two or more sets of CTs, depending on the number of transformer windings. In the days of electromechanical relays, transformer differential protection used to be an

Figure 16.5 Transformer protection functional diagram. MFM, multifunction metering.

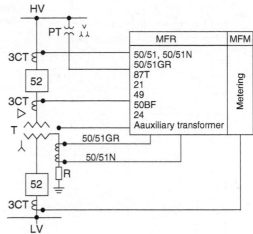

impossible task of matching CT ratios on different voltage sides of transformers. Another issue was that the CTs had to be wired to star or delta depending on the protected winding configuration.

CTs were wired to the star on the delta winding and delta to the star transformer winding. Also, one had to pay attention to the CT polarity orientation.

Well, none of that applies any more when employing digital MFRs. The relay allows the programmer to define and enter the actual CT setup on hand, including the CT ratio, winding setup, and polarity. The relay algorithm takes it from there.

The transformer protection includes the protective elements which detect faulty situations on the transformer as well as on the grid. These may include the overloads or short circuits on the windings as well as on the faults on the external line. On a generator unit system where the transformer is directly connected to the generator, each piece of equipment includes its own differential protection. An additional differential relay 87U is added to cover the overall zone of both the generator and the transformer together as a back up to the individual protections. The overall differential protection is connected to separate CTs from those used for the other protective elements. Typically, transformer differential protection is provided for the transformers greater than 5 MVA.

The 50/51GR is the restraint ground fault protection designed for the Y winding of the transformer either on the HV or LV side of the transformer. The 50/51N is the back up to the 50/51GR.

The transformer protection must include the alarm and fault detection on the auxiliaries including the oil level, oil, Bucholtz, cooling water and winding temperature, carbonization of oil, cooling fan or pump failures, etc.

The generator protective scheme (Figure 16.6) includes two groups of protections A (primary) and B (back up). The A group includes the protective

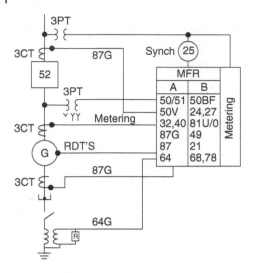

Figure 16.6 Generator protection functional diagram.

elements considered to be critical equipment faults, while B group includes the elements related to the grid dynamics that may affect the generator operation. The group A includes the protective elements for immediate actions, going directly to the lockout trip relay 86. The group B includes the protections that may include some time delays. It operates via a self-reset trip relay 94. The group A and B protections are independent and usually fed from separate sets of CTs. For simplicity, this was not shown in the diagram. Breaker failure relay 50BF must also be included in group B for larger units.

16.5.2 Motors

There are several relay models available for motors, depending on the motor size and importance on the system. The functional diagram in Figure 16.7 shows the most common protective functions.

The key item in this diagram is the (49) thermal protective element, which replicates the thermal image and behavior of the motor based upon the CT inputs to the relay. This protection is like an overload relay, but instead of being based on the current only flowing through the motor, it also acquires a thermal "learned" image of the motor. A wide choice of thermal curves can be selected to suit the specific motor characteristics. The RTDs imbedded within the windings, usually two per phase, can be slow acting, and consequently they are usually used to bias the motor current based thermal model algorithm. The motor differential protection is connected to the CTs to include the leads on both sides of the generator or motor as shown.

Same as for the generators, larger motors may include bearing RTDs and other sensors for oil temperature detection and oil level. These sophisticated

Figure 16.7 Motor protection functional diagram. WNDGs, windings; BRNGs, bearings.

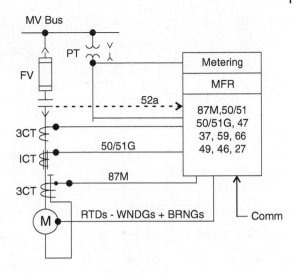

multifunction digital relays are not used for all the motors, but rather for motors ≥100 kW (≥150 HP) and higher, each coming with selectable options for the protective functions.

Smaller motors are generally protected by the MCC starters with appropriate breakers and overloads.

The motor MFR may include other elements, some of which may be applicable for larger motors only. For instance, the motor differential element 87M is used for motors 1500 kW and higher.

16.5.3 Transformers Current Reflections

Figure 16.8 shows the per unit phase and sequence component currents seen on the Δ and Y sides for various Y side faults, on a transformer. This diagram, is extremely important in diagnosing a faulty situation during commissioning.

Note that positive and negative sequence magnitudes are the same on both sides of the transformer but shifted in opposite directions.

The 0.577 factor for ground faults occurs because ground faults are single phase events; hence, the current is multiplied across the transformer by the transformer turns ratio, rather than by the line–line voltage ratio (the turns ratio on a delta/wye bank is $1/\sqrt{3}$ relative to the line to line voltage ratio).

16.5.4 Synchronizing and Synchrocheck Relays: What is the Difference?

The synchronizing relay (25) is the relay that allows to synchronize a generator unit to the grid and to the other units. Synchronizing relay is an active device

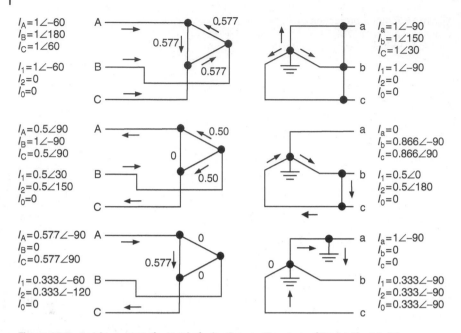

$I_A = 1\angle-60$
$I_B = 1\angle180$
$I_C = 1\angle60$

$I_1 = 1\angle-60$
$I_2 = 0$
$I_0 = 0$

$I_a = 1\angle-90$
$I_b = 1\angle150$
$I_c = 1\angle30$

$I_1 = 1\angle-90$
$I_2 = 0$
$I_0 = 0$

$I_A = 0.5\angle90$
$I_B = 1\angle-90$
$I_C = 0.5\angle90$

$I_1 = 0.5\angle30$
$I_2 = 0.5\angle150$
$I_0 = 0$

$I_a = 0$
$I_b = 0.866\angle-90$
$I_c = 0.866\angle90$

$I_1 = 0.5\angle0$
$I_2 = 0.5\angle180$
$I_0 = 0$

$I_A = 0.577\angle-90$
$I_B = 0$
$I_C = 0.577\angle90$

$I_1 = 0.333\angle-60$
$I_2 = 0.333\angle-120$
$I_0 = 0$

$I_a = 1\angle-90$
$I_b = 0$
$I_c = 0$

$I_1 = 0.333\angle-90$
$I_2 = 0.333\angle-90$
$I_0 = 0.333\angle-90$

Figure 16.8 Δ side currents for Y side faults. Source: Courtesy of Basler Electric [3].

that acts directly on the incoming generator exciter and governor to raise/lower its voltage and speed to match that of the grid and close the breaker. A single relay can be transferred as the synchronizer to several units, one at a time individually.

A synchrocheck relay (25CH), like Basler BE1-25 [3] on the photo (Figure 16.9), is a passive device which monitors the voltages on both sides of a circuit breaker and determines that proper phase angle and voltage magnitude exist to allow the breaker to be closed.

Figure 16.9 Synchrocheck relay. Source: From [3].

A (25CH) contact is typically placed in the breaker closing circuit as a permissive interlock. The relay is used for closing ring main lines, interconnecting busbars, tie lines, and connecting generators to the network.

16.6 Relay Coordination

The majority of the relay coordination work [4] is done with overcurrent relays by selecting and aligning the overcurrent time–current curves at different voltage levels. The other protective elements are configured as a choice between specific settings to suit the equipment ratings and characteristics.

Protection principles must be followed to limit the damage to the equipment and operators due the overloads and short circuits on the principle of inverse trip characteristics. The higher the fault current, the faster is the relay activation. This is achieved by trip settings that limit damage and activate selective equipment isolation to the smallest area possible with proper communication of the action taken. These actions are taken to enable the system to be restored as quickly as possible.

The following steps are recommended when conducting a coordination study:

- Obtain or create a complete one line diagram with equipment ratings and characteristics.
- Obtain transformer data, including kVA rating, inrush points, primary and secondary (and tertiary and quaternary, as applicable) connections, impedance, damage curves, primary, secondary, and tertiary, as applicable, voltages, type(s) of cooling.
- Obtain motor data, including full load currents, kW/HP rating, voltage, type of starting characteristic (i.e. full voltage non reversing (FVNR), full voltage reversing (FVR), reduced voltage and RV (reduced voltage) type–split winding, auto-transformer, solid-state, etc), type of overload relay (Class 10, 20, 30).

And, as applicable

- Obtain fuse characteristics, including fuse types/classes, minimum melt curve, total clearing time curve.
- Obtain circuit breaker characteristics for LV breakers, including time/current coordination curves.

Discrimination is the process of locating a fault, type, and magnitude. It is an important factor of relay selectivity. Discrimination is achieved by several different methods:

- Time
- Current magnitude
- Distance (V/I)

- Time + current magnitude
- Time + distance
- Time + direction of current
- Use of communication
- Use of other quantities: negative sequence, harmonics, etc.

The relay coordination work used to be done on a graph paper by drafting the curves one next to the other. This process was rather tedious as the settings had to be adjusted, by hand, to suit and readjusted over and over again. Fortunately, relay coordination has become rather easy by using appropriate coordination software. Now, one can select any inverse curve from the software catalog and adjust it to fit the object to be protected and the adjacent upstream/downstream protective curves. If the curve does not fit (i.e. it crosses another protection element's curve), one can call out other curves, with different time dial settings or curve characteristics, from the family of curves available in the software, until you find a curve that fits into the coordination scheme. Once all the curves are safely placed on the chart, engineer can read the final settings and write them into the book as the proposed group of protection and control settings.

16.7 Protection Function Elements

16.7.1 Overcurrent (50/51, 50/51N)

This is the principal protection in most of the relays. Two elements form a trip action: pickup + time delay. The time delay is initiated by the onset of pickup. The relay will activate a trip when the current magnitude reaches its pickup point and after the corresponding time delay, based on the time–current $(T-C)$, characteristic selected in the coordination study. The $T-C$ characteristics for the 51 element vary from definite time to extremely inverse. These characteristics are available in every MFR either as IEEE or IEC family of curves as shown in Figure 16.10.

Each $T-C$ curve is further presented as a set of curves ranging from 0.5 to 10 in time increments of 0.1 vertically on the $T-C$ chart. This makes it easier to selectively fit the 51 element to coordinate with the upstream/downstream relays.

The IEEE and IEC curves are virtually identical. Use them as you wish, but once you select IEEE, for instance, stay with IEEE curves for the rest of the coordination process. The 50 element is the instantaneous value position as a vertical line from the selected 51 curve down to the zero time line on the chart. The pickup setting is adjustable and marked as I_n current. This setting is typically 1.25–1.5 of the maximum normal operating current. As an example,

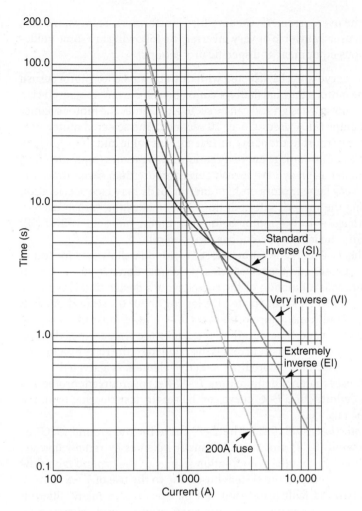

Figure 16.10 Coordination with fuse.

let us select pickup current as $I_p = 250$ A for $I_n = 200$ A. The value of 250 A becomes the tap $= 1$ in the relay coordination language.

The tap selection moves the top of the overcurrent T–C curve left–right on the coordination chart. The time dial setting from 0.5 to 10 moves the bottom of the T–C curve vertically up and down. All this work is done automatically by the relay coordination software by the cursor moving the tap and time setting on the coordination sheet to suit. Once a satisfactory curve shape and timing is positioned, the software records the final tap and time selection for that particular relay setting.

The shape of the overcurrent $T-C$ curve for transformers, feeders, motors, and generators is usually inverse or very inverse. The following are some guidelines for the various equipment and specific protections.

Motors: The $T-C$ curve must be selected with due allowance for motor inrush current and the motor's thermal damage curve. Inrush current magnitude is a function of the motor's design and can be six to eight times the motor's nameplate full load amps (FLA) lasting over 10 seconds. The selected motor protection $T-C$ curve must fit the above inrush current value and time duration and below the thermal damage curve. Premium class motors (high efficiency motor, see Chapter 14) may have inrush current more than seven times the motor's FLA. Very large motors on high inertia loads may face acceleration times exceeding the motor's safe stall time. This makes the Inrush current potentially damaging to the motor before the motor has achieved its full speed and before the inrush current has dropped off.

Transformers: The $T-C$ curve must be selected with due allowance for transformer inrush current and the transformer's thermal damage curve. Inrush current magnitude is a function of the transformer's design and is typically 10 times the transformer's nameplate FLA lasting about 0.1 second. As with motors, the selected transformer protection $T-C$ curve must fit *above* the inrush current value and time duration, but *below* the transformer thermal damage curve. Often the inrush current is overridden by the relay by allowing a time delay of 0.5 seconds before the protective element is activated.

Bus protection: It uses mainly a definite time $T-C$ (flat) overcurrent protection curves. A bus definite time $T-C$ curve can be set for fast clearing of all the breakers on the bus.

Ground fault protection (50/51REF): The $T-C$ curve for the ground (51G) or neutral overcurrent (51N) protection is usually taken as extremely inverse. This is due to the fact that the coordination strings of the ground fault flow are relatively short between the system buses due to the use of Y−Δ or Δ−Y transformers. Ground fault protection can be executed in many different ways, as shown in the following for transformers, motors, and generators.

The diagram (Figure 16.11) shows two types of ground fault protection for transformers. REF, often called REF protection for the transformer Y winding and the neutral ground fault 50/51N as a back up to the REF protection. All the relays (R) shown including phase protection are a single MFR with the desired protection elements. While the (50/51N) protection sees a ground fault through and downstream of the transformer, the REF protection sees earth faults only within the transformer Y winding. This is accomplished by having three CTs connected to the phases and one CT to the transformer neutral in such a way that creates a restricted zone of protection. The CT inputs serving these protective elements are wired directly into the transformer MFR to provide the desired protection.

Figure 16.11 Transformer ground restricted protection, 50/51GR.

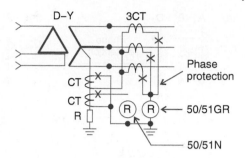

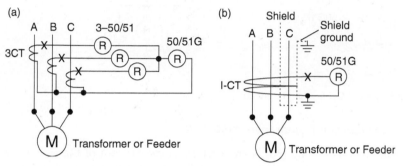

Figure 16.12 (a, b) Motor ground fault protection.

The diagrams in Figure 16.12a,b are different approaches for protecting a motor, feeder, or transformer. In the method (a), the MFR uses phase readings to calculate the residual ground fault current, assumed to be flowing in the neutral. The problem with this method is that the phase CT ratios selected for the phase fault conditions are relatively high. Due to the high CT ratio, the settings tend to be at the lowest end of the relay protective range, say 0.1 A of the 0.1–10 A range.

The method shown in Figure 16.12b is more effective as the CT ratio fed to the MFR can be selected to suit the ground fault conditions. In particular, this is applicable to the resistance grounded systems. The CT is a single window type that allows passage of all three phases. If the cable shield is passed through the CT, it must be pulled back through the CT from the termination to cancel the shield fault current in the relay assessment.

The functional diagrams Figure 16.13a,b show the method of ground fault detection and protection for generators by MFRs.

Method (a) uses a single CT to read the current passing through the generator neutral resistor. This fault current is in the order of 5–10 A. Thus, the CT can be of low ratio to suit.

Method (b) uses a single phase distribution transformer to read the voltage rise of 0–120 V generated in case of a ground fault. The method of protection and setting selection is presented in Chapter 5.

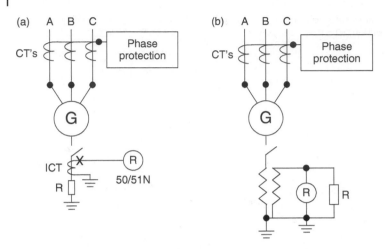

Figure 16.13 (a, b) Generator ground fault protection.

16.7.2 Overcurrent Instantaneous (50, 50N)

The instantaneous setting (50) is an addition to the bottom of the (51) T–C curve setting. Where the (51) element provides overload protection, the (50) element provides short circuit protection. It is usually set at 10–$12 \times I_n$.

When setting the (50/51) function for a transformer on a primary side of a step down transformer, the (50) instantaneous element must be set high as not to pass through the transformer and not interfere with the protection on the secondary side of the transformer. The protective logic of this protection for a generator step-up transformer may be different, though.

Here is an example of 50/51 settings for a transformer:

Transformer kVA rating: $1000\,\text{kVA}, 13.8\,\text{kV}$–$480\,\text{V}, Z = 5.75\%$.

$$\text{Transformer primary current: } I_n = \frac{1000\,\text{kVA}}{\sqrt{3} \times 13.8\,\text{kV}} = 42\,\text{A}$$

Set (51) element at: between 50 and 60 A (Tap)

$$\text{Transformer through fault current: } I_{tf} = \frac{1000\,\text{kVA}}{\sqrt{3} \times 13.8\,\text{kV} \times 0.0575} = 728\,\text{A}$$

Set (50) element at: 1500 A to be considerably higher than 728 A

The (50) setting will act to quickly remove transformer from service for faults within the transformer, but will not interfere with the protection on the LV side.

Overcurrent voltage restraint (51 V): This is a special function used for generator protection. The overcurrent protection (51 V) is additionally conditioned and accelerated by a voltage component. The voltage must fall below its setting to allow the overcurrent element to trip. The lower the voltage sensed by the generator protection relay, the faster is the relay conditioned to act.

Directional overcurrent (67, 67N): This function is used mainly in the line and feeder protection. The relay sensing is in one protection only; front or back. The direction to the protection is given by an input from the bus voltage transformers.

Differential (87), generator (87G), transformer (87T), motor (87M), line (87L): Differential protection [5] is one of the most widely used protection elements. It requires at least two sets of CTs. Due to the cost constraints, (87) is generally applied to generators >1.5 MW, transformers >5 MVA, and motors >1.5 MW.

The differential element is usually set at 20–30% mismatch within the protective zone covered by the CTs on both sides of the protected object. MFRs must include filtering of the second harmonic in the transformer inrush current. Transformer inrush current contains a large content of second harmonics which arise immediately following the breaker closing. The 87T relay filters it to override its impact. A careful investigation must be made of the CTs such that the CTs are properly rated to avoid possible saturation and malfunction. CT saturation may happen on one side only which would also lead to a malfunction.

A biased differential percentage setting chart is shown with two or three slopes, 20%, 25%, and 30% (see Figure 16.14). The higher the ratio of the mismatch current I_{diff} over the restraining current I_{stab}, the faster the relay operates. The 87 trip function is always directed to a lockout trip relay (86) to force the operator to manually reset the protective and trip relay after carefully inspecting the protected equipment.

Pilot differential (87LP): In this protection scheme, a communication circuit is used to compare the system conditions at the opposite ends of a transmission line. This method provides selective high speed fault clearing anywhere within the protection zone. The communication could be a wire circuit, microwave transmission, leased telephone channel, or nowadays fiber

Figure 16.14 Differential relay 3 slope setting.

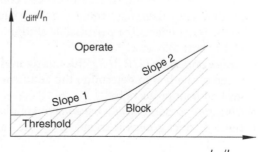

optic channel, all of which may be included as options with the protective relays. The communication channel can provide a trip blocking interlock, line trip, or sending a breaker transfer trip as a breaker failure signal.

Under voltage (27): It is used as a backup protection. An under voltage situation may signify other faulty conditions on the system, such as breaker trip, generator trip, etc. The trip function is usually configured as a self-reset condition (94).

Under/over frequency (81U/81O): Frequency excursions are mostly relevant to the islanded generations. A rise in frequency may result from a loss of load while a drop in frequency may mean a sudden loss of generation on the system or a sudden large load applied to the system. A drop below a certain threshold may be used to initiate quick start up of new generators or load shedding.

Breaker failure (50BF): This protection supervises faulty breaker operation. A separate parallel timed circuit is employed by a (50) element to monitor if current is still flowing through the breaker following a trip activation. After a set time interval, usually 0.5 seconds following the initial trip, tripping of the other adjacent breakers may be initiated.

Generator ground fault (64G): Refer to Figure 16.13b. It is used for generators as ground fault detection on the generator stator. It is measured as an overvoltage across the generator's high-resistance grounding resistor (see Chapter 5 for details).

Generator reverse power (32): Should the generator's mechanical driver fail, the generator will start motoring (driving the motor or turbine). This is a highly critical condition which must be cleared by the reverse power protection element as quickly as possible. A backup relay is also required.

Loss of field (40): This function is used for generators and synchronous motors to indicate a failure of the excitation circuit. The voltage element is set to drop out at a specific voltage, below which the system cannot recover during loss of excitation. This may be caused by a diode or thyristor failure or a loss of DC supply.

Negative sequence (41): This protection is used mainly on all the rotating electrical equipment to indicate an unbalanced condition due to a loss of phase or a ground fault. These fault conditions may cause overheating of the rotor due to the $(I_2)^2t$ effect. The permissible setting must be based on the generator $(I_2)^2t$ chart tolerance.

Distance protection (21P, 21G): This multizone protection is based on the ohmic value of the line to determine the zone coverage, separately for phase and ground fault protection.

Setting groups: Some MFRs have up to six distinctive setting groups. Each of these distinct groups is triggered by the various relay inputs (e.g. day and night, heavy or light loading, one or two lines input, or bus tie closed). For instance, closing the bus tie breaker on switchgear drastically changes the operating conditions on the switchgear for which a different set of protective operating logic would be warranted. Having two transformers on a single incoming feeder transformer causes larger magnetizing current input into the relay. In order to avoid the nuisance trips caused by the higher inrush currents, the bus tie breaker closure would be a trigger input to move the settings to a different group, which would allow higher inrush current and additional loading on the transformer feeder.

Communications: The MFRs provide advanced communications technologies for remote data and engineering access, making it easy and flexible to use, and integrate into new and existing infrastructures. Direct support for fiber optic Ethernet provides high-bandwidth communications, allowing for high-speed file transfers of relay fault and event record information. Ethernet and serial port options support a wide range of industry standard protocols and capability for a direct integration into DCS and system control and data acquisition (SCADA) systems on DNP3.0, IEC 61850, IEC 62439/PRP.

Metering and disturbance analysis: MFR offer power quality monitoring for fault and system disturbance analysis, by providing monitoring and recording of harmonics measurement for currents and voltages. The transient recorder gives the user ability to capture long disturbance records critical for the applications, both in digital points and 16 analog values, user assigned.

Data logger with fault reports provides the recording of analog values selected from the analog values calculated by the relay over a specified time interval for prefault and fault conditions.

Event recorder chronologically lists all triggered elements with an accurate time stamp over a long period of time.

Flexlogic: FlexLogic is the relay programming logic used by GE-Multilin that provides the ability to create customized protection and control/operation schemes, thus minimizing the need and associated costs of auxiliary components and wiring. Other relay manufacturers provide similar programmable software. It operates on the basis of AND and OR gates to join and combine miscellaneous, inputs, outputs, analogs, interlocking, and virtual logic outputs by drawdown method for a specific trip/close or other action. Once programmed, the relays display the logic in a relay logic block diagram for review. The drawdown is built up automatically while the relay is being programmed.

Figure 16.15 shows a case of flex logic prepared for managing a MVAR flow on the 69 kV, 250 km line from Mayo to Dawson City in Yukon, by switching ON/OFF reactors at the Mayo end.

Open Breaker MR1 Logic

```
Virt Out10 --------|
                   |-- AND(2)---|
FxE 4 PKP ---------|            |
                                |--- OR(2)---|--- AND(2)--- === Virt Op 11 ------>> Cont Op 7 IOn {Open MR-1}
                  AND(3)---|    |
Virt Out 9 ON ----|        |    |
FxE 3 PKP --------|        |
Virt Out10 ON(NOT)---|
Cont Ip H7c On ------------|
```

{2 units are on}
{Vars OUT > 2.0 MVAR}

{Vars OUT >1.0 MVAR}
{2 units not on}
(FxE Inhibit Contact)

Close Breaker MR1 Logic

```
Virt Out 10 ON------|
                    |--AND(2)--|
FxE 1 PKP ----------|          |
                               |--- OR(3)----|--AND(2)--------->>Cont Op 8 IOn {Close MR-1}
                  AND(3)---|   |
Virt Out 9 ON ----|        |   |
FxE 2 PKP --------|        |
Virt Out10 ON (NOT)---|
Phase UV2 OP ------------|
Cont 1p H7c On------------|
```

{2 units are on}
{Vars IN > –1.0 MVAR}

(1 Unit On)
{Vars IN < –0.5 MVAR}
(2 units not On)
(Close on Loss of V for 3 sec.)
(FxE Inhibit Contact)

Figure 16.15 Flex logic for controlling MVAR flow on line.

16.8   Time Grading

The idea here is to grade the operating times of the relays in such a way that the relay closest to the fault operates first. Time-graded protection is best suited for Radial Distribution Systems. Radial distribution is those systems which start with a power source at the top and finish with a load at the bottom.

Time-graded protection (Figure 16.16) is implemented by using overcurrent relays with either definite time or inverse time characteristics. The operating time of definite time elements (relatively flat magnitude curve) do not depend on the magnitude of the fault current, while the operating time of inverse time relays is made shorter as the magnitude of the fault current increases. The inverse time protection is used in radial networks where variations of short-circuit power due to changes in network configurations are small or where the short circuit current magnitude at the beginning and of feeder differs considerably. In these cases, the use of inverse time characteristics can speed up the operating time of the protection scheme for higher fault current magnitudes closer to the source of power.

Time grading coordination with fuses is also easier to obtain by employing inverse time characteristics. While the definite time characteristic is relatively flat, the inverse characteristics vary by their steepness and are termed inverse, moderately inverse, very inverse, and extremely inverse.

The protective devices are placed first on the chart, shown in Figure 16.17a, curves 1 and 2 . The protected equipment's thermal capability must be laid out to the right of the first curve. All the other protection curves must be placed further to the right without crossing, not less than 0.4 seconds distant from each other, just ahead of their thermal characteristics for transformers and cables. The minimum time interval between the relay characteristics is calculated to include the following operating times: Relay activation + circuit breaker operation + relay retardation + safety margin ≥ 0.4 seconds.

Figure 16.16 Inverse overcurrent protection.

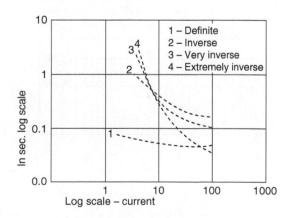

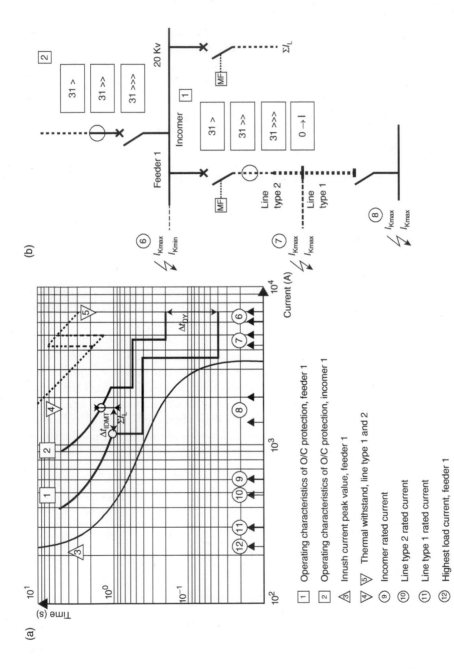

Figure 16.17 Time graded relay coordination. Source: Courtesy of ABB [6].

The numbers 6–8 (increasing from 8 to 6) at bottom of the graph indicate the fault levels (A) along the one line diagram shown in Figure 16.17b. That helps one select proper setting based on the fault magnitude. The numbers 9–12 indicate the equipment ratings, which must be safely protected.

16.9 Time–Current Grading

Figure 16.18 shows a situation when multiple generators can produce different fault levels depending on the number of units in service [6] at the moment of fault. For example, a fault on one generator's terminals will create different fault contribution from the generator at fault (1G) and contribution from healthy generators in service (1G, 2G, or 3G). This is like having a third dimension in the coordination process. A similar situation occurs during ground fault conditions when there is a grounding transformer on each bus side, each capable of contributing 200 A to a ground fault. Once the bus tie closes the ground fault level increases from 200 A on a single bus to 400 A on the tied bus.

The study of the time grading with multiple generator feeders is straight forward if the operating characteristic of the protection of the other generator feeders are combined in a single operating characteristic of a so-called equivalent generator feeder. This is obtained by multiplying the current values of the relay operating characteristic of a single generator by the number of generators in use at any time, up to 3G.

Inverse current characteristics are suited for this situation. From the diagram, it can be seen that when a fault occurs on generator feeder 4, the total fault current fed by the plant and the other feeders reaches the level indicated by 4. Thus, the operating time of the protection can be even shorter than 100 ms. The fault current fed by the equivalent generator is at least on the level indicated by 2. It can clearly be seen that in this way a reliable time grading is obtained between the generator feeders also in cases where the fault current fed by the network is particularly low or if one generator is out of operation.

The "group settings" feature included with the most of the MFRs allows one to make six different groups of settings for particular protected equipment. For instance, the case of four generators in Figure 16.18 can easily be resolved by providing four different group settings for the faults on the generators conditioned by the number of generators in service at the time of fault. The current count of the generators in service is imputed into the relays to select an appropriate operating setting group for the relay.

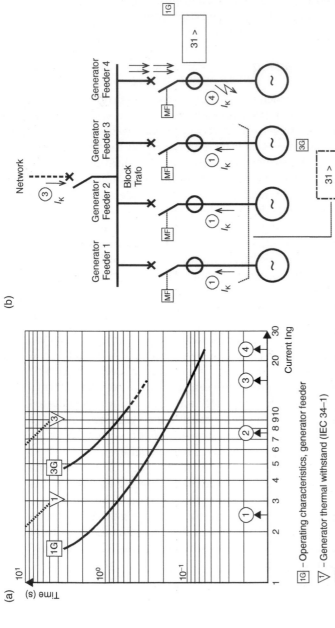

(a)

(b)

1G – Operating characteristics, generator feeder
▽ – Generator thermal withstand (IEC 34–1)
3G – Operating characteristics, equivalent generator (3 generators)
▽ – Equivalent generator thermal withstand (IEC 34–1) (3 generators)
① – Fault current supplied by one generator
② – Fault current supplied by equivalent generator (3 generators)
③ – Fault current supplied by external network
④ – Total fault current supplied by external network and equivalent generator

Figure 16.18 Time–current graded protection. Source: From [6].

16.10 Reclosing

The MFRs for lines and feeders include a reclosure element. A large majority (over 90%) of overhead line faults are ground faults, transient in nature, caused by lightning, conductor swinging, or tree branches falling on the line, all of which can be cleared by momentarily de-energizing the line. It is therefore feasible to improve service continuity and network stability by automatically reclosing the breaker following a relay trip activation. The reclosing system can be used with protective relay schemes that trip only one breaker pole for single line-to-ground faults, and all three poles for all multiphase faults. The reclosing system may provide one or two reclosing operations with selection of different high speed reclosing times for single-phase and multiphase faults. The second reclose operation is always time-delayed.

16.10.1 Breaker Duty Cycle and Interrupting Capability

The standard duty cycle for a circuit breaker is a close–open (CO) operation, a dead time of 15 seconds for slow or 0.3 seconds for fast reclosing; and another CO operation after 3 minutes. This is defined in IEEE Std. C37.04-1999 clause 5.8 [7].

In an area of high lightning incidence, most **transmission line** breakers will successfully reclose with a single try (shot). Multiple reclosing shots on transmission lines are not warranted.

At the **subtransmission level**, reclosers may use multishot activity. The first shot may be delayed if there are motors and generators on the system. Multishot reclosers are common on radial circuits as there are no stability issues. Often some form of circuit checking before time delayed reclosure is required to ensure the synchronism is present or one circuit is dead.

Multiple shot reclosers are common on the distribution circuits with significant tree presence, based on the experience that tree branches are causing the problems. Success of a reclosure on a transmission circuit is highly dependent on the speed of the line tripping. The faster the clearing, the less fault damage (due to arc ionization) the less of a shock to the system on reclosure. A line pilot protection is a fast and the most effective method of the line tripping and most commonly used to initiate successful reclosure on single shot activation.

Some dead time is required to deionize the fault arc prior to reclosing. The following formula based on experience is used for establishing the minimum dead time:

$$t = 10.5 + \frac{kV}{34.5} \text{ (cycles)}. \tag{16.1}$$

This works out for different voltages:

69 kV → 12.5 cycles,
138 kV → 14.5 cycles,
220 kV → 16.4 cycles,
345 kV → 20.5 cycles.

The reclosing relay must offer multiple options for reclosing, including synchro-check between the two sides to insure the voltages on the two sides are in the exact synchronism and within the acceptable window of angular difference to minimize the shock on the system upon closing the breaker. The most common reclosing options are live line dead bus (LLDB) and live bus dead line (LBDL).

16.11   Load Shedding and Automatic Quick Start of Generators

In a stable operating system, the generators normally operate at a frequency of 60 Hz in North America (50 Hz in other regions of the world), providing a perfect match between generation and mechanical drive input. The rotating masses of the generator rotors act as repositories of kinetic energy of the system. Any significant upset to the balance will cause a frequency change, one way or the other, depending which part of the balance is upset. When the electrical load exceeds the generation (mechanical input), the rotors slow down (frequency drops) by supplying the kinetic energy to the system in an effort to maintain the balance. In opposite direction, the rotors speed up to absorb the excess kinetic energy.

Generator unit governors act on the small changes in speed to adjust the mechanical input to return to the normal frequency. Sudden load changes, loss of a major generator, or loss of a transmission line can produce a severe load–generation imbalance resulting in a rapid frequency change. If the governors, hydroturbines and boilers cannot respond quickly enough, the system stability may be adversely affected. Rapid, selective dropping of loads can make recovery possible and avoid a system collapse and prolonged process of system power restoration.

In industrial plants with islanded generation, the most likely scenario is a loss of generation. In this situation, multistep frequency relays are assigned to monitor the frequency change and the rate of change. If the frequency drops, the relay starts removing some of the loads in steps until a balance is achieved and the fall of the frequency arrested. Frequency relays are installed at the main switchgear distribution centers. Procedure for setting up the load shedding scheme may include the following considerations:

- Number of loading steps.
- Load kilowatt to be shed/step.
- Frequency settings: The first step may be at 59.5 Hz.
- Time delay for each step.
- Load location of the frequency steps.

Power utilities are reluctant to drop industrial plants upon a loss of generation. They prefer to run the system with sufficient capacity and plenty of spinning reserve and add generation to quickly re-establish the balance. Based on their experience of the regional system operation, power utilities will operate the system with a mix of generation. Multistage frequency relays will initiate quick acting hydro units to soften the frequency drop and allow the slower starting units to be brought into service after.

One of the biggest utilities in Asia, with over 10 GW of installed capacity, operates its power system at 49.5–50.5 Hz. Active remedial actions are taken in case of major disturbances if the frequency reaches the limits at 47 and 52 Hz.

16.12   (86) Lockout and (94) Self-Reset Trip Relays

Trip relays (Figure 16.19) receive signals from the relay protective elements like 50, 32, etc. to initiate tripping of the electrical breakers. This action must be fast, expected within a ½ cycle (8 ms).

16.12.1   Lockout Trip Relays

Contacts of this relays remain in the operated position after the controlling quantity is removed. They can be reset either by hand or by an auxiliary electrical element.

Critical operating actions like differential, reverse power, loss of field protections are directed to these trip relays, often called master trip relays. They will be manually reset only after a field review and acceptance that there was no damage made to the protected equipment.

16.12.2   Self-Reset Trip Relays

The contacts of this trip relay (Figure 16.20) remain operated only while the controlling quantity is applied and return to their original condition when it is

Figure 16.19 Trip lockout (86) relay.

Series 24
Manual reset LOR

Figure 16.20 Self-reset trip relay.

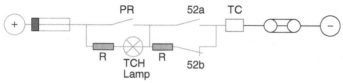

Figure 16.21 Trip supervisory relay.

removed. Typically, protective elements such as undervoltage and other transitional abnormal states are directed to the self-reset trip relays. Once the relay is reset, the equipment can be restarted and placed back into operation.

16.12.3 Trip Supervision Relay

Most of the MFRs have a trip supervision element (Figure 16.21), which monitors the state of the protective relay contact. What is the point of having a protection relay if you do not know if the other circuit elements are healthy: fuses, coils, contacts, etc. In addition, the switchgear trip coil must be supervised as shown in the previously provided sketch. A small current (insufficient to trip) is made to flow through a resistor R to monitor the continuity of the circuit. TCH (trip circuit healthy) lamp indicates the status of the main trip circuit in both, breaker open or closed conditions. This supervisory element can be arranged as a remote alarm.

References

1 ANSI. Standard Electric Power System Device Function Numbers acc. to IEEE C.37.2-1991.
2 GE Multilin: https://www.gegridsolutions.com/multilin/brochure/index.htm.

3 Basler Electric, (2018) *Transformer Protection Application Guide.* https://
www.basler.com/Product/BE1-11t-Transformer-Protection-System.

4 IEEE Southern Alberta Section PES/IAS Joint Chapter Power System Relay
Coordination.

5 Thompson, M.J. (2014). *Percentage Restrained Differential, Percentage of
What?* Schweitzer Eng'g Labs, Inc.

6 ABB (2011). Relay Coordination, Section 8.2. In: *Distribution Automation
Handbook,* 3–12. ABB.

7 IEEE Std. C37.04 Rating Structure for AC HV Circuit Breakers.

17

Plant Automation and Data Networking

CHAPTER MENU

17.1 Plant Control, 379
  17.1.1 Relay Logic, 380
  17.1.2 Programmable Logic Control (PLC), 381
  17.1.3 Distributed Control System (DCS), 385
  17.1.4 Interposing Relays, 385
  17.1.5 Input/output (I/O) Cards, 385
  17.1.6 Digital I/O, 386
  17.1.7 Analog I/O Channels, 387
17.2 Motor Controls Integration, 388
17.3 Human Machine Interface (HMI), 389
  17.3.1 Screen Elements, 390
  17.3.2 Colors, 391
  17.3.3 Critical Alarms, 391
17.4 PLC or DCS: What Is the Difference?, 391
17.5 Data Networking, 393
17.6 Means of Communication, 396
  17.6.1 Transfer Data, 396
  17.6.2 Physical Media, 397
  17.6.3 Logical Schemes and Arbitration, 399
  17.6.4 Open Industry Standards, 399
  17.6.5 Open Protocols, 400
17.7 Web-based HMI, 402
  17.7.1 Work Place, 402
  17.7.2 Creating Interface Screens, 403
17.8 SCADA Applications and Communication Protocols in Power Industry, 403
References, 406

17.1 Plant Control

Electrical engineers work closely with other engineers on the project including the control engineers. Their work and responsibility for the plant process and automation is highly related to the electrical engineering. On many projects,

Practical Power Plant Engineering: A Guide for Early Career Engineers, First Edition. Zark Bedalov.
© 2020 John Wiley & Sons, Inc. Published 2020 by John Wiley & Sons, Inc.

the plant automation falls under the electrical discipline. Clearly, as an electrical engineer, you will not be doing their work, but you will need to know what it takes to create a plant-wide control system and what plant interfaces are needed to implement it.

17.1.1 Relay Logic

During the last 30 years, the technology has advanced and left behind my favorite equipment: the hard-wired relay panels. They were replaced with something much better and more imaginative: programmable logic controllers (PLCs) and distributed control systems (DCSs).

The old relay logic generally consisted of two different parts located: (i) partly within each individual motor starter or valve circuit, and (ii) the rest of the logic was in a separate relay control cabinet as in Figure 17.1. The cabinet included an "overhead" circuit with hundreds of relays wired to issue start/stop commands to the plant drives, motors, and valves. Some of us enjoyed preparing wiring diagrams with relays and timers to generate the desired interlocking and operating logic. It was possible to interlock the conveyor system of several conveyors to operate the system by pressing a single push button on a relay panel. That was something. This would start the last conveyor in sequence followed by the next one, second to last, and so on, in reverse order of the ore flow until the first conveyor, which would then be loaded when all the conveyors were running. Or stop them in a proper order to insure that the conveyors belts were left empty. Similarly, for water flows and tank filling, by sequencing the valves and pumps.

Figure 17.1 Old relay logic panel.

It was not easy to prepare schematic diagrams. To wire the panel, it was even harder. To make a change in a field, it was a nightmare and a plant down time. How do you follow the wires in a bulky relay control panel? Not easy. One thing was certain; we were extremely busy. Each relay had a limited number of normally open (NO) and normally closed (NC) contacts and inevitably one would run out of them. So one had to chain the interposing relays to multiply the contacts. When commissioning, it was not easy to figure out which relays were energized with your actions. Well, you would poke a megger instrument into each one and place stickers on them if energized. Then someone invented a little light on the top of relay to indicate it was energized. That was a huge progress as the electrical engineers finally got something that was visible, and it is following their actions. The technological progress just commenced.

In comparison to the current automation with PLC and DCS it really was the dark ages. There were skeptics, who thought the PLC logic will become so complicated that it would smother itself.

17.1.2 Programmable Logic Control (PLC)

By current standards, relay logic was rather poor engineering, though we were working with the automation at that time and thought we were at the top of the class. Remember that Apollo flight to the Moon in 1969 was designed based on the relay logic. Just stop and think how many miniature relays were needed to generate some reasonable flight operating logic.

Then suddenly the world for electrical engineers changed dramatically. Programmable logic [1] was introduced. The only hardware needed was a computer and a number of I/Os, with input and output cards. The little relays had gone out forever? Not really. Actually, they came back in different ways to help the DCS. Later on that we started programming with more imagination the PLCs in a language resembling ladder logic diagrams. I could not ask for a better job. Relays were used in the PLCs also, but they were created out of nothing. They were virtual relays. The program generated as many relays and NO and NC contacts as one wished to have. It was just a matter of memory in the central processing unit (CPU). The ladder rungs represent the wiring between the different components which in the case of a PLC were living in the virtual world of a CPU. So if you understood how basic electrical circuits work, then you could understand ladder logic, except that it was much simpler.

There were some doubts about the PLC initially. Some plants insisted on having duplicate controls for each motor and valve: one in relay logic and the other in ladder logic. The thinking was that the virtual ladder relays and contacts would not last. Or that a bit of a voltage spike may erase the memory and make the conveyors run backwards or nowhere. That thinking didn't last long. The PLCs have proven themselves and are here to stay.

Figure 17.2 PLC with I/O cards by Rockwell.

Now you can program your logic and operating plant sequence, and then revise it off line if conditions change. Just as a temporary measure only, if needed. Furthermore, the best part is that you can test the circuit off line by forcing various operating conditions. Testing on PLC could be arranged to be more inclusive and conducted much faster than you could in the real world. There were less chances of damaging the equipment. All of that without a need to rewire anything. Wow? Is not that wonderful? With the relay panels, one had to continuously work to add more relays and more wiring as the plant operating sequence changed.

You wonder, what took them so long to come up with this wonderful innovation? Now, this really has become black box engineering. You wire inputs and outputs into the black box, one drive after the other in an uniform wiring fashion: say, four inputs and two outputs per motor, even though the logic for each motor may be totally different. Instead of calling relays CR1, CR2, ... , now you identify them as I/O numbers from the PLC I/O cards for each drive. These numbers are retained in the PLC memory, assigned, and wired to that specific motor. Therefore, each drive has its own specific I/O numbers, consisting of inputs or outputs and the terminals on an eight point I/O cards, as shown in Figure 17.2.

Nowadays, the systems are becoming ever smaller. The I/O cards can have far more I/O points. The I/Os can now be programmable to be either input or output. However, while the I/O terminals have become smaller, the outputs contacts have become less powerful to drive big motor contactors.

On a motor schematic, an input is an operating relay, for instance **I0805** representing an interlock input from another drive or an input from a stop button from its own drive. On the same schematic, an O output is shown as contacts driven by those virtual relays ( ), which are activated by the PLC logic to start the motor or become a permissive for another interlocked drive. The contacts are called digital input (DI) and digital output (DO) for the discrete entries and AI and AO for the analog inputs and analog outputs. On the ladder logic, you can rename them to suit the actual activating sensors and call them a Start or Stop. The processing sequence is shown in Figure 17.3.

Figure 17.3 PLC processing sequence.

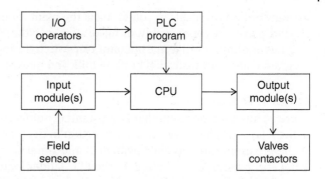

Once everything is assigned and wired, the rest is to program the black box, actually a big black box that includes logic for hundreds of motors, valves, and sensors. Bingo! Now even mechanical engineers can understand the electrical schematic diagrams: here is the motor, and right there is the black box into which you can write any logic you desire for the drive to follow. How simpler can you get?

In the ladder logic (Figure 17.4), two contacts in series is an AND condition, while two contacts programmed in parallel are an OR condition. The software program allows you to add "comparisons," "if" conditions, "skips," and even use math, routines, analog values, etc. Anyone can write ladder logic as there are

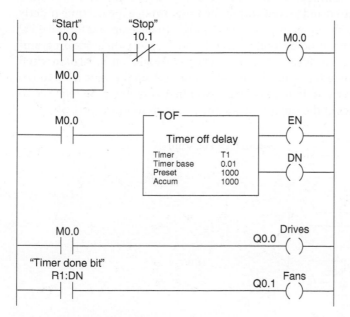

Figure 17.4 Part of PLC ladder program.

so many options available to you to make the plant interactions to automate a bottling plant or a car assembly, or make a flight to the Mars.

If the conditions change, the operator can switch the logic from program A to program B and now instead of bottling milk you may be bottling orange juice.

17.1.2.1 PLC Scan

There is, however, one item that is substantially different from the relay logic. In the relay logic, there was no time direction in the logic. The logic progresses as the drives go through their sequences and timers, valves and motors were being energized or de-energized. In the PLC ladder logic, the PLC (computer) scans the I/O status and the applicable logic instructions rung by rung from top to bottom and again and again within milliseconds. It applies the logic as read in sequence inclusive of the changes and the events taking place in the plant process. The program does not realize a status change until it returns to it in the next scan as shown in Figure 17.5.

While the scan passes the ladder rung, it freezes it in that position until it returns back to this rung in the next scan. So when the next scan arrives and sees an I/O status change (input no longer energized, output is now giving a permissive, or an analog input has reached the setpoint), the controller initiates a different action in the successive rungs (but not yet in those behind) as dictated by the program logic. Those rungs behind may be affected and will be updated when the scan reaches them in the next pass. Therefore, the software rungs must be written and placed within the program in a logical time orderly sequence one after the other as you want the logic to develop and for the CPU to scan it in that specific order. That is an additional dimension in the programming that the relay logic did not have. The relay panel logic was a real hard wired thing; it held an operating sequence, but it did not sweep the logic from top to bottom or any other way. In the PLC logic, you may sometimes be required to reposition the order of the rungs to make the program to work correctly.

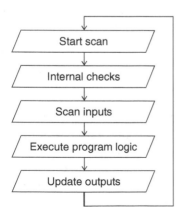

Figure 17.5 PLC scan.

17.1.3 Distributed Control System (DCS)

The technological progress did not stop with PLC. Evolution in the automation came up with something even more sophisticated in processing input/output (I/Os), DCS. DCS is a more powerful control system with much larger memory and hardened hardware than PLC. The main difference between the two systems is the handling of the analog values. While PLC can also utilize analog values for water levels, pressures, and temperatures to initiate control actions, the DCS handles analog values and control loops in a functional way which is a more transparent method to handle these variables.

The DCS uses function blocks for its programming. The function blocks are preprogramed routines for every possible control function. It is basically a matter of connecting the lines between the function blocks from/to their specific nodes to activate final control elements (motors, control valves, etc.). As electrical engineers, we like ladder logic with relays and normally open and closed contacts. That looks like a true electrical schematic. We read that much faster. One can force (simulate) a contact to check what happens as a consequence of your forced action. One can do it in DCS too, but in a less transparent way.

The two systems are evolving and getting closer in their performance. PLCs execute its actions in milliseconds, while DCS due to their larger memory takes one to two seconds to initiate an action and bring back a status feedback. For that reasons, PLCs will always be present for the specific machines in the plant, while the DCS will run the plant. The PLCs will be connected and be fed into the plant DCS. In some industries, milliseconds do matter.

17.1.4 Interposing Relays

It used to be that the PLC or DCS outputs were directly wired into the motor control circuits as permissives or as initiating contacts. The contacts had the ability of switching up to 10 A in a motor control circuit. However, since the present I/O cards have increased a number of inputs and outputs on the cards, their switching capability has dropped. Instead, the outputs are now driven to small interposing relays (IR), which have more hardened contacts to be wired directly into the motor control circuits. On a recent project, thousands of these relays were noted (Figure 17.6).

The relays came back, did they not? I found them extremely useful during the plant commissioning, because they were visible when energized.

17.1.5 Input/output (I/O) Cards

The applications for industrial process-control systems are diverse, ranging from simple-level control to complex electrical power grids. The intelligence of these automated systems lies in their measurements by the control units

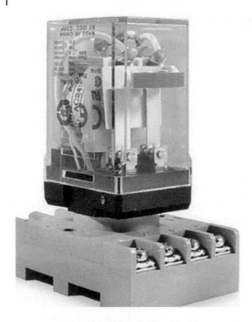

Figure 17.6 Interposing relay (IR).

within the PLCs and DCSs systems: comprising power supplies, CPUs, and a variety of analog-input, analog-output, digital-input, and digital-output modules (cards). The I/O system provides the physical connection between the operating floor equipment and the PLCs/DCSs. Looking at an I/O card one reveals terminal strips where the external field devices connect. There are many different kinds of I/O cards that serve to condition the type of input or output, so the CPU can use it for its logic. It is simply a matter of determining what inputs and outputs are needed, filling the rack with the appropriate cards and then addressing them correctly in the CPUs program.

Input devices may consist of digital or analog devices. A DI card handles discrete devices. An analog input card converts a voltage or current (e.g. a signal that can be anywhere from 0 to 20 mA) into a digitally equivalent number that CPU understands. Examples of analog devices are pressure transducers, flow meters, and thermocouples for temperature readings.

17.1.6 Digital I/O

DI inputs on the PLC cards are binary either on or off, representing field push-buttons, limit switches, relay contacts, proximity switches, photo sensors (on/off), pressure switches, and more. DIs cards are available in both DC as well as AC to accommodate the external variables.

DO outputs on PLC cards represent and control the field devices including control relays, motor starters, interposing relays, and solenoid valves.

Figure 17.7 Analog controls.

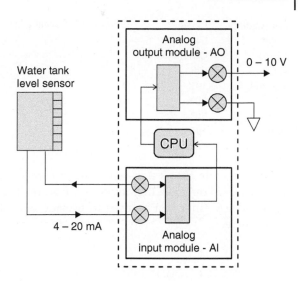

17.1.7 Analog I/O Channels

The standard ranges of analog variables are dominated by 4–20 mA, 0–5, and 0–10 V metering ranges. The most popular Analog I/O range is 4–20 mA [2]. In practice, most modules have multiple channels and configurable ranges. The resolution of input/output modules typically ranges from 12 to 16 bits. The accuracy can be 0.1% over an active range.

Let us look at the sketch (Figure 17.7) displaying the action of a little system using analog cards. A water level sensor measures the water level in a tank and transmits the information to the PLC central point, CPU.

The control unit consists of an analog input module AI that takes a reading form the tank sensor and presents it in the 4–20 mA range to the PLC CPU. A corresponding analog output module AO receives it at 0–10 mV and controls the required system variable or alarm. In other words, a level transmitter representing the water level is continually passing a standard 4–20 mA signal over a current loop in CPU and transmits it to an analog AO module. The output is converted and read at 0–10 V. At a certain point, depending on the 0–20 mA input, the AO activates an alarm or other action: WL → AI → CPU → AO → alarm/increase water feed.

17.1.7.1 Scaling

It is important to understand the scaling (Figure 17.8) of the 4–20 mA range. The total range is 16 mA for a 0–100% level or fill. Therefore, 4–20 mA = 0–100%. We can calculate that for every 1.6 mA segment within the 4–20 mA range it is 10% of the actual level. The key here is the 4 mA as the starting point and not 0.

mA reading	Level
0 mA	Card fault
1.6 mA	Calibration fault
4 mA	**0%**
5.16 mA	10%
7.2 mA	20%
12 mA	50%
20 mA	**100%**
>20 mA	Over range

Figure 17.8 Scaling list.

If the module shows 0, it represents a card fault. No other module, such as 0–20, 0–24 mA, has that important feature. The current readings outside the 4–20 mA range can be used to provide fault-diagnostic information, as shown here.

17.2 Motor Controls Integration

The plant motor, variable frequency drive (VFD), and other field devices are integrated and made functional within the plant control system. This integration takes the form of a looped DeviceNet, Control Net (FieldBus), ModBus, or Fast Ethernet 100 Mb/s network interconnecting the plant motor control center (MCC) starters, VFDs, and field devices to facilitate and enhance the plant communications and process interlocking logic. Fast Ethernet over F/O strands seem to be gaining ground here due to their speed of data transfer capability, large number of nodes, and encoding capability, which is discussed later in this chapter.

Chapter 1 discusses the typical motor schematic and shows that following inputs and outputs were networked directly by a DeviceNet cable as a system node:

Output 1	Device 42: Starter motor main coil
Output 2	Not used
Input 1	Ready/stop
Input 2	Start
Input 3	Motor running
Input 4	Fault/overload

The plant motors and other devices are controlled locally by start–stop local control stations (LCS) and remotely from the central control room. The Loc/Rem selector switch is generally located as a software item on a DCS screen, which may be showing a flow diagram of a part of the process; for instance a mixing plant. This selector switch transfers the mode of operation from the control room to the field (local). When in remote mode (control room and default position), the field devices are part of the overall plant operating sequence, conditioned by the field interlocks and process logic. When in local mode, the motors and field devices are operated locally on an individual basis for maintenance purposes only. In this situation, safety interlocks are included, but the process interlocks are not included. The stop button is operable in both modes: Rem and Loc.

Some plants insist on having Loc/Rem switches as part of the field push buttons LCS. In that case, the field operator must receive a permission from the control room operator to allow him to change the switch position to "Loc." He may be allowed to do a switchover on a standby pump, but not on a running pump. While switching the running pump may cause a plant interruption, it may be allowed if the tank level is high enough to allow for a temporary interruption.

For motors 25 kW and larger, motor starters may be provided with intelligent overloads, which will indicate the present motor loading as well as operate on an overload condition.

Starters for the drives operated on as-needed basis, such as roof fans, are operated locally (start–stop) only without going through PLCs.

In specialized situations, the supplier's packaged equipment is supplied with integral PLCs that are networked directly to the plant DCS over a plant wide Ethernet, ModBus, or similar link.

17.3 Human Machine Interface (HMI)

Human machine interface (HMI) system gives plant operators a way to interact with and manage the process in the plant. HMI platform is included with DCS. PLC can be installed without an HMI, but if the process/machine is regularly monitored and controlled, a HMI becomes necessary, and it is ordered as an extra.

The operator plant interaction HMI is developed through a graphical user interface (GUI), by allowing the plant status, measurements, plant control, and information exchange on the graphical screens. The graphic screens have to be developed using standard features and symbols.

Often, different sets of screens are developed for the electrical one-line diagrams and process, starting from the overall view, which allows navigation to other screen representing individual parts of the plant or process. The screens must be simple, readable, and intuitive. For the electrical systems, the first screen is a simple overall one-line diagram, leading off to the individual lower voltage diagrams for each section of the plant. Mechanical systems must include screens for the individual plant process departments: air, water, steam, chemicals, ore conveying, fire protection, and HVAC. Figure 17.9 shows a

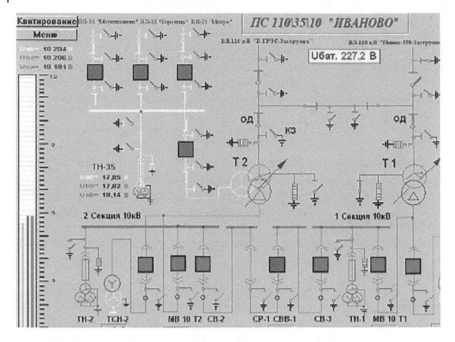

Figure 17.9 An HMI screen for an electrical one-line diagram.

graphic operating screen for an electrical single-line diagram and this one is from Roccia.

Selecting HMI software from different vendors typically starts with an analysis of product specifications, features, and previous usage. The key considerations can include the system architecture, performance requirements, speed of interaction, integration, and cost of procurement and operations. The vendor must be known and be in business for at least five years.

17.3.1 Screen Elements

In order to allow the operator to supervise and control the process, depending on the complexity of the process, the screens may include in distinctive colors the following operating elements:

- Equipment layouts and material flows based on P&ID and one-line diagrams with means of navigation between the plant subsections.
- Equipment status, expressed in different colors.
- Readiness to start.
- Control buttons to activate certain operating actions. Control actions are typically arranged as certain checkouts and handshaking interrogation

procedure. The emergency buttons are direct, generally allowed by holding two different keys at the same time.

- Flow, pressure, level, and temperature measurements.
- Pre-alarms and alarms.

The pre-alarms may include the display bars to indicate the current levels and the operating margins as well as features to be able to look at the historic data of the important plant-operating parameters. This is particularly important for the situations when the plant is forced to operate in some part outage conditions.

17.3.2 Colors

The selection of the colors on the screens for the various elements generally follow the operating company standard colors previously used in their operating plants, so the operators can navigate and operate in the familiar environment. Power plant utilities typically dictate the color selection for the screen elements.

17.3.3 Critical Alarms

A major plant failure will likely cause hundreds of associated failures, stoppages, and alarms. When a major failure occurs, an operator may be observing a screen that is unrelated to the failure. The alarm handling procedure must allow the operator to deal with the root cause of the failure. The supervisory logic must bring the critical alarm to an immediate attention of the operator. Selected critical alarms must pop up accompanied with a buzzer action onto the screen that the operator is handling during the critical plant failure.

There were situations where the operator was unaware of the major occurrences in the plant, because he was away from his desk or viewing different screens.

Typical critical alarms for a hydroelectric project may include

(1) Generator circuit breaker trip
(2) Intake gate trip
(3) Spillway level high
(4) Fire alarm
(5) Plant flooding.

17.4 PLC or DCS: What Is the Difference?

Briefly, PLC rules machine control, while DCS dominates process control. To manufacture plastic widgets, use PLC. For producing chemicals, DCS is your best friend.

Analog values are extremely important to batch and chemical plants; for example to add additives at certain temperatures, slow down the flow at certain liquid levels, open valves and direct activities or change flows when pressure rises or come down. That is the process area where DCS is used. On the other hand, PLCs are more efficient with the discrete values, open/closed/status. For a large plant, both technologies are required to make use of PLC fast response and DCS for the analog and plant organizing capability, as discussed later.

Today, the two technologies share the automation territories as the functional borders between them continue to overlap. PLCs still dominate high-speed machine control. DCSs prevail in complex continuous processes requiring proportional integral derivative (PID) loop control. PLCs were born as replacements for multiple relay logic and are used primarily for controlling discrete manufacturing processes and standalone equipment.

If integration with other equipment is required, connecting HMIs and other control devices is needed. The DCS, on the other hand, was developed to replace PID controllers and is found most often in batch and continuous production processes, especially those that require advanced control measures. As users demanded more production information, PLCs gained processing power and networking became common. At the same time, the DCS spread out to incorporate PLCs and PCs to control certain functions and to provide reporting services. The DCS now supervises the entire process, much like the conductor in an orchestra.

Which one to use? Less-expensive PLC or more functional DCS? Like most things in the world of automation, it depends on the needs of the application. Here are the key factors to consider:

(1) *Response time*: PLCs are fast. That is a huge asset. Response times of one-tenth of a second make the PLC an ideal controller for near real-time actions such as a safety shutdown, firing control, or machining. A DCS takes much longer to process data, so it is not the right solution when response times are critical. In fact, safety systems require a separate controller.

(2) *Size*: A PLC can only handle a few thousand I/O points or less. It is just not as scalable as a DCS, which can handle many thousands of I/O points and more easily accommodate new equipment, process enhancements, and data integration. If you require advanced process control, and have a large facility or a process that is spread out over a wide geographic area with thousands of I/O points, a DCS makes more sense.

(3) *Redundancy*: Another problem with PLCs is redundancy. If you need power or fault-tolerant I/O, do not try to force those requirements into a PLC-based control system. You will just end up raising the costs to equal or exceed those of a DCS. Redundancy in processing and networking favors DCS.

(4) *Complexity*: The complex nature of many continuous production processes, such as oil and gas, water treatment, and chemical processing, continue to require advanced process control capabilities of the DCS. Others, such as pulp and paper, are trending toward PLC-based control.

(5) *Frequent process changes*: PLCs are best applied to a dedicated process that does not change often. If your process is complex and requires frequent adjustments or must aggregate and analyze a large amount of data, a DCS is better suited.

(6) *Historical data and trending*: Process historian. Operator features and security. Organization and consistency. All of these lean toward DCS.

(7) *Programming*: The biggest benefit in PLCs is that it is easier for plant personnel to implement and configure internally than a DCS. Many technicians and engineers have experience with ladder logic, and you may decide to take care of your processes in-house.

(8) *Cost*: If you have a large process or a process that you hope will expand, consider paying more upfront for a DCS, and it will save you money over time. You will pay your integrator less in engineering time because the alarming, security, tag logging and trending, block icons, and face plates, and even the logic blocks are already developed, tested and, for some packages, used in plants all over the world.

(9) *Installation*: Purchasing a PLC allows you to buy only the software with the features you need. If you have a simple application or a standalone skid system, a PLC (with a small HMI) might be all you need. A PLC can be installed, programmed, and ready to go very quickly.

(10) *Use*: PLC is best suited for machine automation (quicker processing time): skids, stand-alone systems, ore handling and conveying, pulp and paper, smaller hydro power plants. DCS is best suited for large process chemical plants, batch management, and large power plants for SCADA applications.

17.5 Data Networking

So how is the motor operation and sensor data being transferred to the conductor (operator) of the orchestra, i.e. to the PLC or DCS and HMI to let the operator monitor, supervise, and operate the plant from his command post, and at the same time, allow the management observe and read power generated or gadgets produced from a remote location?

The answer is networking [3] at different levels from a simple sensor on the factory floor to the Internet cloud, as shown on the distributed network architecture in Figure 17.10. Every protocol seems to have its territory within the architecture, but with a demand for ever higher speed of data transmission, distance to be covered, and the number of connections (nodes) to be made,

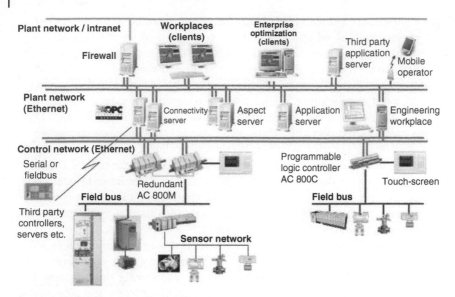

Plant network / intranet

Workplaces (clients)

Enterprise optimization (clients)

Firewall

Third party application server

Mobile operator

Plant network (Ethernet)

OPC

Connectivity server

Aspect server

Application server

Engineering workplace

Control network (Ethernet)

Serial or fieldbus

Redundant AC 800M

Field bus

Programmable logic controller AC 800C

Touch-screen

Field bus

Third party controllers, servers etc.

Sensor network

Figure 17.10 Plant automation network architecture. Source: Courtesy of Rockwell Automation [3]. Copyright 2018, Rockwell Automation Inc.

the user must assemble the system architecture to include all the factors as well as desired functionality and cost. Several of the most popular methods are reviewed here.

A communication protocol must be selected to define how information flows on a network composed of sensors, MCC starters, valve controllers, and PCs. The motor starters and the field sensors used in the process will be networked to PLC and/or DCS.

An industrial network integrates discrete, analog, and relays/actuators and signals over a large plant area for monitoring and control of various applications. Industrial networks transfer bits of information serially with a few wires to exchange data between the nodes. All of that is to insure the data is transferred reliably error-free and securely between the network components.

The industry has developed a number of suitable protocols for different levels of communications including ControlNet, ProfiBus, DeviceNet, AS interface, FieldBus, to name a few for the plant floor levels and fast Ethernet for the higher hierarchy levels. Table 17.1 illustrates the characteristics and capabilities (2017).

The DeviceNet and AS Interface may be the low-cost favorites for small projects. Control Net and ProfiBus may suit larger plants where critical high-speed activity is needed. Industrial Ethernet 10BaseT and 100BaseT with low-cost unshielded twisted pair (UTP) Cu cables appear to be the candidates

Table 17.1 Machine floor networking.

	ControlNet	ProfiBus	DeviceNet	AS-I	Industrial Ethernet
Media format	RG6	Twisted Pr	Thick/thin C	UTP, F/O	UTP, F/O
Message format	Peer to peer	P to P	Polling	M/S	Peer to peer
	Multi-master	M/S		Strobing	
Distance	250 m to 5 km	0.1–24 km	100–500 m	100–300 m	50 km
Nodes	99	127	64	32	1024
Baud rate	5 Mb/s	9.6 Kb/s 12 Mb/s	125–500 kb/s	167 Kb/s	10–2.4 Gb/s
Message bytes	500 bytes	244 bytes/scan	8 bytes/node	8 bytes/node	46–1500 bytes

The baud rate falls with the number of nodes connected. Thus, it may be recommended to string the nodes not to exceed 50% of its node capacity per each network card. It may be recommended to have one or two MCCs per network card.
Some systems covering more than 1 km will need repeater stations.
Source: From [4].

for large plants and to be integrated with the 100BaseF and GigBaseF media links essential at the higher plant operation level.

The industrial automation can be complex with respect to the amount of data to be handled, varied applications, and the need to know by the various parties operating the plant from the operators to the management. In principle, three hierarchical levels are in play: floor level, control room, and remote offices, as follows:

(1) *Plant floor level*: At this level, field devices such as sensors, switches, actuators, and analog devices are connected and feed data to the PLCs, DCS. Most of the devices are of low intelligence, while the actuators process the data and receive directives to operate valves and pumps. Here appropriate Field Buses such as DeviceNet and ControlNet and others can be employed to function at this level. Industrial Ethernet with UTP cables seems to be also encroaching into this area.

(2) *Control room level*: At this level, communications are configured for peer-to-peer networks between PLCs, DCS, HMI, and other computers for historical archiving supervisory control as well as to coordinate between the manufacturing cells. Ethernet TCP/IP is the most adopted for this area.

(3) *Remote management*: The plant-level management collects information from all area levels to manage and plan the overall automation process. Here, Ethernet local area network (LAN) and wide area network (WAN) are employed for production planning and management information

exchange. This also is a gateway to the Internet cloud and other industrial networks.

In many cases, engineers' work is to update or renovate the operating plant. This includes the connections of dissimilar networks and technologies for which other specific equipment will be required. Some of that may not be evident at the outset and may be decided during the equipment replacement. Many old technologies are no longer supported as the new technologies based on fiber optic systems command their presence in the plants. The following additions may be required:

- *Gateways and bridges*: to interface dissimilar systems of different characteristics and protocols to make them operate across the gateways.
- *Routers*: to redefine the paths and to channel the communication packets within the networks.
- *Repeaters*: to extend the networks to handle more nodes greater distances between parts of the network.

17.6 Means of Communication

17.6.1 Transfer Data

There are several techniques used to transmit information [4]. Nearly, all data network systems in use today use binary digits (bits), a series of 1s and 0s, to send information, but there is a difference in the methods of carrying the bits across the network. Here are the keys words to understanding the networking:

Baseband involves the use of the entire bandwidth of a channel to transmit a single signal, using one carrier frequency.

Broadband divides multiple analog signals into different frequencies and transmits them simultaneously.

Bandwidth is the range of frequencies that a given carrier is able to effectively transmit. The rate at which data can be sent depends on the bandwidth of the cable and is expressed in "baud."

Baud rate is actually the rate of signaling events (changes in frequency, amplitude, etc.).

Bits per second are not necessarily the same as the baud rate and should not be confused, even though the terms are routinely used interchangeably. The baud rate is the same as the b/s rate if and only if each signal element is equal to one bit exactly.

Messages are packets of data in bytes with formatting and addressing information, along with the data.

Parity is a part of message, a simple error-checking method, to determine if the byte was received correctly.

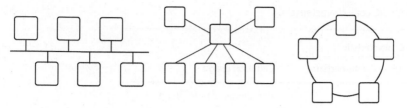

Figure 17.11 Network topology [4].

Simplex transmissions are only in one direction, all of the time.

Half-duplex is bidirectional communication allowed in one direction at any given time.

Full duplex is bidirectional transmission in both directions simultaneously.

Synchronous (clocked) transmissions are timed so that both devices know exactly when a transmission will begin and end.

Asynchronous (unclocked) transmissions must mark the beginning and the end of messages. Synchronous transmission is usually faster than asynchronous, but the timing issue between two remote machines can introduce problems causing asynchronous transmission to be simpler and less expensive, and therefore more widely used. Asynchronous transmission does, however, introduce extra control bits into a message, which slows the rate that actual data can be transferred.

Network topology is the arrangement of the devices on a network (called nodes) called the topology. There are several types (Figure 17.11) that are commonly used, including the following: Bus, Star, and Ring, as shown later. Combinations of the basic types are used, depending on the application.

17.6.2 Physical Media

There are several types of media used in industry: coax, twisted pair, and optical fiber are the most common [4].

Coax: In coaxial cable (coax), RG stands for "Radio Guide." If a terminating end resistor with a value equal to the characteristic impedance is placed at the end of a cable run, it approximates an infinite cable, and minimizes signal reflections (see Table 17.2).

Shielded twisted pair (STP): A cable with a number of individually insulated Cu conductors, twisted into pairs, and surrounded by a shield to reduce EMI/RFI. The pairs may be individually shielded or have one overall shield, or both. STP is more expensive than UTP, but is more resistant to interference, making it the cable of choice for most industrial networks.

Unshielded twisted pair (UTP): A cable with Cu conductors without any shielding (see Figure 17.12). It is commonly used in informational Ethernet

Table 17.2 RG coax characteristics.

	Characteristic	
Coax type	Impedance	Common usage
RG-6	75	Broadband, carrier band (drop)
RG-8	50	Thick Ethernet
RG-11	75	Broadband, carrier band (trunk)
RG-58	50	Thin Ethernet
RG-59	75	Broadband drop

Source: From [4].

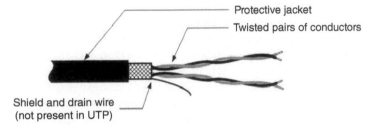

Protective jacket

Twisted pairs of conductors

Shield and drain wire
(not present in UTP)

Figure 17.12 UTP cable.

applications because of its less-expensive construction and ease of use. Cat stands for category (see Table 17.3).

Optical fiber cable: Optical fiber has the ability to transmit higher data rates for longer distances than copper media. It costs a lot more and is usually more difficult to work with. Fiber optic cables have the added important benefit of being completely immune to EMI and RFI. Optical fiber systems are nonspark-producing so that they can be more easily used in explosive areas. The optic fibers are roughly the size of a human hair. There are typically several fibers 2, 4, 6, 12, 24, 48 in cables, called Cores, all surrounded by an outer jacket and sometimes in a metallic armor for mechanical protection.

There are two types of optical fibers: single-mode and multimode. Single mode fiber is smaller in diameter (usually 8.3 μm core) than multimode fiber (usually 62.5 μm core).

Single-mode fiber is intended for a single (frequency) signal, to transmit a higher power signal. Single-mode optical applications typically operate in the 1300/1550 nm (nanometers) range.

Multimode fiber can handle multiple frequencies at once, but the signals can interfere with one another if their power level is too high. Multimode equipment normally operates around 850/1300 nm.

Table 17.3 UTP cable application.

Cable type	Data rate (Mb/s)	Common usage
Cat 1	N/A	Voice grade analog only
Cat 2	4	Digital voice
Cat 3	10	10BaseT
Cat 4	16	Token ring
Cat 5	100	100BaseT
Cat 6/(5e)	1000	1000BaseT

Source: From [4].

Therefore, a single-mode signal can be sent a far greater distance than a multimode signal for transmissions over 2 km. Within a plant, one would use multimode fiber.

17.6.3 Logical Schemes and Arbitration

How the systems pick and transmit data?

(1) *Client/server and polling (master/slave)*: In a client/server arrangement, a client makes a request to the server, which responds to the client. Servers may not speak until spoken to. There may be many clients and many servers; however, if more than one client is present, another arbitration method (usually physical) is required among them. Roles of devices may change, and a node that at one point is a client making a request to another, may later be a server to that very node.

(2) *Polling* is a similar access method in which one device makes a request from another, but the connotation is that of a more regular, timed interval than that of client/server.

(3) *Master/slave arrangement* is similar in nature, in that the master makes requests of the slave. The difference is that the roles do not change: masters remain masters and slave devices remain slaves.

(4) *Peer-to-peer*: In a peer-to-peer network, no one device is the controlling authority. Each is equal in priority, that is to say, they are peers. Devices take turns writing to, or making requests of one another.

17.6.4 Open Industry Standards

The Electronic Industries Association (EIA) has developed several "Recommended Standards" (RS-XXX) to aid in the ease of connection [4].

(1) *RS-232C*: The most widely used and versatile physical standard. It was developed mainly for the interface between data communication equipments (DCEs), typically a modem, and data terminal equipments (DTEs). Since RS-232C is an unbalanced system, the distance between devices is limited to roughly 15 m (45 ft, at 19.2 kb/s), which is its largest downfall, but the inherent flexibility makes it a popular choice for many users.

(1) *RS-422:* Being a balanced system, RS-422 makes longer distances and faster transmission rates than RS-232C possible. It is capable of 10 Mb/s at distances of 4000 ft (1200 m). The RS-422 standard recommends a 24 AWG twisted pair cable with a 100 Ω characteristic impedance. Cat 5 cables used in Ethernet meet these requirements and is widely available, making it a good choice for RS-422 installations. In keeping with the practice of terminating the cable with the equivalent of the characteristic impedance, a 100 Ω resistor should be connected across the "A" and "B" lines.

(2) *RS-485*: A serial interface standard which (like RS-422) is a balanced system, except that it uses a tri-state line driver. It permits even faster rates over longer distances – 100 kb/s at up to 4000 ft (1200 m). It can operate on one pair of twisted pair in half-duplex mode, or on two pairs in full-duplex mode. The full-duplex mode is limited to a master/slave arrangement only.

17.6.5 Open Protocols

Ethernet IP is the most widely used network protocol. The most widely used LAN technology is the Ethernet and it is specified in a standard called IEEE 802.3. It is rapidly being accepted in the plant control applications down from the signal level up to the Internet cloud and executive level. The advent of switches and routers has made Ethernet the most popular industrial networking protocol in full duplex mode.

Major manufacturers are introducing industrially hardened Ethernet interfaces, switches, and connectors (Figure 17.13), as well as industrial control message protocols (Modbus/TCP, Ethernet/IP, Profinet, Fieldbus HSE, and more). Ethernet utilizes a baseband signal. The most common configuration is over twisted pair wires (T) in either a bus or star topology or by optical fiber (F) (see Table 17.4).

Figure 17.13 Patch box for Ethernet IP.

Table 17.4 Ethernet connections.

Signal name	ID	Data rate	Maximum distance (m)	Cable media
10Base5	Thicknet	10 Mb/s	500	50 Ω thick (RG-8) coax
10BaseT		10 Mb/s	100	Cat. 3 or 4 twisted pairs
10BaseF		10 Mb/s	2000	Fiber optic cable
100BaseT	Fast E	100 Mb/s	100	Cat. 5 twisted pairs
100BaseF	Fast E	100 Mb/s	400	Fiber optic cable
1000BaseF	Gig E	1 Gb/s	220	Fiber optic cable

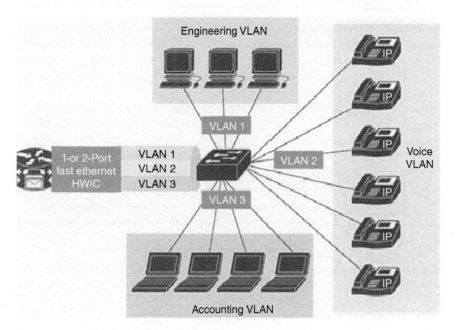

Figure 17.14 Ethernet TCP/IP network.

The most common protocol used with Ethernet is TCP/IP, and this is the route that the new industrially hardened Ethernet implementations are following (see Figure 17.14). Transmission Control Protocol (TCP) defines the source port and destination port numbers that allow data to be sent back and forth to the correct application running on each device. Internet Protocol (IP) is responsible for inserting routing information into the header of each message on the network layer composed of 32 bits.

The Cat-5e twisted pair cable supports Gigabit Ethernet, where all four pairs of twisted wires in the cable are used to achieve the high data rates. Cat-5e or higher cable categories are recommended for network video systems. Most interfaces are backwards compatible with 10 and 100 Mbs Ethernet and are commonly called 10/100/1000 interfaces.

Modbus Plus is an RS-485 based, peer-to-peer network protocol, which uses the Modbus data structure. It transmits at a rate of one mega baud and allows up to 32 nodes on a segment of up to 455 m. A repeater allows an additional segment to be attached for a maximum of 64 nodes, over 900 m. Schneider calls this protocol: Ethernet ModBus.

17.7 Web-based HMI

The innovation marches on. This is an HMI available directly on your own computer or mobile phone anywhere in the world wherever you can access Internet. One can now watch tennis match on your laptop and switch from time to time to check on the process on the factory floor for the number of car parts produced and the alarms to be attended to.

It uses a Web browser, such as Microsoft Edge, Internet Explorer, Google Chrome, Safari, or Firefox, to provide an interface with vendor's graphics, trending and alarming applications (applications within the HMI/SCADA suites). There is no need to install any software onto the remote clients or to export or convert displays.

The main complaint by the regular HMI users is that they cannot keep up with the advance of technology. As soon as the system is installed and configured, the technology may already be outdated. Well, here is the solution. Web-based HMI is continuously being updated as your home PC.

The speed of refreshing the screens may be lacking.

One user told me that this new HMI approach may not be secure and can be hacked. A large Asian utility has just implemented this type of system and tells me it fits well into their security requirements.

17.7.1 Work Place

Conventional supervisory control systems generally require operator desktop computer or operator panel, fixed in the area where personnel work. It is a special device, which is designed and configured to perform this specific task. The operator can organize to work on the project over the Internet wherever there is Internet available. The main way to access the system for control, setup, configuration, or just observation – is a web browser. Any browser would do.

User is not limited to the used device and the location for access to the controlled equipment. This can be any stationary or portable computer, smartphone, connected via a local network, or a wireless access point to WebHMI.

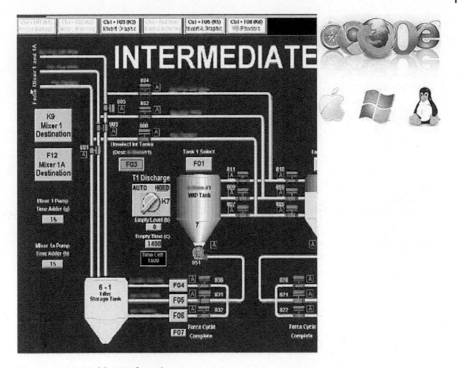

Figure 17.15 Webb HMI faceplate.

There is no limit on the number of users working in the system simultaneously. The access rights and role credentials are, however, assigned to different users. Anyone who is online at a moment sees all the configuration changes at once, in real time.

17.7.2 Creating Interface Screens

If the function is to supervise and operate only, the system will transmit the screens that have been prepared already. Like a regular HMI, WebHMI has a built-in graphic editor with standard elements to create a working interface process screens. Graphic elements created in other specialized or animated graphics program can be added to the image library (see Figure 17.15).

17.8 SCADA Applications and Communication Protocols in Power Industry

Supervisory Control and Data Acquisition HMI (SCADA HMI) is used by utilities; power generation and distribution, electrical grids, water distribution,

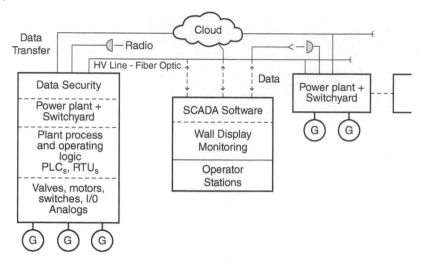

Figure 17.16 SCADA powerplant basic architecture.

traffic control, airports, oil and gas supply, waste management, etc., to integrate and provide economical, functional, and safe operation of the whole operating region. The data is collected over transmission lines using fiber optic, microwave, and other redundant means of communication over an entire operating area.

In case of power plants, the plants of an area are integrated for monitoring and support each other in case of emergencies, economic load sharing, and import/export of power. For instance, SCADA systems for a US state may be equipped to monitor and control individual power plants and switchyards as well as to be integrated with the neighboring states and the overall operating area. Not every SCADA system is programmed to control and monitor every detail of every plant, but on the basis of their regional responsibilities.

SCADA is a specific integrating software for centralized data acquisition, control and monitoring based on the operation and programmed logic of the plant PLCs and remote terminating units (RTUs). SCADA is similar to a plant DCS control system, except that it is integrated with other power plants over the entire area. This particular difference leads to the requirement for greater safety and security and the use of specific communication protocols [5], as noted in Figure 17.16. The IEC Technical Committee 57 has generated the following companion transmission protocol standards for SCADA:

IEC 60870-5-101	Companion standards especially for basic telecontrol tasks.
IEC 60870-5-102	Companion standard for the transmission of integrated totals in electric power systems (not widely used).
IEC 60870-5-103	Companion standard for the informative interface of protection equipment.
IEC 60870-5-104	Network access for IEC 60870-5-101 using standard transport layering profiles.
IEC 61850	Used for electrical utilities, switchyards, and SCADA applications.
DNP3	Widely used in North America for electrical utilities and SCADA applications.

The most common transmission protocol presently used is IEC101. There seems to be a transition developing toward IEC104 for F/O cable bandwidth layering and used for transmission between plants.

IEC 60870-5-101 (IEC101) is a standard for power system monitoring, control, and associated communications for tele-control, tele-protection, and associated telecommunications for electric power SCADA systems. It uses standard asynchronous serial tele-control channel interface between DTE and DCE.

IEC 60870-5-104 (IEC 104) protocol is an extension of IEC 101 protocol with the changes in transport, network, link, and physical layer services to suit the complete network access. The standard uses an open TCP/IP interface to network to have connectivity to the LAN and routers. There are two separate link layers defined in the standard, which is suitable for data transfer over Ethernet and serial line (PPP – Point-to-Point Protocol). However, the security of IEC 104 by design has been proven to be problematic, as SCADA users witness encryption problems to prevent attacks. Unfortunately, due to the increase in complexity, vendors are reluctant to roll this standard out on their networks.

Figure 17.17 SCADA with DNP3 transmission.

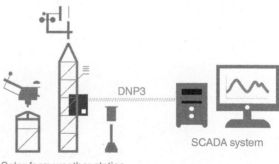

DNP3

SCADA system

Solor farm weather station

DNP3 is a set of communication protocols used between intelligent electronic devices (IEDs) components in process automation systems and is widely used in North America. Its main use is in utilities such as electric and water companies.

Specifically, it was developed to facilitate communications between various types of data acquisition and control equipment (Figure 17.17). It is based on the **IEC61850** standard, also used for power utilities and electrical substations.

IEC 61850 is an international standard widely used over Ethernet TCP/IP for electrical substation automation, in particular for connections to protective relays and measuring instruments. It provides a comprehensive model for how power system devices should organize data in a consistent manner across all types and brands of devices.

References

1 Rockwell Automation (2017). "ControlNet PLC-5 Programmable Controllers.
2 Slattery, C., Hartmann, D., and Ke, L. (2009). *PLC Evaluation Board Simplifies Design of Industrial Process-Control Systems*. Analog Devices.
3 Rockwell Automation (2019). *Plant Networking Systems*. Rockwell Automation Pre-engineered Network Solutions.
4 Chet S. Barton, PE – *Introduction to Control Networks in an Industrial Setting*. Course IC-3001.
5 Clarke, G.R. et al. (2004). *Practical Modern SCADA Protocols: DNP3, 60870.5 and Related Systems*. Newnes. ISBN: 0-7506-5799-5.

18

Generation

┌───┐
│ **CHAPTER MENU**   │
│   │
│ 18.1 Types of Generating Plants, 407   │
│   18.1.1 Power Plant One-Line Diagrams, 411   │
│   18.1.2 Generator Capacity Curve and Limits, 413   │
│   18.1.3 Generator Excitation and V Curves, 415   │
│   18.1.4 Short Circuit Ratio (SCR), 416   │
│   18.1.5 Generator Impedances, 418   │
│   18.1.6 Generator Transient Conditions, 419   │
│   18.1.7 Generator Stator Core Test, 420   │
│   18.1.8 Generator, Motor Testing after Installation, 421   │
│ 18.2 Governors, 422   │
│   18.2.1 Digital Governors, 424   │
│   18.2.2 Auxiliary Equipment, 426   │
│ 18.3 Excitation: Control for Voltage and Reactive Power, 427   │
│   18.3.1 Generator (Motor) Magnetizing, 427   │
│   18.3.2 Static or Brushless Excitation, 428   │
│ 18.4 Generator Circuit Breaker, 431   │
│ 18.5 Generator Step-up Transformers, 432   │
│ 18.6 Heat-Rate Curve, 432   │
│ 18.7 Hydraulic Turbine Cavitation, 433   │
│ 18.8 Generator Cooling, 434   │
│ 18.9 Plant Black Start, 434   │
│ 18.10 Synchronous Motor, 436   │
│   18.10.1 Damper Winding, 436   │
│ 18.11 Plant Capacity and Availability Factors, 437   │
│   18.11.1 Typical Capacity Factors for Power Plants, 438   │
│ References, 439   │
└───┘

18.1 Types of Generating Plants

There exist a number of conventional ways to produce electrical energy, some by burning fuel and others by exploiting the renewable natural resources, such as wind, sun, water, and geothermal sources. The natural resources are not

Practical Power Plant Engineering: A Guide for Early Career Engineers, First Edition. Zark Bedalov.
© 2020 John Wiley & Sons, Inc. Published 2020 by John Wiley & Sons, Inc.

always close to the locations of the load centers. In most cases, the renewable resources are located away from the load, and the power produced is transported to the load centers by long transmission lines. On the other hand, the generating plants using fuel are built closer to the load centers, and the fuel resources are transported to the plants.

The following critical points have to be evaluated in order to determine the type of plant to be used: site location, space required, T&D energy transport, initial cost, running cost, maintenance cost, cost of fuel transport, environmental issues, unit starting quickness, standby operation losses, and overall efficiency.

Let us briefly describe the principal characteristics of the major sources of generating power:

1. *Photovoltaic farms*: This is the simplest form of generating electricity (see Chapter 25). Large number of photovoltaic (PV) panels are laid out on a flat ground and directed toward the Sun. The plates absorb the sun's rays and generate electricity. The plates are connected in a way to produce the desired voltage, which is then converted from DC to AC in power inverters and led to transformers to convert to a higher voltage and transmit the power to the users at some distance from the PV farm. To boost the output, the sun trackers are employed to follow the sun from sunrise to sunset. Plants 1000 MW installed capacity already have been built.
   Advantages: Low cost production and easy to maintain.
   Disadvantages: High capital cost per kWh produced, large area required, remote from load centers, availability of power cannot be predicted, low capacity factor, operates only during day hours, large output losses, generates no reactive power, and wind issues.

2. *Concentrated solar plant (CSP)*: This solar plant uses mirrors to concentrate solar energy to heat water or oil to operate a thermal cycle with low pressure turbines and condensers (see Chapter 25). Plants employ parabolic mirrors, Fresnel mirrors, and power tower technology to concentrate the sun's heat. Sun trackers are employed to follow the sun from sunrise to sunset. Plants in excess of 300 MW have already been built.
   Advantages: Low cost production and more efficient than PV plants. It can generate reactive power.
   Disadvantages: High capital cost per kWh, remote from load centers, environmental problems on power tower plants, availability of power cannot be predicted, low capacity factor, operates only during day hours, large output losses, and wind issues.

3. *Wind farm*: It employs 1–7 MW wind towers. Due to the wind variability, the generator design employs wound rotor asynchronous generator model that gets its excitation from the grid through AC/DC to DC/AC converters. Unit transformers are a chain connected at 33 kV and led to the main substation

for connection to large transformers for power transmission to remote load centers (see Chapter 26).

Advantages: Low cost production.

Disadvantages: Requires large area, power not predictable, used only when wind available, Difficult transport of blades. High installed capital cost per kWh. Remote from load centers. Low capacity factor. Large output losses. Wind turbulence issues.

Requires import of reactive power at the main and unit stations.

4. *Hydroelectric plant*: A hydroelectric generating station is a plant that produces electricity by using water to propel the turbines and drives the generators. There is a variety of plants employed, reservoir/dam, run of the river, pumping storage. All the plants are powered by the kinetic energy of flowing water as it moves downstream.

The plant power is dependent on the product of water head (height) and water flow.

$$P = 9.81 \times W\,Q\,H\,\eta \times 10^{-3}\,\text{kW} \tag{18.1}$$

Where P – power, W – specific weight of water in kg/m^3, Q – rate of flow in m^3/s, H – the height of fall (head) in meters, η – the overall efficiency of operation.

The plants use Kaplan turbines for low head/high volume sites, Francis for 10–600 m head/1–10 m^3 middle range volume and Pelton turbines for high head/low volume sites.

Advantages: Low cost production, can be used as base or peaking power, and quick start within one minute. The resource is adaptable to changing demand, also used as spinning reserve, can be designed to work as synchronous condenser to provide MVARs, simple operation and maintenance, and efficiency of 80–85%.

Disadvantages: Large capital cost for civil works in creating the catchment and main dam, capacity factor ranging from 15% to 50%, and water inflow not predictable from year to year.

5. *Diesel engine plant*: These plants use liquid or gaseous fuel, mixed with atmospheric air as a working medium. During the suction stroke, the fuel is compressed and ignited inside the cylinders. The thermal energy is converted into the mechanical energy to move the pistons and spin electric generator (see Chapter 20).

Diesel engine plants are built for remote communities and ships. The plants usually operate as base load generation, 24 hours a day to supply power to industrial facilities and the camp or villages. The engines for base load are of slow speed 400–900 rpm type, operated on continuous duty. Waste high and low heat can be recovered and used for winter heating and for industrial plant usage, thus raising the plant efficiencies from 40% up to 80%.

Advantages: Low capital cost, quick installation, high efficiency, waste heat recovery, quick start, suitable for small remote communities, and suitable for peaking duty.

Disadvantages: High fuel cost, environmental emissions issues, and noisy.

6. *Gas turbine plant*: A plant that uses a combustion engine that converts gaseous or liquid fuels to mechanical energy. Natural gas contains 80% methane. The gas turbine engine compresses air and mixes it with fuel that is then burned at high temperatures, creating a hot gas to make the turbine blades spin and turn a generator as it exhaust gas to atmosphere. The burning energy is converted into electricity. The efficiency expected is <30%.

    Plants are built with multiple GT units of 10–250 MW each with a possibility of adding combined cycle thermal generation on every pair of turbine units by using the exhaust gas temperature reaching 540°C.

    Advantages: Low capital cost, quick plant construction, waste heat recovery possibility, quick engine start for peaking duty, and suitable for small remote communities.

    Disadvantages: High fuel cost and environmental emissions issues.

7. *Thermal plant*: A fossil-fueled power station that burns coal, natural gas, waste, or petroleum to produce high pressure and temperature steam in a boiler for steam turbines to drive electrical generators. Thermal power plants are designed on a large scale for continuous operation and base loading. The majority of the plants use one or two 400–600 MW units. Coal-fired plants have large infrastructure for coal handling and storage. EPA requirements dictate a need for having exhaust scrubbers for pollution control. Scrubbers are one of the primary devices that control gaseous emissions, especially acid gases. The pollution control adds to the station load and causes a significant reduction in plant efficiency.

    Advantages: Base load operation, low capital cost per kWh, high capacity factor: 60–70%.

    Disadvantages: Pollution control, slow to start, and high standby operational losses.

8. *Nuclear plant*: A thermal power station with a nuclear reactor as the heat source using uranium as fuel for the chain fission reaction. The fuel is fed into the reactor core and the heat is produced through a controlled (moderated) nuclear fission. The heat generated is cycled in a heat exchanger to generate pressurized steam and generate electricity in a regular thermal cycle.

    The nuclear chain reaction is moderated in the reactor core depending on the type of reactor by regular water H_2O, solid graphite, or heavy water D_2O. Spent fuel taken from the reactor is highly radioactive and stored in water basins within the plant.

    Advantages: Low production cost, base load operation, and no pollution or CO_2.

Disadvantages: Long planning and regulation process, high construction cost, radioactivity, fuel and core containment requirements, spent fuel storage.

18.1.1 Power Plant One-Line Diagrams

Figures 18.1 and 18.2 demonstrate different types of one-line diagrams for power plants, each having two generating units. The first one is used for

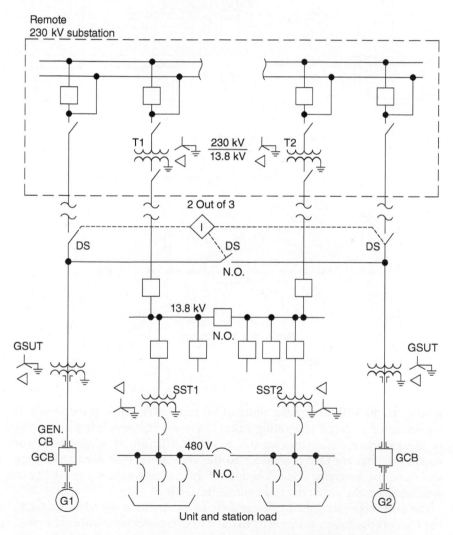

Figure 18.1 Power plant, smaller units. GSUT, Generator step-up transformer; UAT, unit auxiliary transformer; SST, station service transformer; GCB, generator circuit breaker.

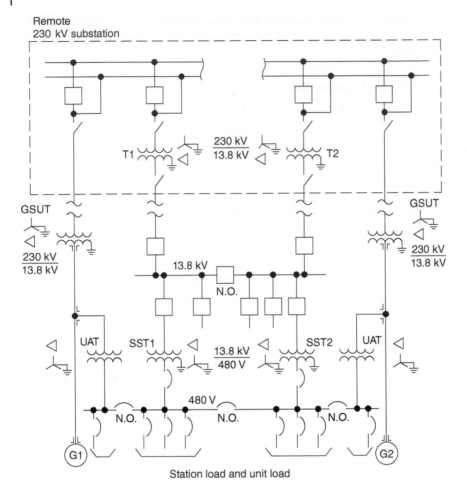

Figure 18.2 Power plant, larger units.

smaller plants with generating units of up to 60 MW, while the 2nd one is implemented for larger generating units. The main difference is in the location of the generator circuit breaker (GCB). The GCBs are expensive pieces of equipment that are placed adjacent to the generators, thus occupying large spaces on the generator floor. A benefit of this arrangement is a capability to synchronize right within the plant across the GCB.

The other approach used for larger generator currents is not to have a GCB, but connect the generators directly to the output transformers within the plant and connect the transformers at HV to the remote switchyard. In this case, the generator synchronizing and switching is done by remote switchyard breakers.

This concept saves on the cost of breakers, but complicates on the breaker remote control wiring and commissioning.

A plant one-line diagram for a diesel generating plant is completely different and shown in Figure 20.3.

18.1.2 Generator Capacity Curve and Limits

Early in my carrier as I was teaching classes, I asked students to tell me the key factor in building the electrical equipment. I heard many answers. Generally, they were all surprised when I told them it was the temperature. Understanding the capability of the system generation is one of the keys into the power plant and systems engineering.

It is desired that electrical engineers understand the significance of the equipment operating limits, impedances in all the operating conditions; steady state and transient, as well as the generator excitation and governor droop.

The chart (Figure 18.3) represents the operating capacity of a hydro generator. The unity power factor is the straight vertical line in the middle. The operating limits for this hydro unit are listed on the chart.

For the hydroelectric generators, there is also a rough operating region between 20% and 40% of power capability due to cavitation and increased

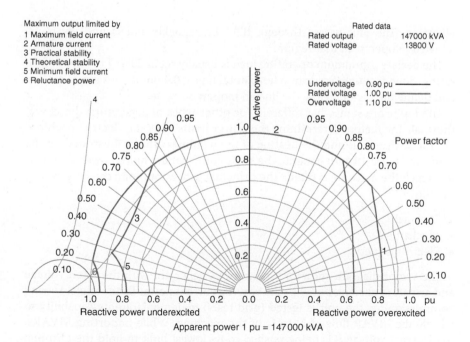

Figure 18.3 Hydro generator: 147 MVA (1 pu), 0.85 lag/0.9 lead power factor.

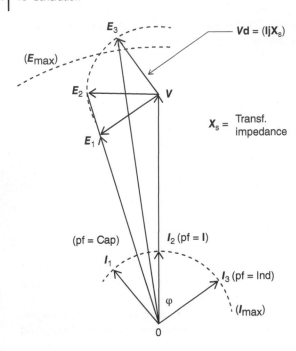

Figure 18.4 Generator steady state limits.

vibration. The unit passes through this zone quickly and is unlikely to be allowed to operate in that region.

The declared minimum operating load is usually set at 50% of the nameplate rating. So in fact one can draw a horizontal line at 0.4 point on the vertical line and define the region below the line an inoperable zone.

The limits are somewhat different for other types of generators: diesel and thermal. The factors determining the power limits of all the electric machines are the shaft torque and the heating of the windings. This is illustrated on the diagram Figure 18.4: In general, the shaft could handle more power than that for which the machine is rated. The principal generator limits can be grouped as follows:

• Thermal
• Mechanical
• Electromagnetic

The vector diagram shows a synchronous generator with the rated voltage and armature current, but with the varying current power factor angle. During commissioning, the unit is tested (grid tests) to demonstrate its capability to handle the MVAR flow of I_1 to I_3 of the machine. While importing MVARs, the internal voltage E is being pushed to its lowest limit to hold the terminal voltage V from collapsing with the help of transformer tap changers.

The steady-state power limits are determined by the heating in the field and armature windings of the machine. Since there is a maximum value given for I_F and E, there is a minimum acceptable power factor of the generator when it is operating at the rated MVA. Since E is the vector sum of V and jX_S I, it is evident that there are some current angles for which the required E exceeds E_{max}. If the generator is operated at these power factors and the rated armature current, the field windings will be overheated. The angle of I results in the maximum allowable E, while V that is at the rated value determines the generator-rated power factor. Therefore, the generator can be operated at a lower power factor (more lagging) than the rated value, but only by reducing the MVA output of the generator.

18.1.2.1 Thermal Limits

Thermal limits set the maximum temperature rise due to the maximum power loss in the parts of the machine. The major cause of losses on the rotor is the field winding resistive loss. If the field current is above a certain level, the I^2R losses will be too high, possibly causing over-temperature in the field winding. In order to manage thermal conditions by limiting the field current, the induced excitation $|E|_{max}$ in the armature winding will be limited. Similarly, on the stator, excessive heating limit sets armature current $|I|_{max}$. In steady state, this limit is the rated current of the machine.

18.1.2.2 Mechanical and Electromagnetic Limits

Mechanically, the input power to the generator is limited by the physical capacity of the prime mover. As a result, the prime mover limit sets the maximum output power from the generator, $P_{max} = |E \sin \delta|_{max}$.

The final limit is related to the mechanical input and the ability of the generator to electromagnetically create a torque equal and opposite to the desired mechanical torque. The torque is a product of two electromagnetic fields or a function of the sine of the angle δ between V and E. At a given excitation, if the mechanical torque increases, the rotor speeds up, increasing δ and the electromagnetic torque. This negative feedback continues until the electromagnetic and mechanical torques balance.

The static stability limit is reached at $\delta = 90°$.

$$P = \frac{VE \sin \delta}{X_s} \tag{18.2}$$

18.1.3 Generator Excitation and V Curves

The following two additional and useful generator characteristics indicate how the generator performs. Once the excitation curve (Figure 18.5) starts turning (saturating), more field current is required per volt I_f/V. For some generators subjected to large voltage variations, the magnetizing curve is made to be

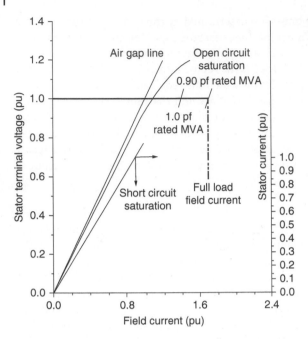

Figure 18.5 Open and short circuit graph.

linear until $V = 1.5$ pu. For this operating requirement, larger magnetic core is required. Generator V curves (Figure 18.6) show the relationship between the stator armature current I_A against the field current I_f for the generator under-excited on the left and overexcited on the right for various loads from 25% to 100%.

18.1.4 Short Circuit Ratio (SCR)

Short circuit ratio (SCR) is a measure of generator stability. It is the ratio of field current I_f required to produce rated armature voltage V_{rated} at open circuit to the field current I_f required to produce the rated armature current I_A at short circuit (see Figure 18.7).

The SCR can be calculated from the **open circuit characteristic** (O.C.C) at rated speed and the **short circuit characteristic** (S.C.C) (Figure 18.5) of a synchronous machine. The SCR is approximately equal to the reciprocal of the per unit value of the saturated synchronous reactance X_d as follows:

SCR $= I_f$ (for V_{rated} at open circuit rated)$/I_f$ (for I_A rated at short circuit) $= 1/X_{d\_sat}$ in pu.

Therefore, a large SCR corresponds to low X_d value and physically smaller size of generator.

SCR affects the operating characteristics and the generator stability performance. Low value of SCR of a synchronous generator would cause large

Figure 18.6 *V* Curves; field current for varied pf.

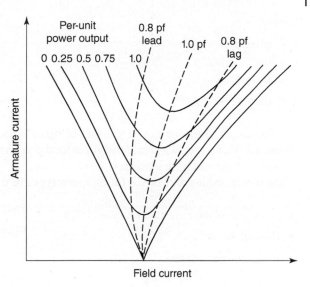

Figure 18.7 Short circuit ratio.

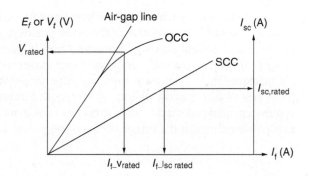

variation in the terminal voltage when a change in load takes place. To maintain a relatively constant terminal voltage, the field current I_f must vary over a wide range.

A synchronous machine with a higher value of SCR performs better with respect to voltage regulation. It also demonstrates higher steady-state stability limit. Higher SCR demands larger size and increases the cost of the machine. Utilities will demand higher SCR machines and higher ceiling excitation voltage or their generators. The SCR is directly proportional to the air gap magnetic reluctance R (magnetic resistance) or the air gap length. Reluctance of a uniform magnetic path is

$$R = \frac{L}{\mu A} \tag{18.3}$$

Where

R = reluctance

L = length of the gap

μ = permeability

A = cross section of path

If the length of the air gap or number of turns is increased, the SCR can be increased. This requires a greater height of field poles and, as a result, the overall diameter of the machine increases.

The typical values of the SCR for different types of machines are as follows:

- For cylindrical rotor machine, the value of SCR   0.5–0.9.
- Salient-pole machine   1–1.5 and
- Synchronous condensers   0.4

18.1.5 Generator Impedances

A cylindrical rotor synchronous machine has a uniform air-gap. The reactance remains the same, irrespective of the rotor position. However, a synchronous machine with salient poles (hydro generators) (Figure 18.8) has a nonuniform air-gap. Therefore, the value of its reactance varies with the rotor position.

Consequently, a cylindrical rotor machine possesses one axis of symmetry (pole axis or direct axis), whereas salient-pole machine possesses two axes of symmetry: field poles axis, called **direct axis** or ***d*-axis** through the poles, and axis passing through the center between the poles at electrical 90°, called the

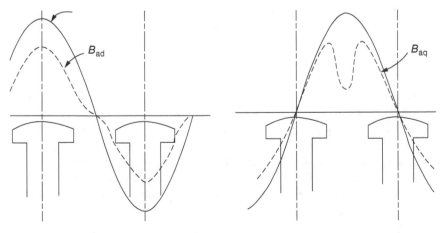

Figure 18.8 Generator impedances X_d, X_q.

quadrature axis or **q-axis**. Since the air gap is nonuniform, large flux is produced along the d-axis and a minimum flux along the q-axis.

This implies that the quadrature flux has significantly more air in its path than the direct axis flux. Therefore, the quadrature flux path reluctance (magnetic resistance) is higher than direct axis flux path reluctance, so $L_d > L_q$, $X_d > X_q$. Also, the sinusoidal magnetic force mmf produces somewhat distorted and reduced flux along the q-axis. For a typical hydro generator, the reactances are $X_d = 1.15$ pu, $X_q = 0.75$ pu. Reluctance in magnetic circuit is analogous to resistance in an Ohmic circuit.

18.1.6 Generator Transient Conditions

During a three-phase short circuit on a generator, the fault current passes through a transition from a subtransient to a steady-state value as shown on the charts for the average on Figure 18.9a and individual phases Figure 18.9b. [1]

The subtransient current during the first two–three cycles attains a value of 10–18 times the full load current. The same transitions apply to the induction motors. The fault currents in the corresponding transition periods are calculated as follows:

$$I'' = \frac{OC}{\sqrt{2}} = \frac{E_a}{X_d''}, \quad I' = \frac{OB}{\sqrt{2}} = \frac{E_a}{X_d'}, \quad I = \frac{OA}{\sqrt{2}} = \frac{E_a}{X_d} \tag{18.4}$$

Where;

I'', I', I	Fault rms current, subtransient, transient, and steady state.
X_d'', X_d', X_d	Direct axis reactances, subtransient, transient, and steady state.
E_a	Induced emf per phase.

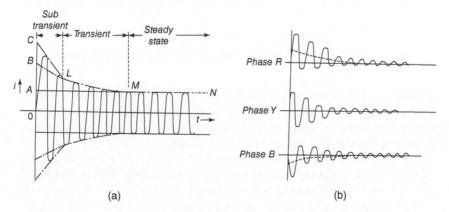

(a) (b)

Figure 18.9 (a, b) Generator transient conditions. Source: Courtesy of Electrical baba. [1]

The DC components in the three phases are different at the instant of fault, thus the waveforms of the phases are not identical. During the fault the current follows the voltage by 90°. If the voltage of the middle phase as above is at maximum during the fault, the resulting DC increment of the short-circuit current is at zero value, etc.

→We engineers, this is important to note. In the short-circuit calculations, the subtransient values are used for the selection of the circuit breaker interrupting capability. For the system stability studies, transient values are used.

The breaker contacts start separating during the transient conditions after an action of protective relay. The r.m.s. value of the current at the instant of the breaker contact separation is called the breaker *breaking* current in kA. However, if the breaker closes on fault, it will be operating on the fault current of the first cycle, i.e. the highest value. This fault current is called the breaker *making* current, which is higher than the breaking current.

18.1.7   Generator Stator Core Test

Stators of large hydro generators are assembled *in situ* starting from stacking the iron laminated core to assembling the stator bars to complete the windings. This work is conducted in a challenging environment of active plant construction going on at the same time around the stator assembly. The stator assembly must be conducted in a strongly controlled and protected environment within an *in situ* tent and heating provided inside the tent to maintain a moisture-free constant temperature.

This early test is required following the completion of the core stacking to detect possible damage to the inter-laminal insulation for possible damage during fabrication punching and assembly, thus causing local hot spots which may spread out and also damage the winding stator bars.

According to Siemens, the following three inspection methods are used after the core stacking [2]:

- *ELCID*: Electromagnetic core imperfection detection
- 60 (50) Hz ring flux test; Core test method for large turbo generators.
- 500 Hz ring flux test.

All the methods inject a tangential flux into the core by means of an additional winding wrapped around the core to simulate the flux operating conditions of the core. The test must simulate the magnetic operating conditions as reasonably as possible and thus induce the rated voltage.

What is the difference between the tests?

ELCID: Simple test, easy setup at low supply, and energy level at 4% of rated induction, corresponding to an induced longitudinal voltage of approximately 5 V/m. A Chattock coil is routed between pairs of adjacent core slots to cause a magnetic potential difference between to contact points. The test is not conducted under realistic conditions and is subject to interpretation of the results.

Ring flux tests: A high energy level of up to 85% of rated induction, corresponding to 1.3 T flux and 100 V/m with a wrapped around coil to cause internal vibration and heating images.

A test of 60(50) Hz is preferred if conditions allow it for set up. It requires large source of power at 2–3 MVA, 400–600 V right at the generator simulating the operating conditions. It takes several days to execute. On the other hand, ELCID may be more appropriate for underground conditions where a large power source may not be available during construction and for this test. It takes only 8–10 hours to execute.

Ring flux test of 500 Hz gives comparable results to the 60(50) Hz test, but with smaller set up, easier transportation, and lower power requirements.

18.1.8 Generator, Motor Testing after Installation

Typical tests for rotating machines and MV cables in power plants are executed in the following order:

1. *Megger test* for a quick determination of the status of insulation and a possibility of some metallic object left behind within the winding.
2. *Polarization index*: This is a test of the insulation using a megger at DC voltages depending on the voltage of the winding, from 500 V to 5 kV. For instance a 13.8-kV machine would use 5 kV megger. A single 10-minute test is made by applying the test voltage on each phase from insulation to ground. The insulation resistance index is calculated by comparing the readings of 10 minutes over the reading of one minute ($R_{10}/R_1 \geq 2$). A reading of 2 is considered acceptable. If the one-minute insulation resistance is above 5000 MΩ, the calculated PI may not be meaningful. In such cases, the PI may be disregarded as a measure of winding condition.

   PI was developed to make interpretation of results less sensitive to temperature. PI is the ratio of two IR at two different measuring times. Insulation resistance is always higher at 10 minutes as the leakage current is lower. If the calculated index is around 1, additional testing after 48 hour of heat drying must be done to verify the insulation. Temperature of the winding does not rise during the test period of 10 minutes. So it is fair to assume that both R10 and R1 are measured at the same winding temperature.

   The timing of the two readings is significant. At one minute, it is considered that the capacitive current has disappeared, while at 10 minutes it is considered that the insulation molecules are fully polarized and oriented in line with the electrical field, therefore, the polarization current is also disappeared.
3. *Hipot Test* [3]: Hipot or dielectric withstand test is generally performed to assure that the winding insulation has a minimum level of electrical strength

to survive electrical stresses in normal service. Hipot tests may be performed with any of three types of voltages: AC at the power frequency, DC, and very low frequency (VLF) at 0.1 Hz.

The hipot test level is $(2V + 1)$ kV for the power frequency AC test and 1.7 $(2V + 1)$ kV for DC test for new stator windings with rated line-to-line voltage V. For instance for $V = 13.8$ kV, the AC test is 29 kV and for DC test is 49 kV.

An AC test is preferred for better fault detection and similarity with the normal operation, while the DC test is preferred due to their smaller testing apparatus required.

The test is conducted by connecting one side to a phase conductor and the other to ground. The test lasts one minute upon gradually reaching the target test voltage.

Repeat this test procedure for all circuit phase conductors.

18.2 Governors

In isolated systems, the governors control frequency. In large systems, governor is a part of the load operation control (Figure 18.10). A governor is a combination of devices that monitor speed deviations in a turbine and converts the speed variation into a change of wicket gate servomotor position, firing rates, to manage the load/generation transients. The governing system is the main controller of the turbine power.

For Governor droop and performance, see Chapter 19. The governing system must include the following:

- PID (proportional – integral – derivative) measurement and logic
- *Redundant speed sensing elements*: electrical, mechanical, and centrifugal
- Governor control actuators to work on the servomotors
- Hydraulic pressure supply system
- Turbine control servomotors: normally supplied as part of a turbine
- Redundant communication modules
- Air compressed system
- Adjustable permanent droop 0–10% and temporary droop 0–100%.

The primary functions of the hydraulic turbine governor are

- Start, maintain, and adjust unit speed for synchronizing with the grid.
- Maintain system frequency after synchronization by adjusting turbine output to load changes.
- Share load changes with the other units in a planned manner in response to a system frequency error.
- Adjust output of the unit in response to operator or other supervisory commands.

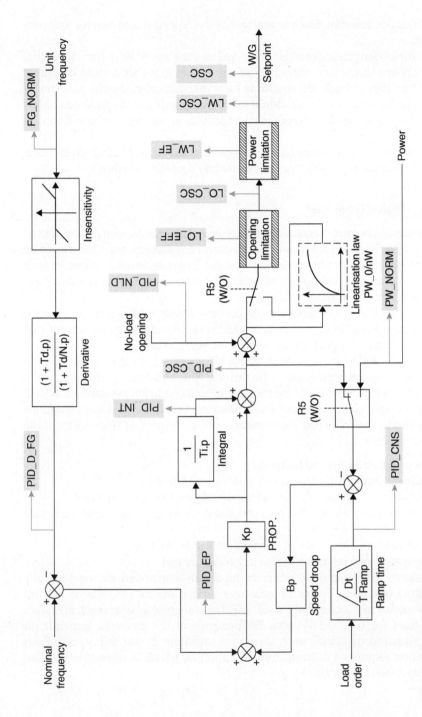

Figure 18.10 Governor frequency – power control, hydro unit.

Perform normal shut down or emergency overspeed shut down for unit protection.

Deadtime: Governor deadtime is defined as the elapsed time from the initial speed change to the first movement of the wicket gates for a rapid change of more than 10% of load. The deadtime for a mechanical-hydraulic governor is 0.25 seconds, whereas the deadtime for an analog or digital governor is less than 0.2 seconds enabling the governor to provide accurate stable speed control sooner.

Deadband: The maximum band between the values inside of which the variation of controlled variable does not cause any detectable action.

18.2.1   Digital Governors

With their lower cost, versatility, and software programmability, they are the favored governors today for new installations or replacements. One should be aware that the product life cycle of digital governors is relatively short, as with most computerized technology of today. Therefore, over time, spare parts can become difficult to procure.

The software and the hardware can become obsolete in as little as 10 years. A best practice would be to choose a well-known reputable manufacturer that will be around to support the equipment for years to come.

Tests: The fundamental performance test for a governor is by reviewing the quality of its speed regulation of the turbine. The measured performance of a governor is a major indicator for the operational condition assessment. ASME PTC-29 [4] specifies procedures for conducting tests in dry and wet conditions to determine the following performance characteristics of hydraulic turbine speed governors

- Droop – permanent and temporary
- Deadband and deadtime – speed, position, and power
- Stability index – governing speed band and governing power band
- Step response gain (PID) – proportional gain, integral gain, and derivative gain
- Setpoint adjustment – range of adjustment and ramp rate
- Overspeed, load rejection, frequency response test

  Actuators: These can be either hydraulically controlled or mechanically (motor) operated. Hydraulic actuators with pressure oil systems and oil servomotor are commonly used. Mechanical (motor operated) actuators are used for large MW units. Performance of the governor depends on the situation on hand: start, shutdown, dynamic power follow, etc. Each function requires a different transfer function, which in operation is called channel (see Figure 18.11.)

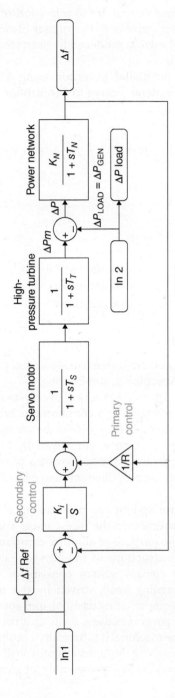

Figure 18.11 Governor typical transfer function.

On hydro turbines, the final output is the activation of the vanes (wicket gates) to allow more or less water onto the runner blades. On a steam turbine, activation of the fuel valve is produced to generate more or less steam as dictated by the governor.

Controls: Electro hydraulic digital governor using PID controllers are present-day practice. The general form of this controller is presented on the following equation [5]

$$M(t) = P + I + D = K_p\, E(t) + K_i \int E(i)\mathrm{d}t + K_d\, \mathrm{d}E/\mathrm{d}t \qquad (18.5)$$

Where

$M(t)$	= Controller output deviation	
$E(t)$	= Error as function of time	
K_p	= Proportional constant	Gain adjustment 0–20%
K_i	= Integral constant	Gain adjustment 0–10%
K_d	= Derivative constant	Gain adjustment 0–5%

18.2.2 Auxiliary Equipment

Oil pressure system: The oil pressure system consists of oil pumps, oil tanks, oil sump, and the necessary valves, piping, and filtering.

Flow distributing valves: The distributing valve system varies in design depending on the type of governor. For a common mechanical governor, the system consists of a regulating valve (that moves the servomotors) controlled by the valve actuator, which in turn is controlled by a pilot valve. These valves coupled with the oil pressure system provide a power amplification in which small force movements are amplified into large force movements of the servomotors.

Control system: The control system can be mechanical, analog, or digital depending on the type of governor. In the truest sense, the control system is the heart of the governor. The purpose of all other components in a governor system is to carry out the instructions of the governor control system. For mechanical governors, the control system consists of the fly-ball/motor assembly (ball-head or governing head) driven by the permanent magnet generator (PMG), linkages, compensating dashpot, and speed droop device.

Speed sensor: Mechanical governors use a PMG as a rotating speed sensor which is driven directly by the machine. It is basically a multiphase PMG that is electrically connected to a matching multiphase motor (ball head motor) inside the governor cabinet that drives the fly-ball assembly (or governing head) which is part of the control system.

Analog and digital governors use speed signal generator (SSG) driven directly by the unit which provides a frequency signal proportional to the unit speed usually through a zero velocity, magnetic pickup monitoring, rotating gear teeth, or through generator frequency measured directly by a potential transformer (PT).

Wicket gate position feedback: The restoring mechanism is a "feedback" device that feeds back the current wicket gate position and the post movement command position to the control system. In a mechanical governor, this is typically a pulley cable system. With digital governors, it may be a linear potentiometer or linear noncontact electrical positioning system.

18.3 Excitation: Control for Voltage and Reactive Power

18.3.1 Generator (Motor) Magnetizing

Where does the reactive power come from in the generator?

That is a good question (see Figure 18.12). Electrical energy cannot be converted directly into useful mechanical work (rotation of motor). Or the other way around, it is not possible to take the turbine input and convert it to electrical energy. To convert mechanical energy into electrical energy or the way around, magnetic field has to be created in the gap of stator and rotor of a generator (motor). Hence, some amount of energy has to be used in creating the magnetic field.

The power that is needed to create magnetic field is known as reactive power. Reactive power therefore is not actually a loss, but a transmission system that creates magnetic field serving to convert mechanical turbine energy in rotor into electrical energy in stator.

If underexcited, the generator will take in (import) reactive power with leading power factor because the reactive power is insufficient to create the magnetic field of the generator.

If overexcited, generator will deliver (export) reactive power with lagging power factor in addition to the reactive power needed to magnetize the generator. The generator must shed (export) some reactive power to work with the same flux.

Regardless of the reactive power flowing in or out, generator will always deliver real power, including even if there is no flow of reactive power at power factor equal to unity. The source of reactive power is in the generator exciter that magnetizes the generator armature. By means of the exciter DC current, the exciter develops the controlled generator flux and then the terminal voltage on the armature (stator). The voltage will then push the current to the connected load. The DC current is increased/decreased by

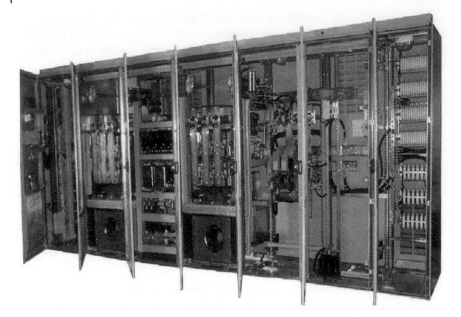

Figure 18.12 Generator static exciter board.

varying the exciter thyristor cutting (firing) angle to let more/less of the AC input through the thyristors. Therefore, when the DC current input is reduced (underexcited), the generator must import (absorb) some reactive power to maintain a constant flux and consequently to meet the voltage set point. As the terminal voltage changes due to the load change, the exciter must provide the reactive output to match the change and to keep the voltage constant.

The thyristor bridges are 100% redundant and automatically switchable in case of a failure. The AC/DC conversion process generates a lot of heat, which has to be channeled out of the excitation cubicles and out of the plant control room by redundant forced ventilation. A failure of the ventilation will lead to a trip of the generator unit.

18.3.2 Static or Brushless Excitation

The principal function of the excitation system is to furnish power in the form of DC current and voltage to the generator field for creating magnetic field. Furthermore, the exciter automatic voltage regulator (AVR) regulates the generator electrical output. Most of the large utility generators use redundant AVRs fed from external potential and current instrument transformers, as shown in Figure 18.13. Fast response of the excitation current is desired to preserve unit stability and to control its voltage.

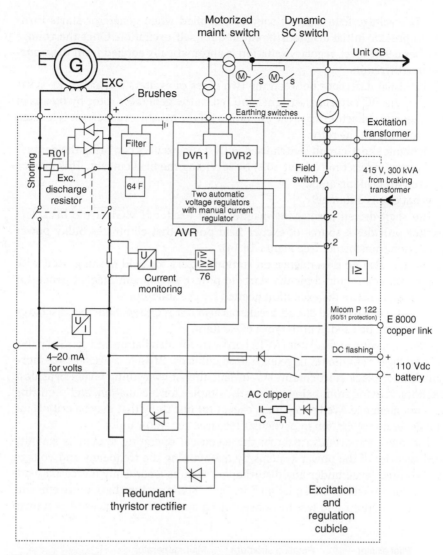

Figure 18.13 Static excitation diagram (with brushes).

Ceiling voltages range from 1.5 to 3 pu. Going over 2.0 pu may be subject to excessive magnetic core saturation and may not be practicable from the economic reasons.

In a generator with static exciter, shown in Figure 18.13, DC excitation current is transferred from the exciter panel to the generator rotor poles by carbon brushes to generate the magnetic flux and the electro-magnetic force (EMF) voltage.

A DC voltage field flashing source is applied when generator starts turning to provide initial magnetism required for self-excitation. Once the voltage reaches a 90% level, regular excitation is automatically applied from the transformer/thyristors.

The visual difference between the two types of exciters is that for the static exciter, the DC current is external and fed to the generator rotor by means of brushes and collector rings. The excitation for the brushless generator is located and rotating on the rotor.

Dynamic short-circuit switch used with static exciters can bring the generator to stop in less time (about 30%) by reversing the field current. This may be desired for peaking units.

What else is different?

The shaft-driven permanent magnet pilot exciter (PMG) of the brushless exciter provides a source of exciter field power that eliminates bulky power excitation transformer (see Figure 18.14).

There is also no dependence on station battery for field flashing used with static exciters, required for the start-up power. Permanent magnet generator (PMG) can initiate the excitation needed for the start-up.

Large expensive field circuit breaker is also not required. Shorting discharge resistors can be used in both types of exciters.

Compact voltage regulator (AVR) hardware for installation and control panel or switchgear eliminates large excitation cubicles. Brushless exciters are more suitable for black start conditions. In fact almost everything favors brushless exciters, starting from being less costly, simpler for installation, and requiring less maintenance. Well everything, except for one thing that may be critical for the generator operation in particular for large generator units.

The most important aspect in the electrical operating system is stability and security of the power systems by maintaining the frequency and voltage at a desired level, under any disturbance, such as a suddenly increased load, generator outage, or transmission lines fault. Static exciters have more efficient transfer functions and fast response during system disturbances. A response

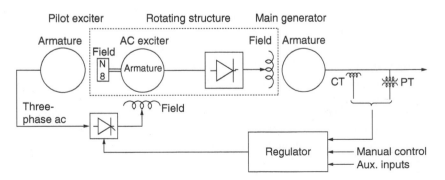

Figure 18.14 Brushless rotating exciter with PMG.

time of 50 ms is now achievable by brushless generators, while static generators response is even faster and to a higher ceiling voltage of 2 pu and over.

Power system stabilizers (PSSs) are also better suited to operate with static exciters. The basic function of PSS is to extend the stability limit by modulating generator excitation to provide the positive damping torque to power swing modes. The PSS generates a supplementary signal, which is added to control loop of the generating unit to produce a positive damping.

Smaller generators up to 100 MW for industrial applications favor the generators with brushless excitation. Diesel engine generators always use brushless, due to their small layout requirements and their black start capability. On the other hand, the utilities by their codes will demand on having generators with static exciters to better ride them through system disturbances.

18.4 Generator Circuit Breaker

Circuit breakers for generators as illustrated in Figure 18.15 [6, 7] are not ordinary switchgear breakers, but specially made for the generator duty to meet the requirements specified in IEEE Std.37.013.

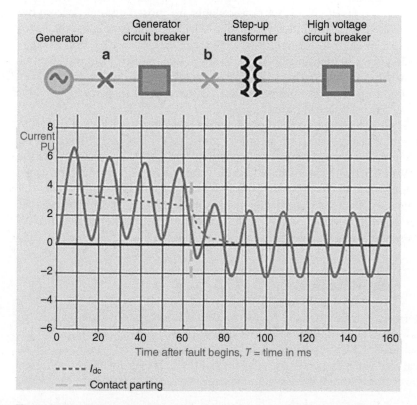

Figure 18.15 Generator circuit breaker. Source: Courtesy of Eaton and ABB.

Generator-fed faults can experience delayed current zero condition, where the high inductance to resistance L/R ratio of the system can cause the DC component of the fault current to exceed 100%. Furthermore, the breakers must possess extremely high short-circuit interrupting ratings for the fault locations at "a" and "b" shown in Figure 18.15. Generator breaker ratings at 5 and 15 kV, 4–5 kA continuous up to 75 kA interrupting are available for installation in the plant switchgear by Eaton Corp. The breaker offers a relatively economical solution for generators up to 60 MW, 13.8 kV, 0.8 power factor. For even larger generators over 30 kA, ABB breakers [7] are used. These can be fitted in line with the isolated phase bus (IPB) installation to serve for the generator protection switching and synchronizing.

18.5 Generator Step-up Transformers

Large GSUTs are typically directly linked to individual generators to operate on a unit basis. Often, an overall differential protection is zoned around both pieces of equipment (G + T), in addition to the protection zones provided for each item separately.

The MVA ratings of GSUTs for power plants are chosen differently from the transformers used in the industrial plants. We can recall from Chapter 2 that the industrial plant transformers are selected with their base MVA ratings capable of carrying a ½ plant bus load, i.e. to share the load with the transformer paired on the same bus. In case of a transformer failure, the healthy transformer is supposed to pick up the full plant load and carry it on its forced cooled oil natural air forced (ONAF) rating.

GSUTs work as single units; thus, these transformers are rated to carry the full generator MVA output on its forced cooled MVA rating. For instance, on our recent project, a 125 MW, 0.8 pf generator had a GSUT rated at 156 MVA (125 MW/0.8 pf) with three out of four oil/water coolers operating. One water cooler is always held as spare and the generator is generally loaded to its full MVA capacity. Even the transformer impedance for this transformer was based on the cooled rating of 156 MVA. In this case, the transformer impedance on 156 MVA base was 13%.

18.6 Heat-Rate Curve

This term is commonly used by mechanical engineers in power stations to express plant or unit efficiency. The heat rate is the heat energy in British thermal unit (BTU) required per kWh of electrical output generated. Thus, the heat rate curve indicates the efficiency of the unit over its operating range. The lower this number, the less input energy is required to produce each kWh of electricity. The heat rate numbers will determine which unit is chosen to operate and how to operate it in the most efficient way.

18.7   Hydraulic Turbine Cavitation

Cavitation [8] is a problem in hydraulic reactive turbines that negatively affects their performance and causes substantial vibration and serious damages. It generally happens at a particular operating zone below 40% loading. Cavitation erosion problems (Figure 18.16) are found in the turbine runner and other parts of the turbine system as a result of rapid changes in pressure, which leads to the formation of small bubbles or cavities in the liquid.

It occurs when the static pressure of the liquid falls below its vapor pressure, the liquid boils and large number of small bubbles of vapors are formed. These bubbles collapse near the metal surface at a high frequency in such a way that the elastic shock wave created is able to cause erosion of the adjacent materials. These bubbles mainly due to low pressure are carried by the stream to higher pressure zones where the vapors condense and the bubbles suddenly collapse, as the vapors are condensed to liquid again.

This results in the formation of a vacuum cavity and the surrounding liquid rushes to fill it. The streams of liquid coming from all directions collide at the center of cavity giving rise to a high local pressure whose magnitude may be as high as 7000 atm. Formation of cavity and high pressure are repeated many thousand times a second, causing pitting on the metallic surface of runner blades or draft tube. It manifests itself in the pitting of the metallic surfaces of turbine parts. The material then fails by fatigue, added by corrosion.

Reaction Francis turbines and propeller/Kaplan turbines are suited for medium and low head hydropower sites.

Figure 18.16 Result of cavitation.

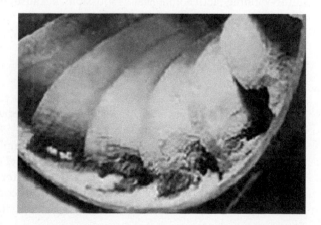

In general, there are two ways to reduce the cavitation damage. One involves optimizing the hydraulic design of equipment; the other involves developing coatings for the substrates of wetted parts, which can prolong the overhaul interval of hydraulic components.

18.8 Generator Cooling

Equipment cooling is essential to maintain the design thermal limits. Cooling in generators can broadly be classified into two types: open and closed circuit.

Open circuit cooling is suitable for small generators. The air is drawn into the generator by means of fans and circulated inside. The air is then released back into the atmosphere.

Closed circuit cooling is used in large alternators. Hydrogen and water are used as the cooling media. Hydrogen and/or water can transfer heat better than air as they have higher specific heat rates. Hydrogen has a low density that results in reduced windage losses for the alternator rotor resulting in lower generator dimensions and weights and better generator efficiencies.

Cooling performance is greatly improved through the use of hydrogen-cooled gas in place of air as coolant for the rotor winding and stator core. Hydrogen is circulated through the radial and axial ducts of the generators by means of pumps. Though, water has a better cooling capacity as compared to hydrogen, the circulation equipment for water is more expensive. Special systems for the purification of water are also required.

Generator winding cooling system is closed circuit ventilation, with circulating air passing through air/water heat exchangers. The generators are normally designed to operate continuously at rated power with one cooler out of service.

Large generators over 1000 MVA are hydrogen-cooled for rotor and water cooled for stator windings. An increased output of per unit volume of stator-winding active material is possible because of the higher thermal conductivity and specific heat of water.

18.9 Plant Black Start

The principle of a black start is that should there be a total shutdown and isolation of the plant from other power sources, the power plant must have a capability to restart itself from its own resources.

Black start in an *industrial plant* may be very simple. When the lights go down and everything comes to a stop, there is a diesel standby generator that will immediately get energized and prepare the plant for a new start. The diesel engine will kick start based on sustained (three seconds) loss of volt logic. Diesel engine must be self-energizing by means of its own 24 Vdc battery.

Obviously, the diesel standby generator will not be required to run the whole industrial plant. It must, however, have the capability to restart the essential loads, charge DC batteries, provide lighting, heating, and assist in critical chemical processes that may be solidified if held in its down state for too long. The plant will have station batteries and UPS to ensure the plant critical circuit breakers and other control systems are always ready to reclose as soon as the main power is restored. The UPS and the station battery must be of sufficient capacity to run the control system for four to eight hours.

In a *power plant*, the criterion is a bit different. The black start is a power plant application used when the grid due to unexpected failures or climatic storms is broken into several nonfunctional islands.

Here again, standby generator and station battery play a key role to maintain the system in readiness to restart. The standby generator must be capable of powering all the auxiliary (essential) loads of a single power plant unit to bring it to operation and readiness to synchronize on the HV grid. A 100 MW hydro unit would typically need about 400 kW of operating auxiliary load to be able to bring the power plant to produce power. Once it is connected to the grid, it can supply its own load and standby generator is no longer needed. A 1 MW diesel generator can easily provide that load and have no trouble starting the biggest motors in the plant. The rule of thumb here is that the standby diesel engine must have MW capacity at least as large as 1.5 times the rating of the largest pump motor and enough capacity to power all the auxiliaries for several hours.

As the standby generator kicks in, the power plant auxiliary power distribution system must be realigned and restored. The bus incoming breakers must close in a proper sequence as directed by the plant control system. All the essential plant breakers including the plant control system are DC controlled from the station battery. It allows the essential auxiliaries to be restarted and help the generating unit to bring itself to a position of synchronizing to the grid to serve its customers.

Generally, the power plants feed the grid via HV switchyards with each generator connected to double busbars. The goal is to reintegrate the overall grid one plant at a time into an operating system for the area. Once the first generating unit is energized, it is then synchronized to its adjoining switchyard dead bus. This can be a bus alone or a bus with a line to an adjacent remote power plant. The synchronizing is being done via a procedure called "line charging," whereby the generator breaker is closed on a dead bus and then the voltage is raised to the generator's nominal value through the GSUT.

During this process, the governor switches from droop to Isochronuous mode (ISO) with a 0% droop. In this mode, the generator is capable of rapidly raising its frequency (speed) and voltage to facilitate synchronizing at a remote bus. This adjustment is being done from a remote source calling for more/less speed/ voltage to match the remote source. Once the match is accomplished, the remote HV breaker will close and rejoin the two power plants together.

When the generator load reaches 10% of its nominal MW rating, the governor automatically switches to the generator droop 2–4% mode to properly share the load.

18.10   Synchronous Motor

Unlike the asynchronous (induction) motor, this motor rotates at the synchronous speed, thus the two magnetic fields (rotor and stator) tend to be aligned with each other.

One advantage of the synchronous motor is that the magnetic field of the machine can be produced by the direct current in the field winding so that the stator windings need to provide only a power component of current in phase with the applied stator voltage, i.e. the motor can operate at unity power factor. This condition minimizes the losses and heating in the stator windings.

The power factor of the stator electrical input can be directly controlled by adjustment of the field current. If the field current is increased beyond the value required to provide the magnetic field, the stator current changes to include a component to compensate for this over-magnetization. The result will be a total stator current that leads the stator voltage in phase. It provides reactive volt-amperes needed to magnetize other apparatuses connected in the plant such as transformers and induction motors. Operation of a large synchronous motor at such a leading power factor is an effective way of improving the overall power factor of the plant electrical loads.

The field current may be supplied from an externally controlled rectifier through slip rings, or, in larger motors, it may be provided by a shaft-mounted rectifier with a rotating transformer or generator.

18.10.1   Damper Winding

A synchronous motor is not self-starting. This apparatus needs be given a push by an induction pony motor or a damper winding to reach its full synchronism. At any speed other than synchronous speed, its rotor would experience an oscillating torque of zero average value as the rotating magnetic field repeatedly passes the slower moving rotor. The solution is a short-circuited damper winding similar to that of a rotor of an induction machine, which is added to the rotor of the synchronous machine to provide its starting torque. Once the motor is brought to about 95% of synchronous speed, the field current is then applied and the rotor pulls into synchronism with the revolving field winding. [9]

The damper winding has an additional property of damping out any oscillation that might be caused by sudden changes in the load on the rotor when in synchronism. Adjustment to load changes involves changes in the angle by which the rotor field lags the stator field and thus involves short-term changes in instantaneous speed. This causes currents to be induced in the damper windings, producing a torque that acts to oppose the speed change.

18.11 Plant Capacity and Availability Factors

Capacity factor of a power plant is the ratio of its actual energy output over a period of time to its potential output at full nameplate capacity continuously over the same period of time. It is a utilization factor, which includes all the plant forced and planned outages, seasonal factors, availability of fuel, and dispatch requirements of the daily load cycle. It can be expressed in terms of day, month, and most often in terms of a year.

Annual capacity factor $\quad C_{Fa}$ = Generated MWh/Nameplate MW plant capacity $\times$ 8760 h

Annual availability factor $\quad A_{Fa}$ = Time plant is available to produce/8760 h

$$C_{Fa} < A_{Fa}$$

Availability factor of a power plant is the ratio of the amount of time that it is able to produce electricity over a certain period, to the amount of the time in the period. Occasions where only partial capacity is available may or may not be included. Where they are included, the factor is called: *Equivalent availability factor* (EAF). The capacity factor for a period will always be less than the availability factor for the same period. The difference depends on the dispatch choice to run the plant or not. If the unit is ready or on spinning reserve, then it is considered as available.

The availability of a power plant greatly depends on the age of the plant, type of plant, fuel, and how the plant is operated. Everything else being equal, plants that are run less frequently have higher availability factors because they require less maintenance. Most thermal power stations, such as coal, geothermal and nuclear power plants, have availability factors between 70% and 90%. Newer plants tend to have significantly higher availability factors. Gas turbines as peaking units tend to have relatively high availability factors, ranging from 80% to 95%.

Plant capacity factor is one of the key factors used to estimate a costing structure for a power plant. The plant efficiency does not matter much if the fuel (water, sun, wind) is not available. Or if the fuel is expensive and must be burned only if no other energy resources are available. The capacity factor can be estimated separately for each unit to determine the favored units to be used. Herewith, in this chapter, one can get a feel for the utilization of the various types of power plants and generation.

The following power plants were reviewed with respect to their annual operating capacity factors: thermal, hydro, nuclear, solar PV, concentrated solar, and wind.

Table 18.1 Capacity factors in the UK.

Year: Plant type	2007	2008	2009	2010	2011	2012	2013	Average (%)
Nuclear	59.6	49.4	65.6	59.3	66.4	70.8	73.8	61.9
Gas turbine, Comb. Cycle	64.7	71.0	64.2	61.6	47.8	30.3	27.9	56.6
Coal fired	46.7	45.0	38.5	40.2	40.8	56.9	58.4	44.7
Hydro	38.2	37.4	36.7	24.9	39.2	35.8	31.7	33.7
Wind	27.7	27.5	27.1	23.7	29.9	29.0	32.3	27.5
Solar photo voltaic	9.9	9.6	9.3	7.3	5.1	11.2	10.2	8.6

In some cases, the plants may have dual utilization as a source of electrical power and a source of heat by utilizing the waste heat. In that case, the plant should be judged on the basis of two capacity factors: primary one for electrical power and the other on the basis of the waste heat delivered and used. The waste heat factor is seasonal and highly dependable as a by-product of the usage of the electrical power.

"Renewable energy is marching on. It's a self-reinforcing cycle," they say. As more renewables are installed, coal, oil, and natural gas plants are used less. As coal and gas are used less, the cost of using them to generate electricity goes up.

The following figures in Table 18.1 were collected by the Department of Energy and Climate Change on the capacity factors for various types of plants in United Kingdom grid [10] (Table 18.1):

18.11.1 Typical Capacity Factors for Power Plants

1. *Thermal power plant*: Expected capacity factor: 60–65% Base load generation.

   Thermal power plant capacity 4000 MW (4×1000 MW Units) produces annual energy of 21 725 000 MWh.

   Calculate annual plant capacity factor C_{Fa}. 365 d $\times$ 24 h = 8760 h/yr.

   Capacity factor is dependable on the daily load cycle, and plant planned and forced outages.

   $$C_{Fa} = \frac{\text{Energy MWh Produced}}{\text{Plant MW Nameplate} \times 8760} = \frac{21\ 725\ 000}{4000\ \times\ 8760} = 0.62 \rightarrow 62\%$$

2. *Hydroelectric power plant*: Expected capacity factor: 38–50%. Peaking power.

Nameplate capacity: 200 MW (4 × 50 MW Units). Annual energy production: 700 800 MWh.

Capacity factor is dependable on seasonal water availability. Generate electricity intensely during wet seasons to avoid water spilling and sparingly during dry spells. Often, used for spinning reserve to stabilize the system due to their rapid start capability. Planned outages can be scheduled during annual dry periods.

$$C_{Fa} = \frac{700\ 800}{200 \times 8760} = 0.40 \rightarrow 40\%.$$

3. *Nuclear power plant*: Capacity factor expected: 80–93%. Base load generation.

Plant nameplate capacity: 2400 MW (4 × 600 MW), Annual energy production: 16 800 000 MWh

Capacity factor is dependable on the daily load cycle, and plant planned and forced outages.

$$C_{Fa} = \frac{16\ 800\ 000}{2400 \times 8760} = 0.80 \rightarrow 80\%.$$

4. *Solar (photo voltaic) plant*: Capacity factor range expected: 13–19% . Solar dependable.

Nameplate capacity: 3 MW, Annual energy production: 450 000 MWh. Capacity factor dependable on the plant locations and periods of high insolation. Power storage limited.

$$C_{Fa} = \frac{450\ 000}{3 \times 8760} = 0.171 \rightarrow 17.1\%$$

5. *Concentrated solar plant (CSP)*: Tube and tower type. Capacity factor range expected: 25–35%. Plant nameplate capacity: 350 MW, Annual energy production: 9 100 000 MWh

$$C_{Fa} = \frac{9100\ 000}{350 \times 8760} = 0.297 \rightarrow 29.7$$

6. *Wind power farm*: Capacity factor range expected: 20–40%, Wind farm capacity: 80 MW (40 × 2 MW), Annual energy production: 180 600 MWh. Capacity factor is wind dependable, which is also a seasonal factor. High wind and low winds are not useful.

$$C_{Fa} = \frac{180\ 600}{80 \times 8760} = 0.315 \rightarrow 25.7\%$$

References

1 Electrical Baba (2016). Electrical Concepts. https://electricalbaba.com/concept-of-subtransient-transient-steady-state (accessed 15 July 2019).

2 Guo Qiang (2017). *Application of ELCID Test and Ring Flux Test on a Stator Core.* Xi'an: APT technology co., ltd.

3 Gupta, B.K., Stone, G.C., and Stein, J. (2009). *Stator Winding Hipot (High Potential Testing).* IEEE.

4 ASME PTC-29-2005 Governor Test Procedure.

5 R. Thapar *Chapter-6 Hydro Turbine Governing System.*

6 Eaton: VCP-WG, 15 kV Generator Breaker, Catalog Brochure BR01301001E.

7 ABB: Generator Circuit-Breaker, Catalog Brochure.

8 Khuranaa, S. (2011). Navtejb and Hardeep Singh – effect of cavitation on hydraulic turbines. *International Journal of Current Engineering and Technology* 2 (1) Inpressco.

9 Jadric, M. and Francic, B. (2004). *Dynamics of Electrical Machines.* Issued by Graphis in Croatian.

10 Department of Energy and Climate Change, UK Annual Capacity Factors.

19

Power Dispatch and Control

CHAPTER MENU

19.1   Plant and System Operation, 441
     19.1.1   Load Variability, 442
     19.1.2   Plant Scheduling, 442
19.2   Load – Frequency Control, 444
     19.2.1   Moment of Inertia H, 445
     19.2.2   Governor Control, 446
     19.2.3   Deadband, 446
     19.2.4   Speed Droop, 446
     19.2.5   Generator Droop Control, 447
     19.2.6   Generator Operation, 449
19.3   Voltage Reactive Power Control, 452
     19.3.1   AVR Droop Characteristic, 452
     19.3.2   Generation Operation Modes, Reactive Power Sharing, 453
19.4   Line Transfers, Import/Export Power, 456
References, 459

19.1   Plant and System Operation

Power dispatch and generation control are some of the major daily activities in the power plant and system operation. It is important for young engineers to learn the basic concepts of the operation of power systems to be able to insert themselves into the ongoing conversations. There are two principal activities handled by dispatch centers responsible to operate the power system on a 24-hour basis to generate, transmit, and deliver power to factories, residences, services, and roads on a timely, stable, and economic manner. These are

(1) Load and frequency control – governor action
(2) Reactive and voltage control – excitation action

Practical Power Plant Engineering: A Guide for Early Career Engineers, First Edition. Zark Bedalov.
© 2020 John Wiley & Sons, Inc. Published 2020 by John Wiley & Sons, Inc.

Figure 19.1 Regional dispatch center.

19.1.1 Load Variability

Electricity is a unique commodity that must be produced at the instant it is needed. It cannot be economically stored in large quantities using today's technology. The electrical load changes in a daily cycle. The tariffs paid for energy used (kWh) can slightly change the shape of the consumption curve. Namely, on the consumer side, the demand can be "stored" for the periods of lower tariffs. For example washing dishes can be postponed (see Figure 19.1).

Normally, the load is the lowest at night and highest during the day. While in the most part, the changes can be predicted; often, there are load changes that are sudden and unexpected. The operators try their best to anticipate and maintain a balance on the grid to match the generation with the load. However, major power system upsets do occur: unit outages, transmission line failure, and system short circuits. All of these may cause sudden changes to the system frequency and a possibility of instability and breakages of the system. The grid frequency must be maintained within ±0.5 Hz.

19.1.2 Plant Scheduling

To accommodate the load cycle, the utilities like to possess a mixture of generation, including diesel engines, hydro, gas, coal, wind, solar, and nuclear. The best operating power systems are those with multiple generators, none of the units greater than 3–5% capacity of the overall system, units spread around, and concentrated close to the load centers to minimize line losses.

Each type of generation is assigned its own particular duty of peaking or base load generation. Hydro power units, for instance are low-cost operating units. They can be used for base load at times of high waters as well as to serve as standby spinning reserve units capable of responding relatively quickly and fully loaded within a minute. Thermal units will take longer to start and accept load due to their limitations in turbine rate of rise in metal temperature (MW/min).

The utilities normally try to schedule and operate the generators at a given power level known to be the most efficient for the season and time of day. The generation dispatchers decide which unit is the most economical to increase power output, or bring additional units on line to maintain the frequency at its proper level. Water resources must be used wisely to insure adequate water levels for efficient power production and never fall under the minimum operating level (MOL).

Load factor: A utility's load factor is its average load as a percentage of its peak load. It is far more economical to serve large industrial loads having high load factors (>85%) in comparison to let us say convention centers with load factors of <10%.

Base load: Utility must generally supply some minimum base load 24 hours a day, every day of the year. The generators that are the most economical to operate are used to supply this base load. They are usually loaded or dispatched close to their maximum power level.

Intermediate plants: These units generally have their power output raised every morning and throughout the day, and then shut down every evening. They run considerably more hours than peaking units, but fewer than the base-load units. Utilities typically meet intermediate loads with older generating units that were once base-load plants, but have recently been replaced by newer, more efficient units.

Peak loading plants: The peak load is daily, seasonal, and annual demand that requires careful planning. To meet the varying system, load dispatchers decide which generators to run and at what power level and in which district.

Throughout the day, dispatchers stack up generators into a predicted load curve. Factors that the dispatchers weigh in include each generator's specifics, such as

- Cost of operation – production cost
- Maximum capacity
- Maintenance schedule
- Environmental emissions
- Spinning reserve needed

Production costs: The cost of electricity from each generator includes both fixed and variable costs. Fixed costs include costs of financing plant construction, leases, and taxes. Variable costs are those expenses that increase roughly proportional to the use of generator. Oil and gas units are less costly to build but costly to operate due to their high cost of fuel, so it makes sense to use them rarely and for peaking duties.

Generation reserves: This is the additional capacity 15–20% beyond the actual maximum load to allow for maintenance outages and operating flexibility. It is expressed as a percentage of generating capacity available above the utility's peak load.

Spinning reserve: The system dispatcher must plan for the possibility that one or more generators may fail to produce power at any time. The dispatcher must have several units partially loaded at all times so they can be brought up to full load quickly if and when needed. It is common to have 15–20% spinning or rolling reserve to provide reliable service in case some units trip, break down, or demand suddenly surges.

Maintenance schedule: Besides occasional forced outages, generators require periodic planned outage for scheduled maintenance. Most of maintenance is planned during seasons when the load is relatively low, typically spring and autumn.

Environmental emissions: Increasingly stringent environmental regulations are forcing utilities to move from economic dispatch procedure to what is called environmental dispatch. For example generators are dispatched to minimize air pollution on smoggy days even if it results in higher generating costs. Generation plants are sometimes limited in their annual release of emissions.

19.2 Load – Frequency Control

The system generators operate in synchronism with each other on a system's common frequency. If the load exceeds the generation, the system generators will slow down and the system frequency will fall. A change in frequency of ± 0.5 Hz is manageable. A fall in frequency of 1.5–2.5 Hz is a sign of a serious disturbance that may lead to a fragmentation of the system. A major effort must be undertaken by the machine governors to restore as much generation as quickly as possible.

Initially, any change in frequency due to an increased load must be initially matched by the kinetic energy of the generator masses until the governors are ready to act to increase fuel, steam, or water flows to reaccelerate the machines. A similar situation occurs in the reverse direction for a loss of load. The governors act based on their operating software, by choosing an appropriate governor channel and transfer function dependent on the severity of the situation for rapid or slow response.

A sudden loss of load may lead to a severe overspeed on the generators during the governor dead band, until it catches on and starts regulating; in this case, slowing down the machine. Each machine is designed to withstand a certain amount of overspeed. To insure a safe operation of the machines and to protect it against a failure of the governor, the turbines are equipped with ultimate safety trip mechanisms, both electrically and mechanically operated, to shut down and close the fuel valve or water wanes and stop the unit. For a steam turbine, a 10% overspeed would cause a trip. On a recent project, we tested the mechanical trips on two high inertia hydro units at 145% of nominal speed. The electrical safety trip was set just over the mechanical setting, at 150%.

19.2.1 Moment of Inertia *H*

The change in speed naturally depends on the moment of inertia of the machine, or the *H* stored energy constant in kWs/kVA. Let us look at what happens to a turbine generator running isolated at nominal speed when subjected to a load change:

Unit capacity	100 MW, running at low load
Nominal speed	1800 rpm at 60 Hz
H constant	4.5 MWs/MVA
Governor dead band	0.3 second before the steam valve starts regulating
Sudden load increase	30 MW, or loss of generation
Stored energy	$= 4.5 \times 100$ MW $= 450$ MW-s (stored energy is proportional to speed^2)
Loss of stored energy	$= 30$ MW $\times 0.3 = 9$ MW-s
Temporary loss of frequency	$= 60 \times \sqrt{450 - 9/450} = 59.4$ Hz (loss of 0.6 Hz, 1% drop)

Well, of course, this is an academic case; a single generator on the system. A loss of 30 MW on a large system would not be felt as severely. A system having a larger connected capacity during this event would be defended by multiple units, with a range of inertia constants, dead bands, and transfer function responses.

In many a situation, generators are intentionally made with large diameter rotors and high inertia constants. For instance, in a large nickel smelter plant in Indonesia, we built three hydro generating plants for a total of 450 MW produced by seven generators. Each 65 MW generator was built with a 9 m diameter rotor for a high inertia constant of about 9 MWs/MVA. New generators were installed as new smelters were added to the plant production. Smelting process causes huge swings in reactive and active power demand and consequently large swings in the frequency. Initially, when the plant was running on three generators, the frequency swings during plant disturbances were in the range of 9 Hz. As the plant expanded and two more generators were added, the swings were reduced to a 2 Hz range and finally with seven generators, the swings were reduced to about 0.5 Hz.

19.2.2 Governor Control

Speed turbine governors [1] vary prime mover output (torque) automatically for the changes in the system speed (frequency). The speed sensing device is usually a flyball assembly for mechanical-hydraulic governors and a frequency transducer for electro-hydraulic governors. The output of the speed sensor passes through signal conditioning and amplification (provided by a combination of mechanical-hydraulic elements, electronic circuits, and/or software) and operates an actuator to adjust the prime mover output (torque) until the system frequency change is arrested. The governor action arrests the drop in frequency, but does not return the frequency to the pre-upset value (50 or 60 Hz) on large interconnected systems. Returning the frequency to the nominal frequency is the job of the automatic generation control (AGC) system. The rate and magnitude of the governor response to a speed change can be tuned (transfer functions) to suit the characteristics of the governor controls and the power system to which it is connected.

Therefore, if a decrease in system frequency occurs due to a loss of generation or an increase in load, the shaft speed of each connected synchronous generator will also decrease. This speed decrease is transmitted to a mechanical or electro-hydraulic governor causing the valve to move up and allow more flow (fuel, steam, water, etc.) to the prime mover. Thus, the output power (torque) of the controlled prime mover will increase and help arrest the frequency change.

19.2.3 Deadband

There are two types of deadbands in speed governing systems: inherent and intentional. Inherent deadband is very small (<0.005 Hz) on most governors connected to the power system and can be neglected. Intentional deadband is used by manufacturers and generation operators to reduce activity ("hunting") of controllers for normal power system frequency variations and may be large enough (about 0.05 Hz) to affect the overall power system frequency control performance.

19.2.4 Speed Droop

The droop setting indicates the percentage amount the measured quantity must change to cause a 100% change in the controlled quantity. For example a 4% frequency droop setting means that for a 4% change in frequency, the unit's power output changes by 100%. Or, in other words, if the frequency falls by 1%, the generator with a 4% droop setting will increase its power output by 25%.

Figure 19.2 Two unit plant.

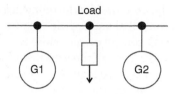

The droop line formula is an equation of a straight line (see Figure 19.2 for a graphical presentation).

$$y = dx + q \tag{19.1}$$

where,

d = droop.

In general, the percent movement of the main prime mover control mechanism can be calculated as the speed change (in percent) divided by the per unit droop. A governor tuned with speed droop will open the control valve a specified amount for a given disturbance. This action is initiated from a feedback from the main prime mover control mechanism. If a 1% change in speed occurs, the main control mechanism must move enough to cause the feedback through the droop element to cancel this speed change. Thus, for a 1% speed change, the percent movement of the main control mechanism will be the reciprocal of the droop (i.e. if the droop is 4%, the movement will be $1/0.04 = 25\%$).

If the governor is tuned to be "isochronous" (ISO) (i.e. 0% droop), it will keep opening the valve until the frequency is restored to the original value. This type of tuning is used on small, isolated power systems. On large, interconnected systems, ISO setting would result in excessive governor movement. Therefore, speed droop is used to control the magnitude of governor responses for a given frequency change of all generators to produce a shared response after a disturbance. The lower the droop setting, the faster is the governor action. Hydro units generally are set with a 2% droop.

19.2.5 Generator Droop Control

Governor droop allows synchronous generators to run in parallel, to share load among the generators in proportion to their power ratings. The droop action causes the unit to adjust its valve, gate, etc., positions to raise/lower the power output. The major advantage of using droop in paralleling is that it allows dissimilar MW machines to be paralleled without concern for their load sharing interface.

The droop characteristic of the unit governor is selectable on the machine for the engine output in the range of 0–10%. The units having different droop characteristics will share the load inversely proportional to their droop characteristics and based on their ratings. The higher the percentage droop setting on the unit, the less the unit will assume of the overall load. The ISO droop is a flat control characteristic with 0% droop. The standby units typically run with an ISO control or very close to it. The standby units, typically diesel engine units, will accept all the load that is left connected as part of the "essential" load. The nonessential load will be reconnected after the main power restoration.

The governors used for thermal, hydro, gas, diesel generation, etc., are similar in their applications. The difference is in the type of actuators acting on the mechanical parts to regulate valves, gates, and solenoids as well as in the transfer functions (software) implemented for each type of generation.

The droop characteristic is not necessarily always linear. For peaking units like hydro, the starting characteristic from 0% to 100% will certainly be different from that engaged during the normal operation. The starting droop will initially approach the ISO mode and then change as the unit approaches the full speed. For that reason, governors may have several acting channels with different percentage droop settings and the characteristic shapes to suit the type of generator and its application.

A speed reference as percentage of actual speed is set in the droop mode. Since the actual prime mover speed is fixed by the grid, this difference in speed reference and actual speed of the prime mover is used to increase the flow of working fluid (fuel, water, steam, etc.) to the prime mover, and hence the power output. The amount of power produced is strictly proportional to the error between the actual turbine speed and speed reference. As the difference increases, fuel flow is increased to increase power output, and vice versa. This type of control is referred to as "straight proportional" control. If the entire grid tends to be overloaded, the grid frequency and hence actual speed of the generators is decreasing. All units will see an increase in the speed error, and therefore will increase fuel flow to their prime movers to generate higher power output.

Machines [1] in an isolated plant set to *the same droop* will share the load proportionally to their capacities. This is applicable to any droop setting set equal on the units. Assume G1 = 100 MW, G2 = 200 MW, plant load $P = 200$ MW (see Figure 19.3). The machine loading will be

$$G1 \rightarrow P\frac{G1}{G1+G2} = 200 \times \frac{100}{100+200} = 66.6 \text{ MW},$$

$$G2 \rightarrow P\frac{G2}{G1+G2} = 200 \times \frac{200}{100+200} = 133.3 \text{ MW}$$

If the droop settings on the two machines are different, the loading will not only be based on the machine MW ratings. The machine with lower (flatter) governor droop will assume more load. Governors using speed droop require a sustained change in system frequency to produce a sustained change in prime mover control mechanism or generator power output. Therefore, governors alone cannot restore the power system frequency to the pre-disturbance level.

Figure 19.3 Load sharing on droop.

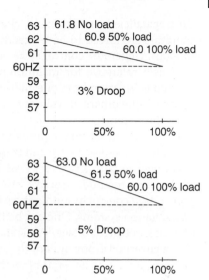

An important feature of a governor system is the device by which the main prime mover controls mechanism and hence the generator power output can be changed without requiring a change in system speed. This is accomplished by the speed reference (or speed adjustment). This adjustment called supplementary regulation is typically made remotely from the dispatch center by the AGC system.

19.2.6 Generator Operation

Real power kW sharing [2] on an electrical grid is fully automatic, and it depends on speed and fuel rate control between the generator sets based on percentage of kW load. When multiple synchronous generators are connected in parallel to an electrical grid, the frequency is fixed by the grid, since individual power output of each generator will be small compared to the overall load on a large grid. Synchronous generators connected to the grid run at the same frequency, but at various speeds due to their difference in the number of poles.

When generator sets are operating together on an isolated (islanded) bus, they are commonly provided with controls to allow each machine to operate at the same percentage of load as the percentage of load on the total system. This is termed a "load sharing control system." Hopefully, the operating control systems are of the same suppliers. In many cases, the control systems (governors) are not compatible with each other. Although it may be possible to integrate systems from different manufacturers, preferably generator governors and load sharing controls should be of the same manufacturer and model to avoid conflicts in responsibility for proper system operation.

In a paralleled arrangement, the frequency and voltage outputs of the generators are forced to have exactly the same values when connected to the same bus. Generator control systems cannot simply monitor bus voltage and speed as a reference for maintaining equal output levels, as they do when operated in isolation from one another. If a generator is regulated to a different speed than the others, it will not share kW load properly with other generator sets in the system.

Successful load sharing requires addressing of both kW and kVAR load sharing, under both steady state and transient conditions. The generating systems always require both kW and kVAR load sharing, but they do not both need use the same method. Reactive kVAR power sharing is primarily dependent upon voltage control by automatic voltage regulator (AVR) through the generator excitation systems. One can be ISO and the other can be droop and of the different % setting. Voltage/kVAR sharing is dealt with later in this chapter.

For a governor droop system to function correctly, the following conditions must exist for all the generators on the common bus:

(1) The generators must have the same no load frequency when operating disconnected from the bus.
(2) The generators must be set to droop at the same rate from no load to full load.

When trying to interface dissimilar load sharing units from different suppliers, it is possible to configure the system so that some of the generators in the system operate at a base load level, and others operate in a load share state. The base load machines operate at a constant load, while the generators operating in ISO load sharing mode will "float" with the balance of the available load.

19.2.6.1 Single Generator Operating in Parallel on Grid

The frequency is preselected for the generator [3]. If the generator speed setting is even fractionally higher than the frequency of the utility, the governor will go to full load in an attempt to increase the bus speed. Since the utility speed cannot be influenced, the engine will remain at full fuel.

Equally, ISO (flat frequency control) operation in parallel with utility is impractical. If the governor setting were even a bit above of the utility frequency, the unit would go full fuel, since it cannot reach the reference speed. Similarly, if the setting were slightly below actual speed, the fuel racks would go to fuel-off position. Therefore, governors should not parallel in ISO with any operating system, which is too large to be able to influence its speed.

Droop solves the problem. It causes the governor speed reference to decrease as load increases.

When paralleled with a bus, the load on the engine is determined by the reference speed setting of the governor droop. Increasing the speed setting, while not able to change the speed of the bus, it will cause a change in the load the engine is providing. The graph in Figure 19.3 indicates that the amount of load is determined by where the droop line intersects the speed of the bus. If the location of this line is moved, either by changing the reference speed or the droop angle in the unit, the amount of load will also be moved.

19.2.6.2 Multiple Engines on Isolated Bus

In these plants, the engines are capable of changing the frequency (speed) of the bus in line with a change in load. Multiple engines can also be paralleled on an isolated bus with all engines on droop and one in ISO mode control. This arrangement can provide a stable situation at constant speed. The droop engines are sharing base load, while the ISO unit picks up the slack in either direction (Figure 19.4), providing the load change does not exceed the MW rating of the ISO engine. If the load increases beyond the ISO unit capability, the whole plant would slow down for the droop units to resettle to the new droop – speed positions, while the ISO unit would remain overloaded to the point where it would be unable to meet the governor reference speed.

Peaking generators typically have special governor droop characteristics, usually 2%, to quickly reach their nominal target load point. This is declared as minimal loading, like 50% or 100%. The generator must reach these load points in specific times as specified by the utilities. The droop characteristic may not be linear droop or flat ISO, but something of a mixture to ensure the unit reaches that desired loading without overshooting or undershooting. The characteristic includes certain transients as not to overshoot when approaching the load target.

Figure 19.4 ISO and droop operation.

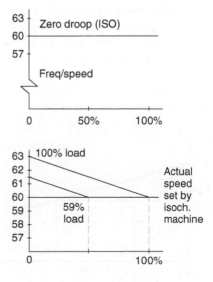

Once they are synchronized and loaded on the grid, the frequency is dictated by the grid based on the performance of mixture of loads and generators. The generator loading is dictated from the grid control board by dictating the set points and scheduled targets. Hundreds of AC generators are running synchronously with the power grid acting like an infinite sink.

The operators know their daily schedule and loading pattern, thus, they add/remove the units in a projected generation daily pattern to match the expected grid operated load.

Frequency droop control is useful for allowing multiple generating units to automatically change their power outputs based on dynamically changing loads. In case of a significant loss of a large generating unit, if the system does not collapse, all the other units on the grid must respond quickly to pick up the slack, at a frequency value **below** its nominal value (for example 49.7 Hz (or 59.7 Hz) along their droop lines.

If a large load is tripped, then generators will be off loading, and the frequency will settle at a steady-state value **above** its nominal value (for example 50.5 or 60.5 Hz). Supplementary frequency controllers are therefore necessary to bring the frequency back to its nominal value (i.e. 50 or 60 Hz).

19.3   Voltage Reactive Power Control

19.3.1   AVR Droop Characteristic

The excitation and voltage control of a synchronous generator is done by AVR that use generator voltage and/or current as inputs in order to control its voltage output to a pre-set value. With that comes the reactive power control. AVRs assume different control approaches to optimize the generator performance depending on whether the generator is connected to the grid, or operating in an island mode (see Figure 19.5).

Droop compensation is a control technique designed for the generator operation on the grid. It is not required when one generator is in island mode.

An interpretation of the graph is that as the reactive power demand from the generator increases, the generator terminal voltage decreases. The set point in

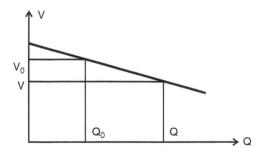

Figure 19.5 Voltage droop.

the AVR is chosen so that when the generator reactive power Q supplied is zero, the generator voltage is equal to the nominal voltage.

The reactive power generated is calculated from the generator voltage and current signals fed back to the AVR. Droop compensation is set as percentage drop of the nominal voltage V_N for maximum reactive power Q generated. Droop setting can be given values from 0%, which effectively disables the droop, to a maximum of usually 10%. Typically, a setting of 4–6% is chosen.

19.3.2 Generation Operation Modes, Reactive Power Sharing

The following are the generator operating scenarios:

- Island mode as a stand-alone generator.
- Synchronized with grid.
- Operating in island mode and in parallel with other generators.

19.3.2.1 Island Mode, Single Generator

This is the simplest case in terms of AVR control (Figure 19.6) [4]. A single synchronous machine operating in island mode is only responsible for two actions:

- Control the busbar voltage to the required nominal level.
- Supply the load with the required reactive power and respond fast to any load changes to meet the demand at any time.

The AVR in this case does not require droop compensation to control its output. Set the droop setting within the AVR to 0% and disable (close) the current compensating switch.

19.3.2.2 Single Unit, Synchronized with Grid

In this case (Figure 19.7), the AVR needs droop compensation in order to control its output [4]. The grid voltage V_G is fixed by the grid and cannot be controlled by the unit AVR. Any requirement for reactive power from the generator will result in the AVR internal voltage set point V and corresponding power factor to change to meet the new demand. The increased reactive power demand

Figure 19.6 Island mode, single unit.

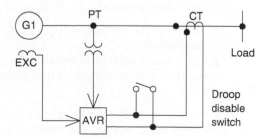

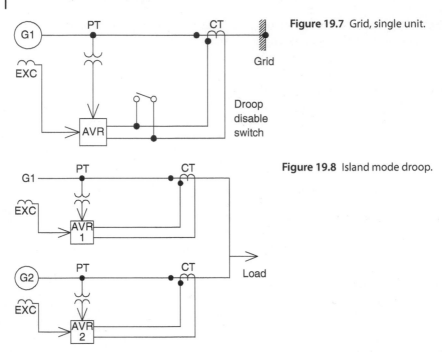

Figure 19.7 Grid, single unit.

Figure 19.8 Island mode droop.

Q_L causes the AVR set point to increase from V_G toward V_L because of the droop compensation control.

Therefore, the droop can be enabled or disabled, allowing the flexibility to disable it when in island operation and to enable it before connecting it to the grid. This eliminates the undesired effect of lower-than-nominal voltages when in island operation.

19.3.2.3 Island Operation with Paralleled Generators

For the case of island mode operation with at least two generators connected in parallel to supply the load, the control of the voltage and the reactive power requirements has to be equally shared between the generators in parallel (Figure 19.8). There are two control methods for the generator AVRs to accomplish equal sharing by either: control with droop compensation or with cross-current compensation.

(1) *Control with droop compensation* [4]: In this case, the following conditions need be satisfied:
- The generators must be of equal size.
  The AVRs must have the same droop characteristic and the same setting applied.

In the simplest case, the AVRs operate in droop compensation mode. The two generators share equally the reactive load connected according to the

Figure 19.9 Cross current compensation.

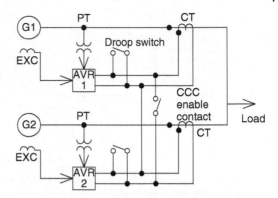

droop characteristic of the AVRs and the setting applied. This control mode is ideal in the grid situations. In an island mode, it results in the voltage output changes dependent on the reactive power demand. So as the requirement for reactive power increases (kVARs exported), the output voltage from the generators increases due to the droop compensation.

(2) *Cross-current compensation (CCC)* [4]: Figure 19.9 is a method that allows two or more paralleled generators to share equally a reactive load, given that the following assumptions are satisfied:
- There is no grid connection, i.e. the generators operate in island mode, the generators are of equal size.
- The AVRs have the same droop characteristic.

CCC is one of the methods to facilitate sharing kVARs between the units running on the same bus. CCC is the operation of paralleled generators without intentional voltage droop in particular when the excitation on the units is dissimilar. This is achieved by the insertion of a current transformer (CT), usually on "B" phase of each generator, and paralleling the CTs together to provide an identical voltage bias to each AVR in the system.

The system works best when the voltage regulators are all of the same manufacturer and model, which is not always the case. This may require changing of all the voltage regulators in the system to a new model. Using CCC results in no intentional droop in voltage from no load to full load on the system, so it is considered to be superior to a reactive droop compensation system from a performance perspective.

This system includes both the droop-compensation and the CCC enable-disable contacts. When in CCC, the same current develops through the compounding CTs of all the generators in parallel, since they are identical, and when the CCC contact closes, it stops flowing through the AVRs, but flows only through the CTs. The configuration allows flexibility by switching

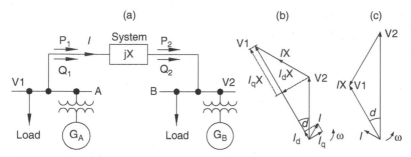

(a) System of two generators interconnected.
(b) Phasor diagram when V1>V2, and I_d and I_Q are components of I.
(c) Phasor diagram when V1<V2.
Weedy [6].

Figure 19.10 Line power transfer. Source: From John Wiley & Sons.

the contacts to enable or disable CCC or droop, to operate in grid or as island, as desired.

19.4 Line Transfers, Import/Export Power

There is a close relationship between the absolute voltage difference $|V_1 - V_2|$ between two nodes (power plants) and the flow of reactive power in the inter-connecting tie line, as explained by B.M. Weedy [5] (see Figure 19.10). The voltage difference between two nodes connected by a transmission line, having X and R characteristics is

$$\Delta V = \frac{RP + XQ}{V_2} \tag{19.2}$$

Also, it can be shown that the transmission angle is proportional to active and reactive power flow P and Q, as follows:

$$\delta V = \frac{XP - RQ}{V_2} \quad \rightarrow \quad \delta = \sin^{-1} \frac{\delta V}{V_2} \tag{19.3}$$

The power transfer relation is given as follows:

$$P = V_1 \times V_2 \frac{\sin \delta}{X} \tag{19.4}$$

X is the line reactance. If line $X \gg R$, as it is the case in most of power circuits, the MVAR and MW power flows can be considered proportional to ΔV, the voltage difference and line angle δ, respectively:

$$\text{MVAR} \rightarrow \Delta V = \frac{XQ}{V_2}, \quad \text{MW} \rightarrow \delta V = \frac{XP}{V_2} \tag{19.5}$$

In essence, it can be said that Q will be directly proportional to ΔV, while P will be proportional to δV.

In a network, for $V_1 > V_2$, the current will flow from V_1 to V_2. Based on the aforementioned, one can conclude as follows:

(1) The direction of the MVAR flow will be from the bus with higher voltage.
(2) The flow of P power will be proportional to the line power angle δ between the two voltage vectors. In fact both rotor angles are negative. To make the power transfer one node must be less negative. That station will be the power sending end. The direction of the MW flow will be from the bus which is operating with a rotor angle advanced to the other rotor.
(3) If the power transfer is arranged over a number of lines of different paths, the power flow will be inversely proportional to the path impedances.

The aforementioned facts are of fundamental importance to the understanding of the operation of power system.

The voltage difference is due to the operator actions on the exciter or transformer tap changers to direct the Q flows, both in magnitude and direction.

The angular difference is due to the operator actions on the governors at the plants to direct the P flows, both in magnitude and direction.

The power flows P and Q are nondependent on each other. The flows of either MW or MVAR will stop or reverse when the angle and voltage difference closes or reverses, respectively.

All of the above looks like a lot of work for an engineer. Fortunately, lately none of this work is done by hand. The engineer may use a Power Flow software program and tweak the voltages and angles on a preplanned plant electrical diagram and observe the changes and flow adjustments on the screen.

Consider the Figure 19.10 above. G1 is in phase advance of G2, while V_1 is greater than V_2 as shown on the phasor diagram above. Evidently, I_d and hence P is determined by angle δ, while I_q and hence Q mainly by $|V_1 - V_2|$. In this case, $V_1 > V_2$, and reactive power is transferred from G1 toward G2. If operators change the situation and arrange for $V_1 < V_2$, the direction of Q flow reverses. These two operations are independent of each other.

At some point within a network, there may be a deficiency of reactive power, hence, the voltage at that point is low and the reactive power is being imported. Reverse operation is also valid. Therefore, the voltage can be controlled by importing or exporting reactive power VARs into the network by acting on the generator exciter.

The generator excitation control is directed by a digital AVR, which controls the firing of thyristors to supply the generator with variable field current and stable output voltage. If the internal generator voltage EMF is lower (higher) than the grid voltage, the generator will import (export) reactive power and operate with leading (lagging) power factor.

An overexcited machine, the one with greater than normal excitation, generates (exports) reactive (inductive) power while operating on the power factor lagging side of the generator capability curve. An underexcited machine absorbs reactive power and operates on the leading side of its capability curve. It generates negative or leading VARs. The same is valid for the synchronous motors or condensers. The generator operating at the power factor leading side, while importing reactive (capacitive) power must lower the internal voltage E. In some instances, the generator may be in danger of falling out of synch due to the low voltage generated. In this case, the generator stability curve is lowered.

Local or dispatch operator selects on the AVR the set point for the generator operation, either to import or export VARs or to operate at power factor 1, with a goal of maintaining the end voltage stable. This type of action is effective for small systems and short supply lines. For long lines, the lines must be compensated by series capacitors. If the load is highly inductive, the load has to be compensated with shunt capacitors placed at the load buses.

Generator capability is tested during generator tests to prove its reactive power limits. Often, the generators are not at the load centers and are unable to prove their full capability in the grid as the test may raise the grid voltage. In this situation, the test can be performed in the same plant by exchanging MVARs between two plant units, by making one generator export (lagging pf), while the other is made to import (leading pf) (see Figure 19.11).

The MVARs are regulated by the generator exciter AVR. The AVR of G1 is adjusted to be underexcited (to import), while G2 is made to be overexcited. By exporting MVARs, the voltage of G2 raises toward its limit of 105% of V_n, while the voltage of G1 falls toward its low voltage operating limit of 95% of V_n. If this arrangement is unable to meet the MVARs flows without going over the voltage limits, the generator transformers OffLTC taps could be adjusted

Figure 19.11 MVAR exchange.

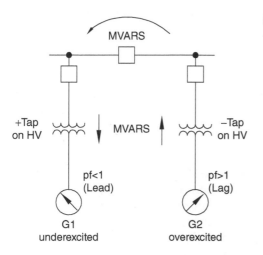

on both units, by holding G1 voltage up and G2 down. The transformer T1 is tapped up (+) and T2 down (−).

Therefore, the onload tap changer (OnLTC) on generator transformers enhances the unit capability to better manage the MVAR flows. If the transformers do not have OnLTCs the taps would have to be preset ahead of the test. After the test, the taps are returned to their regular positions.

References

1 Hoa D. Vu & J. C. Agee; Tutorial on Speed Governors, June 2002, Self published.
2 Gary Olson, White paper, Cummins; Paralleling Dissimilar Generators: Part 3 – Load Sharing Compatibility, Published by Cummings.
3 Woodward (1991); Speed droop and power generation, Application Note 013202.
4 Stefanos Spanopoulos, Power Engineering Consultants; Voltage Control – Droop and Cross-Current Compensation.
5 B.M Weedy, 1967; John Wiley and Sons, Book: *Electric Power Systems*.

20

Diesel Engine Generator Plant and Standby Power

CHAPTER MENU

20.1 Gen-Set Ratings and Classifications, 461
20.2 Plant Design, 463
20.3 Unit Performance, 465
    20.3.1 Example 1, 465
    20.3.2 Example 2, 466
20.4 Plant Electrical One Line Diagram, 468
20.5 Waste Heat Recovery (WHR), 469
20.6 Engines for Ships, 473
References, 473

20.1 Gen-Set Ratings and Classifications

Diesel engine generation is the generation of choice in industrial plants for remote areas where there is no connection to the electrical grid. For remote or isolated locations, diesel generation is reliable and economical in both initial capital and operating cost. The remote plants must be operated as efficiently as possible because transporting diesel fuel is costly. Diesel engine maintenance can be costly as well. In many cases in the Arctic, diesel fuel transport is only possible during winter months over frozen lakes and rivers.

Medium speed units consume about 0.225 kg of fuel per kWh produced. Slower base load engines consume about 0.192 kg of fuel per kWh produced. When the engine waste heat is recovered for the plant process and space heating, the fuel efficiency may raise from 40% to about 80%. Diesel engines run most efficiently when loaded at 85% or more of their ratings. Prolonged engine running at lower power levels makes them less efficient, more prone to clogging and failing.

Practical Power Plant Engineering: A Guide for Early Career Engineers, First Edition. Zark Bedalov.
© 2020 John Wiley & Sons, Inc. Published 2020 by John Wiley & Sons, Inc.

Diesel engines are rated for their mode of operation depending on the application as follows:

Standby: Generator output is available with varying load for the duration of the interruption of the normal power source. Average output is at 70% of the standby power rating. Typical operation is 200 hours a year with maximum expected usage 500 hours/yr, according to ISO 8528.

Prime: Generator output is available with varying load for an unlimited time. Average load is 70% of the prime power rating. They are capable of demand of delivering 100% of **Prime** rating ekW with 10% overload for emergency use for maximum of 1 hour in 12 hours. Overload cannot be more than 25 hours/yr. ISO 3046 [1].

Continuous: Generator output is available with nonvarying load for an unlimited time. Average output is 70–100% of the continuous rating. Typical peak demand is 100% of the continuous rating ekW for 100% of operating hours. ISO 3046.

Engine power ratings: standby > prime > continuous.

Firm power: Diesel engine generation plants, that are intended to be operated as base load generation, must be built with a firm capacity to cover the whole plant load at the most critical time of year. For Northern regions, this is during the winter months. The number of engine-generator units N must be able to supply the full load with some available spare capacity. This is called the plant firm power. Actually, the plant must have $N + 1 + 1$ units, where N is the number of units providing the firm power, +1 unit on ready standby, and +1 unit on maintenance or repair. This is calculated on the basis of the N units running at 85% to allow for some spinning reserve.

Typical examples of "firm power" selection could be as follows:

- For a plant with a peak load of 15 MW, an appropriate design basis would be to include $(4 + 1 + 1)$ units of 4 MW each.
- For an industrial plant load of 33 MW, the unit make up could be $(6 + 1 + 1)$, with 6 MW units, or $(8 + 2 + 1)$ with 4.5 MW units, where the total generating plant capacity would be 48 and 49.5 MW, respectively.

For a large power plant (Figure 20.1), suppliers should be requested to offer their proposals for a 48–50 MW plant, based on the ratings of their machines, with the individual unit capacity of 4.5–6 MW capacity for each engine-generator set. The quote request will usually stipulate that there shall be one unit on repair for the whole plant and one on standby, ready to start, for every four running units needed to meet the firm power. For Northern climates, plant owners will usually require several engines to be equipped with waste heat recovery (WHR) capability at both levels; low heat from the jacket water and turbocharger and high heat from the lube oil and exhausts.

Figure 20.1 A remote DG power plant.

The technical evaluation by the project engineers must determine the best packages suited for the industrial plant by looking at providing firm power and WHR with respect to the projected annual MWh needs.

The waste heat boilers and heat exchangers should be designed to recover heat from several units to meet the requirements at the most critical time of year. The low heat would be used for the camp and residences, while the high heat would be used in the plant process. Supplementary boilers with fuel burners are also required to supplement the heat during the periods of low generation, system failures, and for extreme cold. It is desired to run the basic plant production in shifts, around the clock, to insure adequate diesel engine generation is used, and it guarantees sufficient waste heat is made available for the plant services during 24 hours.

20.2 Plant Design

The generators are typically brushless excitation design with permanent magnet, Class H insulation with Class F temperature rise 105 °C over 40 °C ambient.

Smaller units, up to 2 MW, can be designed to be installed into transportable containers. Typically, the container comes complete with all the associated electrical, control, fire protection, and mechanical equipment enabling the diesel generator (DG) set to operate on its own. Several containerized DG sets can be placed and cabled together to form a larger power plant.

These containerized mobile DG plants come complete with the switchgear assembly in a separate container. It typically consists of one cell for each generator, two breakers cells for the station service transformers, four to six feeders for power distribution, and a tie breaker. The station service load is fed from the station service transformers and LV motor control centers (MCCs). The overall plant auxiliary load can be estimated at about 5% of the total plant generating capacity.

Plants with four to six and more engines are common. The spacing between the engines is 6–7.5 m (20–25 ft), center line to center line, depending on the engine width. A loading/maintenance bay is allowed at the main gate entry side. The engine auxiliary bay is located at the rear of the engines, while the electrical space is allowed on the generator side, with a walkway or truck driving space laid in between the generators and MCCs. The main medium voltage (MV) switchgear is located in a separate room adjacent to the control room, both of which are located a floor above the main floor.

The cabling (usually middle voltage) is marshaled in trenches or cable trays in such a way to minimize cable crossings between the generators. Cables for the engine auxiliaries are laid on overhead cable trays at the rear side of the engines.

Generator grounding: Each generator is high resistance grounded for 5 A fault current by means of a single phase distribution transformer (see Chapter 5).

Plant automation: In-house generating plants are usually operated automatically as an islanded operation, without manpower intervention. The control system will automatically add and remove units, based on the ongoing power demand in the plant, taking into account the most efficient operating regions of the units (80–110%). The units will generally operate with a droop of 3–4% to equally share the load between the units.

Fuel storage requirements: Fuel storage tanks must be designed with a total capacity of fuel proportional to the annual consumption, considering the fuel source and the plant locations. In remote areas with difficult access, consider for the seasonal turnaround having at least three months spare fuel capacity. As we noted earlier, large gensets generate about 5 kWh for each kilogram of fuel. For the fuel estimation purposes, the true reading of the consumption with varied load including the auxiliaries one should use a figure of 3.5 kWh generated for each kilogram of fuel.

Industrial plants with their own in-house generation with multiple units do not need designated standby unit(s). It is unlikely that all the operating units will fail or trip at the same time. All the units must be black start capable, by either a DC battery or by air compressors battery operated. The black start is an operating condition applicable to standby units in large industrial or power plants. Upon loss of all the main generation or a loss of utility connection, the standby unit has to have sufficient capacity to initiate the plant restart, run all the essential auxiliaries and make the units ready for resynchronization to the utility supply.

Diesel gensets typically run with 42% efficiency at 100% load and about 37% at 50% load, not considering the auxiliary plant load. With waste heat recovered from the engines, the overall efficiency approaches 80% in the winter months. During summer days, some low heat may have to be discarded into atmosphere.

Day tank: This fuel tank rated for a day supply of fuel is located within or just outside the engine plant. This tank does not require explosion proof design or special fire protection installation.

Stack: Each engine comes with its own 30 m stack. Combining of stacks for several units in a single stack should be avoided. It reduces unit efficiency as it increases back pressure in the stack.

Two-stroke/four-stroke engines: A two-stroke engine is less costly, smaller, and mechanically simpler. It develops more power for the same speed and piston displacement because of elimination of two idle strokes. A four-stroke engine is more efficient, less noisy, has simpler cooling, and is more suited for waste heat exploitation.

Supercharging vs. turbocharging: Both are power boosters. The forced air charge to achieve better fuel economy can be increased either by reducing the air temperature or increasing air density or pressure, which is more effective and preferred. Either form of forced air induction into the engine cylinders can be used successfully. One can even use both on the same diesel engine, but unlikely.

Superchargers use belt-driven compressors by the engine crankshaft to draw more air into the engine's combustion chamber and deliver a power boost to the engine. The increased airflow to the engine allows more fuel to be combusted and can increase the power by about 40–50%.

Similarly, turbochargers use high temperature exhaust gas to drive turbine/compressors for injecting more air into the engines.

Though the superchargers are easier to maintain, more fuel-efficient turbocharger operation is preferred for the diesel engines to boost power.

After cooling: This is a heat exchange cooling device used on turbocharged and internal combustion engines to improve their volumetric efficiency by increasing intake air-charge density through isochronic cooling. A decrease in air intake temperature provides a denser intake charge to the engine and allows more air and fuel to be combusted per engine cycle, increasing the output of the engine.

20.3 Unit Performance

This information provides an idea of a typical diesel engine genset used for standby, prime, and continuous duty.

20.3.1 Example 1

Let us review the ratings, efficiencies, and heat balance of the machines.

The following data is from a major genset supplier for an engine operating at 60 Hz, 1200 rpm, 0.8 pf, at 90 °C coolant temperature, turbocharged, after-cooled (see Tables 20.1 and 20.2).

Table 20.1 Diesel engine fuel consumption.

Rating	Prime 1450 ekW	Continuous 1325 ekW	Prime 1450 ekW	Continuous 1325 ekW
% Loading	Fuel consumption (l/h)		Fuel consumption (kg/kWh)[a]	
100	354.2	348.3	0.222	0.239
75	284.3	260.1	0.237	0.238
50	193.8	178.7	0.242	0.245

a) Calculated based on ekW.

Table 20.2 Coolant distribution.

Heat rejection, kW at 100% loading	Prime	Continuous
To coolant	630	591
To exhaust	1445	1296
To atmosphere	207.8	205.8

The test fuel is distillate lower heating value (LHV): 42 780 kJ/kg, 18 390 British thermal unit (BTU)/lb, 1 L = 0.838 9 kg.
The use of heavy oil fuels (Bunker C) is considered 9% less efficient.
ekW = power measured at the generator terminals.

20.3.2 Example 2

Let us review the kilowatt ratings for the specific engines. We tabulate data for the 50 and 60 Hz machines (Figure 20.2). This tabulation provides information on the performance of the units used for base generation when operated at 50–110% of its nominal rating (see Table 20.3).

Specific losses and balance for the largest model above 16 cylinder, 900 rpm, 60 Hz, turbocharged, after-cooled are listed and presented as an energy tree in Figure 20.5 (see Table 20.4).

For the remote areas, medium speed 1200–900 (750) rpm engines are used.

Slow 400 rpm, heavier, but more efficient engines that require less maintenance are used for large power producers. Cost of these engines can run 4× the 900 rpm units. Plant standby units use 1800 rpm (1500) engines.

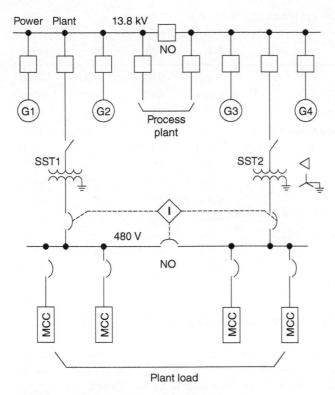

Figure 20.2 DG plant basic one line diagram.

Table 20.3 Engine performance, base load.

Model ratings (kW)	Standby	Prime	Continuous	Standby	Prime	Continuous
	720 rpm, 60 Hz			*900 rpm, 60 Hz*		
6 Cylinders in line	1680	1525	1375	2000	1820	1650
8 Cylinders in line	2220	2020	1830	2660	2420	2200
12 Cylinders in V form	3360	3050	2750	4000	3640	3300
16 Cylinders in V	4440	4040	3660	5320	4840	4400
	750 rpm, 50 Hz			*1000 rpm, 50 Hz*		
6 Cylinders in line	1730	1570	1420	2150	1940	1760
8 Cylinders in line	2280	2080	1890	2860	2600	2350
12 Cylinders in V	3460	3140	2840	4300	3880	3520
16 Cylinders in V	4580	4160	3780	5720	5200	4700

Engines 1375–5720 kW, turbocharged, after-cooled, four stroke cycle.

Table 20.4 Engine percentage of loading performance.

Loading (%)	110	100	75	50
Engine power (bkW)	5 060	4 600	3 450	2 300
Generator power (ekW)	4 840	4 400	3 300	2 200
Engine efficiency, nominal (%)	42.8	42.8	42.2	39.4
Fuel consumption, nominal (kg/bkW-h)	195.7	198.1	201.3	212.5
Exhaust gas mass flow (kg/h)	35 711	32 228	22 864	14 782
Fuel input, LHV (kW)	11 812	10 753	8 182	5 835
To jacket water (kW)	1 029	971	819	647
To oil cooler (kW)	510	485	425	360
To exhaust, 25 °C (kW)	3 663	3 379	2 704	2 016
To exhaust, 177 °C (kW)	2 792	2 481	1 623	1 006
To after-cooler (kW)	1 201	951	387	94
Atmosphere (kW)	329	345	382	407

20.4 Plant Electrical One Line Diagram

Remote generating plants are usually built as multiple engine plants for 10–30 MW installed capacity. The generator voltage is determined based on the total load to be distributed and the distance to be distributed.

The diesel engine generators rated at 1 MW and higher are built to any standard voltage; 3.3, 4.16, 6.6, 11, and 13.8 kV; 50 or 60 Hz, depending on the location.

If the plant is erected close to the load, the generator voltage will be selected 4.16 or 13.8 kV (North America) to suit the load without having any major transformation at the power plant, except for the station service load. In a typical industrial plant, the operating load is mainly at 480 V for the process equipment and 4.16 kV for the crushers and mills.

In view of the above, the power plant one line diagram (Figure 20.2) will comprise a 5 or 15 kV metal-clad switchgear with the following number of fully equipped breaker positions:

$N + 1$, generator cells, complete with protective relay and metering panel,
2 cells, for two station service 4.16 (13.8) kV–480 V, 500 kVA transformers,
1 cell, for bus tie breaker, and
3 cells, for feeder breakers for plant distribution.

The generator cells are typically full height, while the other cells can house two breakers per vertical section. The cells shall be equally distributed around the tie breaker, which will be held normally open.

Often, I have encountered the following situation. In a remote location of a power plant, a spillway gate is operated by a 100 HP, 480 V motor, which is to be powered by a transformer fed from an 11 kV line. In addition, a standby diesel generator must be provided to insure a safe operation in case of an outage. The installation may have a few additional smaller motors, lighting, and heating. The main drive is a concern due to its inrush current during a start.

So how big transformer does one need and how big a diesel engine is required for this application?

Without doing the calculation, my ball park recommendation for safely starting a 100 HP motor, I would use a diesel engine of 100 kW (that is in excess of 35% over the pump rating) and a transformer of about 200 kVA. Why this difference?

Well, a diesel engine generator has an automatic voltage regulator (AVR) which will adjust the voltage almost immediately to ensure the operating voltage is at a safe level. The transformer connected to a long line, on the other hand, will be subjected to a large voltage drop during a motor start, thus a larger transformer will be required to insure a safe voltage level for the motor during the start. A larger kVA transformer will have a relatively smaller impedance on the motor kVA base than a smaller transformer, thus the voltage drop on the transformer would be smaller.

20.5   Waste Heat Recovery (WHR)

The diesel engine generator is a relatively efficient electricity producer, but more than 60% of the input energy is wasted and thrown out through the stacks and in the cooling water. Waste heat from the diesel engine represents a potential source of energy that can be used for heating and cooling purposes, or, when coupled with a Rankine cycle,[1] can generate even more electrical power. See Figure 20.3.

Figure 20.4 shows realistic values of the potential waste heat use from a 16-cylinder, 4.5 MW unit. It is contributed by Galmea Consultants Ltd. (Spain) [2]. It is based on a Genset technical data sheet at 100% load, 60 Hz, 300 rpm, 100 m above sea level (ASL), and 25 °C.

The presentation is approximate, subject to change due to site conditions for each project.

Large part (40%) of waste heat is produced during the diesel engine's combustion cycle. As this heat heats up the cylinder walls, cooling is required to prevent the lube oil from evaporating and damaging the engine. Jacket water is

---

1 The Rankine cycle as shown previously is an idealized thermodynamic cycle of a heat engine that converts heat into mechanical work. The heat is supplied externally to a closed loop, which usually uses water as the working fluid.

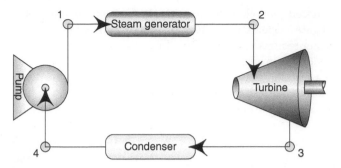

Figure 20.3 Rankine cycle.

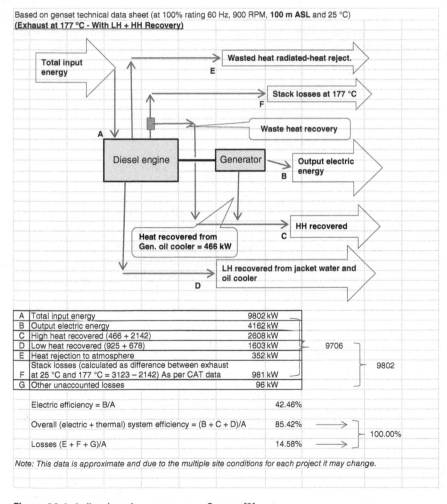

A	Total input energy	9802 kW	
B	Output electric energy	4162 kW	
C	High heat recovered (466 + 2142)	2608 kW	
D	Low heat recovered (925 + 678)	1603 kW	9706
E	Heat rejection to atmosphere	352 kW	
F	Stack losses (calculated as difference between exhaust at 25 °C and 177 °C = 3123 – 2142) As per CAT data	981 kW	9802
G	Other unaccounted losses	96 kW	

Electric efficiency = B/A	42.46%
Overall (electric + thermal) system efficiency = (B + C + D)/A	85.42% →
Losses (E + F + G)/A	14.58% →

100.00%

Note: This data is approximate and due to the multiple site conditions for each project it may change.

Figure 20.4 A diesel engine energy tree. Source: [2].

pumped in for engine cooling. To avoid excessive thermal stresses in the engine, a temperature difference of about 5–8 °C is recommended between the inlet and outlet water temperatures. Instead of using fans and outdoor radiators to cool off the jacket water, this waste heat can be recovered for the heating and cooling through heat exchangers.

The rest of the waste heat, 20%, is exhausted at the end of the expansion stroke. The temperature of the exhaust gases vary from 350 to 700 °C. With turbochargers and WHR units, the exhaust gas temperature is greatly reduced to as low as 177 °C. The 177 °C limit is established as the temperature below which some condensation and corrosion occurs.

If waste heat is taken from the engine and exhaust, the corresponding electrical load (pumps, fans, exchangers, etc.) will not be needed to be generated in the form of electricity and taken from the generator. In view of that, the engine load will be reduced as well as the waste heat. That seems to be a counter purpose operation and needs be taken into account. If the jacket water waste heat is not sufficient for the space heating and cooling, then exhaust heat must be added to cover for the deficiency. There are two systems that can be employed for recovering exhaust heat, as follows: (i) pressurized water system and (ii) two-phase, steam/water system.

Of the two recovery systems, the pressurized water system (Figure 20.5) is the simpler and less costly plus it needs no modifications to the engine. It uses a mixture of water and glycol suitable for lower temperatures. It works in temperatures between 82 and 92 °C and uses a gas/water heat exchanger. If higher water temperature, closer to 100 °C, is required, the system must be further pressurized to avoid flashing in the engine.

The steam/water system, option (Figure 20.6), also uses the jacket water, but it needs to bypass the turbocharger to reach higher temperatures needed

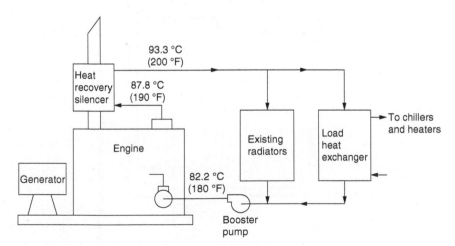

Figure 20.5 Pressurized water system. Source: [3].

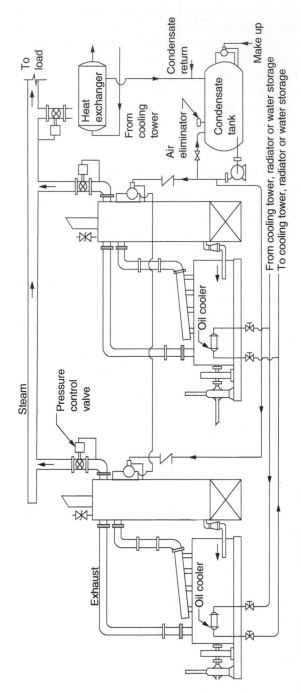

Figure 20.6 Steam/water WHR. Source: [3].

to saturate and superheat the steam coming out of the water cooling system. The operating temperature here is in the range of 110–120 °C and needs to be pressurized to avoid flashing in the engine. It also needs a condensate tank in addition to the heat exchanger.

The economy of the WHR will greatly depend on the industrial users and the daily cycle of the plant.

WHR definitely will be less in demand during summer months and during plant low operating period.

20.6 Engines for Ships

Large ships use diesel engines for driving the propellers and for generating electricity for other users. Shipboard diesel engines tend to have relatively constant load and a long life cycle. The load is reduced when the ship is in harbor and tied to shore power. Based on their steady load, WHR can augment the electrical generation by adding a steam turbine cycle and consequently add about 5% more electricity to the ship. The steam cycle module is relatively costly, and the ship builders estimate a payback of five years.

The turbocharger added to the exhaust pipe reduces the temperature in the exhaust. The steam cycle needs higher temperatures, thus a bypass is needed to capture higher temperature. This option can add 5–10% to the electricity production and reduce exhaust gas emissions.

MAN (supplier) claims that to a DG plant of 27–80 MW, a 2–9 MW steam turbine package can be added.

References

1 ISO International Standards Organization.https://www.iso.org/standard/28330.html.
2 Jose Galindo (2018), Galmea Consultants Ltd., Waste heat expert.
3 Chai, V.W. (1979). *Comparison of the Total Energy Systems for a Diesel Generation Plant*. ResearchGate.

21

Reliability Considerations and Calculations

CHAPTER MENU

21.1 Definitions, 475
21.2 Basic Reliability Engineering Concepts, 477
    21.2.1 Basic Mathematical Concepts in Reliability Engineering, 477
    21.2.2 Failure Rate and Mean Time Between/to Failure (MTBF/MTTF), 478
21.3 Different Failure Rates vs. Time Scenarios, 479
    21.3.1 The "Bathtub" Curve, 479
    21.3.2 The Exponential Distribution, 480
21.4 Estimating the System Reliability, 483
    21.4.1 Series Systems, 483
    21.4.2 Parallel Systems, 484
    21.4.3 Parallel and Serial Elements, 485
    21.4.4 Piper Alpha Rig Explosion, 488
21.5 Common Mode Failure, 489
    21.5.1 Space Shuttle Program, 490
21.6 Availability, 490
    21.6.1 Inherent Availability, A_i, 491
    21.6.2 Operational Availability, A_o, 492
    21.6.3 Availability Calculations, 492
    21.6.4 Availability in Series, 493
    21.6.5 Availability in Parallel, 493
References, 494

21.1 Definitions

Reliability engineering deals with the longevity and dependability of parts, products, and systems. Reliability engineering incorporates a variety of analytical techniques designed to help engineers understand the failure modes and failure patterns of these parts, products, and systems. Every engineer and manager must learn the terminology and usefulness of the reliability principles [1, 2] given below.

As a young engineer, I was lucky our company sent me on several seminars related to the reliability engineering. This was followed by me lecturing others.

This chapter refers for definitions, formulae, and some figures of the book *Reliability Engineering Principles*, best summarized and written by Drew Troyer, an utmost authority on reliability matters. Of course, the formulae and methods have been widely known and used by the industry before the book was issued.

- *Availability*: A measure of the degree to which an item is in the operable and committable state at the start of the mission, when the mission is called for at an unknown state.
- *Common mode failure*: The failure that can be assigned to be cause of failure of many components on the machine. For instance a lube pump failure can be a cause of failure or critical wear of several components.
- *Failure*: The event, or inoperable state, in which an item, or part of an item, does not, or would not, perform as previously specified.
- *Failure mode*: The consequence of the mechanism through which the failure occurs, i.e. short, open, fracture, and excessive wear.
- *Failure, random*: Failure whose occurrence is predictable only in the probabilistic or statistical sense. This applies to all distributions.
- *Failure rate*: The total number of failures within an item population, divided by the total number of life units expended by that population, during a particular measurement interval under stated conditions.
- *Maintainability*: The measure of the ability of an item to be retained or restored to specified condition when maintenance is performed by personnel having specified skill levels, using prescribed procedures and resources, at each prescribed level of maintenance and repair.
- *Maintenance, preventive*: All actions performed in an attempt to retain an item in a specified condition by providing systematic inspection, detection, and prevention of incipient failures.
- *Mean time between failure (MTBF)*: A basic measure of reliability for repairable items: the mean number of life units during which all parts of the item perform within their specified limits, during a particular measurement interval under stated conditions.
- *Mean time to failure (MTTF)*: A basic measure of reliability for nonrepairable items: the mean number of life units during which all parts of the item perform within their specified limits, during a particular measurement interval under stated conditions.
- *Mean time to repair (MTTR)*: A basic measure of maintainability: the sum of corrective maintenance times at any specified level of repair, divided by the total number of failures within an item repaired at that level, during a particular interval under stated conditions.

- *Mission reliability*: The ability of an item to perform its required functions for the duration of specified mission profile.
- *Reliability*: (i) The duration or probability of failure-free performance under stated conditions. (ii) The probability that an item can perform its intended function for a specified interval under stated conditions.

Engineers responsible for designing various engineering projects must incorporate reliability engineering into their plans and designs, including plant flow diagrams, procurement, operations, and maintenance. The most relevant and practical methods for plant reliability engineering, include

- Basic reliability calculations for failure rate, MTBF, availability, etc.
- Exponential distribution.
- Field data collection.

21.2 Basic Reliability Engineering Concepts

A basic understanding of the fundamental and widely applicable methods [2] can enable the engineers to improve the manufacturing/production process by acknowledging the critical part of the production, the most failure of the prom equipment, and part of the line where the blockages occur, their nature and root causes, and impact on the production process. Following that understanding, the engineer will ascertain which equipment needs redundant drives, more spare parts, better protection from environment, or a change in the production technology to suit the raw materials used to make products.

Let us look at one of the simplest and the most widely used reliability concepts and cost-benefits of the reliability calculations.

Suppose, a pumping system that is *critical* to your production is in a failed state 10% of time. It presently operates with a single pump. What would be an improvement if an additional pump is added in parallel with the present pump? If the system fails now at 10% of time and the system is made redundant, the failure rate would fall down to $0.1 \times 0.1 = 0.01$ (1%), which is an improvement of: $(0.1/0.01) \times 100 = 1000\%$. Well, it would be crazy not to improve the production line for an expenditure of $500. On the other hand, you would not want to spend money on a *noncritical* installation where you can tolerate the failure and make repairs at your convenience.

21.2.1 Basic Mathematical Concepts in Reliability Engineering

Many mathematical concepts apply to reliability engineering, particularly from the areas of probability and statistics. Likewise, many mathematical distributions can be used for various purposes, including the Gaussian (normal) distribution, the log-normal distribution, the exponential distribution, the Weibull distribution, and a host of others. We will discuss the exponential distribution, which is the most widely applied in the reliability engineering.

21.2.2 Failure Rate and Mean Time Between/to Failure (MTBF/MTTF)

The purpose for quantitative reliability measurements is to define the rate of failure relative to time and to model that failure rate in a mathematical distribution for the purpose of understanding the quantitative aspects of failure. The most basic building block is the *failure rate*, which is estimated using the following equation:

$$\lambda = r/t \tag{21.1}$$

where

λ = failure rate
r = total number of failures occurring during the investigation period
t = total running time/cycles/miles/etc., during an investigation period for both failed and nonfailed items, usually taken as one year.

For example, if five electric motors operate for a collective total time of 50 years with five functional failures during the period, the failure rate is 0.1 (10%) failures per year. Therefore, one motor failure occurs in this plant every year. This data one cannot acquire from your new plant. For the typical failure rate data, one has to search through the publications of the institutions collecting and sorting this type of information.

Another basic concept is the mean time between/to failure (MTBF/MTTF). The only difference between MTBF and MTTF is that we employ MTBF when referring to items that are repaired when they fail. For items that are simply thrown away and replaced, we use the term MTTF. The computations are the same. The basic calculation to estimate MTBF and MTTF is simply the reciprocal of the failure rate function. It is calculated using the following equation.

$$\theta = t/r \tag{21.2}$$

where

θ = mean time between/to failure
t = total running time/cycles/miles/etc., during an investigation period for both failed and nonfailed items, usually taken as one year
r = the total number of failures occurring during the investigation period.

The MTBF for our industrial electric motor example is five years for each motor, which is the reciprocal of the failure rate for the motors. Incidentally, we would estimate MTBF for electric motors that are rebuilt upon failure. For smaller motors that are considered disposable, we would state the measure as MTTF (see Figure 21.1).

The failure rate is a basic component of many more complex reliability calculations. Depending upon the mechanical/electrical design, operating context,

Figure 21.1 Failure rates. Source: From [2].

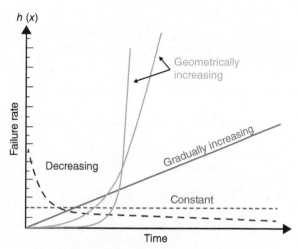

Depending upon machine type, the failure rate may decrease, remain constant, gradually increase or geometrically increase as a function of time

environment, and/or maintenance effectiveness, a machine's failure rate will vary significantly.

21.3 Different Failure Rates vs. Time Scenarios

21.3.1 The "Bathtub" Curve

The Gaussian distribution [2] has its place in evaluating the failure characteristics of machines with a dominant failure mode, but the primary distribution employed in reliability engineering is the exponential distribution. When evaluating the reliability and failure characteristics of a machine or a product line, we must begin with the "bathtub" curve, which reflects the failure rate vs. time from the installation to the point in time when it is totally worn out and replaced. That time is when you say: "It's no longer worth fixing it, as it will never work as new."

The bathtub curve effectively demonstrates three basic failure rate characteristics: declining, constant, and increasing. Regrettably, the bathtub curve has been harshly criticized in the maintenance engineering literature because it fails to effectively model the characteristic failure rate for most machines in an industrial plant, which is generally true at the macro level. Most machines spend their lives in the early life, or infant mortality, and/or the constant failure rate regions of the bathtub curve. We rarely see systemic time-based failures in industrial machines (see Figure 21.2).

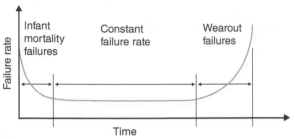

Figure 21.2 Bathtub mortality. Source: From [2].

Often criticized (unjustly) the bathtub curve is more of a conceptual tool than a predictive tool.

Despite its limitations in modeling the failure rates of typical industrial machines, the bathtub curve is a useful tool for explaining the basic concepts of reliability engineering. As an engineer, you must view every piece of equipment as being subjected to the bathtub mortality. Initial "bugs" must be weeded out to arrive at a more stable plant operation.

Look at the failure rates on a macro as much as on a large scene. The initial failures are generally within the machines and software. Once the software is debugged or adjusted and failed machine parts, bearings, gaskets, relays are removed and replaced, the view widens on the review of the whole production line to verify the production constraints: insufficient flows, pressures, and temperatures.

The human body is a good example of a system that follows the bathtub curve. People, and other organic species, tend to suffer high failure rate (mortality) during their first years of life, but the rate decreases as the child grows older. Assuming a person reaches puberty and survives his or her teenage years, his or her mortality rate becomes fairly constant and remains there until age (time)-dependent illnesses begin to increase the mortality rate (wear-out).

These factors can be metaphorically compared to factors that influence machine life. Design and procurement is analogous to genetic predisposition, installation and commissioning is analogous to prenatal care and mother's nutrition, and lifestyle choices and availability of medical care is analogous to maintenance effectiveness and proactive control over operating conditions.

21.3.2 The Exponential Distribution

The exponential distribution, the most basic and widely used reliability prediction formula, models machines with the constant failure rate, or the flat section of the bathtub curve. The industrial machines spend most of their lives in the constant failure rate, so it is widely applicable. Beneath is the basic equation for estimating the reliability of a machine that follows the exponential distribution, where the failure rate is constant as a function of time.

$$R(t) = e^{-\lambda t} \tag{21.3}$$

where

$R(t)$ = reliability estimate for a period of time, cycles, miles, etc. (t)

e   = base of the natural logarithms (2.718)

λ = failure rate (1/MTBF, or 1/MTTF), taken from statistics.

In our electric example for motors, if you assume a constant failure rate of 0.1/yr, the likelihood of running a motor for six years without a failure, or the projected reliability, is 55%, calculated as follows:

$$R(6) = 2.718^{-(0.1 \times 6)} = 0.5488 = \sim 55\%$$

In other words, after six years, about 45% of the population of identical motors operating in an identical application can probabilistically be expected to fail. It is worth reiterating at this point that these calculations project the probability for a population. Any given individual from the population could fail on the first day of operation, while another individual could last 30 years. One motor may fail twice within that period, also. That is the nature of probabilistic reliability projections. Preventive maintenance on the machine, lubrication, and monitoring can influence the durability of the plant and equipment and extend their life.

The failure rate of 0.1/yr we assumed for the motors was taken from a wide population not necessarily from the motors operating at similar circumstances as our plant. Our motor may be operating at a high load factor, often over-loaded, and hard to reach to properly maintain. It may fail sooner than pre-dicted. But in our plant, we may have 20 similar motors operating at different conditions and loading factors. That large family of motors may closer resemble to the overall population of the motors. Based on the projected failure rate, we may say that two motors will fail every year in our plant and will need be replaced or repaired.

If a failure rate for oil transformers is one in 17 years, and if our plant has 17 transformers, we will expect to have some repair work on one transformer every year. Or that in 17 years, all transformers will have a major failure that will affect our production. Or that in 17 years, there will be approximately 17 failures, some of them having several failures, while others may not have any. Natu-rally, the operating conditions, equipment redundancy, and maintenance will influence the failure rate on the individual units and on the whole population.

Any of Gaussian, Exponential, and Weibull distribution will determine the reliability of the product being reviewed, all based on some relevant failure rate, at various degrees of accuracy that we cannot be sure which one is more accurate of the three methods. The fundamental block in every calculation is the failure rate λ, with random accuracy of within 50–60%, depending on the source it came from and under what kind of mode of failure it occurred.

Based on the aforementioned, we may not be interested to view the plant reliability through several different methods of calculations, because the error margin is high. What we are interested in is a calculation which will direct us in our assessment of better reliability of the options available to us to implement in the design of the project. All the methods mentioned above will likely point toward the same alternative as the most reliable and very likely with the same margin in comparison of one against the other. That is what matters and not by how much, when the margin of error for the key item (failure rate) in the calculations is statistical and not dependable. In fact, the failure rate number may be irrelevant. We know that the motor will fail, but do not know how often at our operating conditions (failure mode) and at any other operating condition.

At a reliability conference this author attended, we went into a lot of details of calculating reliability, modes of failure, fault three analysis, and possible improvements to the power generation, etc. Finally, one engineer from India stood up and said: "I don't see anything here to improve my life when my village gets power only on every Tuesday morning and Friday afternoon."

The probabilistic calculations will point to us the way and the rest is on us engineers to design the highest possible plant reliability and availability within the cost restraints by using proven products of high quality that have been proven to that it will perform well for this specific service. Furthermore, we will make sure that the critical units are placed in the plant in such a way that can be well maintained and/or repaired or replaced for the greatest level of the production line reliability, and most of all, the plant availability.

Let us look at the plant control system that operates and supervises the plant. The batch plant software activates hundreds of interposing relays to initiate the plant operation. One interposing relay, which activates a little solenoid or a machine to move a product in one or other direction as directed by the commanding software, may fail one day and stop the production line. The bottles on the line will not be capped or filled or labeled. The whole line will stop, and it will be difficult to figure out which relay out of several hundred working this line had failed and stopped the packaging line. Hopefully, our intuition will guide us to focus on a few relays most likely causing this type of line stoppage. The faulty relay will be located, thrown out, and replaced with a new one within an hour or two to allow us to restart the production line. Hopefully, the other lines feeding to this line had a storage capacity (a bin), so they continued with their part of the production.

In this production aforementioned, we cannot duplicate the individual components. Everything is lined up in a series. When one item fails, everything stops and fails. The company will likely have several lines operating in parallel, so the production in some reduced capacity will continue.

Here are the things we can do to improve the line reliability. These include

- Access and maintainability
- Spare parts availability
- Reduced time for repairs
- Alarm indications on the control system
- Failure time analysis by control system.

21.4 Estimating the System Reliability

Once the reliability of components or machines has been established relative to the operating context and required mission time, plant engineers must assess the reliability of a system or a process. The production line can be laid out by having the components installed in series, parallel, and shared-load redundant system (m/n systems).

21.4.1 Series Systems

Reliability block diagrams map a process from start to finish [2]. For a series system, Subsystem A is followed by Subsystem B, and so forth. In the series system, the ability to employ Subsystem B depends upon the operating state of Subsystem A. If Subsystem A is not operating, the system is down regardless of the condition of the other Subsystems functioning (see Figure 21.3). To calculate the system reliability for a serial process, you only need to multiply the estimated reliability of Subsystem A at time (t) by the estimated reliability of Subsystem B at time (t), and so on. The basic equation for calculating the system reliability of a simple series system over the same period of time is

$$R_s(t) = R_a \times R_b \times R_c \times \cdots \times R_n \tag{21.4}$$

where

$R_s(t)$ = system reliability for given time (t)
R_{a-n} = subsystem or sub-function reliability for given time (t).

So for a simple system with three subsystems, having an estimated reliability as noted at time (t), the system reliability is calculated as follows:

$$R_s = 0.90 \times 0.80 \times 0.90 = 0.576, \text{ or about } 57\%.$$

That is not very healthy. The more components are chained in series, the worse it gets. Here comes the famous statement: "**You are as strong as your weakest link.**" The components chained in series must be extremely reliable to render a

Figure 21.3 Block diagrams.

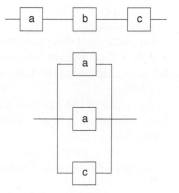

reliable system. Those interposing relays we mentioned earlier are all chained in the same single chain, thus rendering a suspect system requiring a strong and experienced maintenance crew. The interposing relays may not be connected in series electrically, but they are in series from the reliability point of you, as failure of any, fails everything.

21.4.2 Parallel Systems

Often, design engineers will incorporate redundancy into critical systems [2]. Reliability engineers call these parallel or redundant systems. These systems may be designed as active parallel systems or standby parallel systems. The block diagram for a simple three component parallel system in given time is shown and calculated as follows:

$$R_s(t) = 1 - [(1 - R_a) \times (1 - R_b) \times (1 - R_c) \times \cdots \times (1 - R_n)] \qquad (21.5)$$

where

$R_s(t)$ = system reliability for given time (t)

R_n = subsystem $(a-n)$ or subfunction reliability for given time (t).

To calculate the reliability of an active parallel system of three subcomponents, where all units are running, use the following simple equation for the subsystems having reliability of 0.9, 0.8, and 0.9:

$$R_s(t) = 1 - [(1 - R_a) \times (1 - R_b) \times (1 - R_c)]$$
$$= 1 - [(1 - 0.9) \times (1 - 0.8) \times (1 - 0.9)] = 0.998 \qquad (21.6)$$

The reliability of the parallel system is considerably greater than that of the series system.

The reliability for a single pump operation is composed of a serial combination of all the components, shown in Figure 21.4. To improve on the reliability, a standby pump must be added and switched on. The calculation of standby systems requires knowledge about the reliability of the switching mechanism to transfer from one pump to the other. The reliability will be improved if the design arrangement is made to switch on the second unit as soon as the first one is tripped or switched off. This can easily be accomplished by the plant control system.

For instance if this concerns two redundant 100% flow rated pumps feeding a tank, a short outage may not be detrimental to the plant operation. The factor of availability of the parallel system will likely be greater than the factor of reliability.

The value of 0.9 assumed for reliability of a subsystem is not unreasonable. Each subsystem a, b, or c may be a combination of serial components for each pump and motor.

Consider a parallel system of two $2 \times 100\%$ capacity pumps. In this case, each parallel subsystem would include a motor, pump, electrical starter, circuit

Figure 21.4 Single pump operation. STR = starter, SOL = solenoid valve.

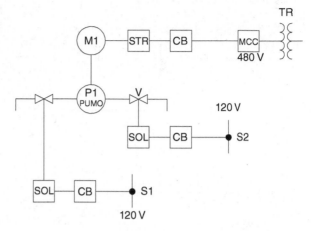

Figure 21.5 Serial/parallel operating system.

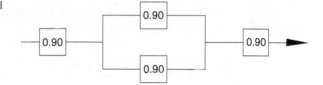

breaker, valve (2), solenoid (2), etc., each of which would be a serial component that may fail and cause a failure of the whole subsystem. With respect to the reliability, all the devices operating on a subsystem are serial components.

21.4.3 Parallel and Serial Elements

Often, the systems contain parallel and serial subcomponents as shown in Figure 21.5 [2]. In this system, the series components are the weak links. Evidently, the parallel components within the chain are as solid as rock. The overall system reliability can be calculated as follows:

$$Rs = R_1 \times [1 - (1 - R_2) \times (1 - R_3)] \times R_4$$
$$= 0.9 \times [1 - (1 - 0.9) \times (1 - 0.9)] \times 0.9 = 0.8019 \sim 80\%$$

The overall reliability will always be lower than the reliability of the worst serial component (m out of n systems (m/n systems)) [2]. An important concept to plant reliability engineering is the concept of m/n systems as in Figure 21.6. These systems require that m units from a total population of n connected in parallel be available for the system to be successful. An example is a large transformer cooled with four oil/water heat exchangers of which any three modules must be in service (three out of four), to allow the transformer to be fully loaded. Fewer cooling modules are required if the transformer is lightly loaded.

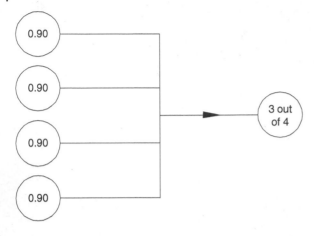

Figure 21.6 "*m* out of *n*" system.

Our concern is the scenario of the fully loaded transformer. The reliability calculation for an *m/n* system can be reduced to a cumulative binomial distribution calculation according to the following formula:

$$R(t) = \sum_{k=m}^{n} \frac{n!}{k!(n-k)!} (e^{-\lambda t})^k (1 - e^{-\lambda t})^{(n-k)} \tag{21.7}$$

where

$R(t)$ = system reliability given the actual number of failures (k) is less than or equal the maximum allowable $k \le m$

n = the total number of units in the system

$P = e^{-\lambda t}$ = the probability of survival, or a component considered to be equal for a given time (t)

k = the actual number of failures

m = the number of units required for success.

Here is a different and more understandable approach to calculate this concept [2]. An example of such a system might be an air traffic control system with *n* displays of which *m* must operate to meet the system reliability requirement.

For the sake of simplicity, let us assume that the units are identical, they are all operating simultaneously, and failures are statistically independent. Then,

R = reliability of one unit for a specified time period, assumed 0.9 (90%).
$Q = 1 - R$ = unreliability of one unit for a specified time period, assumed $1 - 0.9 = 0.1$ (10%).

$$\text{For } n \text{ units} : (R + Q) = 1 \tag{21.8}$$

$$(R + Q)^n = R^n + nR^{n-1}Q + \frac{n(n-1)}{2!}R^{n-2} Q^2 + \frac{n(n-1)(n-2)}{3!}$$
$$R^{n-3} Q^3 + \dots + Q^n \tag{21.9}$$

This is the familiar binomial expansion of $(R + Q)^n$. Thus,

$$P\,(\text{at least } n - 1 \text{ survive}) = R^n + nR^{n-1}Q \tag{21.10}$$

$$P\,(\text{at least } n - 2 \text{ survive}) = R^n + nR^{n-1}Q + \frac{n(n - 1)}{2!}R^{n-2}\,Q^2 \tag{21.11}$$

$$P\,(\text{at least } n - 2 \text{ survives}) = 1 - Q^n \tag{21.12}$$

Let us look at the specific case of four coolers that meet the previously mentioned assumptions. If the reliability of each cooler for the same time t is 0.9, what is the system reliability for time t if three out of four coolers must be working? R represents healthy state, while Q represents a failed state.

$$(R + Q)^4 = R^4 + 4R^3Q + 6R^2Q^2 + 4RQ^3 + Q^4 = 1$$

from which:

$P_4 = R^4$ (all four will survive), OK! $\rightarrow$ 0.6561
$P_3 = 4R^3Q$ (three will survive), OK! $\rightarrow$ 0.2916, Total (for three or four): **0.9477 or 95%**
$P_2 = 6R^2Q^2$ (two survive, assumed system failed)
$P_1 = 4RQ^3$ (one survives, assumed system failed)
$P_0 = Q^4$ (all fail, assumed system failed)
$R_s = R^4 + 4R^3Q = 1 - 6R^2Q^2 - 4RQ^3 - Q^4$ (at least three survive)
$R_s = R^4 + 4R^3Q + 6R^2Q^2 = 1 - 4RQ^3 - Q^4$ (at least two survive)
$R_s = R^4 + 4R^3Q + 6R^2Q^2 + 4RQ^3 = 1 - Q^4$ (at least one survives)
$R_s = R^4 + 4R^3Q = (0.9)^4 + 4(0.9)^3\,(0.1) = 0.6561 + 0.2916 = 0.9477$

So the likelihood of completing the mission three out of four within time (t) is 0.9477 (0.6561 + 0.2916), or approximately 95%.

Sometimes, an engineer must decide between choices of using a system with $3 \times 50\%$ capacity components against a system with $2 \times 100\%$ fully redundant units.

Let us consider a case of an air blower system for an ore cyclone separator. A separator uses a process method of creating an air vortex inside the cyclone to separate fines from the coarse ore. The fines are separated at the top, while coarse ore is discharged at the bottom.

The engineer wants to weigh in the reliability and cost of having a case 1 with $3 \times 350\,\text{HP}$ against a case 2 with $2 \times 500\,\text{HP}$ electrical motors (see Figure 21.7).

Case 1: System fails if two motors fail. A 50% operation is not acceptable.
Case 2: System fails if both motors fail.
Case 1: $3 \times 50\% \rightarrow R_s(t) = R^3 + 3R^2Q = 0.9^3 + 3 \times 0.9^2 \times 0.1 = 0.972$, or 97%
Case 2: $2 \times 100\% \rightarrow R_s(t) = 1 - [(1 - R_a) \times (1 - R_b)] = 1 - 0.1 \times 0.1 = 0.99$, or 99%

Evidently, the case 2 ($2 \times 100\%$) is a close reliability winner and if the cost warrants it, it should be chosen.

Figure 21.7 Cyclone separator

But that may not be the full story. The components can be repaired or replaced and arranged to return the system back to its full capacity. How long does it take to repair or replace one motor or one blower in either case and make the systems fully available again? The reliability calculation does not offer this answer. The availability calculations take into account the repair time and may be more appropriate method for this decision-making process, as discussed later in this chapter.

21.4.4   Piper Alpha Rig Explosion

On 6 July 1988, a catastrophic explosion occurred in North Sea resulting in 167 people losing lives. It started on the condensate pumping system extracting gas to the shore. It was a redundant $2 \times 100\%$ parallel pump system that was supposed to be highly reliable. On this particular day, a discharge valve of Pump A failed and was disabled and left for a routine maintenance.

Therefore, for this particular case, this particular day, we have converted the system from a parallel to a serial. Watch out! (see Figure 21.8). The pipe was loosely flanged off, but the pump was *not padlocked and not tagged off* in the control room. During the night shift, the Pump B failed and was giving alarms. Shutting down the pumps would cause a loss of production. The operator started the A pump that was not tagged off. The pump started and pushed the gas through the flange and caused the explosion.

This accident tells us that even a parallel and reliable system can fail. The likelihood that the second pump will fail while the other pump is on a brief maintenance is minuscule, but it does happen. Tragically, in this situation, a series system with one pump would cause severe loss of production due to

Figure 21.8 Piper alpha
rig catastrophe.

periodic failures, but it would have saved lives. Perhaps a reliable parallel system has made the operators to be careless and caused them to fail tag off the pump that was on maintenance.

21.5   Common Mode Failure

The term "common mode failure" is a center piece of every reliability discussion. Shown is a critical system within the plant, failure of which will cause stoppage of the plant production. It is a task of the engineer to determine the common mode components within this critical system. Failure of such component may cause a failure of the whole system of two pumps (see Figure 21.9).

Consider again a redundant $2 \times 100\%$ capacity pump system. The pumps work in parallel, both electrically and mechanically. They are piped together to a common header and each fed through its own valve. Also, the motors are fed from a common transformer. In terms of reliability, the devices associated with Pump 1 are serial components of the P1 chain, similarly, for the P2.

Failure of any of those components will disable its own related pump, but not the whole system. However, if a failure or planned outage occurs on the motor control center (MCC) supply transformer (common mode component), it will disable both pumps and stop the downstream plant operation. With such an arrangement, the reliability of this critical system is badly compromised, and the overall reliability lowered to be less than the reliability of the common mode device.

Common mode components may be acceptable in some noncritical systems that can tolerate lengthy outages, but for the important dependable systems, the

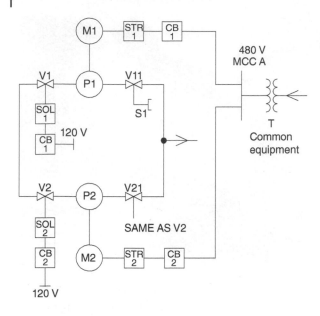

Figure 21.9 Common mode parallel system.

number of common components must be reduced to a minimum or eliminated. In this case, P1 pump should be fed from MCC1, and P2 from MCC2, thus removing the common mode failure.

21.5.1 Space Shuttle Program

The space shuttle orbiter was the most complex machine ever built. It comprised 2.5 million moving parts and a huge amount of other parts like heat shield plates.

On *Challenger*, an O-ring, a perfectly designed part, eroded on the rocket booster and caused a destruction of the orbiter during its lunch due to unexpected cold weather.

Columbia was destroyed because of damaged thermal protection from foam debris that broke off from the external tank during ascent. It proved that even a light feather can make a massive damage. The shuttle was assembled of too many parts. Many of them were of common mode nature; failure of each could destroy the shuttle. At the end, its complexity proved to be an insurmountable problem and a reason for the space program dissolution.

21.6 Availability

Reliability is a function of time, defined as the conditional probability that the system will perform correctly throughout the interval, given that the system was performing correctly at the time t_0. *Availability* is considered as a function of time, defined as the probability that system is operating correctly and is available to perform its function at the instant of time "*t*." The major difference between these two terms is in the time, which is considered as an interval in former and instantly in the latter.

21.6.1 Inherent Availability, A_i

Inherent availability is the steady-state availability when considering only the corrective maintenance (CM) downtime of the system. This classification is what is sometimes referred to as the availability as seen by maintenance personnel. This classification excludes preventive maintenance downtime, logistic delays, supply delays, and administrative delays. Since these other causes of delay can be minimized or eliminated, an availability value that considers only the corrective downtime is the inherent or intrinsic property of the system. Many times, this is the type of availability that companies use to report the availability of their products (e.g. computer servers) because they see downtime other than actual repair time as out of their control and too unpredictable.

The CM reflects the efficiency and speed of the maintenance personnel, as well as their expertise and training level. It also reflects the characteristics that should be of importance to the engineers who design the system, such as the complexity of necessary repairs, ergonomics factors, and whether ease of repair (maintainability) was adequately considered in the design. For a single component, the inherent availability can be computed by

$$\text{Reliability}: e^{-\frac{\text{Time}}{\text{MTBF}}} \tag{21.13}$$

$$\text{Availability}: A_i = \text{MTTF}/(\text{MTTF} + \text{MTTR}), \text{ or } A_i$$

$$= \text{MTBF}/(\text{MTBF} + \text{MTTR}) \tag{21.14}$$

MTBF = uptime/number of system failures
MTTR = CM downtime/number of system failures

The higher the MTBF value is, the higher the reliability and availability of the system.

MTTR affects availability. This means that if it takes a long time to recover a system from a failure, the system is going to have a low availability.

High availability can be achieved if MTBF is very large compared to MTTR.

One should keep in mind that until steady state is reached, the MTBF calculation may be a function of time (e.g. a degrading system). In such cases, before reaching steady state, the calculated MTBF changes as the system ages and more data are collected. Thus, the above formulation should be used cautiously. Furthermore, it is important to note that the MTBF defined here is different from the MTTF (or more precisely for a repairable system, MTTFF: mean time to first failure).

21.6.2 Operational Availability, A_o

Operational availability is a measure of the "real" average availability over a period of time and includes all experienced sources of downtime, such as administrative downtime, logistic downtime. The operational availability is the availability that the customer actually experiences. It is essentially the *a posteriori* availability based on actual events that happened to the system. The previously discussed availability classifications are *a priori* estimates based on models of the system failure and downtime distributions. In many cases, operational availability cannot be controlled by the manufacturer due to variation in location, resources, and other factors that are the sole ownership of the end user of the product.

Operational availability is the ratio of the system uptime to total time. Mathematically, it is given by

$$A_o = \text{Uptime/operating cycle} \tag{21.15}$$

where the operating cycle is the overall time period of operation being investigated and uptime is the total time the system was functioning during the operating cycle. If it concerns a generator, the loading on the generator does not enter the picture. (*Note*: The operational availability is a function of time, t, or operating cycle.)

The concept of operational availability is closely related to the concept of operational readiness. In military applications, this means that the assigned numbers of operating and maintenance personnel and the supply chain for spare parts and training are adequate. In the commercial world, a manufacturer may be capable of manufacturing a very reliable and maintainable product (i.e. very good inherent availability). But what if the manufacturer has a poor distribution and transportation system or does not stock the parts needed or provide enough service personnel to support the systems in the field? The readiness of this manufacturer to go to market with the product is low.

Logistic planners, design engineers, and maintainability engineers can collaboratively estimate the repair needs of the system, required personnel, spares, maintenance tasks, repair procedures, support equipment, and other resources. Only when all downtime causes are addressed will you be able to paint a realistic picture of your system's availability in actual operation.

21.6.3 Availability Calculations

System availability is calculated by modeling the system as an interconnection of parts in series and parallel in a similar mode as the reliability calculations [3].

If failure of a part leads to the system becoming inoperable, the two parts are considered to be operating in series. If failure of a part leads to the other part taking over the operations of the failed part, the two parts are considered to be operating in parallel.

21.6.4 Availability in Series

As stated earlier, two parts (a) and (b) are considered to be operating in series if failure of either of the parts results in failure of the combination (Figure 21.3). The combined system is operational only if both part (a) and part (b) are available. From this, it follows that the combined availability is a product of the availability of the two parts. The combined availability is shown by the equation beneath:

$$A_s = A_a \times A_b \times \cdots \times A_n, \text{ or } A_s = \Pi A_i \qquad (21.16)$$

where

Π = product of multiplication
i = parts 1–n

Example

Device	Availability	Downtime
Mill	0.95	18.25 days → 0.95 × 365 = 18.25 days
Ore transport	0.97	10.95 days
Overall	0.92	29.2 days

This table shows the availability and downtime for individual components and the series combination. The implications of the above equation are that the combined availability of two components in series is always lower than the availability of its individual components.

21.6.5 Availability in Parallel

As stated earlier, two parts are considered to be operating in parallel if the combination is considered failed when both parts fail.

The combined system is operational if either is available. From this, it follows that the combined availability is 1 − (both parts are unavailable). The combined availability is shown by the equation beneath:

$$A = 1 - (1 - A_i)^2$$

or,

$$A = 1 - \Pi U_i = 1 - \Pi(1 - A_i) \qquad (21.17)$$

A = availability
U = unavailability
Π = product of multiplication
i = parts 1 to n

Device	Availability	Downtime
Pump 1	0.95	18.25 days ($0.95 \times 365 = 18.25$ days)
Pump 2	0.95	18.25 days
Overall	0.9985	0.91 days

Evident from the above Eq. (21.17) is that the combined availability of two components in parallel is always much higher than the availability of its individual components and must be used for all the critical components. The components operating in parallel are typically identical because they share the same product, both at the input and the output. Two pumps piped together are working in parallel, generally of identical reliability parameters.

Consider a system with N components where the system is considered to be available when at least $N-M$ components are available (i.e. no more than M components can fail). The availability of such a system is denoted by $A_{N,M}$ and is calculated, as follows:

$$A_{N,M} = \sum_{i=0}^{M} \frac{N!}{i!(N-i)!} \; A^{N-1} (1-A)^i \tag{21.18}$$

References

1 MIL Standard 721 – Reliability Engineering Terms (2019).
2 Troyer, D. (2011). *Reliability Engineering Principles*. Noria Corporation.
3 EventHelix.com (2019): System Availability Calculations.

22

Fire Protection

CHAPTER MENU

22.1 Plant Fire Protection System, 495
    22.1.1 Fire Detection and Monitoring, 496
    22.1.2 Smoke and Heat Detectors, 498
    22.1.3 Main Fire Alarm Panel (MFAP), 499
22.2 Fire Sprinkler Systems, 500
    22.2.1 Sprinkler Heads, 500
    22.2.2 Wet-Pipe Sprinkler System, 501
    22.2.3 Dry-Pipe Sprinkler System, 501
    22.2.4 Preaction Systems, 502
22.3 Gas Flooding Suppression, 503
    22.3.1 Large Generators, 503
    22.3.2 Control Rooms, 504
22.4 Fire Hydrants and Standpipes, 506
    22.4.1 Standpipes, 506
22.5 Portable Fire Extinguishers, 507
22.6 Fire Safety Dampers and Duct Vents, 507
22.7 Deluge Systems, 508
22.8 Fire Water Supply System, 510
    22.8.1 Water Storage Tanks, 511
    22.8.2 Fire Pumps, 511
    22.8.3 Pump Controllers, 512
    22.8.4 Fire Pump Sizing, 513
22.9 Cables and Conduits used for Fire Protection Circuits, 514
22.10 Fire Detection and Notification Witness Testing, 515
References, 516

22.1 Plant Fire Protection System

Fire detection and fire suppression systems are comprehensive protection systems for industrial and power plants. As a fully integrated system spread over a wide area, it includes fire detection, notification, and suppression. This is a "mechanical project" with substantial participation by the electrical engineers

Practical Power Plant Engineering: A Guide for Early Career Engineers, First Edition. Zark Bedalov.
© 2020 John Wiley & Sons, Inc. Published 2020 by John Wiley & Sons, Inc.

through the detection and notification and later as commissioning engineers they take charge of it to prove the system performance.

The following specific fire protection systems may be employed in an industrial or a power plant:

- Gas flooding, VESDA, or sprinkler systems for large generators and plant control rooms.
- Automatic wet or dry sprinkler systems for general offices, stores and cable spreading areas.
- High velocity water (HVW) deluge for large transformers (>10 MVA).
- Fire hydrant and standpipe hose cabinets in all areas as manual secondary protection.
- Portable fire extinguishers in all areas, serving as the initial activation by personnel.

22.1.1 Fire Detection and Monitoring

The main function of the fire alarm and detection [1] system is to serve as a completely supervised fire alarm reporting system that enters into an alarm mode by activation of any of the alarm initiating devices. The system remains in the alarm mode until the initiating device and the fire alarm control panel are reset and restored to normal. All the connections necessary for making the system functional must be in accordance with the NFPA guidelines and relevant governing local laws.

The proposed fire detection and alarm system is intelligent, addressable, and microprocessor based. The system includes various types of notifying devices (Figure 22.1) and fire detectors strategically positioned and connected in loops in all the areas of the plant building. The goal is an early detection of outbreak of fire and to provide plant operators with a warning so that remedial action can be initiated. The plant fire detection system typically would include the following subsystems/equipment to supervise the proper functioning of

Figure 22.1 Fire notifying devices.

- One (1) Main Fire Alarm Panel (MFAP) located in the central and accessible location, fed from the plant UPS and 24 V_{dc} internal battery.
- Two (2) peripheral mimic panels in the remote main substation and the camp.
- Ionization type smoke detectors and thermal detectors complete with necessary cabling distributed throughout the plant.
- Smoke detectors located at strategic points inside the housing of major equipment.
- Manual pull stations and notification devices such as fire horns/bells, strobe flashing lights distributed strategically throughout the plant for manual intervention and audible and visual alarms.

The detector types are specialized and selected to suit the environment in which they are installed. A mixture of various types of fire detectors are employed to best serve the area protected, based on the type of equipment, human presence, and the amount of combustible materials. A complete list of rooms must be prepared as part of the fire protection design criteria, indicating the type and number of detectors to be located in each. The detectors must not be mounted in air pockets on the ceilings where air circulation may be blocked by deep structural beams.

All the detection devices operate from 24 V_{dc} power supplied by the MFAP with a limitation of >85% of voltage to the last device on the circuit.

Addressable fire alarm control panels require programming to establish a database of all the devices the fire alarm panel is supporting. End-of-line (EOL) resistors are not required in the addressable fire alarm systems. When an addressable fire alarm panel is working properly, it constantly sends messages down its addressable two-wire circuits to verify each detector is working properly. Each device responds to the request with a short status report. In case of a wire break or if a device fails, the MFAP annunciates a trouble condition indicating that one or more of the devices have failed and identifies the failed devices.

The benefits associated with addressable technology are significant over the conventional nonaddressable technology, as follows:

- Simplified wiring – the wiring does not require additional two return wires for a string of devices like in Class A wiring.
- EOL resistors are not required like in conventional Class B wiring.
- Event reporting is precise for every individual device on the system and its location.
- Field testing of the devices is simplified, quicker, and less confusing.
- On a conventional fire alarm panel, the terms "zone" and "circuit" are often interchanged (one circuit covers one zone). On an addressable panel, the zones are created in software. Any device can be assigned to any zone.

22.1.2 Smoke and Heat Detectors

Ionization detectors: The detector, without moving parts, detects products of combustion using an ionization chamber. The pilot light (red light-emitting diode) indicates the alarm status of the detector. Each detector can be tested for its sensitivity.

Ionization smoke detectors are responsive to *flaming fires*. These detectors have a small amount of radioactive material between two electrically charged plates. The radioactive material ionizes the air between the plates and causes current to flow between the plates. Smoke entering the chamber disrupts the flow of ions, reducing the current flow, which activates the alarm.

Photoelectric smoke detectors (Figure 22.2) are generally more responsive to fires that begin with a long period of smoldering (called "smoldering fires"). Photoelectric-type devices aim a light source into a sensing chamber but at an angle away from the light sensor. Smoke entering the chamber deflects light onto the light sensor, triggering a fire alarm.

These addressable detectors are made to work in the environment's air velocity, altitude, humidity, temperature, and color of smoke. They are plug-in types with a detector base containing terminals for making connections. The plug-in base must be compatible for all detectors irrespective of the detector's principle of operation. For the best protection, it is recommended that both (ionization and photoelectric) technologies be used. In addition to individual ionization and photoelectric devices, combination devices that include both technologies in a single device are also available.

Fixed temperature heat detectors: Fixed temperature detectors are the most common and operate when the heat sensitive element reaches the set-point temperature and changes state from a solid to a liquid. The most common fixed temperature point for electrically connected heat detectors is 58 °C (136.4 F). The newest detectors activate at 47 °C (117 F), increasing the available reaction time and safety margins.

Figure 22.2 Smoke detector.

Figure 22.3 Pull station.

Rate-of-rise heat detectors: Rate-of-rise heat detectors operate on a rapid rise in element temperature of 6.7–8.3 °C (12–15 F) increase per minute, irrespective of the starting temperature. This type of heat detector can operate at a lower temperature fire condition than would be possible if the threshold activation were fixed. It has two heat-sensitive thermocouples or thermistors. One thermocouple monitors the heat transferred by convection or radiation and the other responds to ambient temperature. The rate-of-rise heat detector responds when the first temperature increases relative to the other.

Rate-of-rise detectors may not respond to low energy release rates of slowly developing fires. To detect slowly developing fires, a combination of detectors adds a fixed temperature element that will ultimately respond when the fixed temperature element reaches the design threshold.

Beam detector: Beam detectors are often employed to cover large areas and work on the principle of light photo-sensitivity obscuration due to smoke developing between the beam transmitter and receiver.

Manual pull stations: They are installed in accordance with suitable standards and codes at all points where the personnel are able to safely initiate an alarm. Manual fire red pull stations (Figure 22.3) work on the break-the-glass principle to push the button and initiate an alarm. The manual call points are addressable and indicated on the fire panels with its number and location.

22.1.3 Main Fire Alarm Panel (MFAP)

The microprocessor-based fire alarm panel (Figure 22.4), with internal 24-hour chargeable battery, provides automatic, address supervised, detection and

Figure 22.4 Fire alarm panel.

alarm. The control modules within the MFAP supervise the alarm annunciation lines, detection and notification circuits, release of sprinkler and deluge systems, release of gas flooding system, and panel internal faults and wiring issues.

A physical live mimic diagram is often integrated into the main fire alarm panel to show the fire zones and associated alarms and indications. If an alarm, fault or release occurs, the pertinent light glows. After an acknowledgement, the alarm horn or bell is silenced, but the MFAP light remains steady and is turned off only after resetting the MFAP.

The mimic panel may also show the flow diagram along with the required indication including ventilation fire dampers, plant stops, staircase pressurization, release commands, evacuation signal release for sections, and general plant evacuation release. The alarms can be arranged to be heard in the specific zones and/or the complete plant.

22.2 Fire Sprinkler Systems

22.2.1 Sprinkler Heads

Each sprinkler system [2] consists of an isolating valve (with auxiliary contact for remote indication of its status), normal wet or dry distribution piping, a flow switch and glass type, and temperature-activated sprinklers (Figure 22.5). The temperature rating of a bulb/sprinkler is at least 30 °C greater than the

Figure 22.5 Sprinkler head.

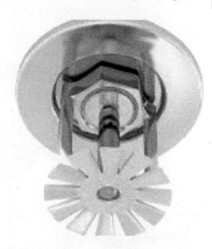

highest expected ambient temperature of the location. Sprinkler heads for the wet and dry systems are identical. Sprinklers on an average discharge 75–150 l/min (20–40 gpm), depending on the water head.

Sprinklers are generally installed in all the areas except those covered by gas flooding, such as offices, workshops, stores, cable spreading areas, and tunnels. The emphasis is given to the protection of cables, diesel generator room, and fuels storage.

22.2.2 Wet-Pipe Sprinkler System

This type of system is the most common and is typically the easiest to design, install, and maintain. Wet-pipe systems, with a series of closed sprinkler heads (Figure 22.6), contain water under pressure at all times in the overhead piping. If a fire occurs and produces a sufficient amount of heat to activate one or more sprinkler heads, water immediately discharges from the affected sprinkler(s). Wet-pipe systems are more inherently reliable and less costly to maintain than dry systems. Wet pipe systems, however, should never be considered for the areas, where there is a possibility of freeze up in the piping systems or for the sites where temperature can go below 3 °C. Also, some industrial facilities will not allow wet systems piping to run over the electrical and electronic equipment (e.g. LV or MV switchgear, MCCs, DCS I/O cabinets).

Wet pipe sprinkler systems require the least amount of effort to restore itself following activation. In most instances, sprinkler protection is reinstated by replacing the fused sprinklers and turning the water supply back on.

22.2.3 Dry-Pipe Sprinkler System

These are the systems for the areas having substantial chance of freezing. The piping does not contain any water. The piping is charged with air or occasionally

Figure 22.6 Wet pipe sprinkler.

nitrogen under pressure and is connected to automatic, normally closed sprinkler heads. If one or more of sprinkler heads opens due to fire causing lowering of the air pressure in the pipe, the remote main or branch valves will activate and open to allow water flow to the activated head(s). The glass fuse of the sprinkler head actually melts and starts releasing air.

Dry-pipe systems are more complex than wet-pipe systems. They require a reliable air supply source and, because of the delay associated with water delivery from the dry-pipe valve to the open sprinklers, are subject to certain design limitations on the areas allowed to be covered.

22.2.4   Preaction Systems

Preaction systems are typically found in spaces containing computer or communications equipment, museums, electrical switchgear, and other facilities where inadvertent water damage from system piping is a concern. It is a dry-pipe system, with an additional advance initiation by the electrical fire detection system. The preaction system comes in three variations, as follows:

(1) *Single Interlocked preaction system*: This system has a preaction valve, rather than the dry pipe valve, to hold back the water. The electrical addressable system will initiate the valve opening to release water into the piping under some double cross zone or equivalent fire protection

activation criteria. Water is released only after a number of sprinkler heads open due to the heat caused by fire.

(2) *Double-Interlocked preaction system*: This system has characteristics of both the single-interlocked system and the dry-pipe system and typically consists of a dry-pipe valve mounted on top of the preaction valve. For water to enter the system piping, both the supplemental detection system and the sprinklers on the system must operate. The double-interlocked system is the most common in freezer facilities where accidental valve operation can result in the immediate freezing and system piping damage.

(3) *Noninterlocked preaction system*: This variation of the preaction system initiates water flow release through the sprinklers either on activation of the supplemental detection system or the opening of a single sprinkler.

All the wet and dry pipe systems must include the alarms connected to the fire alarm control panel, fire suppression initiation reporting, zone (valve) initiation, water flow, and system operation readiness when required. Preaction and dry-pipe systems may require additional effort to reset the control equipment for system restoration.

22.3   Gas Flooding Suppression

22.3.1   Large Generators

Gas flooding the generators as a method of fire suppression is not common in the North American power plants. Water spray is more common. A new type of generator winding insulation, made of glassy materials, is not affected by water applied in case of fire. Once spun and loaded, the generator insulation dries rapidly. However, this approach of water spraying is not acceptable in some Asian countries. Argonite gas is a preferred option (Figure 22.7). Other gas systems like Halon are also used in different parts of the world.

Figure 22.7 Gas cylinders.

The Argonite gas system (AGS) provides total flooding within the generator enclosure and maintains the gas concentration for at least 20 minutes after the first gas discharge. For that reason, each generator uses two banks of cylinders. The main bank provides for rapid release of gas and the reserve bank for slow release of gas. These two banks maintain an effective concentration of Argonite during the deceleration of the generator. The system conforms to the NFPA Code No. 2001.

The generator louvers close during the firefighting operation. Exhaust gases and smoke are later exhausted by the HVAC system.

Fire detection circuit includes a combination of smoke and heat detectors within the generator and within the generator external enclosure. The gas flooding system control panel alarm is initiated if any detector is activated. The following cross zone logic is used for automatic gas release:

(1) One smoke detector and generator electrical protection (87G or 64G) activated.
(2) One heat detector and generator electrical protection (87G or 64G) activated.

Manual release of Argonite is available by operating the valve trip mechanism at each generator. The fire detection equipment is self-contained for each generator and completely independent from other generators and from the rest of the plant fire detection and alarm system.

22.3.2 Control Rooms

The fire detection includes a mixture of heat and smoke detectors within the subfloors and above the electrical/control cabinets. An automatic release of Argonite requires concurrent operation of a smoke detector together with a heat detector. The gas flooding system provides for total flooding and maintaining the necessary Argonite concentration for at least 20 minutes after the first gas discharge.

The pipework, nozzles, and valves convey gas from the cylinders to the associated rooms. The control room piping and fire suppression system runs above the plasterboard sections of the suspended ceiling to the distribution nozzles placed to provide flooding of the rooms. Locking pins are provided for the cylinder release mechanisms to ensure a release of Argonite into the protected area can be positively prevented whenever required. The Argonite indicating light stations are located above each room entrance with one red and one green indicating light, corresponding to the condition within the protective area.

22.3.2.1 VESDA Systems
Alternatively, the fire detection can be done by means of VESDA detection system as in Figure 22.8. The name VESDA is an abbreviation for **Very Early Smoke**

Figure 22.8 VESDA installation.

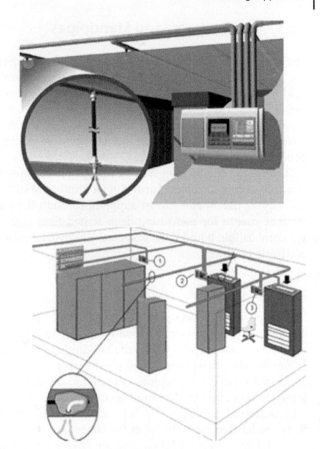

Detection Apparatus. It is a new and advanced approach to the fire detection. A VESDA detector works by continuously circulating air into a distributed pipe network via a high-efficiency aspirator. The air sample then passes through filters and finally a laser detection chamber to evaluate the smoke concentration.

It comprises an external monitor and a number of aspirating pipe circuits with an adjustable detection sensitivity range of 0.005–20% obscuration/m. The signal is then processed and presented via a bar-graph display, alarm threshold indicators, and/or graphic display.

In power plants and industrial applications, this concept of fire detection is used for the highly valued control rooms and large generators in enclosed environments. Serving as a primary detection source, it can easily be mixed with the other type of detectors to permanently indicate the level of smoke particles in the protected environment. Once the particle level reaches a certain concentration level, it provides an alarm and firing of the gas cylinders regularly in correlation with some other detection media.

22.4 Fire Hydrants and Standpipes

Fire hydrants (Figure 22.9) are a backbone of the firefighting systems in a building or industrial plant. The purpose of the fire hydrant system is to provide a readily available source of water to any point throughout or around the building. It is a water distribution system consisting of a water tank, suction piping, fire pumps, and a distributed piping system. The distributed piping system establishes connectivity throughout the building through fire hydrants, hoses, and standpipes.

Through the fire hydrant system, firefighters are provided with a reliable and versatile system capable of providing a number of different methods for controlling the fire. Water can be supplied through the fire hydrant system as a straight stream for combating deep seated fires and as a spray for combating combustible liquid fires. A typical fire hydrant discharges through a hose 900 l/min (250 gpm).

22.4.1 Standpipes

It is a type of rigid water piping that is built into multistory buildings to which fire hoses can be connected, allowing manual application of water to the fire. It serves the same purpose within the buildings as hydrants on the street.

Wet standpipes with hoses are filled with pressurized water at all times. In contrast to the dry standpipes, which can be used only by firefighters, wet standpipes can be used by building occupants.

In industrial plants, a hydrant loop is placed around the buildings with hydrants located every 30–50 m for coverage with a 50-m hose in all parts of the buildings and external equipment. Water supply for fire hydrants comes from a plant water supply system and pump house. Pressure reducing stations are installed in order to limit pipe pressure, if required.

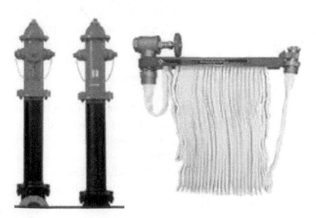

Figure 22.9 Hydrants and fire hose.

22.5   Portable Fire Extinguishers

The main function of this system is to provide initial protection against fire to various equipment and working areas. Portable fire extinguishers shall comply with the requirement as per the provisions of NFPA 10 [3] and International standards (IS) for the fire classification expected, as follows:

Class A: for solid combustible materials
Class B: for liquid combustible materials
Class C: for energized electrical equipment

The extinguishers (Figure 22.10) are located at strategic locations with easy access. Portable fire extinguishers operate manually. The number and location of hand fire extinguishers and fire buckets are indicated on the fire protection drawings during the detail engineering.

22.6   Fire Safety Dampers and Duct Vents

The following fire safety rules must be adhered to in the plant design:

- Fire transition from one room to the other must be prevented.
- Plant ventilation must be shut down.
- Motorized wall dampers and duct vents (Figure 22.11) must be automatically closed following a fire outbreak in the area. The dampers must have at least a three-hour fire rating.

Figure 22.10 Hand extinguishers.

Figure 22.11 Vent louvers.

- Fires must be starved of oxygen.
- Wall cable openings for piping and cable transition must be sealed with approved one to two hour fire-rated materials.
- Cable trays will not pass from one room to the other through wall openings. This is to limit the heat on the metal to be transferred to the other side of the wall.

22.7 Deluge Systems

Deluge systems, as the name implies, deliver large quantities of water over specified areas in a relatively short period of time. These systems are used to protect against rapidly growing and spreading fires, generally on oil-filled transformers.

A deluge valve (DV) (Figure 22.12) controls the system water supply and is activated by a supplemental fire detection system. Typically, sprinklers used in a deluge system may not contain thermally sensitive operating elements and, as a result, they are referred to as open sprinklers. Because open sprinklers are employed, system piping is at atmospheric pressure. As water reaches each sprinkler in the system, it is immediately discharged from the system. The nature of this system fire suppression makes it appropriate for facilities in which significant amounts of highly combustible materials are present. It is also used for situations in which thermal damage is likely to occur in a relatively short period of time. Aircraft hangars are one area of application of deluge systems. Additionally, they are used for exterior protection

Figure 22.12 Deluge system.

Figure 22.13 Deluge system in operation.

of high-value equipment essential to the continuity of operations such as long-lead-time-to-replace equipment (see Figure 22.13).

Transformer Deluge Water Spray System is employed for relatively large transformers (>10 MVA). A HVW spray is produced creating a misty emulsification that cools, smothers, and extinguishes oil burning fires. Refer to photos for a 150 MVA transformer. My presence was on the second photo, taken during a Deluge commissioning test.

The HVW spray system when mixed with oil forms an emulsion that cannot burn. In addition, the fine droplets striking the hot surface absorb heat, imparting a cooling and smothering effect thereby extinguishing fire.

Each automatic HVW spray system over transformer is provided with two separate pipe networks as follows:

(1) A wet detection pipe network over the protected equipment with temperature-sensing detectors (sprinklers) located at strategic locations. The sprinkler head consists of a quartzoid bulb (QB) that bursts when the temperature reaches its limit, to release water. These QBs are color coded in red, yellow, and green. The rating of the red bulb is 68 °C, yellow is 80 °C, and green is 93 °C.
(2) The fire water pipe network is installed around and over the protected equipment, equipped with spray nozzles to spray water onto the equipment.

Both of these piping networks end up at the DV provided for each transformer. Water supply up to DV comes through main header from the fire water network. Detection pipe network between DV and detectors is kept pressurized with water tapped at the upstream of DV. In case of fire and the activation of the sprinkler sensor circuit, the DV valve opens automatically and releases water from the header to the pipe ring mains with nozzles and sprays over the protected equipment.

A water alarm gong is provided to sound a continuous alarm while the spray system is in operation. Shutdown of the water spray system is manually operated by closing the motorized valve.

Motorized valve at the inlet of DV and manual butterfly valve at the outlet of DV is provided for isolation and maintenance.

22.8 Fire Water Supply System

This system provides water to the plant suppression system. It includes the following basic elements:

- Water storage tanks or water reservoir (e.g. lake or river)
- Pump house and pumping
- Piping distribution system with hydrants, standpipes, and sprinklers

The shutoff pressure (churn pressure) is the highest pressure point on the pump curve. Most sprinkler components are rated for 175 psi. When selecting the pump and designing the system, the pump head must be lower than 175 psi. Should maximum pressure exceed 175 psi, pressure reducing valves are required according to NFPA. Sprinkler operating pressure varies with the distance from the water source. To make the discharge and pressure more uniform throughout an area and within 40–70 psi, sprinklers are laid out in branches, rather than in one radial string.

Large industrial plants need fire water tank capacity for at least two hours supply to meet the demand of the largest fixed fire suppression system plus the maximum hose stream demand. Pressure reducing stations are required on all major branches in the pipe network for achieving the desired pressure required by the hydrant/sprinkler system. Hydro test pressure of water pipe shall be 1.5 times the maximum working pressure of each section.

22.8.1 Water Storage Tanks

The tank capacity is defined by the hazard assigned to the project facility as follows (see Table 22.1).

In hydroelectric power plants, if the power house is located below the dam, water is often tapped off the penstocks. One penstock must be kept full at all times. For the surface part of the plant, a two-hour concrete water tank may be required.

Conversion: 1 bar $(kg/m^2) = 100$ kPa $= 14.05$ psi $(lb/in.^2)$.

22.8.2 Fire Pumps

A fire pump house is required in each industrial plant and power plant as well as in commercial and residential areas. In an industrial plant, a pump house feeds fire water to the sprinkler and hydrant system from a water tank or other water supply sources. The pump house is built close to and below the water supply tank to benefit from water delivery by gravity.

An industrial or a power plant application will include three pumps:

- *Jockey pump*: Maintains water pressure in the water delivery system. It is engaged when the system pressure falls below 7 bar. It is shut off for pressure at 8 bar.
- *Duty pump*: Electric motor driven and sized to the kW (HP) capacity based on the extent of the water distribution system. The operating system is held pressurized to 10–12 bar. The pump is re-engaged when the pressure falls to 6 bar.

Table 22.1 Water tank capacity.

Duty	Application	Capacity (h)
Light	Residential and office buildings	1
Normal	Industrial plants, factories, warehouse	2
High	Refineries, chemical and paint factories, aircraft hangars, stores for explosives and flammable materials.	3–4

Figure 22.14 Diesel fire pump.

- *Standby pump*: Diesel engine driven (Figure 22.14) and of the same capacity as the electrical pump. The engine must be DC battery start capable. It is initiated when the pressure falls down to 4 bar or loss of power supply. In this situation, all three pumps may be operating if power supply is available.

Depending on the project, the fire water system pressure varies at the pump house. On our latest power plant project, the pressure was from 95 to 105 psi (6.75–7.45 bar).

The pump's operation is controlled by the pressure switches as follows:

- The **Jockey Pump** is activated by a fall in pressure on the system typically set at about 7 psi to maintain the system pressure.
- The **Duty Pump** is initiated on the system pressure falling to about 65 psi. The jockey pump is kept off when the duty pump is in operation.
- The **Standby Pump** kicks in if the pressure falls to the set point due to the duty pump failing, or the jockey pump being unable to maintain the pressure on the system, or the plant electricity system fails.

22.8.3 Pump Controllers

The controllers are built for installation and use in accordance with **ANSI/ NFPA 20, Standard for the Installation of Stationary Pumps for Fire Protection** [5]. Fire pumps are listed for fire pump application, which means they are tested and certified by accredited laboratories and listed by authorized institutions such as Underwriters Laboratories (UL) or Factory Mutual (FM).

Fire pump controller panels (Figure 22.15) contain electrical components such as circuit breakers, starters, switches, relays, and other devices dedicated

Figure 22.15 Pump controller.

to the operation of fire pumps. For large pump motors, Delta/Star, or Soft Start contactors are employed. The fire pump controller performs all the control functions for monitoring system pressures and operating pumps and valves. A local/remote switch is provided on the panel for controlling duty, jockey, and standby pumps manually (local) or automatically (remote).

22.8.4 Fire Pump Sizing

The total head required for the fire protection system is the head required by the standpipe or sprinkler system, whichever is higher. The first step in sizing a fire pump is to determine the head demand, which is the fire protection system head demand minus the head available at the pump suction side. Head loss due to friction is also required to be included to the total head required for the sprinkler/standpipe.

The pump capacity demand depends on the maximum flow rate for the entire system. For a combined system (standpipe and sprinkler), NFPA 14 Section 7.10.1.3 [4] states that only the highest demand between both systems needs to be accounted for in the pump sizing calculations. Should the combined system not have an area fully sprinkled system, then both demands (standpipe and sprinkler) are counted in for sizing.

Once the required capacity and head have been determined, a pump can be sized, based on the NFPA 20. Table 4.8.2 [5] shows the standard pump capacity list.

22.9 Cables and Conduits used for Fire Protection Circuits

NEC Article 760 covers "...the installation of wiring and equipment of fire alarm systems including all circuits controlled and powered by the fire alarm system." These systems are defined in the NEC as follows: "The portion of the wiring system between the load side of the overcurrent device or the power-limited supply and the connected equipment of all circuits powered and controlled by the fire alarm system."

Large industrial installations will employ fiber optic cabling between the major plant control panels and remote areas. Within the plant, one may use conventional conduit and wire installations to the devices.

Generally, most of the fire protection contractors insist on using special non-flammable red-colored conduits for the fire protection power and signaling wiring. The idea here is that the special conduits will endure fires for a bit longer than ordinary conduits. The wiring system also must be water proof and include NEMA 4 enclosures.

Having worked abroad in the tropics for a number of years, I'm still waiting to see a properly water proofed, well-gasketted fire protection outdoor installation. There is absolutely no chance for a conduit system to work properly in this environment. I have seen fire protection devices in NEMA 4 enclosures full of water after rain storms. Water generally enters from the connecting conduits, irrespective whether the entry conduit is at the top or bottom of the NEMA 4 enclosure. Naturally, the devices filled with water are failing and are not responsive. Sometimes they come back to life when the water is drained from the enclosures and conduits; that is, until the next rainstorm. Furthermore, the external conduits also bring water to the indoor installations, thus the indoor devices fail as well. Well, of course, after a serious rainfall, fire alarms are heard everywhere.

This was happening so often that something different had to be done. A similar situation was evident with the "waterproofed" conduit installation for external lighting. To make conduits more waterproofed is an illusion. On another project, we had an international company working on the lighting in a leaking tunnel under a dam for two years. The conduits were weatherproofed, and weatherproofed but without success. Ground fault circuit interrupters (GFCI) breakers were continuously sensing moisture in the conduits and blocked the circuits to be energized. The lighting was held permanently off. To deal with the massive number of alarms, the only solution we found that worked was by replacing the conduit/wire installation with armored cables and water-tight Teck connectors. We did that on the external fire system and the tunnel lighting as well. One can even remove the outer PVC jacket from the cables to make it more adaptable to the fire system installations. In cases where we could not replace the conduits, we simply drilled a small drain hole in the NEMA 4 enclosures to continually drain it. It worked.

It just happened that I passed by this power plant recently and looked at all these fixes made 15 years earlier. People have changed. They did not know what I was talking about. But they told me that they had not heard a nuisance fire alarm from that side of the plant, and the tunnel lighting was also operational.

Using specialized cables for fire protection systems certainly helps in some exceptional situations.

22.10   Fire Detection and Notification Witness Testing

Field devices: All the detectors are functionally checked in the field. Smoke is used for smoke detectors, manual activation for call stations, and heat gun for heat detectors. In order to pass, the device must trigger an audible alarm, be correctly identified at the MFAP display screen, and be successfully silenced and cleared.

Fire panel interconnections: Samples of each device are randomly selected and are functionally checked in the field. In order to pass, the device must trigger the light at the mimic (if applicable), be identified as an alarm at the local fire panel, main panel, and be communicated to the distributed control system (DCS). Same applies for broken wires.

Flow switches: The devices are triggered by manually opening the respective sprinkler's head drain valve. In order to pass, the device must trigger a local panel light, display it on mimic panel, correctly identify on the respective fire panel, and be communicated to the DCS.

Deluge valve operation: Each deluge system must be tested for its automatic and manual functioning. Water pressure must be tested to display proper water spray and mist distribution. Insufficient pressure will cause water to drip instead of creating a mist. Automatic operation is tested by a heat gun applied to one of the heat detectors on the sprinkler circuit. Manual operation is tested directly by operating a manual pull station on the DV.

Gas flooding: Operation of the Argonite gas "directional valves" must be demonstrated, but without triggering the gas discharge. For the gas to be delivered to a room, it requires both gas pressure and pilot valve triggering to activate.

The generator lockout trip (relay 86) must be disabled during testing, but its functioning simulated. To prevent the release of gas, a discharge event must be simulated. Opening of the pilot valve must be witnessed in conjunction with actuation of the "Master Cylinder" plunger for the Argonite canisters (plunger disconnected from the cylinders). The valves must demonstrate that the necessary sequence of operation is working as intended:

(1) 30-second countdown following AGS panel initiation
(2) Simultaneous actuation of "Master Cylinder" plunger and "Pilot Valve Solenoid"

Room fire integrity for gas: The test is conducted to prove the room tightness to hold a sufficient design concentration of gas for >10 minutes, as recommended by NFPA. The test is performed by creating negative and positive pressure into well-sealed rooms and monitoring the pressure (leakage) inside the room. The room is considered satisfactory if the test desired reading exceeds 10 minutes.

References

1 Mike H. (2014) CFAA Technical Seminar – Addressable Notification.
2 NFPA 13 – (2019) Standard for the Installation of Sprinkler Systems.
3 NFPA 10 – (2016) Standard for Portable Fire Extinguishers.
4 NFPA 14 – (2019) Standard for the Installation of Standpipe and Hose Systems.
5 NFPA 20 – (2019) Standard for the Installation of Stationary Pumps for Fire Protection.

23

Corrosion, Cathodic Protection

CHAPTER MENU
23.1 Process of Corrosion and Cathodic Protection, 517
23.2 Galvanized Steel, 519
23.3 Sacrificial Anodes, 519
23.3.1 Galvanic (Sacrificial) Anodes, 519
23.4 Impressed Current Application, 521
23.5 Soil Resistivity, 525
23.6 Cathodic Protection for Ships, 525
23.6.1 Parts Affected on Ship, 525
23.6.2 Protection against Corrosion, 526
23.7 Corrosion due to H_2S Gas, 527
23.7.1 Mitigation, 529
References, 530

23.1 Process of Corrosion and Cathodic Protection

Cathodic protection [1, 2] is an electrochemical practice for preventing corrosion of metallic structures exposed to a natural electrolyte, either submerged or buried. NACE is the worldwide corrosion authority and place to study. The process involves application of DC electrical current to the metal surface from an external source. Cathodic protection essentially means reduction or total elimination of natural corrosion on metal surface by forcing the metal to become cathodic.

Electrochemical corrosion takes place when two different metals come in contact with a conductive liquid (electrolyte). The current flows from the anodic metal through the liquid or moist soil to the cathodic metal. Anodic metal gets to be corroded, while the cathodic metal steel structure is protected from corrosion. Whenever electrical current leaves an anode to enter the electrolyte, small particles of iron are dissolved into solution, causing pitting at the anode. When current enters the cathode, molecular hydrogen gas is formed on the surface to preserve and protect the cathode from corrosion.

Practical Power Plant Engineering: A Guide for Early Career Engineers, First Edition. Zark Bedalov.
© 2020 John Wiley & Sons, Inc. Published 2020 by John Wiley & Sons, Inc.

In other words, by installing the cathodic process, the natural corrosion process is blocked and reversed.

There are three methods of providing cathodic protection:

- Galvanizing the metallic structure
- Sacrificial anodes
- Impressed current

A battery used in our gadgets is a typical galvanic cell. Zinc in the battery container is an anode. The carbon post in the middle is the cathode. The battery case is filled with acidic electrolyte. Once the battery is placed in use, current flows from the zinc (Zn) case (anode) to the carbon post (cathode) through the electrolyte. As oxygen is generated at the face of anode, particles of Zn are dissolved into the solution and hydrogen gas is deposited on the carbon rod. The current will continue to flow until the Zn case is broken.

When a long pipeline is laid in the ground, soil moisture serves as the electrolyte. The anode and cathode are established on the same pipeline, with the pipeline providing the return path. This is due to differences in the acidity of the soil at different places, which causes the same metal to be more active (+) in one location over another, thus causing a potential difference to drive the current between the more active spot to the other (see Figure 23.1).

It is the specific potential difference generated by two different metals in an electrolyte that drives the current in the galvanic cell between the metal to be protected (cathode) and the active metal to be sacrificed (anode). The same metals found in a different electrolyte will generate different voltages. If any two metals are selected loosely joined together and buried in moist ground, they will establish a galvanic cell. The metal that is more positive will become an anode to the other metal, which will become cathodic. The anodic metal will corrode, while the cathodic metal will be protected from corrosion. The greater the difference between selected metals the greater and faster will be the process of induced corrosion. A metal structure placed in soil or ocean is the fourth condition for starting corrosion: anode, cathode, electrolyte, and metal structure.

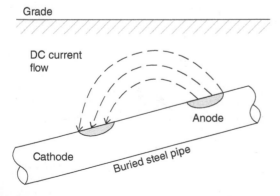

Figure 23.1 Buried pipe corrosion.

Table 23.1 Metal solution potentials.

Anodes (+)	Volts	Cathodes (−)	Volts
Lithium	+2.96	Copper	−0.80
Potassium	+2.93	Silver	−0.80
Barium	+2.90	Platinum	−0.86
Calcium	+2.87	Gold	−1.50
Sodium	+2.71	Carbon	
Magnesium	**+1.75**		
Zinc	**+1.10**		
Aluminum	**+1.05**		
Chromium	+0.56		
Nickel	+0.23		
Tin	+0.14		
Lead	+0.12		
Iron	+0.04		
Hydrogen	0.00		

Bold indicates that the metals are the key materials for anodes.

Here are the solution potentials of some materials as observed in sea water (see Table 23.1).

23.2 Galvanized Steel

Galvanizing generally refers to hot-dip galvanizing that is a way of coating steel, fences, and other exposed metals with a layer of metallic zinc or tin. Galvanized coatings are quite durable in most environments because they combine the barrier properties of a coating with some of the benefits of cathodic protection. If the zinc coating is scratched or otherwise locally damaged and steel is exposed, the surrounding areas of zinc coating form a galvanic cell with the exposed steel and protect it from corrosion. This is a form of localized cathodic protection – the zinc acts as a sacrificial anode.

23.3 Sacrificial Anodes

23.3.1 Galvanic (Sacrificial) Anodes

Galvanic anodes are used for smaller structures and longer painted structures with some breaks in coating. Also, where less current is needed to maintain

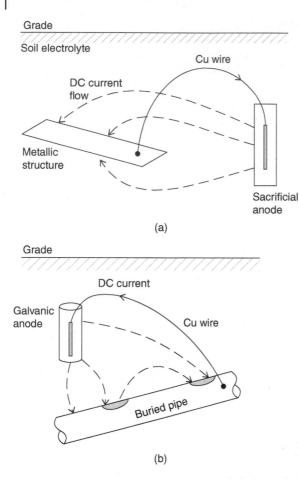

Figure 23.2 (a, b)
Protection by sacrificial
anode.

corrosion protection, sacrificial anodes are used and buried next to the structure. If more current is needed, impressed current is usually a better choice.

Corrosion occurs where current leaves the sacrificial anode. Electrolyte conducts the current to the cathode. No corrosion takes place at the cathode. Refer to the simplified Figure 23.2 of sacrificial anode applications. Cathodic protection is achieved by transferring the corrosion from the metallic structure (cathode) to a sacrificial anode, which is regularly replaced once it is consumed. Current flows to the protected structure at all contact points with the electrolyte along the whole buried pipeline.

Power in galvanic cathodic protection is generated by the anode. The efficiency of the anode is enhanced when installed in a backfill of 75% gypsum, 20% bentonite, and 5% sodium sulfate. This special mixture lowers anode-to-earth resistance and allows electrical current to flow more easily to the targeted structure.

Figure 23.3 Groundbed with sacrificial nodes.

Three metals are commonly utilized for cathodic protection of coated or painted steel, as follows:

• Magnesium	−1.75	Soil and freshwater applications
• Zinc	−1.10	Low resistivity soils and saltwater
• Aluminum	−1.05	Saltwater and limited freshwater applications

A bed of sacrificial anodes is placed in proximity to the protected structure and is connected to the pipe by a thin copper wire. This allows the anodes to distribute galvanic current toward the structure. No outside power is required. Disadvantages of the sacrificial anode application are limitation in the current produced and rapid consumption of the anodes. Anode replacement is needed every five years.

When a sacrificial magnesium anode groundbed is established (Figure 23.3) near and connected to a pipeline, the potential difference of 1.5 V between the two metals will be sufficient to overcome all other naturally created galvanic cells on the pipeline in the immediate vicinity of the anode. The naturally created anodes on the pipes will be converted to be cathode for the new overall galvanic cell.

23.4 Impressed Current Application

Buried, unpainted steel structures require a lot more current than painted structures in order to be fully protected from corrosion. Sacrificial anodes cannot generate sufficient currents to offer adequate corrosion protection. In these situations, impressed current systems must be employed as shown in Figures 23.4 and 23.5.

Impressed current [3, 4] is not used solely for the bare pipes. Painted or insulated gas pipes are protected this way also. Gas lines tend to be large, and their integrity is critical. This technology involves the use of an external source of

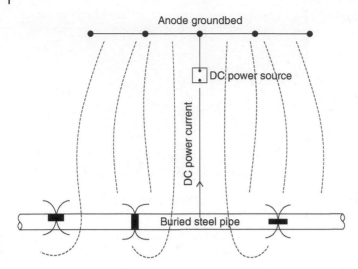

Figure 23.4 Impressed current installation.

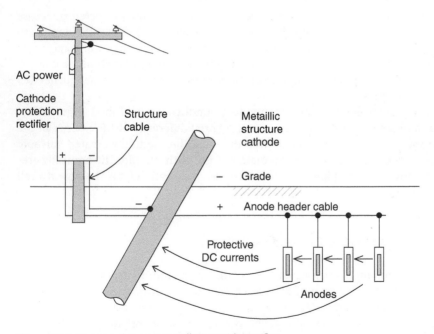

Figure 23.5 Impressed current installation with rectifier.

DC current through long-lasting anodes buried in the vicinity of the protected object.

Impressed current systems can add considerably more current to the structures than sacrificial anodes. Rectifier stations can be placed along a pipeline every km of line depending on the power demand. Potential measurements along the line will determine how much current is needed to create adequate protection. The requirements for more or less current may be seasonal, thus adjustments could be made at the stations to keep the measured potential more negative than −0.85 V.

So what is the criterion for accepting the installation to be well protected from corrosion?

Industry accepted criteria for effective protection is the pipe-to-soil voltage measurement method. The accepted value is **−0.85 V** or **more** negative voltage between the structure and a Cu–CuSO$_4$ electrode placed in the electrolyte near the structure. This is considered the minimum voltage to insure any naturally occurring galvanic cells in the pipelines are overcome. Voltage measurements giving more positive than (**−0.85 V**) are considered not suitable for corrosion protection.

Anodes for impressed current cathodic protection systems are engineered for use in various environments, soils, and structure facilities. Impressed current anodes can be made from graphite or titanium and in the forms of a mesh, ribbon, or tube. Anodic materials with low corrosion consumption rates are selected for long-lasting (20-year) applications. Anode materials proven to be suitable for these applications are graphite, high silicon cast iron, and mixed metal oxides. The anode groupings are called groundbeds. Just like in the sacrificial method, a group of anodes is connected to the pipe by a Cu wire, but this time through a source of DC current. The power source is usually an AC/DC charger receiving power from an adjacent overhead line or solar panels.

The DC source is adjustable to generate the amount of current as needed and as seasonally demanded. Protected structures include buried pipelines, buried storage tanks, offshore structures, above-ground tanks, etc. A single groundbed may comprise 20 or more anodes laid in the ground at various depths and in various configurations.

For example, anodes may be placed vertically in augured holes 10 cm in diameter (4 in.) at 5-m (15 ft) intervals. The anodes are typically 2.5 cm diameter (1 in.), with lengths of 50, 100, or 150 cm.

Anode loading is typically

- 2.5/50 cm anode can be loaded to 5.5 A$_{dc}$,
- 2.5/100 cm anode can be loaded to 11 A$_{dc}$

Significant maintenance effort is required to maintain the electrical circuit in operating condition, including the replacement of anodes and periodic replacement of backfill in which the anodes are buried. The impressed current system

Figure 23.6 Rectifier transformers.

will protect parts of the structure that are buried and in contact with electrolyte, but will not be effective for the parts above ground.

The following must be determined for the application (Figure 23.6) of the impressed current equipment:

- Structure painting or coating
- Soil conditions: moist, dry. Moisture variations from region to region
- Resistivity of electrolyte
- Buried structure coverage
- How much DC current is needed
- Source of power

A typical impressed current installation includes

- Supply transformer, voltage to suit, and connected to a local distribution line. Transformer may be oil filled or dry type. Oil-filled rectifier transformers will be rated for 65 °C temperature rise. Dry type will have Class H insulation for 180 °C (356 F) temperature rise.
- NEMA 4 or 4× enclosure, floor or pad mounted.
- Lightning arrester.
- Main circuit breaker.
- AC/DC rectifier, DC voltage to suit, up to 750 A_{dc} output.
- Controller and monitoring unit.

Available modes of operation include constant current or constant potential control with voltage limiting. The AC input supply from a stepdown transformer is applied to a fully controlled Silicon Controller Rectifier (SCR) full-wave bridge assembly. The controlled and rectified voltage is filtered by a reactor and capacitor combination to yield a low-ripple DC output. The low ripple has the advantages of increased efficiency, no communication interference, and more precise control of the output.

23.5   Soil Resistivity

Equally as in the case of the plant grounding (Chapter 6), soil resistivity is the key factor in selecting a groundbed location. It will define the number of anodes required, the depth of the backfill, and the power rating of the rectifier trans-formers. In general, locations with the lowest and most uniform soil resistivity with relation to depth should be utilized for deep groundbed sites.

Soil resistivity ranges from 150 in swampy lowlands to 2500 Ω-m in high country regions (1 Ω-m = 100 Ω-cm). Carbonaceous backfill is used to bury anode groundbeds. Its resistance must be periodically measured and replaced if greater than 3 Ω. Moisture in the groundbed must be maintained because the anode operation drives the moisture out of the groundbed through an electro-osmotic force, which increases with anode current density.

23.6   Cathodic Protection for Ships

Corrosion is one of the greatest enemies of ships and their machinery [5]. Corrosion takes away the durability and strength of metal. From the main body of the ship to the smallest equipment used in operations, corrosion makes its presence felt in almost every type of equipment used on board a ship. Sea water is one of the best electrolytes.

A ship's hull is continuously in contact with sea water and the moisture-laden winds make it highly susceptible to corrosion. Corrosion increases with hull speed through the water. Corrosion rates are also affected by water salinity and water temperature. For these reasons, corrosion protection in the form of paint-ing, sacrificial anodes, and Impressed Current Cathodic Protection (ICCP) is essential for ships' longevity (see Figure 23.7).

Due to the presence of moisture and water, materials used in ships are contin-uously undergoing chemical and electrochemical reactions. Chemical reactions are generally on a metal's surface, and it is there where the degradation of the metal starts. Generally, the exchange of charge takes place between the reac-tants, the steel surface, and atmospheric oxygen.

23.6.1   Parts Affected on Ship

The main areas of concern are the propeller region, steel plates in the hull, pip-ing systems, and copper plates in heat exchangers. Corrosion also takes place

Figure 23.7 Boat hull protection.

The white patches visible on the ship's hull are zinc block sacrificial anodes.

in and around regions where ice forms (e.g. refrigeration and air conditioning areas), and unpainted areas or the areas where the painting is worn out.

Metals with compositions that are not homogenous may be susceptible to localized corrosion. One such substance is brass, which is an alloy of zinc and copper. On contact with moisture, zinc, which is anodic in nature, dissolves leaving behind copper.

Electrochemical reactions involve compounds that have the capacity to dissolve ions or charged particles in them. These kinds of compounds are acids, seawater, or some metals and gases, which provide mobility to the dissolved ions. Stainless steel is not affected by chemical reaction.

23.6.2 Protection against Corrosion

The following are the three key elements to mitigate the effects of corrosion on ships:

- Painting metal surfaces exposed to water or moisture. Special paints (usually two-part epoxy) are used to protect against chemical reactions.
- Use of sacrificial anodes to reverse the natural flow of galvanic currents.
- Application of an impressed current system to reverse the natural flow of galvanic currents.

Sacrificial galvanic anodes are attached to the hull of small and big vessels. ICCP is employed for larger vessels (see Figure 23.8).

Galvanic anodes use the following materials: magnesium (Mg), aluminum (Al), zinc (Zn). Zn is the most common. Sacrificial galvanic anodes are generally

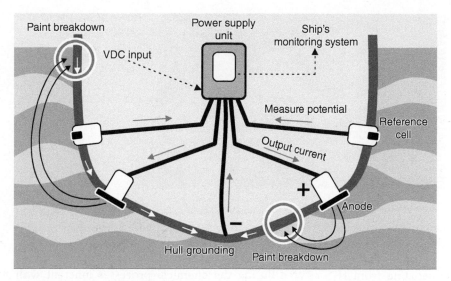

Figure 23.8 Ship impressed current.

placed in the water and fitted flush to the hull to minimize drag. They are also used to protect ballast tanks and heat exchangers. As with all galvanic cathodic protection installations, the effectiveness relies on a solid electrical connection between the anode and the item to be protected. For ICCP on ships, the anodes are usually made of relatively inert materials such as platinized titanium. The anode cables are introduced into the ship via a compression seal fitting and are routed to the ship's internal DC power source.

Smaller vessels with nonmetallic hulls, such as yachts, are equipped with galvanic anodes to protect areas of outboard motors.

A large amount of positive current is passed through the water surrounding the ship. The rectifier is connected to the surface steel on the negative terminal. The positive terminal is connected to the anodes that are fixed inside the ship's surface. As with the galvanic anodes, ship ICCP anodes are flush-mounted outside, minimizing the effects of drag on the ship, and located a minimum 1.5 m (5 ft) below the light load line in an area to avoid mechanical damage.

The current density required for protection is a function of ship size to be considered when selecting the current capacity and location of anode placement on the hull. The impressed current can be in the range of 10–600 A. The voltage can reach as high as 20–30 V.

23.7 Corrosion due to H₂S Gas

Hydroelectric and geothermal power plants are subject to severe corrosive issues emanating from hydrogen sulfide gas (H_2S) [6] created in the lake

Figure 23.9 Cu terminal corrosion.

catchments and reservoirs due to decaying leaves and tree trunks sunk at the bottom of lakes. Tree trunks in the tropics tend to sink. Petrochemical and paper and pulp industries are also subject to severe corrosive issues emanating from H_2S created during the chemical processes inherent with these industries.

H_2S corrosion is a destructive attack on a metal by chemical or electrochemical reaction with its environment. It is a phenomenon that affects materials in most plant systems, including electrical systems. Steel used as tanks are particularly susceptible. Copper, aluminum, and silver are the most widely used materials in electrical conductors and enclosures. Aluminum is resistant to atmospheric corrosion by H_2S, while copper and silver are not (see Figure 23.9).

When H_2S permeates the insulation surrounding a copper conductor, the hydrogen sulfide reacts with the copper to form copper sulfide and causes the conductor to swell. Copper sulfide increases the conductor's resistance thereby reducing the ability of the conductor to conduct electricity. The thickness of electrical terminations on microelectronic boards used for distributed control system (DCS) circuit boards has become so tight that even a minute element of corrosion can cause a board failure.

H_2S corrosion happens in water and air. Due to water turbulence during hydroelectric generation, H_2S gas is released from the bottom of the reservoir and affects metallic structures within the hydraulic water path and the hydro turbine. The gas also gets into the air and affects copper wiring in the electrical and controls installations. If you work in this environment, you will smell it and feel it. Silver chain around your neck will be become black within a day. The corrosion is aggressive and within a year can cause serious damage to the turbine components including runner blades, grounding cables, and wiring in the electrical panels (see Table 23.2).

Air quality in control rooms is defined by the Instrument Society of America ISA-71.01-1985 [7]. A noncorrosive environment is described by NACE as G1 [8].

Table 23.2 Human health effects by H_2S sponsored by WHO (UN).

Symptom	H_2S concentration
Odor threshold	1 ppm
Bronchial constriction in asthmatic individuals	2 ppm
Increased eye complaints	5 ppm
Increased blood lactate concentration, decreased skeletal muscle citrate synthase activity, decreased oxygen uptake	7–14 ppm
Eye irritation	5–29 ppm
Fatigue, loss of appetite, headache, irritability, poor memory, dizziness	28 ppm
Olfactory paralysis	>140 ppm
Respiratory distress	>500 ppm
Death	>700 ppm

NACE Tabulation G1 for air quality in control rooms stipulates the following limits for different gases:

Gas	SO_2/SO_3	H_2S	Cl_2	NO_x
Concentration (ppb)	<10	<3	<1	<50

Human health effects at various H_2S concentrations [9] are given in Table 23.2. Courtesy of Publication [6] sponsored by the UN.

Evidently, H_2S presents an intolerable situation that must be resolved in particular in the areas of high gas concentration. Utilities need to clear their reservoirs of dead trees before filling with water. Removal of H_2S gas through abatement processes is expensive and unlikely to be effectively implemented.

23.7.1 Mitigation

Clean air, free of H_2S, for use in control rooms is not easily obtained as this gas is part of the environment. Concentration of the gas is higher in underground installations in particular in hydroelectric surge chambers. Some means of ventilation and air removal must be established in these areas. Other measures that can be taken to mitigate the H_2S problem are

- Copper conductors and connectors must not be left exposed. All exposed parts must be painted with appropriate paint and repainted again every few years.

- All control wiring must use tinned wires. All control wire terminations must be tinned.
- External grounding cables must be insulated. Though [7] indicated a corrosive attack through the conductor insulation, we had not witnessed this problem on our last project.
- Dissimilar metals placed next to each other in water are an area of massive corrosion. If this is required, the dissimilar metals must be welded together.

References

1 NACE Publications SP-0169:(2013) Standard Practice Control of External Corrosion.
2 NACE International – Publications (current web). www.nace.org/publications
3 Peabody, A.W. (1967). *Control of Pipeline Corrosion*. NACE International, The Corrosion Society.
4 NACE Test Standard for Pipeline Measurements TM0497-2018.
5 Van Dokkum, K. (ed.) (2003). *Ship Knowledge – A Modern Encyclopedia*. DOKMAR.
6 Rivera, M.A. (2007). *Design Considerations for Geothermal Power Plants, Emphasis on H_2S Related Problems*. UN University.
7 ISA 71-01-1985, Standard for Environmental Conditions for Process Control Systems.
8 Vimala, J.S., Natesan, M., and Rajadran, S. (2009). Corrosion and protection of electronic components in different environmental conditions. *Open Corros. J.* 2: 105–113.
9 World Health Organization: Human Health Aspects by H2S (2003), Document 53, page 14.

24

Brief Equipment Specifications and Data Sheets

CHAPTER MENU

24.1 Power Transformers, 531
    24.1.1 Winding Configurations, 532
    24.1.2 Transformer Phase Shift, 532
    24.1.3 Tap Changers, 537
    24.1.4 Dry and Oil Type Transformers, 538
    24.1.5 Transformer Dielectric Tests, 539
    24.1.6 Basic Insulation Level (BIL), 540
    24.1.7 Excitation Current, 540
    24.1.8 Inrush Current, 541
    24.1.9 Autotransformers, 541
    24.1.10 Transformer Parallel Operation, 541
24.2 Motors up to 200 kW, 542
    24.2.1 Design Criteria, 542
    24.2.2 Torque, 542
24.3 Motors > 200 kW (Medium Voltage Motors), 545
24.4 VFD Specification Requirements, 548
24.5 13.8 kV Isolated Phase Bus + PTs (IPB), 549
24.6 Electrical Enclosures, 550
    24.6.1 NEMA Enclosures, 551
    24.6.2 CE and IEC Classifications for Enclosures, 553
24.7 Technical Data Sheets, 553
References, 565

24.1 Power Transformers

The following discussion outlines the options available and helps electrical engineers to determine the criteria in his equipment specifications to suit his specific application.

Practical Power Plant Engineering: A Guide for Early Career Engineers, First Edition. Zark Bedalov.
© 2020 John Wiley & Sons, Inc. Published 2020 by John Wiley & Sons, Inc.

24.1.1 Winding Configurations

The following transformer winding configurations are often mentioned. Only a few of them are used in the plant distribution;

Y–Δ	The best choice for HV grid transformers, with solidly grounded neutral and lower BIL. Yd11 was the configuration used on our last project for the main 275 kV transformers. Yd1 is equally common.
Y–y	Grounded or not grounded. Not used in the power distribution industry, mainly due to their poor performance in establishing stable grounding conditions.
Δ–y	The best transformer choice for MV and LV systems: from 3.3–34.5 kV.
Δ–Δ	Rarely used, mostly for special applications.
A–Δ	Autotransformer, tapped as Yy, with tertiary Δ winding. Used for extra high-voltage (EHV) grid connections. The Δ winding offers stable and balanced grounding conditions to both autotransformer voltages. The Δ tertiary winding, other than stabilizing the neutral points, can also be loaded with capacitors or reactors to provide voltage regulation on the line.
Zg	Single winding. Mainly used as a grounding transformer. Like the Δ winding on a Y–Δ transformer, it allows the flow of **Io** currents with a flux cancelling effect.

24.1.2 Transformer Phase Shift

The Y–Δ or Δ–y transformer produces a natural 30° phase shift, lead or lag, from the primary to the secondary terminals. Considering the counterclockwise (CCW) phase sequence, the lead 30° is often called Dy11 transformer. On the other hand, a 30° lag angle is called Dy1 transformer. If wound around the iron core in the same direction, each phase: primary and secondary are in phase. While the Y winding connects the phases between each phase and neutral, the shift occurs due to the Δ winding connections between two phases.

Due to the 30° phase shift between the windings, various additional configurations can be made to achieve Dy5 (150° lag shift) or Dy7 (150° lead shift), both of which are also popular in the industry. It is a matter of shifting the phases by one phase, to lead or lag, as follows: ABC – abc → ABC – cab, or ABC – bac, respectively.

Let us look at the transformations used in a typical power plant from top (Grid) down to 415 V services (bottom) along the two paths; station services and generator paths, as shown on the phasing diagram Figure 24.1.

Power plants have generally standardized on two winding configurations: 30° lead (11) and 30° lag (1) transformer types. The practice is to string the downstream plant transformers in sequence to go back and forth to maintain the angle from the grid (assumed 0) to the rest of the station service, either +30° or −30°, as shown in Table 24.1 as an actual case for a power plant.

On this particular power plant, the unit services bus is operating at 415 V and shifted by +30° after three transformations from the grid when served from the

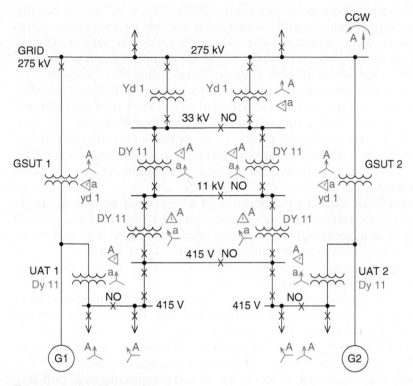

Figure 24.1 Plant phasing diagram.

Table 24.1 Plant phasing diagram.

Path: from grid to 415 V station service				
Transformation		Shift angle	Overall shift	Location
275 kV grid	Assumed 0	0	0	275 kV grid
Ynd1	275–33 kV	+30° lead	+30°	Local distribution
Dyn11	33–11 kV	−30° lead	0	Plant distribution
Dyn1	11 kV–415 V	+30° lead	+30°	Plant services
Path: from grid to generator				
Transformation		Shift angle	Overall	
275 kV grid	Assumed 0	0	0	275 kV grid
Ynd1	13.8–275 kV	+30° lead	+30°	Gen. voltage
Dyn11	13.8 kV–415 V	−30° lead	0	Unit services

grid through station service transformers (SSTs). The same services bus can also be reached from the grid through the generation path via unit auxiliary transformer (UAT) on two transformations. Since the two paths to the same LV bus are not in phase, the transfer SST to UAT, from one path to the other, must be performed out of "sync" as a dead "Break before Make" transfer.

The unit (generator) starts with the unit services being fed from SST. After reaching 90% voltage, the unit load is regularly transferred to the UAT. The unit services are shut down temporarily for <2 seconds to allow the UAT to close. After the breaker closing, the plant control system remembers which drives were in service and reinstates them back to service. If this transfer is not done satisfactorily with all the critical starting loads running within four seconds, the unit is shutdown.

The two 33 kV transformers are placed in this operating scheme to provide for a local 33 kV distribution to nearby communities. If these transformers were 33 kV service not required, the two paths could be arranged to be in phase at the 415 V bus and the synchronizing could be done as live transfer.

Industrial plants generally do not have these problems, as they operate mainly on the basis of radial distribution.

Figure 24.2 shows the most common configurations in the power and industrial plants including the Zg grounding transformer. The individual diagrams show the phasing connections.

Note that the transformers at the bottom right have opposing configurations: Dy11 = Yd1, Dy1 = Yd11, when looked at from the opposing ends.

Transformer impedance: In North America, the impedance is known as Z on pu or % basis of the base MVA rating of the transformer. In Europe, it is known as uk%, and the short-circuit voltage drops. Both are the same thing. The impedance must refer to the transformer base rating, which is the transformer rating before forced cooling.

Turns ratio: The turn ratio on normal taps is equal to the ratio of the voltages for the transformer zero load at the neutral tap. Once the transformer is loaded, the voltage ratio is no longer equal to the turns ratio due to the voltage drop caused by the transformer impedance.

Transformer losses: Transformer efficiencies are high; running at over 99.5% for large transformers. No-load losses usually run at 1 : 6 against the load losses.

No-load losses, also called Iron losses, are all the losses due to magnetization at full voltage applied with secondary circuit open (not loaded). They are measured while performing an open circuit test. It includes eddy current loses, hysteresis losses, I^2R loss due to the exciting current, and dielectric losses. The no-load-losses are relatively constant for any transformer loading and proportional to the voltage up to the nominal voltage. They rise significantly in the core saturation region above the nominal voltage.

Load losses, also called Cu Losses, include I^2R losses caused by the load current I_n flowing through the windings at a rated secondary current. Cu losses

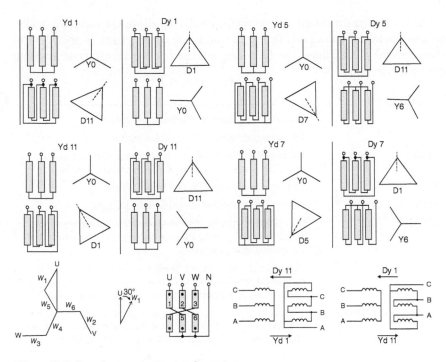

Figure 24.2 Transformer winding configurations.

vary as the square of the load current and are measured while conducting a transformer short-circuit test.

Auxiliary losses are the losses for the auxiliary power used for transformer cooling fans, water pumps, etc. to support the MVA output as the loading approaches the nominal levels. The auxiliary load typically is less than 1/10 of the no load loss for the water pumping systems and 1/20 for the fan cooling.

Transformer cooling: The transformer losses both no-load and load generate heat in the transformers, which must be removed in order to maintain the temperature class of the transformers and to ensure longer life of the insulation.

For air (dry type) cooled transformers, cooling is accomplished by providing adequate ventilation and cooling ducts in the coils. Fans are added to increase heat transfer away from magnetic elements and vulnerable dialectic insulating components.

Cooling of oil-filled transformers is similar. Cooling ducts in the coils must be in sufficient number and size to allow oil to flow through the coils to remove the heat. This fluid moves simply by natural convection, or it can be force cooled by air fans, or oil and water pumping. The tank surface with detachable radiators must be large enough to transfer heat away from the cooling media by a combination of conduction, convection, radiation, and forced cooling.

The cooling banks of fans and/or water pumps usually operate automatically on a temperature rise signal. Typically, one spare cooling bank is provided to allow for one bank failure. The following are the popular transformer cooling designations:

ONAN	Oil natural air natural. The basic self cooled unit
ONAF	Oil natural air forced. Unit cooled with air fans
OFAF	Oil forced air forced. The unit equipped with oil pumps circulating oil and oil being cooled by external air fans.
AN	Dry type transformer cooled by air, naturally
ANAF	Dry type transformer cooled by external fans
OWAF	Oil cooled by means of air cooled water exchangers. This may be the size-saving solution for larger transformers built for limited space locations.
ODWF	Oil directed water forced. Similar to OWAF, but the oil is directed through the special channels in the iron ore and next to the coils, then cooled in the water cooled heat exchangers.
ONAN/ONAF/ONAF	Transformer with this cooling configuration has two stages of fan cooling, each rated 33% of the base rating. For a base 20 MVA unit, this would be 20/26.6/33.3 MVA rated transformer.

Temperature rise (insulation class): It is defined as the average temperature rise of the windings above the ambient (surrounding) temperature, when the transformer is loaded at its nameplate rating.

Liquid-filled transformers come in standard temperature rises of 55 and 65 °C. The 55 °C rated transformer is bigger and more costly.

Dry-type transformers are available in standard temperature rises of up to 220 °C.

These values are based on a maximum ambient temperature of 40 °C. That means, for example that an 80 °C rise dry transformer may operate at an average winding temperature of 120 °C when fully loaded in a 40 °C ambient environment. A hot spot allowance of 5 °C is added above the nominal temperature rise.

The maximum acceptable temperature rise is based on an average ambient of 30 °C during any 24-hour period and an ambient of 40 °C at any time (Table 24.2).

Earlier in my carrier, it was noted that transformer life is reduced by 50%, if operated at 10 °C above its nominal class. It is not known whether this criterion is still valid and by how much with the new type of insulation materials.

Transformer phase markings: Looking from the low voltage side X1–X2–X3 from left to right. The high voltage side will match the low voltage side H1–H2–H3 from right to left.

Table 24.2 Transformer insulation class.

Insulation	Average winding	Hot spot	Maximum winding
Class and rating	Temperature rise (°C)	Temperature rise (°C)	Temperature
A Class 105	55	65	105 °C oil type
B Class 130	80	110	150 °C dry type
F Class 180	115	145	180 °C dry type
N Class 200	130	160	200 °C dry type
H Class 220	150	180	220 °C dry type

Transformer accessories: Breather, oil-level indicator, top oil temperature indicator, oil drain valves, winding temperature indicator, Buchholz relay, etc. specify all standard accessories, with alarm and trip contacts where applicable.

Conservator oil tank: It is a standard equipment for large transformers.

Hermetically sealed tank transformers are oil cooled with fins without conservator expansion tank. The oil expansion due to temperature changes is allowed by a rubber membrane and a nitrogen cushion. The tank operates close to vacuum at 0.5 bar. In these transformers, oil is not exposed to air, and it cannot be contaminated, which extends the life of the transformer.

This type of transformer is generally built for transformers with less oil. Though some transformers of this type were built for higher MVA ratings, it is generally recommended for ratings up to 10 MVA.

24.1.3 Tap Changers

The tap changers can be either On Load (OnLTC) or Off Load (OffLTC). There was even a transformer tap changer with both. The OnLTCs are generally used on larger transformers. The taps are mostly located on the primary side (facing utility) to maintain a stable secondary voltage (facing the plant; see Figure 24.3).

The taps are typically ±10% for power transformers, but often also +5, −10%. Smaller distribution and service transformers up to 2 MVA, go with ±5% taps.

The most common tap changers are ±10% in (0.625%) steps, in comparison to 2.5% steps, typical for the OffLTCs. Multiple steps offer a finer voltage following capability.

The tap changers work on the principle of increasing turns (+taps) and reducing (−taps) on primary winding to decrease/increase voltage on the secondary winding, respectively. The tap switching is performed on a "Make before Break" principle, controlled by a voltage relay.

By lowering the voltage on one side, the MVAR flow turns to the station with lower voltage. If the plant is required to automatically share and regulate MVAR

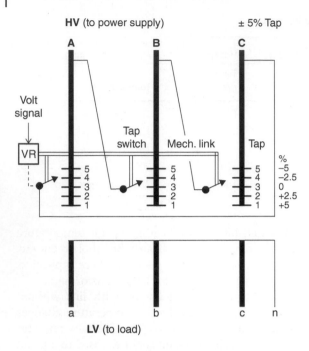

Figure 24.3 Tap Changer on Δ-Y transformer.

flows, the plant will need transformers with on load tap changers. OffLTC is operated manually when the unit is switched off. Transformer Off Load taps are full-capacity taps. The OnLoad taps may be allowed to be of reduced capacity, based on the system requirements.

24.1.4 Dry and Oil Type Transformers

It was told that dry type transformers were three times more expensive than oil type transformers of the same rating. No confirmation have been seen in experience to this fact. Actually, it is firmly believed that the dry type transformers are less costly. They are far simpler to build, enclose, cool, transport, and need far less accessories and instrumentation for monitoring. They are designed for higher classes of insulation, notably Class H – 220 °C, which is about twice that of the oil-cooled transformers. So they are designed to have smaller windings for more current A/mm^2 flow, which causes considerably higher Cu losses and higher impendences.

Dry type transformers are considerably easier to test and place in service. They can also be equipped with cooling fans for additional MVA capacity (see Figure 24.4).

Dry type transformers are usually built to about 3 MVA. Much larger units were built for special situations. Dry transformers are small transformers compared to the oil types and used mostly for indoor unit substations. That exactly

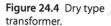
Figure 24.4 Dry type transformer.

is their biggest advantage. They are installed indoors, while the oil-cooled transformers are a major fire hazard and must be placed outdoors with all those specific constraints and requirements for oil containment, disposal, and spacing away from the buildings. Dry type transformers can be placed directly against the LV switchgear to form a unit substation assembly for power distribution to the motor control centers (MCCs) and other distribution boards, thus eliminating the bus ducts and breakers. On the primary sides, they can be furnished with 15 kV (5 kV) load interrupter switches (fused switches) for primary protection. The load interrupter switches are added if the transformers are located >50 m away from the 15 kV (5 kV) switchgear. The load interrupter switches are usually equipped with arcing horns shaped as whiplash horns, which allow the switch to interrupt the magnetizing current of unloaded transformers.

In conclusion, looking at a big picture and the installation requirements, dry type transformers are considerably less expensive than the oil transformers of the same size.

24.1.5 Transformer Dielectric Tests

The purpose of the tests is to prove the transformer was built with quality materials and workmanship to withstand the operating conditions during its operating life. The site impulse tests are usually conducted at reduced values (80%) from those of the factory tests.

The IEEE (Institute of Electrical and Electronic Engineers) standard PC57.12.200 × Table 6 High Frequency Test Tables [1]. The standard defines the following tests based on the equipment BIL level:

- Lightning impulse (BIL), kV Crest, $1.2 \times 50\,\mu s$,
- Chopped wave, kV Crest, $1.1 \times BIL$,
- Switching impulse, kV Crest, $0.83 \times BIL$.

Factory impulse tests are to be conducted in succession:

- A full wave impulse, negative polarity.
- A reduced wave impulse, negative polarity, having crest value of 50–70% of full wave.
- A chopped wave.
- An applied potential test at power frequency between one phase to ground with other phases connected to ground, for a duration of one minute.
- Induced potential test, applied between terminals of one winding voltage is equal to twice the normal voltage of that winding, at twice the nominal frequency for one minute.

24.1.6 Basic Insulation Level (BIL)

Transformers with primary wye (star) windings, secondary delta windings, which are *solidly (effectively) grounded* on the HV side, can use 80% rated arrestors and lower BIL levels on HV side. A transformer with a 900 kV BIL instead of 1050 kV on its 230 kV winding is a huge financial incentive to implement effective grounding at the HV side. Also, the effectively grounded transformer can employ lower-graded BIL level at its neutral point; for instance, 15 kV for a 230 kV transformer. Nowadays, this approach has become less important and may not save money, as the insulations have become more uniform and of higher dielectric strength.

24.1.7 Excitation Current

The excitation current is relatively small, consisting of the magnetizing (reactive) current and smaller active current. The shape of the magnetizing current depends on the saturation of the transformer core. At the nominal voltages, it is proportional to the voltage, but at the voltage levels >100% nominal voltage, the core begins to saturate. The excitation current increases due to the core saturation and becomes severely distorted with harmonics in order to create a sinusoidal flux in the magnetic core. The second and third harmonics are dominant. The third harmonic is the zero sequence current which flows in each phase and is in phase with each other. Under the nominal operating conditions, the excitation current for larger transformers is in the order of <2% of the nominal current.

24.1.8 Inrush Current

During the *first* energizing of the transformer, the initial inrush current may reach a magnitude of *10 times* the nominal current depending on the instant when the transformer is subjected to the voltage on the sinusoidal curve. It is wise to gradually energize the transformer for the first time from the secondary side. In most cases, the transformer has already been energized during the factory tests so that may not be a concern. The transformer inrush current is always present and may trip the incoming breaker if the protection is set strictly for the load current.

When setting the transformer protection, consider setting the overcurrent element (50) to $6 \times I_n$ to $8 \times I_n$ for the inrush current for a duration of 0.5 seconds. The high value of current is due to the high degree of saturation of the magnetic circuit at the moment of energizing. It decays rapidly due to the resistive losses, which provide a damping effect. New numeric relays claim that they have means of suppressing the effects of the second harmonics.

During temporary emergencies, two transformers operating on the same bus may be required to be energized on a single incoming breaker. This operating condition must be verified during the plant commissioning to prove the relay settings are appropriate for this situation to handle the inrush for two transformers. A short ½ second delay can be programmed into the settings to allow for this emergency operating condition.

Transformers seem to be able to tolerate multiple switching. Once we had a situation where a faulty HV disconnect switch was continuously opening and closing (energizing) a 5 MVA transformer every 10 minutes. By the time we arrived to this remote location three hours later, the switch was still doing its number. We observed it for a few minutes and then shut it down. The transformer was not damaged.

24.1.9 Autotransformers

They are rarely used nowadays, though they may be getting a new lease on life on the EHV systems. Autotransformers, one phase or three phase, come as single winding transformers. They are usually wired as Y–Y with solidly grounded neutral. The two voltages are established by tapping the common winding; say 230–13.8 kV. Since a wye grounded auto transformer cannot provide a stable neutral point, a tertiary delta winding is added to allow for the flow of third harmonic currents (zero sequence) to stabilize the neutral point of the auto-transformer. The delta tertiary winding is typically rated at not more than 35% capacity of the autotransformer.

24.1.10 Transformer Parallel Operation

There was no experience of a situation where the transformers were operating in parallel. This is a classical school case, but never used in the industry. The transformers are often connected in parallel on the HV side, but on the LV side, they connect to their own separate busses and loads. The low voltage buses are connected together by a bus tie breaker, only if one of the transformers is taken out of service, thus making one transformer responsible for feeding the whole load.

Of course, if you want to parallel the transformer on your computer, a system study software will allow you to do so. Then change the impedances on the transformers or taps at will and observe the effect of circulating currents and overloading the transformer with lower impedance over the other.

24.2 Motors up to 200 kW

The responsibility for defining the motor operating requirements usually rests with the driven equipment manufacturer or supplier and not the motor manufacturer. The motor requirements must be coordinated with the driven equipment supplier and tabulated on motor datasheets normally submitted to the motor manufacturer as an attachment to the specification.

The motor rating, without considering application of a service factor (SF), must be equal to or greater than the shaft power of the driven equipment when operating under all operating conditions. The motor rating, at the rated service factor, shall be no less than 100% of the power required to operate the driven equipment at its maximum capability (i.e. pump run-out conditions or fan test block conditions).

24.2.1 Design Criteria

Motors shall be constructed of first quality materials and designed in conformance with applicable standards stated in the specification suitable for the site altitude and operation on the plant operating low voltage system.

Motors shall be severe duty, premium efficiency, and induction type (380 V, 50 Hz), (460 V, 575 V, 60 Hz), three phase for totally enclosed fan cooled (TEFC) , Class F Insulation for Class B temperature rise, site rated. (Other enclosure types may be applicable).

Motors intended for operating with variable frequency drives (VFDs) shall be designed, insulated, and certified for inverter duty, with insulation rated for >1500 V.

The insulation shall be nonhygroscopic, fully moisture sealed.

24.2.2 Torque

Unless specified otherwise on the datasheet, motors shall be National Electrical Manufacturer's Association (NEMA) Design B or as determined by the supplier. Whichever NEMA torque design is specified, the motor shall meet or exceed the locked rotor and breakdown torques specified in NEMA standards for the design and rating specified.

Locked rotor currents shall not exceed NEMA maximum values for the specified NEMA design and rating.

Motors shall be rated for continuous duty, full voltage starting. Motors shall be able to start without exceeding their safe stall time, with voltage in the range of 80–100% of nominal. When driven equipment requires extended acceleration period, the WR^2 value of the equipment shall be specified on the datasheets. The motor must be capable of accelerating the load without overheating (see Figure 14.1).

The moment of inertia WR^2 of a pump is its resistance to changes in angular velocity as it rotates about its shaft. The inertia is the product of the rotating weight and the square of its radius of gyration. The total moment on inertia for a pump is the sum of the moment of inertia for each rotating component: motor, pump impeller, shaft and flywheel, if any.

Pumps with large rotating mass have higher inertia and therefore take longer to spin down on loss of power and longer to reach full speed during start-up. This is often beneficial for controlling transient pressures as the pump will slowly decelerate after a trip, continuing to move the fluid and minimize rise in pressure in downstream piping. It is for this reason that flywheels are often installed to increase the overall pump moment of inertia.

Enclosures for motors including frames, end brackets, conduit boxes, and external fan covers shall be made of cast iron.

The external fan covers of all TEFC and explosion proof motors shall meet NEMA MG-1 [2] requirements for a fully guarded machine.

All totally enclosed motors shall be furnished with drain holes and rotating shaft seals. Drain holes shall be provided with stainless steel breather/drain located at the low point of the motor.

External cooling fans shall be of low inertia, bidirectional, nonsparking, made of inert materials, and impervious to chemicals. Fan clamps shall be made of noncorrosive material.

Bearings shall be of the antifriction type, double sealed.

Windings and rotors: All motor windings shall be constructed of copper, wound for a single voltage source. Stator windings shall be form wound and braced to eliminate coil vibrations during all operating conditions.

Rotors shall be brazed copper construction, epoxy protected against corrosion on external surfaces. Rotor shall be free from inherent axial thrust and be within standard MG1-12.05 vibration limits. If balanced weights are added, they shall be permanently secured by welding or other approved method.

Space heaters as required → recommended. Energized when motor de-energized.

RTDs as required, → recommended, for motors 100 kW and larger.

Motor sound levels shall be tested according to IEEE standards. Maximum allowable sound pressure level at 1.5 m (5 ft) from any point on the motor surface shall not exceed 85 dBA.

Corrosion protection: All external surfaces shall be properly cleaned and primed with a zinc chromate red oxide primer and finish coated with a full gloss epoxy enamel paint that shall protect against acid and alkaloid fumes, salt air, solvents, and moisture, and have superior resistance to the effects of sunlight and outdoor weathering without chipping or cracking.

Lifting: All motors frame 182T and larger shall have permanent means for lifting.

Fans: Fans shall be of a nonsparking corrosion-resistant type. Bi-directional fans are preferred. If unidirectional fans are used, the fan rotation shall be indicated by a permanent legible marker and shall match the direction of rotation of the driven equipment.

Stainless steel nameplate, stamped according to NEMA standard MG1-10.37, shall be permanently attached to the motor. The motor connection diagrams shall be permanently attached to the motor, either inside the conduit box or on the motor frame, on the same side as the conduit box.

Conduit boxes shall be diagonally split, rotatable in 90° turns, gasketted, cast iron construction with threaded conduit holds. All conduit (junction) boxes for explosion-proof motors shall have a machined metal to metal fit. A ground lug suitable for grounding the motor frames shall be provided inside the conduit box on all motors.

Motor leads: The required number of motor power leads shall be brought to the conduit box. These leads shall have the same insulation temperature rating as the windings. Leads shall be permanently marked in accordance with requirements of NEMA MG1, Part 2.

All motor power leads and auxiliary device leads shall be wired into a motor terminal housing and separated according to the voltage class.

A moisture-resistant barrier shall be provided between the terminal box and the motor cavity; the motor leads shall be potted or sleeve protected to prevent insulation failure due to motor vibration.

Phase rotation: All motors furnished with unidirectional fans shall have the direction of rotation clearly marked on the motor frame or on a nameplate. The motor leads shall be marked for phase sequence T1, T2, T3, to correspond to the direction of rotation and a supply voltage sequence of A, B, C CCW of the supply voltage for North America, clockwise for the International Electro-technical Commission (IEC) countries.

Shafts: Shafts on horizontal motors shall be suitable for direct coupling, belt drive, or chain drive. When specified on the Datasheet, an adjustable sliding base on rails shall be provided for chain- or belt-driven equipment. The driven equipment manufacturer shall be aware of limits established for such drives documented by NEMA Standards MG1-14.07 and MG1-14.42 [2].

Motors furnished separately by equipment manufacturers shall be complete with mounted motor half couplings. If motors are furnished with reducers of the same manufacturer, they shall be mounted with couplings pressed on and securely bolted to a common bed plate prior to shipment.

24.3 Motors > 200 kW (Medium Voltage Motors)

The responsibility for defining the motor operating requirements shall be by the supplier of the driven equipment. These requirements shall be coordinated with the driven equipment and tabulated on the motor datasheets submitted to the motor manufacturer as an attachment to the specification.

The motor rating, without considering application of a service factor, shall be equal to or greater than the shaft power of the driven equipment when operating under all operating conditions. The motor rating, at the rated service factor shall be no less than 100% of the power required to operate the driven equipment at its maximum capability (i.e. pump run-out conditions or fan test block conditions).

The driven equipment manufacturer shall size the motor under "worst-case" conditions during running and starting and shall take into consideration the effect of a motor terminal voltage of 80% of rated nameplate voltage when starting. "Worst-case" conditions shall be defined by the driven equipment manufacturer, and operation beyond these limits may result in damage to the motor.

The load is not to exceed 90% of full load motor capability under conditions stated earlier.

Design criteria: Motor shall be constructed of first-quality new materials and designed in conformance with applicable standards stated in this specification. Motors shall be squirrel cage, induction type unless otherwise specified on the motor datasheet.

Electrical ratings: The motors shall operate on a (three phase, 60 Hz, 4.16 kV) or (three phase, 50 Hz, 3.6 kV) resistance grounded system, applicable to the site altitude.

Torque: Motor NEMA design torque characteristic shall suit the driven equipment.

Motors shall have normal starting torque and shall be arranged for full voltage starting unless specified otherwise on the motor datasheet.

Hazardous locations: Motors approved for Class I or Class II hazardous locations shall be marked with the appropriate T number indicating the maximum safe operating temperature according to National Electrical Code (NEC) Article 500.

Motors for large Ball and Semi-Autonomous Grinding (SAG) mills may be of a synchronous type. In such cases, the motors shall be provided with the controllers suitable for operation and control of synchronous motors.

Figure 24.5 Motor WPI type.

Motor efficiencies: Motors shall have the highest commercially available efficiency. The manufacturer shall state guaranteed efficiencies. The efficiency shall be based on IEEE Standard 112 [3]; Method B and the appropriate (NEMA) index letter shall be inscribed on the motor nameplate.

Service conditions: Motors shall be designed and built for long, trouble-free life in severe industrial service environments and shall be capable of operating successfully under the site conditions.

Enclosures: Motor enclosure, type, and assembly shall be of design to suit the application intended with respect to the ambient, weather, dust conditions, and corrosion requirements, generally as follows:

- *Indoors*: NEMA-WPI; drip proof, weather protected for indoor installations. See Figure 24.5.
- *Outdoors*: NEMA-WPII; weather-protected for outdoor installations.
  Insulation – stator: Insulation shall be Class F, nonhygroscopic, fully moisture sealed, and vacuum-pressure impregnated (VPI), employing double impregnation cycle with a coat of double epoxy varnish after impregnation. Temperature rise shall be Class B at a 1.0 service factor as listed in (NEMA Standard MG1) and based on 40 °C ambient temperature.
  Torques and starting currents: Induction motors shall meet or exceed the locked rotor, pull-up and breakdown torques, with rated voltage, and frequency applied as listed in (NEMA Standard MG1).

Where driven equipment requires extended acceleration periods, the WR^2 value of the equipment shall be specified on the datasheets. The motor must be capable of accelerating the load without overheating and at voltages down to 80%.

Where driven equipment requires repetitive or cyclic starting service, such duty cycle shall be specified on the motor datasheets. The motor must be capable of the required successive starts without overheating, according to (NEMA Standard MG1).

Induction motor starting inrush currents shall not exceed 650% of full load current with rated voltage and frequency applied. Inrush current may be higher for special requirements.

Bearings and lubrication: Motors shall typically be equipped with two end-shield-supported bearings of either the rolling element (anti-friction) or sliding element (sleeve) type. Where necessary, the bearings shall be insulated to prevent shaft currents and related bearing damage.

Anti-friction bearings shall have an average bearing life of 100 000 hours, as defined by anti-friction bearing manufacturers association (AFBMA). Suitable fittings shall be provided to permit conventional positive purging of old grease during regreasing operations. Close running shaft seals shall prevent leakage of grease as well as prevent the entrance of foreign materials, such as water and dirt, into the bearing area.

The sleeve bearings for all two-pole motors and four-pole 1200 kW and higher motors shall be of split sleeve type, self-aligning, and easy to service. The sealing system shall prevent oil leakage along the shaft. Oil ring sight gauge shall be provided.

Stator windings shall be constructed of copper, wound for single voltage service, form wound, and braced to eliminate coil vibrations under all operating conditions.

Squirrel cage rotors shall be closed slot die cast aluminum or brazed copper construction, epoxy-protected against corrosion on external surfaces, and free from inherent axial thrust and be within the vibration limits of the prescribed standards.

The motors intended to be operated by VFDs shall have upgraded insulation to deal with dv/dt stresses and reflected wave phenomena as imposed by the VFDs. As a minimum, the insulation to ground shall be equal to that of line to line.

Space heaters provided as required. In humid or wet climate, it is wise to keep the space (anticondensation) heaters energized at all times even when the motor is energized.

Temperature detection: Motors shall have stator windings equipped with 100 Ω (at 0 °C) platinum resistance temperature detectors (RTDs), two per phase when indicated. Motors above 600 kW shall also have bearings equipped with 100 Ω platinum RTDs when indicated on the motor datasheet.

Motor differential protection: Motors 1500 kW and more shall be equipped with three (3) current transformers (CTs) on the neutral side for motor differential protection when indicated on the motor datasheet. The other set of CTs will be included on the MV controller assembly by others.

Surge protection: Motors 1500 kW and more shall be equipped with suitable surge protection capacitors and arrestors when indicated on the motor datasheet. The surge protection shall be preferably mounted directly at the motor terminals and enclosed in the main conduit box.

Noise: Motor sound levels shall be tested according to IEEE Standard 85. Sound levels shall be as listed in the tables of (NEMA Standard MG1).

Corrosion protection: Motors shall be equipped with corrosion-resistant fittings and hardware throughout.

Stainless steel nameplate: It shall be stamped in accordance with (NEMA Standard MG1) [2] for induction motors and permanently attached to the motor. The motor connection diagrams shall be permanently attached to the motor, either inside the conduit box, or on the motor frame on the same side as the conduit box.

Termination boxes: Motors shall be fitted with an oversized main termination box for the main leads including the stress cones. The box shall be arranged for rotation, so that cable entry from the top, sides, or bottom is possible. Gaskets shall be furnished between the box and the motor frame and between halves of the box. The termination boxes for the motors operated by VFDs shall be sized for one step higher voltage.

Termination box shall enclose the main leads and current transformers, surge capacitors, and arresters as indicated on the motor datasheet. A grounding lug, suitable for grounding the motor frame, shall be provided inside the conduit box.

The power leads shall be permanently marked and brought out to the terminal box. These leads shall have the same insulation temperature rating as the windings. Space heater leads, temperature detector leads, and auxiliary device leads shall be wired into separate cast iron conduit boxes and separated according to the voltage class.

Phase rotation: Motors shall have the direction of rotation clearly marked on the motor frame or on a nameplate to match that of the driven equipment. The motor leads shall be marked for phase sequence T1, T2, T3 to correspond to the phase and supply voltage sequence of A–B–C counter-clockwise for North American applications and clockwise for IEC applications.

Motor shafts on horizontal motors shall be suitable for direct coupling, belt drive or chain drive, as specified required by the driven equipment.

24.4 VFD Specification Requirements

The VFDs used with the plant motors shall have the performance features as listed below. The VFD control circuit shall be microprocessor-based, provided with nonvolatile flash memory to preserve the software settings when power is disconnected. Additional drive requirements are listed in the data sheet, later on in this chapter.

- Overload capacity for normal duty ratings (for Variable Torque Applications) 110% overload capability for up to one minute. 150% overload capability for up to three seconds
- Overload capacity for heavy duty ratings (for Constant Torque Applications) 150% overload capability for up to one minute. 200% overload capability for up to three seconds
- Power loss ride-through of one to two seconds depending on the size of the motor
- Inherent soft start, linear or S-curve function
- Serial communication capability with RS-232-C port or RJ45 port for laptop connection
- Local/remote operation complete with manual speed control
- Adjustable acceleration/deceleration time, torque and current limit, carrier frequency
- Digital diagnostic indication and protection complete with electronic motor overload protection, phase-to-phase short circuit, ground fault, phase loss, and undervoltage protection
- Define regeneration if used, to an external resistor or to grid.

24.5   13.8 kV Isolated Phase Bus + PTs (IPB)

Output voltages for larger generators do not follow the standard voltage ratings. They are selected anywhere from 11 to 28 kV to limit the currents in the connections to the main output transformers. (see Figure 24.6). The connections are typically presented as large ampacity bus ducts, with or without generator breakers.

Thus, the generators are often connected directly to the transformers with bus ducts at the generator voltage, without having circuit breakers in between. In these cases, the transformers/generators are regarded as single switched entity, protected by circuit breakers on the transformer HV side in the 138, 230 kV or higher voltage switchyards.

MV bus ducts can be of three-phase type, enclosed in a common aluminum enclosure, or of an isolated phase bus (IPB) duct type, whereby each phase is

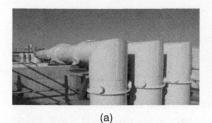
(a)

(b)

Figure 24.6 IPB bus duct (a, b).

Figure 24.7 Phase braid connectors.

fully enclosed in its own aluminum enclosure. This depends on the ampacity of the connection. Separate phase bus enclosures are more common for the higher and medium voltages.

The enclosure also serves as the current return path. The load losses are occurring, both in the conductors as well as in the enclosures. The metallic supporting structures around the bus duct are also conducting the return current. The IPBs rated over 10 kA are generally pressurized and force cooled.

The bus ducts are generally designed for site assembly together with the other associated generator equipment located between the generator and the main output transformer (tap offs), including PTs, excitation and auxiliary transformers, etc.

Bus ducts are designed for high short-circuit ratings, which must be calculated by a short-circuit study. The tap-offs taken from the main bus duct to connect to the auxiliary equipment are usually short-circuit rated twice as much as the main runs, as the taps-offs are subjected to both; contribution from the generator as well as from the grid.

The phase conductor in each phase bus duct is supported on epoxy resin insulators. Flexible copper laminated connectors of required rating are used for connections to the generator and transformers terminals to reduce strain on the bushings (see Figure 24.7). Flexible laminated aluminum or copper connectors of proper rating are used for tap-off connections.

Typical generator IPB bus duct rating characteristics are given later in the data sheet:

Factory tests: The IPB can only be assembled at site. Thus, it is factory tested for individual parts only, including the HV withstand at 50 kV/1 min (for 13.8 kV, IPB), and insulation resistance at 1000 V.

Site: Megger to 5 kV – Insulation resistance test.

24.6 Electrical Enclosures

Electrical enclosures protect sensitive electrical and electronic components from damage caused by elements such as water, wind, dust, dirt, heat, cold, humidity, and chemicals in the operating environment. National standards classify the categories of protection against human touch, equipment damage, and impact of the environment. It is important to understand enclosure ratings and the protection levels they provide. The IEC, NEMA, and UL organizations provide standards to identify the degree of protection enclosures against specific elements, but only UL certifies that enclosures pass specified tests to achieve each rating.

The enclosures are not just small boxes and panels but also the assemblies such as switchgear, MCCs. The standards classify the extent to which an enclosure or assembly must resist the ingress of solid bodies and water under designated tests. The most widely referred to standards are NEMA 250–2003 [4] and IEC 60.529 standards. In this chapter, both NEMA and IEC (IP Code) standards are explained and compared.

While NEMA does not actually test products, it establishes the performance criteria for enclosures intended for specific environments. NEMA standards describe each type of enclosure in general in functional terms, and specifically omit reference to construction details and materials used. In other words, NEMA specifies what an enclosure must do, but not how to manufacture it. This is also true for the EN 60.529/IEC 529 Standard.

24.6.1 NEMA Enclosures

Refer to Tables 24.3 and 24.4 for the usability of the various enclosure types.

NEMA	Purpose
Type 1	General purpose. Protects against dust, light, and indirect splashing but is not dust-tight; primarily prevents contact with live parts; used indoors and under normal atmospheric conditions.
Type 2	Drip-tight. Similar to Type 1 but with addition of drip shields; used where condensation may be severe (as in cooling and laundry rooms).
Type 3	Weather-resistant. Protects against falling dirt and windblown dust, against weather hazards such as rain, sleet, and snow, and is undamaged by the formation of ice. Used outdoors on ship docks, in construction work, and in tunnels and subways.
Type 3R	As 3, but *omits* protection against windblown dust.
Type 3S	As 3, but also operable when laden with ice.
Type 3X	As 3, but for corrosion protection.
Type 3RX	As 3R, X for additional corrosion protection.
Type 3SX	As 3S, X for additional corrosion protection.

NEMA	Purpose
Type 4	Watertight. Must exclude at least 65 gpm of water from 1-in. nozzle delivered from a distance not less than 10 ft for 5 min. Used outdoors on ship docks, in dairies, in waste water treatment plants, and in breweries.
Type 4X	As 4 but for corrosive environments, typically made of fiberglass or stainless steel, used indoors in ore processing plants.
Type 5	Dust-tight. Provided with gaskets or equivalent to exclude dust; used in steel mills and cement plants.
Type 6	Temporary submersible. Design depends on specified conditions of pressure and time; submersible in water or oil; used in quarries, mines, and manholes.
Type 6P	Withstands occasional prolonged submersion. Not intended for continuous submersion.
Type 7	Certified and labeled for use in areas with specific hazardous conditions: for indoor use in Class I, Groups A, B, C, and D environments as defined in National Fire Protection Association (NFPA) standards such as the NEC.
Type 10	MSHA. Meets the requirements of the Mine Safety and Health Administration, 30 CFR Part 18 (1978).
Type 11	General-purpose. Protects against the corrosive effects of liquids and gases. Meets drip and corrosion-resistance tests.
Type 12, 12K	General purpose. Intended for indoor use, provides some protection against dust, falling dirt, and dripping noncorrosive liquids. Meets drip, dust, and rust resistance tests.
Type 13	General purpose. Primarily used to provide protection against dust, spraying of water, and noncorrosive coolants. Meets oil exclusion and rust resistance design tests.

Table 24.3 NEMA type enclosures.

Degree of protection for indoors	NEMA type									
	1	2	4	4	5	6	6P	11	12	13
Incident contact with enclosed equipment	x	x	x	x	x	x	x	x	x	x
Falling dirt	x	x	x	x	x	x	x	x	x	x
Falling liquids and light splashing		x	x	x		x	x	x	x	x
Dust, lint, fibers, and flyings		x	x	x	x	x			x	x
Hose down and splashing water		x	x			x	x			
Oil and coolant seepage									x	x
Oil or coolant spraying and splashing										x
Corrosive agents			x				x	x		
Occasional submerging						x	x			
Occasional prolonged submersion							x			

Table 24.4 NEMA Type Enclosures

Degree of protection for outdoors	NEMA type				
	3	3R	4	4X	6
Incident contact with enclosed equipment	x	x	x	x	x
Rain, snow, sleet	x	x	x	x	x
Windblown dust	x		x	x	x
Hose down and splashing water			x	x	x
Corrosive agents		x			
Occasional submerging			x		

Table 24.5 NEMA/IEC compatibility.

NEMA	IEC	NEMA	IEC
1	IP23	4, 4X	IP66
2	IP30	6	IP67
3	IP64	12	IP55
3R	IP32	13	IP65

NEMA – IP equivalence: This cross-reference table is a close approximation of NEMA and IEC classifications for reference only. Due to the difference in the tests conducted, the enclosure classifications are not fully comparable (Table 24.5).

24.6.2   CE and IEC Classifications for Enclosures

The IEC Publication 60529 Classification of Degrees of Protection by Enclosures (Table 24.6);

24.7   Technical Data Sheets

Included in this chapter are several typical (brief) technical data sheets that accompany the equipment technical specifications for the electrical equipment listed later. The presented data sheets could be used as a base for future engineering work and must be updated to suit the project requirements, site operating conditions, and specific details of the equipment. The suppliers must fill the data sheets. Sometimes, suppliers may decline to provide their

Table 24.6 IP enclosures degree of protection.

First numeral			Second numeral	
IP	Protection of persons	Protection of equipment	IP	Protection of equipment
0	None	None	0	None
1	Protected from contact to body	Protected against objects >50 mm in diameter	1	Protected against vertically falling drops of water or condensation
2	Protection against contact with fingers	Protected against solid objects >12 mm	2	Protected against direct sprays of water up to 15° from vertical
3	Protected against tools and wires >2.5 mm in diameter	Protected against solid objects over 2.5 mm in diameter	3	Protected against sprays to 60° from vertical
4	Protected against tools and wires >1.0 mm in diameter	Protected against solid objects over 1.0 mm in diameter	4	Protected against water sprayed from all directions
5	As 4 above	Protected against dust, limited ingress	5	Protected against low pressure jets of water from all directions. Limited ingress allowed.
6	As 4 above	Totally protected against dust	6	Protected against strong jets of water
			7	Protected against the effects of immersion between 0.15 and 1 m
			8	Protected against long periods of immersion under pressure

Any combination between the first and second numerals is possible.

An IP Example:	**IP45**
First IP Numeral 4	Protection against tools and wires over 1 mm in diameter
	Protection against solid objects over 1 mm in diameter
Second IP Numeral 5	Protection against low pressure jets of water from all directions (limited ingress permitted)

quotations if the data sheets are too extensive and limit the suppliers a freedom of choice.

1. Oil filled transformers
2. Isolated phase bus (IPBs)
3. Motors <200 kW
4. Motors >200 kW
5. VFDs
6. MCCs
7. 15 kV switchgear
8. Unit substation

1. Technical data sheet	Oil filled transformers
	Note: tenderer to fill in the empty spaces (/)
Equipment tag no.	10 UAT-002
Quantity	One (1)
Site conditions	Refer to specification attached with request for quote (RFQ)
Rating	10 MVA, ONAN
Voltage (Primary/secondary)	13.8–4.16 kV, 3ph, 60 Hz
Winding configuration	Dy1
Impedance	7%
Location	Outdoors
Temperature rise	65, 80 °C hottest spot temp, above 40 °C
Parallel operation	No
Tap changer	Off-load, ±2.5%, ±5% primary si de
BIL level; (Primary/secondary)	110/75 kV
BIL level; Bushings	110/75 kV
Bushings orientation	
HV air filled, side cable box for	3 × 1c-750 MCM/ph
LV air filled, side cable box for cable-bus	12 × 1c-750 MCM/ph
LV Neutral	Provide, on top cover
Current transformers	
Primary	None
Secondary	None
LV neutral	(2) – 200/5 A; Class: 2.5C100
Control cabinet enclosure	NEMA 4X, weatherproof, light and switch, lockable
Space heater	Yes, 120 V, 1 ph, complete with thermostat

Nameplate	Stainless steel (anodized aluminum)
Cover bolted/welded	Bolted
Lightning arrestors, primary/secondary	Not required
Cable lugs	By others (Cable-bus supplier)
Audible sound level	ANSI/IEEE Standard C57.12.90
Radiators	Detachable
Power supply for auxiliaries	120 VAC
Neutral Grounding Resistor	Required
Location/mounting	Outdoor/on transformer
Rating (A/s)	100 A/10 s
Material	Stainless steel
Accessories	
Winding temperature detector with alarm contacts	Yes
Cover mounted pressure relief/Bucholz relay	Qualitrol 208.60E or equal
Grounding lugs	Yes
Breather	If applicable
Drain and oil filling valves	Yes
Sampling valves	Yes
Liquid-level indicator with alarm contacts	Yes
Painting	Manufacturer's standard or ANSI # 61 Grey
Primary isolating switch	Not required
Fuses/type/rating	N.A.
Rated current (A)	N.A
Momentary interrupting capacity	N.A
Auxiliary contacts	N.A
Factory inspection by buyer's representative	Required
Efficiency at 50%, 75%, 100%	/ / (To be filled by Tenderer)
No load losses at 100% (V)	/
Total losses at 50%, 75%, 100%	/ /
Transformer shipped filled with oil	Yes/no
Oil type/supplier	/
Oil preservation system	Seal tank/conservator

Weight

With oil (kg)	/
Without oil (kg)	/
Oil quantity (l)	/
Dimensions: H × W × D (m)	/ / /
Forced cooling equipment	Not required
Fans, HP / V / ph	/ / /

2. Technical data sheet	**Isolated phase bus (IPB) for power plant**
Ratings	3 × 1 ph, 6500 A Main/250 A Taps, 13.8 kV, 110 kV BIL
	Short-circuit ratings, main: 1 s: 50 kA/Taps: 100 kA
	Momentary for main: 125 kA, peak, taps 250 kA peak
Standards	IEEE/ANSI C37.23 (2003)
Insulation level withstand	50 kV, Hi – pot, 1 min
Temperature rise	65 °C over 40 °C ambient, Encl. 40 °C over 40 °C Ambient
Pressurization	Air: 5–25 mBar, 240 V_{ac}
Materials	Aluminum alloy for conductors and enclosures
Enclosure painting	Yes, acrylic
Enclosures	IP 44 (IEC)
Taps offs for	UAT, Exc.T, Dynamic Brake, PTs + Capacitors
PTs	13.8 kV/$\sqrt{3}$–110 V/$\sqrt{3}$, star connected.
Arresters	Metal oxide, Station class.18 kV, 15 kV MCOV, 10 kA discharge
Surge capacitors	3 × 0.25 mF, 13.8 kV, sealed, oil filled

3. Technical data sheet	**Motors <200 kW**
Site conditions	Refer to specification, attached with RFQ
System voltage	480 V, 3 ph, 60 Hz
Speed, preferred	1800 rpm
Efficiency	Premium
Enclosure	TEFC
NEMA design torque	B
Insulation class	F for Class B temperature rise, 40 °C ambient
Duty	Industrial, severe
Bearings	Anti-friction type

Space heaters	Motors >99 HP, outdoors, 120 V, 1 ph
RTDs	>200 hp, 1/Phase, PT100, 3 w
VFD operation (if applicable)	Upgrade line to ground insulation

Specific requirements	**Vendor to provide data for each motor**
Motor (kW)	/
Motor application	/
Voltage (V)	460
Speed (rpm)	/
Efficiency at 50/75/100% Load (%)	/
Power factor at 50/75/100% Load (%)	/
Service factor	1.15

Starting characteristic	
Torque	/
Starting (%)	/
Pull-in (%)	/
Pull-out (%)	/
Load Inertia (lb ft^2)	/
Drive coupling	/
Base	/
Sole plate/anchor bolts	/
Frame	/
Enclosure type	/
Bearings type	/
Weight (kg)	/
Stator current (A)	/
Inrush pu rated at 100% (V)	/
Stall time (s) (Hot/Cold)	/
Number of starts/hour (hot/cold)	/
Noise measured at 1.50 m (dbA)	/

4. Technical data sheet	**Motors,** MV, **>200 kW**
	Tenderer to fill in empty spaces (/) for each motor.
Driven equipment	*Refer to Mech. Bid Specification*
Equipment tag no.	*Refer to Mech. Bid Specification*
Site conditions	*Refer to Specification attached with RFQ/request for purchase (RFP)*

Shop tests by buyer	*To be advised*
Service	
Motor operated by VFD	Yes
VFD Type, PWM, CSI	PWM preferred
Isolation transformer	/
Ratings	
Power (kW)	250, 400, 500, 1000, 1500
Power factor	/
Service factor (pu)	1.0
Speed (rpm) preferred	1800
Temperature rise, Class B	80 °C at 40 °C ambient
Voltage/phase/frequency (kV/ph/Hz)	4000/3/60
Drive coupling	/
Load inertia (kg m^2)	/
NEMA design	B
Torque: starting (%)	/
Pull-in%	/
Pull-out%	/
Number of starts/h (cold/hot)	/
Construction	
Sole plates	Provide
Base	Provide
Bearings/type	/
Enclosure	To suit working environment
Painting	Suppliers Standard
Surge protection (motors >1000 kW)	/
CT for differential protection (motors >1500 kW)	/
Space heaters (V)	/
RTDs; two each phase, stator winding	Provide
RTDs; bearings, one/bearing	Provide
Bearings, contact making thermometer >1000 kW	/
Performance	
Stator current (A)	/
Inrush (pu) rated	/
Bearing cooling	/
Frame type	/

Outline drawing (No.)	/
Weight, total (kg)	/
Weight, heaviest piece (kg)	/
*Efficiency % at 75, 100%	/ /
*Losses (kW) IEEE 115	/
Reactance X_d/X_d'/ Xd″(pu)	/ / /
Sound level, IEEE 85/MEMA-MG-1	/

5. Technical data sheet	VFDs
Mounting	Within MCCs/plant floor
System voltage/frequency	480 V, 3 ph/ 60 Hz
Vector control	For drives intended to operate at low speed.
Control voltage	120 V_{ac}/ 24 V_{dc} for DeviceNet
System short-circuit level	50 kA, r.m.s.
Distance to the motors	Less than 15 m
Distance to DCS	10–50 m
Speed range	20–120%
Motor type/NEMA design torque	Induction/NEMA B
Insulation	Inverter duty
Rotation	Reversible/*Non reversible*
Regeneration	Yes/no
Motor thermocouples/RTDs	Provided with motors >100 kW
Topology;	Voltage
Torque design	Variable/constant
Circuit breaker for each drive	Required, 50 kA r.m.s.
Inverter type modulation	PWM

6. Technical data sheet	MCCs
General requirements	**Tenderer to fill the empty spaces!**
Site conditions	Refer to specification attached to RFQ
Voltage	480 V, 3 ph/60 Hz
NEMA (EEMAC) class/type	1B
Phases	3 ph, 3 W, or (3 ph, 4 w)
Enclosure (NEMA) type	1 A (gasketted)
Painting	Manufacturer's standard
Starter control transformers	Individual–120 V, 1 ph

Space heaters with thermostat	Provide: 120 V, 1 ph for each vertical cabinet
Main bus	800/1200 A, Refer individual MCCs
Busbar plating	Tin
Bus bracing	65 kA r.m.s., Asymmetrical
Vertical bus	400 A minimum.
Neutral bus	Not required
Ground bus	1/4 × 1 inch at Bottom (6 mm × 25 mm)
Interrupting devices	Circuit breakers, 150 A frame minimum
Breaker type	Molded case
Provision for padlocking	Required
Breaker interrupting	50 kA r.m.s.
Wiring diagrams	Owner's standard as provided with specification
Cable lugs	Provide
Cable entry plates	Provide
Nameplates	Lamacoid, white on black background
Contactor spare aux contacts	1 NO + 1 NC
Overload heaters	Provide
Control fuses	Provide
Panel accessories	As shown on wiring diagrams
I/O communications to DCS	DeviceNet
DeviceNet I/O voltage, 24 V_{dc} or 120 V_{ac}	24 V_{dc}
DeviceNet module I/O count	$I = 4; O = 2$
Recommended number of nodes/run	/
DeviceNet power supply	/
Trunk cable type and size	/
Max./Min. Baud Rate (kBaud)	/

MCC Specific Details

1.	**MCC-xx1**	**Essential MCC, 480 V, 3 ph, 3 w**
	Cable entry	Top
	Incoming breaker	No
	Bus rating	800 A
	DeviceNet	Yes

2.	**MCC-xx2**	**Building services MCC, 480 V, 3 ph, 3 w**
	Cable entry	Top
	Incoming breaker	No
	Bus rating	400 A
	DeviceNet	No
3.	**MCC-xx3**	**Turbine MCC, 480 V, 3 ph, 3 w**
	Cable entry	Bottom
	Incoming breaker	No
	Bus rating	800 A
	DeviceNet	Yes
4.	**MCC-xx4**	**Boiler MCC, 480 V, 3 ph, 3 w**
	Cable entry	Top
	Incoming breaker	No
	Bus rating	400 A
	DeviceNet	Yes

1. Technical data sheet	**15 kV switchgear**
Equipment tag number	*5300-* SWG-101
Location	Plant, indoors
One-line diagram	5300-47D2-001
Application	Generator breaker, 73.3 MVA, 13.8 kV
Voltage	13.8 kV, 3 ph, 60 Hz
kA interrupting rating	50 kA r.m.s. Symmetry
Enclosure	Metal-clad, NEMA 1 A
BIL level	95 kV
Main bus	2500 A Cu
Control power close/trip	125 V_{dc}
Panel accessories	Breaker control/indication

Circuit breakers	
Make and type	SF6/Vacuum
Rating, incoming	4000 A generator breaker (fan cooled)
Feeder breakers	800 A
Controls	Local/Rem (DCS)
Momentary close and latch	104 kA

Interrupting capacity	50 kA r.m.s. Symmetry
Interrupting time (Cycles)	3
Instruments, make and type	Refer to one-line diagrams
Relays, make and type	Refer to one-line diagrams
Cable entry: power/control	Top/top, cable-bus
Cable lugs, power cables	By others (Cable bus supplier)
Nameplates	Lamacoid, black on white
Space heaters with thermostats	Yes, 120 V$_{ac}$
Key interlocks	Yes, As shown on one-line diagrams
Padlocking	Yes
Surge arrestors	Yes, as shown on one-line diagrams
Revenue metering class CTs and PTs	As shown on one-line diagrams
Protection class CTs and PTs	As shown on one-line diagrams
Rear doors	Required
Inspection by buyer	Yes

8. Technical data sheet   **13.8 kV–480 V Unit substations assemblies**

Data sheet A

Number of assemblies required	Two (2)	One (1)
Assembly tag number	5300-USS-121, -122	5300-USS-623
Site conditions	Attached with RFQ/RFP	Attached
One-line diagrams	5300-47D2-0003, -0004	5300-47D2-0005
Substation type	Single ended	Single Ended
Painting	ANSI 61	ANSI 61
Location	Indoors, Elec. Room	Indoors
Transformer rating (MVA)	2/2.66	1/1.33
Incoming switch, HV side	Not required	5 kV load break
LV Switchgear	LV Switchgear	MCC
Busbars	3200 A	MCC 1600 A Bus
Interrupting (kA)	65	65
Feeder breakers	800 A/1200 A, 3 ph, 3 w, 480 V	MCC starters
Incoming breakers	3200 A	None
Key interlocking	No	No
Bus tie breaker	None	None
Number of feeders	Refer to owner drawing	MCC
Assembly dimensions (m)	/	/

Assembly weights (kg)	/	/
Heaviest piece (kg)	/	/
Owner inspection	Required	Required

Data sheet B	**Dry transformers**	
Tag numbers	5300-UST-111, -UST-112	5300-UST-613
Reference drawing	5300-47D2-0003, -0004	5530-47D2-0005
Transformer type	Dry type	Dry type
Rating (MVA)	2/2.5	1/1.33
Enclosure	Ventilated NEMA 1	Ventilated NEMA 1
Cooling	AA/FA	AA/FA
Voltage (kV/Hz/ph)	4.16 kV, – 480/60/3	4.16 kV, – 480/60/3
Neutral grounding	High resistance	High resistance
Connections	Delta/wye (DYI)	Delta/wye (DYI)
BIL (windings/bushings) (kV)	75/30	75/30
Impedance (%)	5.75	5.75
Lightning arresters	Not required	Not required
Insulation class/Temperature. rise (°C)	H/150	H/150
Tap changers	Off load $\pm$ 2.5 $\pm$ 5% primary	Off load $\pm$ 2.5 $\pm$ 5% primary
Efficiency at 100%/75%/50%:	/	/
Neutral grounding resistor	5A Continuous	5A Continuous
Pulsing ground system	Yes	Yes
Mounting	Within assembly	Within assembly
Current transformer	Not required	Not required
HV Incoming Switch, three pole, fused	N/A	600 A, Load interrupter, Shunt trip operated
Switch momentary rating	N/A	40 kA
Winding temperature monitor	Multilin TS-3 or equal	Multilin TS-3 or equal
Cooling fans	120 V_{ac}/1 ph	120 V_{ac}/1 ph
Connections (HV)	To 5 kV switchgear, provide lugs for 1-3c-350MCM, Top entry	

Data sheet C	480 V switchgear	
	LV switchgear	MCC
Type	Single ended	Single ended
Nominal voltage class	480 V, 60 Hz	480 V, 60 Hz
Main bus rating	3200 A, 3 ph, 3 w	1600 A, 3 ph, 3 w
Main bus/grounding materials	Copper/copper	Copper/copper
Incoming cable entry	Bused to transformer	Bused to transformer
Feeder cable entry	Top, 2c-500 MCM/phase	Bottom, 1c-500 MCM/phase
Enclosure	NEMA 1 A, Indoors	NEMA 1 A, Indoors
Bus tie operation	N/A	N/A
Bus bracing	65 kA r.m.s.	65 kA r.m.s.
Main and feeder cable lugs	Required	Required
Space heaters	Required	Required

Circuit breakers		
Breaker ratings: incomer	3200 A, Shunt Trip	None
Breaker ratings: feeders	800/1200 A, Manual	MCC breakers and starters
Protective devices	Static long, short	/
Interrupting rating	65 kA r.m.s., Symmetry	/
Operation	Spring charged, manual	/
Temperature rise	55 °C at 40 ° ambient	
Metering (incoming breaker)	Digital multifunction	
Metering (feeder breakers)	None	
Test voltage	2200 V	
Control voltage	125 V_{dc}	
Nameplates	Lamacoid black on White	
Auxiliary contacts	4NO + 4NC	

References

1 IEEE Std. PC570 12.00-200x Standard for general requirements for liquid immersed distribution, power and regulating transformers.
2 NEMA MG-1 (2016) Motors and Generators.
3 IEEE 112 (2017) Standard Test Procedure for Induction Motors and Generators.
4 NEMA 250 (2014) Enclosures for Electrical Equipment of less than 1000 V

25

Solar Power

CHAPTER MENU

25.1 Solar Resource, 567
25.2 PV Panel Technology, 569
25.3 Photovoltaic Plants, 571
    25.3.1 PV Equipment, 574
    25.3.2 PV Projects, 576
25.4 CSP, 580
    25.4.1 Power Tower Systems, 580
    25.4.2 Parabolic Trough Systems, 582
    25.4.3 Fresnel Linear Mirrors, 583
    25.4.4 CSP Costs, 585
    25.4.5 CSP Operating Plants and Projects, 586
25.5 Thermal Storage, 591
    25.5.1 Balance of Power (BOP) Plant, 593
    25.5.2 Peaking Storage Plants, 593
    25.5.3 Battery Storage Costs, 594
25.6 Conclusion, 595
    25.6.1 PV Plants, 595
    25.6.2 Concentrated Solar Plants, 596
References, 597

25.1 Solar Resource

Solar power generating facilities fall into two categories: photo voltaic (PV) and concentrated (CSP) solar plants.

Solar power systems offer many advantages. Stand-alone systems can eliminate the need to build expensive new power lines to remote locations. For rural and remote applications, solar electricity can cost less than any other means of producing electricity. Solar plants of PV type require minimal maintenance. These plants are easy to combine with other types of electric generations such as wind, diesel, hydro, or combustion turbines to facilitate better usage of solar

Practical Power Plant Engineering: A Guide for Early Career Engineers, First Edition. Zark Bedalov.
© 2020 John Wiley & Sons, Inc. Published 2020 by John Wiley & Sons, Inc.

resources and other fuels. Solar plants are also easy to expand incrementally by adding more modules as power demand increases.

Solar electric power systems are environment friendly. They run quietly and without polluting. When solar electric technologies displace fossil fuels for pumping water, lighting homes, or running industrial facilities, they reduce greenhouse gases and pollutants emitted into the atmosphere.

Solar power systems make use of direct normal irradiance (DNI) directly from the sun. **DNI** is the amount of solar radiation received per unit area by a surface that comes in a straight-line from the direction of the sun at its current position in the sky. Typically, one can maximize the amount of irradiance annually received by a surface and by keeping it normal to incoming radiation. This quantity is of particular interest to concentrating solar thermal plants and PV installations that track the position of the sun. Sun-tracking PV installations can count on the increased solar exposures.

The two maps (Figures 25.1 and 25.2) mentioned beneath display the solar resource in kWh/m^2/yr in Europe and the world.

Insolation is typically rated as a power density in units of kW/m^2. Daily average DNI is provided in units of kWh/m^2/d. Daily amount of solar energy is seasonal, with the greatest DNI on days close to the summer solstice and the least near and over the winter solstice.

Solar plant resource requires geography with insolation ranging from 5.0 to 6.6 kWh/m^2/d. Areas with low wind activity and low rain precipitation are preferable. For instance, Arizona carries an index of 6.6 kWh/m^2/d or 2400 kWh/m^2/yr.

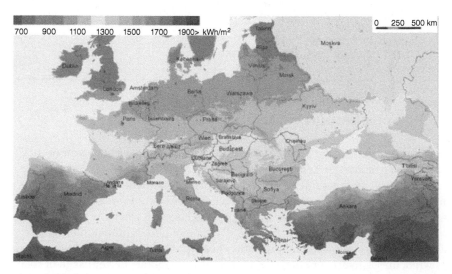

Figure 25.1 Solar resource, Europe [1].

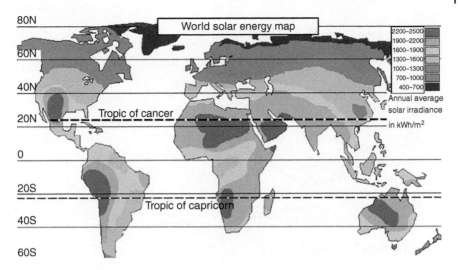

Figure 25.2 Insolation in kWh/m^2/yr. DNI Index [2].

Like wind, sun is an intermittent resource, affected by weather conditions and the position of the sun above the horizon. The angle of the sun's rays relative to the Earth's surface changes during the day and with seasons. Seasonal cloud cover substantially reduces the power output. The maximum annual power output expected from a solar power plant located in an area of high DNI is <30% of its MW installed capacity. In comparison with wind, solar power is more predictable (dependable) and can be planned into the daily operating cycle.

For the power projections, an average time per day that sun will shine range from 4 to 5.5 hours. Globally the accepted factor is 4.8 hours.

25.2 PV Panel Technology

"Quantum efficiency" is a term intrinsic to the light-absorbing material and not the cell as a whole; it refers to the percentage of absorbed photons that produce electron-hole pairs (see Figure 25.3). The energy conversion efficiency is the percentage of incident electromagnetic radiation that is converted to electrical power when a solar cell is connected to an electrical circuit.

This overall efficiency depends on many factors including the temperature, amount of incidention, and the surface area of the solar cell.

The first generation of solar cells, also known as silicon wafer-based photovoltaic, is still the dominant technology today. Single-crystalline and multicrystalline wafers, used in commercial production, promise power conversion efficiencies up to 25%, although the fabrication technologies at present limit them to about 15–20%.

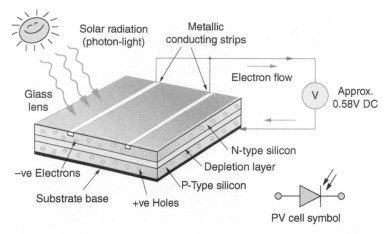

Figure 25.3 Photovoltaic technology [3].

The second generation of PV materials is based on the use of thin-film deposits of semiconductors, such as amorphous silicon, cadmium telluride, copper indium, gallium diselenide, or copper indium sulfide. The efficiencies of thin film solar cells tend to be lower compared to the conventional solar cells, by around 6–10%, but at significantly lower manufacturing costs, thus generating a lower cost in terms of $/watt of electrical output.

The third generation of PV cells has a research goal of increasing efficiency that maintains the cost advantage over the second-generation materials. The technological approaches include dyesensitized nanocrystalline or Gratzel solar cells, organic polymer-based PVs and tandem (or multijunction) solar cell (see Figure 25.4).

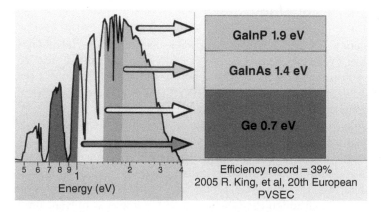

Figure 25.4 Third multijunction photovoltaic. Source: Courtesy of NREL [4, 5].

Multijunction solar cells [4, 5] use a combination of semiconductor materials to more efficiently capture a wider range of photon energies. Depending on the particular technology, present-day multijunction solar cells are capable of generating approximately twice as much power under the same conditions as traditional solar cells made of silicon. Multijunction solar cells have a higher theoretical limit of efficiency conversion as compared to other PV technologies, well above 40%.

Multijunction solar cells use multiple kinds of materials having band gaps that span the whole solar spectrum. The aim is to absorb each color of light with a material that has a band gap equal to the differing photon eV energy.

Figure 25.5 shows the theoretical parts of the spectrum that could be exploited by (a) Si solar cells, and (b) multiple cell layers of distinctive materials.

The spectrum of the sun's light that reaches Earth's upper atmosphere ranges from the ultraviolet to the near-infrared radiation, with peak region (48%) from 400 to 700 nm, which is the visible diapason. PV cells can be defined as p–i–n photodiodes, which are operated under forward bias. They are designed to capture photons from the solar spectrum by exciting electrons across the bandgap of a semiconductor, which creates electron-hole pairs that are then charge separated.

Multijunction solar cells consist of single-junction solar cells stacked upon each other so that each layer going from the top to the bottom has a smaller bandgap than the previous one, so it absorbs and converts the photons that have energies greater than the bandgap of that layer and less than the bandgap of the layer above. Panels with multiple materials harvest solar energy, each layer from a different portion of the light spectrum. Multijunction PVs, as compared to single-junction cells, produce reduced currents because fixed total number of photons is distributed over increasing number of cell layers, owing to the fact that the amount available for electron promotion in any one layer is decreased.

At the same time, the electrons, which are excited, are more energetic and have a greater electrical potential, so the reduction of currents is compensated for by increase in voltages, thus causing the overall power $P = I \times V$ of the cell to increase. Moreover, resistive losses, which are proportional to the I^2 current, are significantly reduced, thus reducing the heat of PV cell.

25.3 Photovoltaic Plants

The two types of power-generating plants which use solar energy are

1. *Photovoltaic (PV) Plants*: The PV technology directly converts solar radiation into electricity through the use of PV panels. These are typically smaller plants for limited commercial operation, located on roofs of the buildings

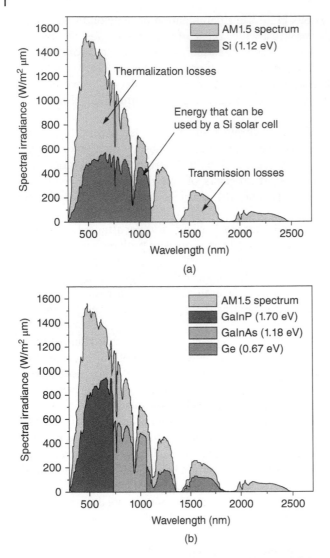

Figure 25.5 (a, b) The AM1.5 solar spectrum and the parts of the spectrum [6].

and in open spaces. The PV panels are tilted to make maximum use of the sun over the year. The module efficiency and power conversion are inefficient. Some of these plants include small battery storage capacity to provide a minimum overnight power reserve.

Larger commercial plants use trackers to benefit from focusing the panels toward the sun during the daylight hours. This also reduces the footprint of the collection area for a given MW output. Tracking adds considerably more energy to the plant production, but it is expensive and high in maintenance. In colder climates, PV panels are more efficient, but issues of snow, cloud, freezing, and lower insolation factor make the installation less desirable.

By the end of 2017, according to the World Energy Council, the PV solar energy installed capacity has passed the 300 000 MW mark. [7]

2. *Concentrated Solar Plants* **(CSP)**: These plants concentrate sunlight using focusing mirrors to generate heat/steam to drive turbines generators, just like a conventional thermal power station. They differ from each other in the method of sunlight concentration. CSP have been operating successfully in California since the mid-1980s. In 2008, a new 64 MW CSP went on stream in Nevada, USA (Nevada Solar I).

   Two new plants 280 MW each have been placed into operation in Arizona and California, 2013 and 2014, respectively.

   CSP may become increasingly attractive as equipment costs in $/kW fall with refinements in the technology and economies of scale.

   According to the HelioSCSP by the end of 2017, [8], the installed capacity of CSP was 5133 MW.

PV technology presents several advantages over the other technologies of power generation (see Figure 25.6). Besides being a clean way to generate electricity and being potentially inexhaustible, it offers the possibility of installing modular and distributed energy appropriate for the areas, in which there is either no electrical network, or grid installation is not cost-effective. PV plants are generally considered for power outputs ranging from 0.5 to 10 MW. The Nellis PV and Alamosa concentrated photovoltaic (CPV) plants are larger at installed capacities of 14.2 and 30 MW, respectively. In China and India, PV plants of 500 and 1000 MW are in operation or planned. These PV plants tend to be less sophisticated, occupying large tracks of land and operate without trackers, thus operating at lower efficiencies.

Figure 25.6 PV panels [5].

Output of a solar PV plant generally matches the load curve of domestic power users until about 4 p.m., at which time its output tapers down and falls below the load demand curve.

PV plants can capture any form of scattered sunlight or ambient light. They produce some power even under cloud cover. Since these types of plants do not offer dispatchability, they are mostly used for remote isolated areas without power lines. A PV plant with a battery storage can offer some degree of dispatchability. There is a possibility of this happening in the future as the cost of battery storage systems are getting lower.

PV solar plants work best when integrated with diesel engine plants (hybrid plants) (sun + diesel/heavy fuel oil (HFO)) to achieve a full level of dispatchability.

25.3.1 PV Equipment

The PV plants include the following basic elements:

- Rows of PV panels, with fixed mountings or placed on solar trackers. The most common panel used for these plants is rated at 200 W peak power, i.e. when exposed to direct solar radiation (see Figure 25.7).
- Cabling to collect DC power.
- Inverters to convert DC to AC power.
- Transformers to step up and step down the power to and from distribution voltage.
- Data acquisition for plant monitoring.
- Reactive power addition.
- Distribution lines and circuits to deliver power to a grid or consumers.

One 200 W panel is expected to generate approximately 220 kWh of energy per year (0.6 kWh/d). The conversion efficiency of a PV cell, or solar cell, is the

Figure 25.7 Dual tracker panels [5].

percentage of the solar energy shining on a PV device that is converted into usable electricity. As noted earlier, the photo/chemical conversion of a panel is in the order of 15–20% of the solar input, depending on temperature and angle tracking and time of day.

Panels convert better in colder temperatures. The operating behavior of the PV cells improves by 4.5% for every $10\,°C$ temperature reduction on the module. The cells get out of focus at $85\,°C$ ($185\,°F$).

The capacity factor of a solar plant is expected to be in the order of 30% of the panel nameplate installed capacity. This assumes the variation in the solar input during the daylight and no input during the night hours. The panels used in America are UL 1703 safety rated for wind, hail, and fire – Class A.

Tracking: Tracking increases power output, by 25% for single, and 35% for double-axis tracking over those of fixed panel systems. The most of the benefit is at the morning and evening hours. This is counteracted by the reduction of the sun collection area, as tracker installations need more space per panel. It increases the cost of maintenance and therefore the cost of kWh produced.

Solar plants are affected by wind. Tracker designs tolerant of wind and sun collection fields of lower profile are desirable. Tracker designs are being made to deal with winds of up to 100 miles/h (160 km/h).

Reactive power: It should be noted that solar panel installations not connected to a grid are unsuitable for loads that require reactive power. The distribution transformers operate at pf = 1.0. Should the load comprise motors, fluorescent lights, etc., that need reactive power, it is necessary to install a capacitor bank near the load.

Capital costs: Solar plants are relatively expensive to build. The current estimates for the capital costs of PV plants are an order of magnitude higher than that of conventional fossil fueled plants. In the solar category, PV plants are costlier per kW installed than CSP. This is mainly a result of their low solar/electricity conversion factors and the need to convert DC into AC power. It is estimated that a typical PV plant with single-axis tracking systems would cost approximately \$11 000/kW installed, or \$11 000 000 for a 1.0 MW plant. The 2008 year cost of Barrick's 1 MW plant in Nevada was \$10 000 000 on an engineering, procurement, construction, management (EPCM) basis.

The estimated installed cost of the Barrick PV plant was broken down as follows:

Plant Barrick

Building, roads, site preparation	1 800 000
PV panels on trackers	7 900 000 (2)
Cables	130 000
Inverters, 2 × 500 kW AC	600 000

Data acquisition	50 000
Water treatment plant for washing	80 000
Miscellaneous	440 000
Total ($)	11 000 000

Power Distribution

- Transformers, 2×1000 kVA (one at each end) $170\,000
- Distribution system, 20 km of 11 kV line $1\,300\,000

Clarifications

(1) The estimate can be reduced by more than $1 500 000 if the distribution line is not required.
(2) A typical 200 W panel: Sanyo 200, Sharp216, GEPV-200 cost about $1000 each.

The size of the panel is 52×35 in., ($1.32\,m \times 0.9\,m$, $1.2\,m^2$). Significant discounts can be obtained for large volume purchases.

Maintenance: Provided that the wind issues are dealt with properly, the maintenance cost of the PV plants is relatively low. It generally consists of washing the panels to maintain their high radiation absorption levels. The plants located in the desert areas are expected to be washed up to 10 times a year. The operating cost is presently estimated at $0.11/kWh for the single-axis tracking plants. The equivalent figure for a plant with fixed panels is lower.

25.3.2 PV Projects

Project name	Alamosa (Figure 25.8)
City	Alamosa, Colorado
Year	2012
Type	CPV, Fresnel Lenses
PV panels	Amonix 7000
Capacity	35 MW
Land	91 Ha
Solar resource (DNI)	2482 kWh/m^2/yr
Number of modules	504
Tracking	Dual axis, with hydraulics
Capacity per module	70 kW
Number of inverters	Solactria: 504×70 kW

Figure 25.8 Alamosa [5].

Project name	Nellis, Nevada
Type of system	PV with tracking
Installed capacity	14.2 MW
Collection area	1000 m × 600 m (140 Acres)
Storage capacity	None
Number of panels	70 000
Panel wattage	200 W each
Tracker model	SunPower T20
Number of panels/tracker	9 and 11
Tracking	Tilted single axis with backtracking
Operating wind resistance	130 km/h
Solar tracking method	GPS
Number of inverters	Xantrex, 52 – GT250, 2 – GT100
Tracking motors per MW	8/1 MW

Project name	Barrick, Western 102, Nevada
Completed	February 2008
Type of system	PV with tracking, Hi Profile
Installed capacity	1.0 MW
Wind gusts	150 km/h (90 miles/h)
Connected to ge grid	No

Project name	Barrick, Western 102, Nevada
Generated power/voltage	1295 kW, 280 V_{dc}
Converted power	1000 kW/3 h, 480 V_{ac}
Collection area	200 m × 180 m (8 Acres)
Storage capacity	None
Number of panels	7404
Panel wattage	175 W, Sharp 175
Number of panels per tracker	9 and 11
Tracking	Tilted single axis 90° reach
Tracker supplier	Thomson, TTI
Tracking motors per MW	5/1 MW
Operating wind resistance	145 km/h
Solar tracking method	GPS
Number of inverters	2 × 500 kW, Satcon, Mississauga, ON
Data acquisition	Draker Labs, California
Total cost (EPCM)	$10 000 000

Owners notes: Significant problems occur with the trackers due to wind gusts. Panel fluttering starts at about 65 km/h, affecting the first three rows of panels. The trackers were guaranteed to safely handle 130 km/h winds. The guarantees were not being met. Other plants with similar trackers are experiencing similar problems.

The recommendations by the operators were to use flat fixed design anywhere where higher wind gusts are expected. This concept would require 25% more panels to achieve the same power output as the comparable single-axis tracked system. Given the high wind speeds and wind gusts, flat fixed panels are the only viable option.

Washing is generally done after rain, with high pressure water spray, followed immediately with squeegees. Washing effort was estimated at four days, up to 10 times/yr.

The Barrick Nevada plant is not grid-connected. The output is connected to the 480 V bus of the plant to feed the plant parasitic loads.

Project name	Sevilla PV, Spain, Lat. 37.2°
Type of system	Low Conc. PV with tracking, Hi Profile
Installed capacity	1.2 MW
Wind gusts	n/a

Project name	Sevilla PV, Spain, Lat. 37.2°
Connected to grid	Yes
Generated power/voltage	n/a
Converted power	n/a
Collection area	400 m × 300 m (29.5 Acres)
Storage capacity	Nil
Number of panels	5544 (12 320 m^2)
Panel wattage	216 W
Number of trackers (Heliostats)	154
Number of panels/tracker	36 (80 m^2)
Tracking	Two axis
Tracker supplier	Abengoa
Tracking motors per MW	n/a
Operating wind resistance	36 km/h (22.5 miles/h)
Solar tracking method	n/a
Number of inverters	n/a
Data acquisition	n/a
Plant capacity factor	12%

Notes: This low concentrated PV plant uses side mirrors placed on heliostats to concentrate more radiation onto the PV panels. In addition the plant uses double-axis tracking to augment its power output. The trackers with panels are 18 m high, thus not suitable for windy areas.

The PV cells and mirrors are cleaned using demineralized water when the power output falls by 1%. High temperature affects the cell performance.

Project name	Tiffany, Whippany, New Jersey, 2006 [5]
Supplier	SunPower, California
Type of system	PV (Roof top)
Installed capacity	0.68 MW
Collection area	9662 m^2
Storage capacity	Nil
Number of panels	3270
Energy produced	782 MWh
Tracker model	None
Number of Inverters	n/a
Plant efficiency of conversion	13%

Notes: In addition to providing energy, the roof top panels reduce the heat dissipation through the roof and extend life of the roof by protecting it from solar UV radiation.

Project	Kamuthi, Southern India
Installed capacity	648 MW, interconnected in one area, Cost $710 million
Area covered	10 km^2
Solar panels	2.5 Million, 576 Inverters, 154 Transformers.

25.4 CSP

CSP can generate both heat and electricity. The heat can be used as a direct energy input to various thermal processes, for example, desalination and process heat. Electricity can be used to power reverse osmosis, or pump water from deep-water wells. Combining CSP and thermal in one place offers opportunities for real cost–saving synergies. CSP technology is highly flexible and still evolving.

The four primary types of CSP technology are described later. Included in the report is a discussion on the concept of thermal storage, which is often associated with CSP.

25.4.1 Power Tower Systems

A power tower system consists of a centralized tower receiver that collects focused solar radiation from a field of moveable heliostats that are placed at the ground level mostly on the north side of the tower (Northern Hemisphere) to allow the heliostats to face south (see Figure 25.9). PV panels are not used. In some areas, solar mirrors are placed all around. Tower operation is based on three main components: heliostats, receptor, and tower. Heliostats are structures with a number of mirrors mounted on single posts. The heliostats must be equipped with double-axis three-dimensional trackers to ensure a perfect focus over the whole day.

The solar energy is concentrated up to 600 times on a receptor located on the upper part of the tower. The plant requires a flat terrain to ensure perfect condition for focusing the heliostats. Tower plants do not perform well in windy environments. Wind causes the mirrors to flutter and the scattered focus may reduce the heat transfer by up to 50%. The centralized receiver transfers solar radiation energy to a heat transfer fluid (HTF), which in turn is used to power a conventional power cycle. Potential temperature of the heating media is over 600–700 °C, which is suitable for smaller plants, as focusing from widely spaced

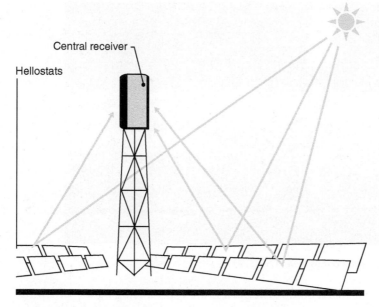

Figure 25.9 CSP tower plant [5].

heliostats becomes difficult during windy conditions. Potentially, a tower plant, providing the wind conditions are right, can produce the highest temperature heating media, owing to the short piping, which is contained solely within the tower and independent heliostats.

Power tower systems are considered the most expensive and arguably most risky of the CSP technologies (see Figure 25.10).

The biggest advantage and attractiveness of this technology to the utilities is the energy storage capability to be able to defer the energy usage to peak demand periods when it is the most needed.

Owing to its short piping, the power tower can use molten salts for heat transfer media, at present the most suitable media for storing energy.

A 11 MW solar power tower presently in operation in Seville, Spain, produces electricity with 624 large movable heliostats, each having 120 m² of parabolic mirrors.

Crescent dune solar near Las Vegas, 110 MW uses 196 m high tower, 10 317 mirror heliostats, three-way tracking, and 10 hours of molten salt energy storage.

Invanpah solar: 377 MW net plant in Mohave Desert, three separate plants. Cost: $2.2 billion ($6000/kW). 170 500 Heliostats, (3) 147 m high towers. After two years of operation, Ivanpah was still unable to meet the production targets.

Figure 25.10 Crescent dunes tower CSP [5, 9].

Environmental issues: Some tower plants are experiencing issues with birds. Birds get to be fried while chasing insects through the sun rays. The plant owners are looking for solutions to this problem by trying to divert insects and birds away from this type of plants. The low cost of natural gas, falling prices of PV plates, production shortfalls, and environmental issues may be a reason for sudden row of cancellations of a number of high-profile projects of this technology in California.

25.4.2 Parabolic Trough Systems

The parabolic trough system [10] is recognized as the proven and installed CSP technology; however, continuing refinements in solar technology may be leading in different directions due to the high cost of the parabolic mirror troughs (see Figure 25.11).

In 1980s, Luz International constructed nine Solar Electric Generating Systems (SEGS) in the California Mojave desert. This still constitutes the world's largest solar power plant with parabolic trough collectors, having a total capacity of 354 MW. Luz's success over the past 20 years has demonstrated the robustness and reliability of the parabolic trough technology. The plant consists of a series of linear parabolic-shaped reflectors that focus the sun's direct radiation onto a linear receiver (tube) located at the focus of the mirror parabola.

The collectors track the sun from east to west (single-axis tracking) during the day to ensure maximum radiation to the linear tube receiver. Oil passing through the receiver tubes is heated to 400 °C as it is circulated through the

Figure 25.11 Parabolic mirror model [11].

receiver pipe and returns to a series of heat exchangers in the power plant, where it is used to produce high-pressure steam to drive turbine generators.

The mirrors are laid down on elevated structures in multiple long parallel arrays to allow for a full rotation of the mirrors, i.e. open during operation and turned down during the off hours or rain. The parabolic mirrors and the glass/steel pipe receivers require specialized fabrication, installation, and focusing and are the costliest components of the plants. Focusing the mirrors to the pipe receivers can also be an ongoing concern in windy conditions.

25.4.3 Fresnel Linear Mirrors

Fresnel linear mirror systems (Figure 25.12) are broadly similar to parabolic trough systems but instead of using trough-shaped mirrors that track the sun, they use long flat mirror strips positioned at different angles that have the effect of focusing sunlight on one or more receiver pipes containing heat-collecting fluid, mounted above the mirrors. As with parabolic trough systems, the mirrors can change their orientation throughout the day so that sunlight is always concentrated on the heat-collecting pipe. The solar energy is reflected onto a tube that contains water, which is heated to steam and sent to storage and turbines. The covered tube orientation partly blocks the sun radiation to the mirror at the bottom.

However, unlike parabolic trough makers, who must secure expensive manufacturer plant capacity to make precision bent glass reflectors, Fresnel mirrors are small flat strips of simple and inexpensive design. When not in use, the mirrors can turn upside down for further protection from the wind, sand storms, or even hail. Lower vulnerability to wind permits the use of lighter and lower profile structures, therefore, reducing the cost per unit area. The simplified mirror concept results in significantly lower installed cost. The relative simplicity of this type of plant means that it is cheaper to manufacture than an equivalent

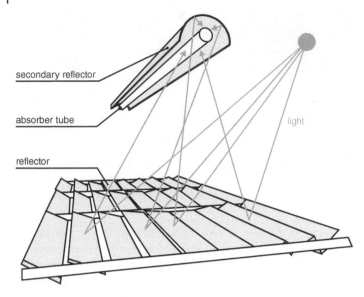

secondary reflector

absorber tube

light

reflector

Figure 25.12 Fresnel mirror model [12].

parabolic trough system, while retaining all the advantages of the parabolic systems.

These systems can be built in smaller areas than either tower systems or trough solar heating systems.

The low profile installation allows for more concentrated packing of the mirrors to produce more MWh generated per km² covered. The current estimates for power concentration generated are 125 MW/km² for a Fresnel system against 60 MW/km² for a parabolic trough plant design. Experts consider this technology the way of the future for the solar technology.

The largest plant using this technology (30 MW) opened in Puerto Errado, Spain, in 2012 (see Figure 25.13). It is only the third commercial scale plant to open using this technology, so less is known about this technology than others in terms of efficiency. According to the operator, the plant has a high efficiency of 38%.

This technology reduces the cost, as follows:

- The reflectors (mirrors) are laid at lower heights on simple stands offering a low profile design for easy maintenance. Mirrors are smaller and significantly less costly than parabolic lenses. Mirrors can direct beams to two receiver tubes, thus allowing for close packing of mirrors.
- The receiver tube is a flat steel, extruded hollow shape, rather than rounded pipe made of steel and glass.
- The heating fluid is water. Fresnel solar eliminates oil, by directly heating water to a temperature of roughly 400 °C at pressure, approximately 50 bars.

Figure 25.13 Solar reserve: crescent dunes solar [5, 9].

It uses the generated steam to drive low-temperature turbines similar to those employed in industrial applications.

One of the advantages of the Fresnel model is that a relatively small amount of capital is required to build an assembly plant. Also important, this product technology lends itself to manufacturing simplicity, requiring little more than well-established engineering materials and processes. As a result, a high-capacity Fresnel plant can be assembled and commissioned in months, rather than years. This makes Fresnel technology rapidly scalable, unlike Solar Towers and wind generators. It has a low visual aspect; it is quiet and can be sited near residential areas.

With joint planning, clear purchasing policies and lead from governments, the turbine manufacturers could build a standardized product, and thus boost capacity rapidly.

25.4.4 CSP Costs

CSP technologies [13] currently offer the lowest-cost solar electricity for large power generation, 10 MWe and larger. Operating costs in 2008 were in the order of 4–5 cents per kWh.

Studies and charts have been developed showing a significant drop in the levelized cost of solar energy over time; however, the actual cost reductions have been lower than anticipated. "Assessment of Parabolic Trough and Power Tower Solar Technology Cost and Performance Forecasts" was produced by Sargent and Lundy LLC, in October 2003 (NREL/SR-550-34440). In this report, the capital cost for CSP in 2008 was forecast at \$3220/kW installed. The actual costs reported for Andasol 1 were \$7280/kW installed with a capacity factor of 41%; however, this system includes thermal storage that comes at a high capital cost.

A present budgetary price for a CSP without thermal storage is \$5000 MW installed with a capacity factor of 30% (this information was provided by Solel, an Israel-based solar energy firm, which purchased their technology from Luz).

An estimated capital cost of a 50 MW plant would be in the order of US\$ 250 000 000.

The annual energy production for the 50 MW Andasol 1 plant, is 158 GWh, at 16% thermal efficiency.

In general, parabolic CSP technology is not economically viable in small-scale solar plants, for example, 1 MW. Typically, the economic sizes start at 50 MW and higher.

There is however, one exception to this rule. One manufacturer of Fresnel mirror CSP (Helio Dynamics) has developed modular systems for hot water heating and air conditioning applications. The installed cost is approximately US\$ 10 000 for an 11 kW module. The units may be mounted at ground level, or on top of buildings and are currently in use in several locations. The roof-top scenario has the advantage of shading the roof which reduces the air condition-ing demand, but at the expense of a higher installation cost. The collector heats the water to a maximum of 180 °C; this water is then used in an absorption chiller to provide air conditioning, and after the chiller, the water may then be used for hot water onsite, for the camp, the mine, or the power plant buildings.

25.4.5 CSP Operating Plants and Projects

The following is a list of some of the plants presently in operation or near the final construction stage:

Project	Crescent Dune Solar, near Las Vegas (Figure 25.13)
Technology	Power Tower, 196 m height, Commissioned, 2014,
Capacity	110 MW
Heliostats	10317, three-way tracking
Mirrors	1196778 m^2
Steam cycle	Molten salt to steam. Salt heated to 566 °C, cooled down to 288 °C
Storage capacity	10 h in molten salt
Salt tonnage	32 000 tons.

Note: During 2017, Dune Solar was on extended maintenance due to a leakage in the molten salt circuit.

Figure 25.14 Puerto Errado, Fresnel model. Source: Courtesy of Novatec Solar [5].

Project	Puerto Errado, Spain (Figure 25.14)
Technology	Fresnel Linear Mirror, Year 2012
Mirrors	28 rows
Area	70 Ha
Solar resource	2095 kWh/m² /yr
Capacity	30 MW, 49 GWh/yr
Efficiency	38%
Heat transfer	Water
Power cycle pressure	55 bar
Solar field inlet/outlet Temperature	140 °C/270 °C, Operating Temperature 270 °C
Cooling	Dry, air cooled condenser
Storage	0.5 h, Single tank, Thermocline

Project	Zhangbei, China
Technology	Fresnel Linear Mirror Concentrators, 50 MW
Storage	14 h

Project	Solana, Phoenix, Az [5].
Technology	Parabolic Trough
Capacity	280 MW, 2 × 140 MW, steam turbines
Year	2013 in Operation
Production	944 GWh/yr
Power agreement	30 yr
Federal loan	US$ 1.45 billion
Capital cost	US$ 2.00 billion
Area	220 Ha
Parabolic collectors	3232 in 808 loops/4 collectors
Heat transfer	Therminol VP-1
Solar field inlet/outlet temperature	293 °C/393 °C
Power cycle pressure	100 bar
Cooling	Wet cooling towers
Fossil backup	Natural gas
Storage	6 h, Molten salt

Project	Mojave, California (Figure 25.15)
Technology	Parabolic Trough
Capacity	280 MW, 2 × 140 MW, Steam Turbines
Year	2014 in Operation
Production	600 GWh/yr
Power agreement	30 yr
Federal loan	US$ 1.45 billion
Capital cost	US$ 1.6 billion
Area	1765 Acres
Heat transfer	Therminol VP-1
Solar field inlet/outlet temperature	293 °C/393 °C
Power cycle pressure	100 bar
Cooling	Wet cooling towers
Fossil backup	Natural gas
Storage	None

Figure 25.15 A row of parabolic mirrors [5].

Project	Andasol 1, Spain [5],
Technology	Parabolic trough
Installed capacity	50 MW, 16% eff.
Solar area	510 120 m^2
Total area	2 020 000 m^2
Solar resource	2136 kWh/m^2/yr
Expected energy production	158 GWh
Heat transfer fluid	Oil, 377 °C, 100 bar
Thermal storage	15 hours, Molten salt
Storage tanks	2 × 14 m height × 36 m diameter
Molten salt	285 T
In operation since	July 2008
Construction period	2 years
Cost	$ 364 000 000, $ 72 000/MW: High
O&M staffing	40

Project	Nevada Solar 1, Las Vegas, Owned by Acciona [5]
Technology	Parabolic trough
Installed capacity	64 MW

Project	Nevada Solar 1, Las Vegas, Owned by Acciona [5]
Number of parabolic troughs	760, 4 m each, with total of 180 000 mirrors
Mirrors	White, 4 mm^2 glass, by Flabeg AG
Receiver tubes	Solel 30%, Schott 70%, total 18 240 of 4 m tubes
Trackers	Hannifin
Turbine	1–70 MW, reheat type, Siemens
Solar area	357 000 m^2
Total area	162 Ha
Expected energy production	130 GWh
Heat transfer fluid	Oil, 391 °C, 100 bar
Thermal storage	30 min, to minimize the effects of transients
In operation since	March 2007
Construction period	2 years
Cost	US$ 260 000 000, or US$ 4060/kW

Project	Kuraymat, Kairo, Egypt [5]
Technology	Parabolic trough
Installed capacity	20 MW
Number of parabolic troughs	n/a
Mirrors	n/a
Receiver tubes	n/a
Trackers	n/a
Turbine	150 MW (hybrid plant)
Solar area	130 000 m^2
Total area	n/a
Expected energy production	33.4 GWh
Heat transfer fluid	Oil, 400 °C, 20 bar
Thermal storage	No
Operation date	End of 2009
Construction period	2 years, Civil works commenced in 2008
Cost	US$ 100 000 000, or US$ 5000/kW

Project	Solar Tres, Spain [5]
Technology	Power tower/Heliostat
Installed capacity	17 MW
Number of heliostats	2590, 115 m^2 each
Trackers	n/a
Turbine	17 Mwe
Solar area	298 000 m^2
Total area	142 Ha
Expected energy production	110.6 GWh
Heat transfer fluid	Molten salt, 565 °C, 20 bar
Thermal storage	Molten Salt, 15 h
Operation date	End of 2007
Cost	US\$ 265 000 000 or ~US\$ 15 600/kW for 74% capacity factor

Project	Sevilla PS10, Spain [5]
Technology	Power Tower, 11 MW
Number of heliostats	624, each with 120 m^2 of parabolic mirrors
Tower height	115 m
Operating temperature	255 °C
Solar area	74 880 m^2
Total area	70 Ha
Expected energy production	23 GWh/yr
Heat transfer fluid	Water
Thermal storage	<1 h, as stored steam
In operation since	March 2007, 2 years construction
Cost	US\$ 47 000.000 or US\$ 4300/kW

25.5 Thermal Storage

Thermal storage [14, 15] is presently cheaper than storage of electricity in batteries, but things are changing (see Figure 25.16). The storage is generally needed to extend the hours of the power availability beyond the sun active

Solana, Az, 280 MW CSP – 6 h molten salt storage tanks

Figure 25.16 Solana AZ, salt storage plant [15].

hours. Thermal storage improves the competitiveness of CSP because it allows the heat to be stored for use during the night or during overcast periods when there is no sun. Thermal storage allows for better usage of generation by delivering at times of daily demand peaks.

An important benefit of thermal storage is its potential to displace the cost of an additional peaking plant. This lowers the cost of the energy produced, and is another reason why utilities are attracted to the technology. By using thermal storage, it is possible to shift the output to make certain demand is met outside the hours when the sun shines.

Each CSP requires a certain small amount of thermal storage capacity, of about 30 minutes to override the effects of thermal transients during the operating daily cycle, i.e. passage of clouds or sudden rain.

This short-term storage is typically accomplished now by the use of oil reservoirs. Molten salt allows the plants to operate at higher temperatures compared to water and oil. Having capability to store solar energy in salt tanks it allows the plant to continue producing energy even after the sun has gone down. In fact, Spain's **Gemasolar** [14] plant became the first solar plant to produce 24 hours of power, because it could continue producing energy from the heat in its salt tanks. This capacity can significantly increase the amount of energy produced by the plant. In the case of Gemasolar, the 19.9 MW plant, 15 hour storage at 565 °C, will produce 110 GWh of energy a year. That is an annual capacity factor of 55%. Pretty good, compared to less than 30% for regular CSP.

Another potential method of energy storage being developed is compressed air in tanks.

25.5.1 Balance of Power (BOP) Plant

Depending on the CSP technology, the fluid flowing between the solar modules is oil, water, or molten salt. A HTF typically flows through the receiver, absorbing heat and returns to a series of heat exchangers in the power plant to generate high pressure steam that in turn drives a conventional steam turbine/generator to produce electricity. The spent steam is condensed in a standard condenser/cooling tower and returned to the heat exchangers via condensate and feed water pumps to be transformed back into steam. Condenser cooling is normally provided by mechanical draft dry or wet cooling towers. The fluid is circulated back from the heat exchangers to the solar collectors.

Power lines are needed to transfer the generated power to consumers. Depending on the size of the plant, the lines can assume any voltage suitable for integration with the external power grid.

25.5.2 Peaking Storage Plants

With the price of large batteries coming down, there are several options developing to better and more efficiently integrate renewable energy with other sources of energy and increase the usage and capacity factors of wind and solar power during the periods of lower solar or wind activity.

25.5.2.1 Hybrids: mixture of battery and gas turbines

Customers want their electricity to be available always on, while utilities want to provide cheaper energy from the renewable resources [16]. But, during the heavy daily load, the wind can weaken and the sun can be behind clouds. To cope with this issue, the renewable energy must be supported by the grid spinning reserves. In California, a new approach has commenced to deal with the power shortfall. It is delivered by quick starting gas turbines called peakers to support the renewables.

In the absence of grid-scale batteries to bridge supply gaps, the natural-gas turbines (GTs) can ramp up and pick up the slack when renewables drop off. But even the fastest gas turbines take several minutes to reach full power, forcing operators to run them at minimum load to keep them ready, burn gas, and put more wear on the machines. This is inefficient combustion that needs extra fuel, costs money, and generates unnecessary greenhouse emissions.

At General Electric (**GE**) **Energy,** they combined the GT peakers and batteries together into a single, efficient package with power management software. With this hybrid system, the turbine is held turned off and the battery will respond instantly. Southern California Edison (SCE) is deploying the solution – the first of its kind in the world – at two sites near Los Angeles. The battery is quick and clean to bridge the energy gap, while the gas turbine is giving the power needed after some delay.

The hybrid assembly includes a GE LM6000 gas turbine that can reach 50 MW in just five minutes – and a 10 MW battery assembled from lithium-ion cells that last up to 30 minutes. When a wind farm output drops, the battery kicks in immediately and gives the turbine the time to start up without cutting off the grid consumers.

25.5.2.2   Kauai (Hawaii) solar plant with litium ion battery storage

Tesla's Kauai solar power facility with a 13 MW SolarCity installation (2016) feeds power to a power storage facility with 52 MWh of total capacity.

It can release power of 5 MW for 10 hours, or 10 MW of peak power for five hours.

The batteries are lithium, identical to those PowerWall batteries used by residential consumers in USA. The storage facility captures energy from the sun during peak daytime production hours, and then holds back the power to be used for peak consumption hours at night.

25.5.3   Battery Storage Costs

The battery storage [17] industry is a key part of the master plan if wind and solar power are ever to be an equal partner on the grid. It is becoming increasingly important in places as South Australia, the region which gets 41% of its

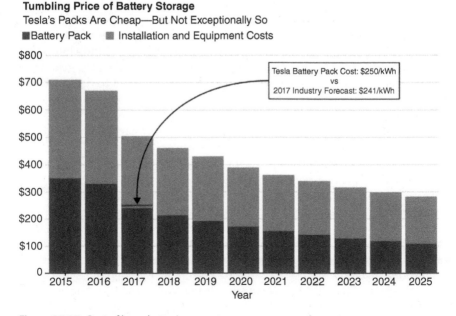

Figure 25.17 Cost of large batteries.

Figure 25.18 Tesla 100 MW storage battery.

electricity from renewable energy, one of the highest penetrations of wind and solar in the world.

This large percentage of renewable coverage can leave a large power shortage when the wind dies down. A magnitude of 100 MW of power was roughly the size of an electricity shortfall suffered by South Australia in a February 2017 blackout (see Figure 25.17).

The South Australia government's current plan is to address its energy shortfall largely through the construction of a state-owned 250 MW gas plant, which would be fired up to support a sudden lack of wind and/or sun. There is a discussion under way for Tesla to supply a 100 MW battery storage installation. Well, the discussion is over, and the project was commissioned in November 2017 in Hornsdale.

The battery actually serves a dual purpose: 70 MW as storage for 10 minutes to provide for the grid stability and 30 MW to serve as a rapid response for sale on the energy market. Instead of running the gas turbine for inefficient maintenance of spinning reserve, the storage part will allow the gas turbines to rest and restart in due time to pick up the load.

The battery life is estimated at 10–15 years. That is still to be seen, as well as how much it will cost to recycle it (see Figure 25.18). [18]

25.6 Conclusion

25.6.1 PV Plants

The following must be taken into consideration, should an investor wish to consider solar PV plants as an alternative source of energy to displace the use of fossil fuel:

- Simple and inexpensive PV plants are taken off in a big way in India and China.

- Each project area under consideration must consider all the factors that maximize the solar resource under the applicable environmental conditions. However, wind and dust concerns may limit applications for solar generation. Wind resource data in the area must be studied, before deciding on an appropriate plant concept. Fixed panel low profile plants may be the only reasonable solution for PV plants in the wind swept areas.
- PV plants for remote consumers, ranging from 500 to 1000 kW seem appropriate.
- PV plants come with a high capital cost. They may be justified either as a source of green power or a source of cheap power over a long period of time. The cost of maintenance can also be high if the area is subjected to high winds or the need for frequent washing.
- PV plants are most suited to desert like areas, which are not subjected to high intensity winds. Deserts usually are. Of all the design concepts, the fixed panel plants are the most suited to tolerate windy conditions. Fixed panel plants require an additional 25–30% of panels to generate the same amount of energy as tracked systems. The overall cost per MW installed may be the same, but the maintenance cost would favor the fixed panel plants.
- A relatively flat terrain is necessary to allow for uniform panel tilt. Paved maintenance roads are needed to limit the dust conditions.
- PV plants are most suited to be placed in the vicinity of small consumers and villages, to make use of low voltage power drawn directly from the inverter switchboards, thus avoiding costly transformers and power distribution circuits. If longer distribution distances are involved, greater than 500 m, transformation to higher voltages is required at both ends of the line.
- Suppliers have been overly optimistic of the tracking systems' capabilities to handle windy conditions in the current early stages of technology development. The use of trackers is beneficial as it adds to the overall efficiency of the plants. Single axis trackers with low-profile layout is the preferred plant design in windy locations. The wind maintenance concerns are likely to be in the order of 1 : 2 : 4 for the fixed, single, and double axis trackers, respectively.
- Use of multijunction solar collectors may deliver unprecedented levels of solar energy production. Efficiencies of up to 45% have already been demonstrated in lab. The theoretical efficiency of the technology is over 80%.

25.6.2 Concentrated Solar Plants

- This promising technology has been hit by some major environmental and technical hurdles that have slowed down their progress.
- The CSP are more adaptable than PV plants to work with storage facilities and become operational and dispatchable over 24 hours, but the cost of energy storage is still high.
- Various technologies are available. Some may be more suited than the others for the specific circumstances and operating environment demanded at

various sites. A number of plants are presently in operation, some of which have been operating for more than 20 years. Others are being planned and will be on stream in the near future. The technology is rapidly evolving. The next several years will see major technological progress clarifying the way forward.

- A CSP at remote sites could be used to displace fossil fuel or imported energy. Thermal storage is still expensive. For now, the industrial process plant in conjunction with other energy sources must be able to absorb all the power generated by the solar plant during the sun active hours. In the event, the output of the solar plant is not required or not available the CSP must be able to reject the solar resource, by turning the mirrors away from the receiver tubes, or some other means.
- The O & M cost, or the cost of power of the CSP is 0.04 $/kW, as per Reference [1], which includes a 5.5% annual solar panel replacement rate. This rate is lower than the cost of the power from the fossil fuel generating plants, when solar energy is available.
- The installed capital cost for the CSP is around $5000/kW (excluding thermal storage), which is higher than for the conventional coal power plants, on an installed MW basis. Furthermore, considering that the capacity factor of solar plants is only 30%, the installed cost in $/kW is very high on the 100% operating base for all the current solar technologies.
- Parabolic trough designs can be used for power plants of 100 MW and larger. However, the technology includes costly parabolic mirrors and receiver tubes, mounted on expensive moving structures. This design type is not suitable for the dusty and windy climates.
- CSP using linear Fresnel mirror designs generate power at lower installed and operating cost per MWh, and use the least amount of acreage per MW of all current solar technologies. The installation is simple, wind friendly and easy to maintain. Within the industry, claims of 50% reduction in costs over the installed cost of parabolic CSP have been made. Refer to the Puerto Errado Plant.

  Most manufacturers of Fresnel CSP designs are targeting the large-scale plant size, 50 MW and above. The small-scale modular Fresnel systems appear to be much more economical.
- The costs of concentrating solar plants (CSP) with storage are expected to fall to less than 6 cents per kWh by 2020. Credit: US Department of Energy Sunshot Initiative.

References

1 British Solar Business: Global Horizontal Irradiation – Europe.
2 OMICS: World Solar Energy Map.
3 Alternative Energy Tutorials (current web). http://www.alternative-energy-tutorials.com/solar-power/photovoltaics.html (accessed 15 July 2019).

4 NREL, Sarah Kurtz (2006) High Efficiency Multi Junction Solar cells.

5 (2006). NREL Concentrated Solar Power Plants. https://solarpaces.nrel.gov (accessed 15 July 2019).

6 Marios Therestis, Nabin Sarmah, Georgios E.Amaoutakis, Tadihg Sean O'Donovan (2013) – *Solar Spectrum dependent thermal Model for HCPV Systems*, Conference paper. www.ResearchGate.net.

7 World Energy Council (2016). Energy Resources. https://www.worldenergy .org/data/resources/resource/solar/ (accessed 15 July 2019).

8 HelioSCSP (2016). Concentrated Solar Power Installed Capacity Increased to 5133 MW by the End of 2017. http://helioscsp.com/concentrated-solar-power-installed-capacity-increased-to-5133-mw-by-the-end-of-2017/ (accessed 15 July 2019).

9 Solar Reserve (2018) Crescent Dunes, Nevada. https://www.energy.gov/lpo/crescent-dunes.

10 NREL/SR-550-34440 (2003). *Assessment of Parabolic Trough and Power Tower Solar Technology Cost and Performance Forecasts.* Chicago, Illinois: Sargent & Lundy LLC Consulting Group.

11 Introduction to CSP Technology (current web). http://www.streammisr .com/l3.php?id=164 (accessed 15 July 2015).

12 Fresnel Mirror Model (current web). https://www.researchgate.net/figure/Principle-of-operation-of-linear-Fresnel-reflector-solar-system-5_fig4_258402998 (accessed 15 July 2019).

13 European Commission, 2007, Concentrating Solar Power – From Research to Implementation.

14 Gemasolar Thermsolar Plant. Molten Salt Storage (current web). www .torresolenergy.com/en/gemasolar/ (accessed 15 July 2019)

15 Renewable Energy World (2016). Susan Kraemer: Commercializing Standalone Thermal Energy Storage. www.renewableenergyworld.com.

16 GE Energy: GE Creates Battery Storage and Gas Turbine Hybrid – California.

17 Bloomberg, Tom Randall: Tesla's $ 169 MM, 100 MW Battery Play – South Australia.

18 Australian Associated Press (2017). *Tesla Powerpacks in Hornsdale.* South Australia, Australian Associated Press.

26

Wind Power

CHAPTER MENU

26.1   Siting a Wind Farm, 600
26.2   Wind Turbine Tower, 602
26.3   Wind Resource, 603
    26.3.1   Wind Gusts, 603
    26.3.2   Wind Rose, 603
    26.3.3   Power, 603
    26.3.4   Wind Tip Speed Ratio λ, 605
    26.3.5   Capacity Factor (CF), 606
    26.3.6   Wind Energy Distribution, 608
26.4   Wind Turbulence, 608
    26.4.1   How Does It Affect the Turbine Performance?, 608
    26.4.2   What Causes Atmospheric Turbulence?, 608
26.5   Wind Turbine Design Classification, 610
26.6   Blade Design for Optimum Energy Capture, 611
26.7   Individual Pitch (Blade) Control (IPC), 612
26.8   Wind Turbine Design Limits, 613
26.9   Wind Turbine Components, 615
26.10 Generators Used with Wind Turbines, 615
    26.10.1   Fixed Speed Wind Turbine Generators, 616
    26.10.2   Variable Speed Wind Turbine Generators, 616
    26.10.3   Synchronous Generator with In-Line Frequency Control, 617
    26.10.4   Doubly Fed Induction (Asynchronous) Generator – DFIG, 617
26.11 Turbine Sizes, 620
26.12 Building a Wind Farm, 623
    26.12.1   Grid Integration Issues, 624
    26.12.2   Utility Stability Requirements for Wind Farms, 627
    26.12.3   Managing Variability and Voltage Regulation at Wind Farms, 630
    26.12.4   Methods of Transient Regulation of Power Generation, 631
26.13 Wind Energy in Cold Climates, 633
    26.13.1   Gaspe Region, 634
26.14 The Effect of Rain on the Wind Turbine Performance, 634
    26.14.1   Conclusion, 634
26.15 Wind Turbines in the Desert Environment, 635
    26.15.1   Blades, 636

Practical Power Plant Engineering: A Guide for Early Career Engineers, First Edition. Zark Bedalov.
© 2020 John Wiley & Sons, Inc. Published 2020 by John Wiley & Sons, Inc.

26.15.2 Rotor Hub, 637
26.15.3 Air Intakes and Exhausts, 637
26.15.4 Bearings, 638
26.15.5 Electronic/Electrical Systems, 638
26.16 Cost, Component Percentage Share, 639
References, 640

26.1 Siting a Wind Farm

A wind turbine is an apparatus that converts the wind's kinetic energy into electrical power. Wind is caused by the sun unevenly heating the atmosphere, the earth's rotation, and surface irregularities. As long as the sun shines, wind will blow, making it a highly variable and sustainable energy source. A cost-effective wind energy storage system has yet to be developed.

On average, wind generated power costs between 0.04 and 0.06 US\$/kWh, at a capacity factor (CF) in the low thirties. Although wind power's price per kWh is cost-competitive, its high initial investment costs pose financial challenges for developers. Sites deemed suitable for wind farm development are often far away from the cities – load centers, thus, transmission lines must be built to bring the electricity into the city, further driving up development and construction costs.

The first step in a wind project development is identifying a site with sufficiently strong winds. See Figure 26.1.

Figure 26.1 Offshore 5 MW turbines.

Accessibility to transmission lines, roads and a ready market are the significant factors when selecting a site. Following an initial assessment, the developer conducts a thorough investigation of the site, collects meteorological data for at least a year, considers environmental and community impacts, and researches siting and permitting. Because most wind farms are built on private land, the developer and landowner must come to a lease agreement before construction begins. The developer then contracts with a local utility to sell the electricity produced and hires an engineering/construction company to build the facility. Upon completion, the wind farm is often sold to an independent operator. In many cases utilities own and operate the wind farms.

Wind turbines are generally divided into a few categories: utility-scale off-shore and onshore, and small wind. Utility-scale wind features turbines are >1 MW. This chapter will not refer to "the small wind."

In Europe, offshore wind farms are typically erected in shallow, coastal waters with depths ranging 5–25 m, with turbines mounted on platforms anchored to the ocean floor. Electricity generated by the offshore turbines is transmitted through underwater cables buried beneath the seabed to a collector station and then a single line connects to the power grid onshore.

To capitalize on the stronger, more consistent winds at sea, most offshore turbines carry higher nameplate ratings from 2 to 7 MW, and more soon, easily doubling that of many onshore turbines. What is more, tower heights typically exceed 80 m and rotor diameters range up to 150 m. Turbines intended for off-shore duty are built to withstand corrosion, storm waves, hurricane winds, and other environmental challenges. In addition to the site location, the following factors have to be determined:

- Wind velocity and duration curve or frequency distribution chart to determine the potential CF.
- Wind turbulence to determine the tower height, blade length, and the wind class.
- Wind rose to determine the wind intensity at varied directions.

Generally, marine locations and exposed hilltops provide the most favorable wind conditions with wind speeds consistently >5 m/s. Turbulent conditions reduce the amount of energy to be extracted from the wind, thus reducing the overall efficiency of the system. This is more likely to be the case over land than over the sea. Raising the height of the turbine above the ground effectively lifts it above the worst of the turbulence and improves efficiency.

It would be a mistake to place turbines right over a cliff to effectively add to the height of the wind turbine tower to benefit from the wind coming from the sea. The cliff at the shore line creates turbulence and it brakes the wind before reaching the turbine rotor. Placing the turbines close to the cliff adversely affects the turbine's performance and life. It is preferable to have a rounded hill in the direction facing the sea, rather than the escarpment.

In most locations around the world, it is windier during the daytime than at night. This variation is largely due to the fact that temperature differences, e.g. between the sea surface and the land surface tend to be larger during the day than at night. It is also advantageous that most of the wind energy is produced during the daytime, since electricity consumption is higher than at night and the tariffs are higher.

On most horizontal smaller wind farms, a spacing of about 6–10 times the rotor diameter is maintained. For large wind farms, distances of about 15 rotor diameters are found to be optimal. This conclusion has been reached by research based on computer simulations taking into account the interactions among wind turbine wakes and turbulence boundary layer.

26.2   Wind Turbine Tower

Horizontal-axis wind turbines with the main rotor shaft and electrical generator at the top of a tower, and the blades pointed into the wind are the most common and commercial turbines today for onshore and offshore applications. Most turbines have a gearbox, which converts the slow rotation of the blades into quicker rotations to be more suitable to drive electrical generators. Solid-state variable frequency drives are now being used to steady the wind variability and better interface the grid transmission system.

Since a vertical tower produces turbulence behind it, the turbine is usually positioned upwind and away from its supporting tower. Turbine blades are pointed into the wind by computer-controlled yaw motors. While some models operate at constant speed, more energy can be collected by the variable-speed turbines. All turbines are equipped with protective features to avoid damage at high wind speeds, by feathering the blades into the wind to cease their rotation and held by brakes.

The MW size and height of turbines is increasing. Land transportation of blades is a concern due to their length. Typically, 40-m long blade is a limit for transportation in the hilly terrain (see Figure 26.2).

Figure 26.2 Single blade in transport.

Offshore 8 MW wind turbines are expected to be in service in a few years. Onshore wind turbines installed in low wind speed areas use higher and higher towers to avoid turbulences. Usual towers of multi megawatt turbines have a height of 70–120 m and in the extremes up to 160 m, with blade tip speeds reaching 80–90 m/s. The higher the tip speeds, the more noise and blade erosion is to be expected.

26.3 Wind Resource

26.3.1 Wind Gusts

It refers to a phenomenon of wind blasts with a sudden increase in wind speed in a relatively small interval of time. In case of sudden turbulent gusts, wind speed, turbulence, and wind shear may change drastically. Reducing the rotor imbalance while maintaining the power output of wind turbine generator constant during such sudden turbulent gusts requires relatively rapid changes of the pitch angle of the blades. However, there is typically a time lag between the occurrence of a turbulent gust and the actual pitching of the blades based upon dynamics of the pitch control actuator and the large inertia of the mechanical components. As a result, load imbalances and generator speed, and hence oscillations in the turbine components may increase considerably during such turbulent gusts. It may even exceed the maximum safe power output level. To ensure safe operation of wind farms during wind gusts, it is highly desired to shut down the turbine operation.

26.3.2 Wind Rose

This is one of the useful tools used to determine the wind power and direction (1 knot = 0.5144 m/s) (Figure 26.3) [1]. The chart aforementioned indicates the wind direction is predominantly northwesterly west north west (WNW). The key is to remove all the obstacles in the direction of wind. At the earth level, roughness to wind is critical. Water surface is considered smoother than long grass and bushes, which will slow the wind down considerably.

26.3.3 Power

Conservation of mass requires that the amount of air entering and exiting a turbine must be equal (see Figure 26.4) [2–5]. Accordingly, Betz's limit gives the maximal theoretical achievable extraction of wind power by a wind turbine as $C_{p,max} = 16/27$ (59.3%) of the total kinetic energy of the air flowing through the turbine. A more realistic value of the power coefficient C_p is in the range of 0.35–0.45.

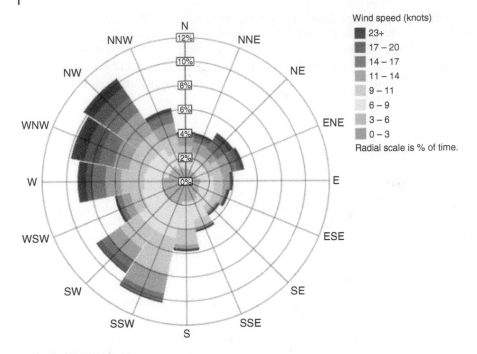

Figure 26.3 Wind rose.

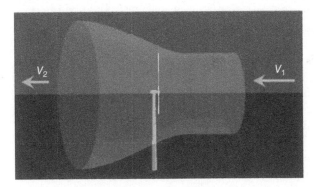

Figure 26.4 Wind tunnel flow.

Taking into account the other losses in the gear box, generator, transformer, the total wind to electricity conversion is in the order of 25–30%. The power output P of a wind machine through the effective swept area A, and the wind velocity v, is

$$P = 0.5 \, C_\mathrm{p} \, \rho \, v^3 \, A \tag{26.1}$$

where, ρ is the air density, 1.225 kg/m^3 at 15 °C at sea level, v is the wind speed in m/s, and A is the swept area of the blades in m^2.

The "wind power density" P/A is called (WPD) and has units of W/m^2.

A cubic meter of air under normal atmospheric conditions has a mass of about 1.2 kg. When it is moving, it possess kinetic energy ($E = \frac{1}{2} mv^2$). Assuming a wind turbine with rotor diameter of 100 m, it sweeps out an area of over 7800 m^2 (πr^2). If, for example, the wind speed is 12 m/s, then the volume of air moving through the rotor area every second is $12 \times 7800 = 94\,000$ m^3. This has a mass of 113 000 kg ($94\,000 \times 1.2$), and its kinetic energy is 7.8 MJ ($E = \frac{1}{2} mv^2$). If the turbine has extracted energy from it, then that mass of air was slowed down on the downwind side to about 9.6 m/s and its kinetic energy was reduced to 5.1 MJ. The turbine has therefore removed 3.0 MJ from the air (8.1–5.1 MJ). The power of the turbine is then (theoretically) 3.0 MW. This is 42% of the original wind power (3/8.1 = 0.37). Maximum turbine efficiencies are generally in the range 40–45%.

Air density and weight variations proportionally affect the power output of wind turbines. Air at −30 °C is 26.7% denser than at 35 °C. The "heavier" the air, the more energy is received by the turbine. At normal atmospheric pressure and at 15 °C air weights some 1.225 kg/m^3. The density decreases slightly with increasing humidity.

At high altitudes, in mountains, the air pressure is lower and the air is less dense. Consequently, for a given wind speed the wind energy will depend on the elevation of the wind turbine above sea level. A turbine placed at 1200 m altitude will lose 15% of its sea level capacity.

26.3.4 Wind Tip Speed Ratio λ

Power coefficient C_p is not constant and static for a turbine [5]. It actually is a function of the ratio λ of the blade tip tangential speed over the actual wind speed, generally known as tip speed ratio (TSR). C_p increases to 0.45 at $\lambda = 8$. Further increases in the tip speed/wind speed ratio, turns C_p down toward 0, as shown in Figure 26.5. Therefore, at $\lambda = 0$ the rotor does not rotate and hence cannot extract power from the wind. At very high λ, the rotor runs so fast that it is seen by the wind as a blocked disc. The wind flows around this "solid," disc so there is no mass flow of air through the rotor. This tells us that the power of a large turbine is limited by the rotational speed of the tip of the blades.

The tip of blade speed (TBS) is calculated if you know the number of rpm and the diameter D, as follows:

$$\text{TBS} = \text{rpm} \times \pi \times D/60 \ \text{m/s} \tag{26.2}$$

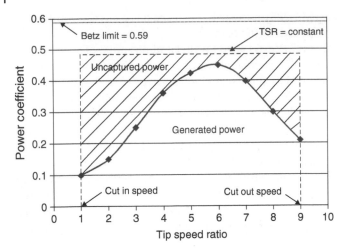

Figure 26.5 Power coefficient C_p, tip speed ratio (TSR) λ.

26.3.5 Capacity Factor (CF)

The capacity factor is simply the wind turbine generator's actual energy output for a given period (usually a year) divided by the theoretical energy output if the machine had operated at its rated power output for the same period.

Therefore, CF = average output/nameplate output ≈ 25–30%.

For comparison, the capacity factor of thermal power generation is between 70% and 90%.

Having CF of 30% or more is a good site and a wind resource.

As illustrated in Figure 26.6, the average (mean) wind speed in m/s is the average value of all the measured wind speeds.

The *modal wind speed* in m/s is the speed of the most frequent occurrence, the top of the chart.

The *weighted average speed* is the speed that allocates the duration of the specific occurrences.

The *median wind speed* divides the curve distributions into two equal parts (Figure 26.6a,b).

What is the average energy content of the wind at a wind turbine site? To determine the potential wind power, one uses the weighted average wind speed of the location. This is to insure that at all speeds are represented weighted with their durations, in particular the high speeds that bring significantly more energy by a power of cube.

A wind distribution such as that illustrated in Figure 26.6 is only valid for the prevailing wind conditions at a particular height above the ground. Average wind speeds usually tend to increase with height than level off, which is why wind turbines are usually installed as high above ground as possible.

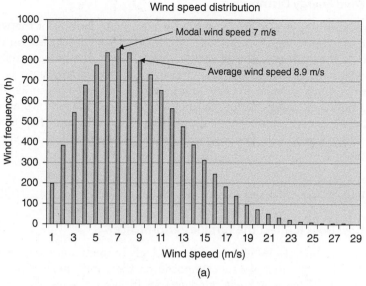

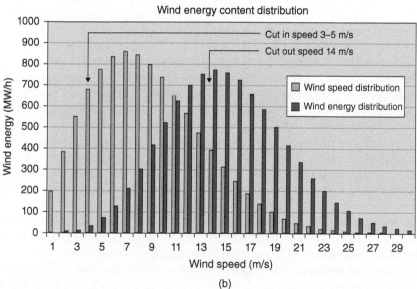

Figure 26.6 (a, b) Wind energy distribution and captured. Source: courtesy of Electropaedia – Energy Technologies [2].

26.3.6 Wind Energy Distribution

The histogram in Figure 26.6b shows the resulting distribution of the wind energy content superimposed on the Rayleigh wind speed distribution that caused it [2–5]. Evidently, not all of this wind energy can be captured by conventional wind turbines. The area between the two distributions is the actual useful wind, defined by the turbine cut-in and cut-out limits, thus capturing only half or less of the available wind energy. A wind turbine will deflect the wind, even before the wind reaches the rotor plane. This means that a wind turbine by Betz' law will never be able to capture all of the energy in the wind.

26.4 Wind Turbulence

26.4.1 How Does It Affect the Turbine Performance?

Atmospheric turbulence [6] is the set of seemingly random and continuously changing air motions that are superimposed on the wind's normal motion. Atmospheric turbulence impacts wind energy in several ways, specifically through power performance effects. It impacts on turbine loads, fatigue and wake effects, and noise propagation.

In the wind energy industry, turbulence is quantified with metric called turbulence intensity – the standard deviation of the horizontal wind speed divided by the average wind speed over some time period, typically 10 minutes. If the wind fluctuates rapidly, then the turbulence intensity will be high. Conversely, steady winds have lower turbulence intensity. Typical values of horizontal turbulence intensity, measured with a cup anemometer, range from 3% to 20%.

When two components of the horizontal-flow fluctuations are measured, they cannot be simply added together to estimate the total horizontal fluctuation. By focusing only on the horizontal standard deviation, the vertical contribution of turbulence is ignored, which can be very strong, particularly during the day, and it likely affects the wind turbine productivity.

For these reasons, turbulence kinetic energy (TKE) is often a more useful metric. Just as kinetic energy is one-half the product of mass and velocity squared, TKE is based on the squares of the variations in velocities.

The standard deviations of all three components of the flow are squared, summed together, and divided by two. Typical values of TKE range from $0.05 \, \mathrm{m^2/s^2}$ at night to $4 \, \mathrm{m^2/s^2}$ or greater during the day. Therefore, both the horizontal and vertical components of the flow are represented in TKE.

26.4.2 What Causes Atmospheric Turbulence?

In the lowest kilometer or two of the atmosphere, known as the planetary boundary layer, turbulence is generated by friction and interactions with the

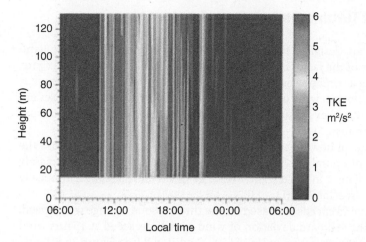

Figure 26.7 Turbulence daily chart. Source: American Meteorological Society.

surface of the earth. On sunny days, solar radiation heats the surface, which warms the air just above, thus causing thermals to rise from the surface. This convectively generated turbulence can be intense. Convective eddies can propagate throughout the entire daytime boundary layer up to a height of 2 km or more. Moving air near the ground slows via friction with the surface.

Faster air flowing over slower air causes wind shear, which, in turn, also causes turbulence. Within the lowest 200 m of the atmosphere, turbulence experiences a daily cycle, as seen in Figure 26.7. Intense daytime solar heating triggers convection, so maximum values occur between 10 a.m. and 6 p.m. In the evening, turbulence drops to low levels.

Characterizing turbulence over differing types of terrain is useful. Off-shore possess a smoother (water) surface than onshore. Exceptions include the occurrence of strong turbulence-producing events, such as hurricanes, tropical storms and deep tropical convection. On the other hand, wind flow in mountainous terrain is usually more turbulent than flow over flat terrain. Turbulence also tends to be higher near the ground surface – where it typically originates – than at higher altitudes.

Measurements of turbulence are often made by using instruments located within the flow or *in situ*, such as with cup, propeller or sonic anemometers. If three-dimensional anemometers are not employed, then the vertical component of turbulence is not measured. The IEC standard for turbine power performance measurements, IEC 61400-12-1, explicitly stipulates that only horizontal components in both directions are to be measured.

26.5 Wind Turbine Design Classification

Wind turbines are designed for the specific site conditions [7]. Turbine wind class is just one of the factors needing consideration during the complex process of planning a wind power plant. Wind classes determine which turbine is suitable for the normal wind conditions at a particular site. Turbine classes are determined by three key parameters – the average wind speed, turbulence, and extreme 50-year gust.

The knowledge of how turbulent a site is of crucial importance because the fatigue failures of a number of major components in a wind turbine are mainly caused by turbulence. In flat terrain, the wind speed increases logarithmically with the height (see Table 26.1).

The extreme wind speeds are based on the three second average wind speed. Turbulence is the standard deviation of wind speed measured at 15 m/s wind speed, based on the definition in IEC 61400-1 edition 2 (see Figure 26.8).

For US waters however, several hurricanes have already exceeded wind Class Ia with speeds above the 70 m/s (156 m/h), and efforts are being made to provide suitable standards.

As an illustration let us review two GE turbine 1.5 MW models built to different classes for different levels of turbulence.

(1) *A Class IIa wind turbine generator (WTG)*; It has a rotor diameter of 77 m and hub heights of 65 m. It is designed for average wind speed at hub height of 8.5 m/s with high turbulence of 18%.
(2) *A Class IIIb WTG*: It has a rotor diameter of 82.5 m and hub height of 80 m. The Class IIIb WTG is designed for lower wind speed (7.5 m/s at hub height) and lower turbulence (16%). The design loads will be smaller.

To increase the rating the blades must be larger and hub height taller.

Table 26.1 Wind classification.

Class	Wind class – turbulence higher 18%/lower 16%	Annual average wind speed (m/s), at hub height	Extreme 50 year gust in m/s and mi/h
Ia	High wind – higher turbulence 18%	10.0	70 (156)
Ib	High wind – lower turbulence 16%	10.0	70 (156)
IIa	Medium wind – higher turbulence 18%	8.5	59.5 (133)
IIb	Medium wind – lower turbulence 16%	8.5	59.5 (133)
IIIa	Low wind – higher turbulence 18%	7.5	52.5 (117)
IIIb	Low wind – lower turbulence 16%	7.5	52.5 (117)
IV		6.0	42.0 (94)

Figure 26.8 Wind standards.
Source: From [7].

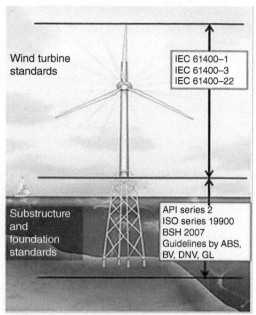

Wind turbine standards

IEC 61400–1
IEC 61400–3
IEC 61400–22

Substructure and foundation standards

API series 2
ISO series 19900
BSH 2007
Guidelines by ABS,
BV, DNV, GL

IEC, API, ISO etc. standards used to certify US offshore wind turbines

Therefore, bigger rotors of Class IIIb WTGs, therefore, capture more wind and yield higher capacity factors compared to Class I or II WTG.

Wind speed conversion from:

m/s →km/h, mph
m/s × 60 × 60/1000 → 3.6 × m/s → km/h
m/s × 60 × 60/1600 → 2.25 × m/s × mph

26.6 Blade Design for Optimum Energy Capture

There is a considerable difference in producing power at 12 m/s wind compared to 6 m/s.

$$P_{12} = P_6 \times (12/6)^3 = 8.0\, P_6$$

Therefore, for a speed change from 6 to 12 m/s, the output power increases by eightfold [2].

Modern, high-power wind turbines used by the utilities in the electricity grid, typically have blades with a cross section similar to the aerofoils used to provide the lift in aircraft wings and with the blade pitch that enhance the capability to increase the blade power.

Just as with aircraft wings, the angle of wind attack increases from 0° to a maximum of about 15° at which point the smooth laminar flow of the air over the blade ceases and the air flow over the blade becomes turbulent. Above this point, the lift force deteriorates rapidly while drag increases leading to a stall. For a given speed of rotation, the tangential velocity of sections of the blade increases along the length of the blade toward the tip so that the pitch of the blade must be twisted to maintain the same optimal angle of attack at all sections along the length of the blade.

Blade twist is thus optimized for a given wind speed. As the wind speed changes, however, the twist will no longer be optimum. To retain the optimum angle of attack as wind speed increases, a fixed pitch blade must increase its rotational speed accordingly. Otherwise, for fixed speed rotors; variable pitch blades must be used.

Yaw control: Wind turbines can only extract the maximum power from the available wind when the plane of rotation of the blades is perpendicular to the direction of the wind. The rotor mount must be free to rotate on its vertical axis and the installation must include some form of yaw control to turn the rotor into the wind.

Large turbine installations have automatic control systems with wind sensors to monitor the direction of the wind and a powered mechanism to drive the rotor into its optimum position. For small, lightweight installations, this is normally accomplished by adding a tail fin behind the rotor in line with its axis. Any lateral component of the wind will tend to push the side of the tail fin causing the rotor mount to turn until the fin is in line with the wind. When the rotor is facing into the wind, there will be no lateral force on the fin and the rotor will remain in position. Friction and inertia will tend to hold it in position so that it does not follow small disturbances.

26.7 Individual Pitch (Blade) Control (IPC)

How can designers build wind turbines with longer lifetimes? Individual pitch control (IPC) plays a key role in compensating loads (adjust wind loads), in particular for larger wind turbines. Recently developed wind turbines are variable speed turbines capable of adapting to various wind conditions. So what is IPC? Any pitch control system allows control of the turbine speed and consequently the power output. It also acts as a brake, stopping the rotor by turning the blades in case of strong winds. Moreover, pitch control, especially an IPC system, has a role in reducing fatigue loads on the turbine structures. Pitch control means the turning of rotor blades between 0° and 90°. When wind speeds are below rated power, typically below 12 m/s, the rotor blades are turned fully toward the wind, which means that the pitch is positioned at 0°.

At increasing wind speeds, the pitch of the blades is controlled in order to limit the power output of the turbine to its nominal value. When wind speeds reach a predefined threshold, typically 25–28 m/s, the turbine stops power production by turning the blades to a 90° position. A reduction of fatigue loads has two considerable advantages: It allows lighter tower and blade designs and translates into longer lifetimes of wind turbines.

Collective pitch control adjusts the pitch of all rotor blades to the same angle at the same time. In contrast, IPC dynamically and individually adjusts the pitch of each rotor blade. This is because wind is highly turbulent flow and the wind speed is proportional to the height from the ground. Therefore, each blade experiences different loads at different rotation positions. Based on current individual loads, this pitch adjustment is carried out in real time, thus benefiting from the reduction of fatigue loads on the rotor blades, the hub, and mainframe and tower structures. In order to compensate these loads, especially symmetric loads caused by inhomogeneous wind fields, the pitch of each rotor blade has to be adjusted independently from the other blades.

Larger turbines in wind parks in remote areas with difficult climatic conditions assume higher expectations on reliability, flexibility, and predictability of electrical power generation. When turbines are getting larger, load reduction, especially for asymmetric loads caused by inhomogeneous wind fields, becomes more and more important. Consequently, the manufacturer anticipates that IPC will play an important role as the most common technology capable of compensating asymmetric loads and reducing fatigue.

26.8 Wind Turbine Design Limits

For safety and efficiency reasons, wind turbines are subject to operating limits [2] depending on the wind conditions and the system design, as follows:

Cut-in wind speed: It is the minimum wind velocity below which no useful power output can be produced from wind turbine, typically between 4 and 5 m/s (10 and 14 km/h, 7 and 9 mph).

Rated wind speed: Also noted as nameplate capacity, this is the lowest wind velocity at which the turbine develops its full power. This corresponds to the maximum, safe electrical generating capacity that the associated electrical generator can handle, in other words the generator's rated electrical power output. The rated wind speed is typically about 12–15 m/s (54 km/h, 34 mph), which is about double the expected average speed of the wind. To keep the turbine operating at nominal rating with wind speeds above the rated wind speed, control systems may be used to vary the pitch of the turbine blades, reducing the rotation speed of the rotor and thus limiting the mechanical power applied to the generator so that the electrical output remains constant.

Though the turbine works with winds speeds right up to the cut-out wind speed, its efficiency is automatically reduced at speeds above the rated speed so that it captures less of the available wind energy in order to protect the generator. While it would be possible to use larger generators to extract full power from the wind at speeds over the rated wind speed, this would not normally be economical because of the low probability of occurrence of wind speeds above the rated wind speed.

Cut-out wind speed: This is the maximum safe working wind speed and the speed at which the wind turbine is designed to be shut down by applying brakes to prevent damage to the system. In addition to electrical or mechanical brakes, the turbine may be slowed down by stalling or furling.

Stalling: This is a self-correcting or passive strategy used with fixed speed wind turbines. As the wind speed increases, so does the wind angle of attack until it reaches its stalling angle. However, increasing the angle of attack also increases the effective cross section of the blade face-on to the wind. A fully stalled turbine blade, when stopped, has the flat side of the blade facing directly into the wind.

Furling or feathering: This is a technique derived from sailing in which the pitch control of the blades is used to decrease the angle of attack which in turn reduces the "lift" on the blades as well as the effective cross section of the aerofoil facing into the wind. A fully furled turbine blade, when stopped, has the edge of the blade facing into the wind reducing the wind force and stresses on the blade.

The cut-out speed is specified to be as high as possible consistent with safety requirements and practicality in order to capture as much as possible of the available wind energy over the full spectrum of expected wind speeds (see Figure 26.6 noted earlier). A cut-out speed of 25 m/s (90 km/h, 56 mph) is typical for very large turbines.

Survival wind speed: This is the maximum wind speed that a given wind turbine is designed to withstand above which it cannot survive. The survival speed of commercial wind turbines is in the range of 50 m/s (180 km/h, 112 mph) to 72 m/s (259 km/h, 161 mph). The most common survival speed is 60 m/s (216 km/h, 134 mph). The safe survival speed depends on local wind conditions and is usually regulated by national safety standards.

Wind turbine noise levels: Sound levels from wind turbines typically is about 55 dB (A) when measured at a distance of about 100 m; this is the same level of sound as you can expect from a car traveling at 60 km/h at the same distance.

Sound levels always decline with distance; the rate of decline (with some exceptions) follows the *inverse-square-law* – twice the distance a quarter the sound, three times the distance a ninth the sound, etc.

26.9 Wind Turbine Components

Basically, there are four major parts to a wind turbine:

- Rotor with hub and blades,
- Nacelle at the top of tower, which contains the generator, variable speed drive and gearbox if used,
- Jaw mechanism,
- Supporting tower.

The rotor blades in wind turbines are composites, made of polyester and epoxy, with glass fibers used as the reinforcing material. The typical modern three-bladed wind turbine can be efficient even if the rotor blades only cover, say, 3% of the swept surface of the rotor because these machines generally have a relatively high rotational speed.

Materials for wind turbine parts (other than the rotor blades, the rotor hub, gearbox, frame, and tower) are made of steel.

Pre-stressed reinforced concrete has been increasingly used for the tower, in particular for the offshore applications.

The power from the rotation of the rotor must be transferred through the drive train, main shaft, the gearbox, and the high speed shaft to a 50 or 60 Hz AC, 3 ph. to the generator. The single gear gearbox brings the shaft speed down to about 20 rpm of the blade rotor.

The mass of the rotor of the generator has to be roughly proportional to the amount of torque it has to handle. So a directly driven generator will be very heavy and expensive. A single gear gearbox converts between slow speed high torque power to high speed 1500 (1800) rpm and low torque power of the generator. This way the generator shaft and weight is significantly reduced.

Generating voltage for the large turbines is 690 V (Siemens 6 MW offshore unit uses 690 V_{ac}).

The power is subsequently sent through a transformer placed next to the wind turbine, or inside the tower, to raise the voltage to somewhere between 11 and 35 kV, depending on the local electrical grid and the number of turbine units.

26.10 Generators Used with Wind Turbines

Grid connected systems are built for average wind speeds 5.5 m/s on land and 6.5 m/s offshore where wind turbulence is lower and wind speeds are higher. While offshore plants benefit from higher sustainable wind speeds [2], their construction and maintenance costs are higher.

Large rotor blades are necessary to intercept the maximum air stream, but these give rise to high tip speeds. The tip speeds, however, must be limited mainly because of unacceptable noise levels and blade tip damage. This results in low rotation speeds as low as 10–20 rpm for large wind turbines. The operating speed of the generator is however much higher, typically 1200 rpm for a 6 pole, 60 Hz generator. Consequently, a gearbox is used to increase the shaft speed to drive the generator at the fixed synchronous speed corresponding the grid frequency.

26.10.1   Fixed Speed Wind Turbine Generators

A typical fixed speed system employs a rotor with three variable pitch blades that are controlled automatically to maintain a fixed rotation speed for any wind speed [2].

The rotor drives a synchronous generator through a gear box in an assembly housed in a nacelle (see Figure 26.9).

Fixed speed systems may however suffer excessive mechanical stresses. Since they are required to maintain a fixed speed regardless of the wind speed, there is no "give" in the mechanism to absorb gusty wind forces resulting in high torque, high stresses, and excessive wear and tear on the gear box.

At the same time, the reaction time of these mechanical systems can be in the range of tens of milliseconds so that each time a burst of wind hits the turbine, a rapid fluctuation of electrical output power can be observed.

26.10.2   Variable Speed Wind Turbine Generators

Variable speed wind turbines handle turbulences better and can capture 8–15% more of the wind's energy than constant speed machines [2].

For these reasons, variable speed systems are preferred over the fixed speed systems.

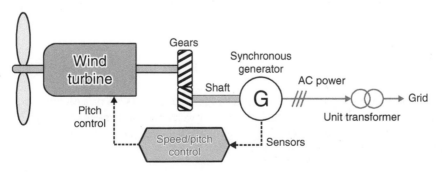

Figure 26.9 Fixed speed turbine generator.

A variable speed generator is better equipped to cope with stormy wind conditions, because its rotor can speed up or slow down to absorb the forces when bursts of wind suddenly increase the torque on the system. Having less stress on the gear box is a great advantage of these units. The electronic control systems will keep the generator's output frequency constant during these fluctuating wind conditions.

26.10.3 Synchronous Generator with In-Line Frequency Control

Rather than controlling the turbine rotation speed to obtain a fixed frequency synchronized with the grid from a synchronous generator, the rotor and turbine can be run at a variable speed corresponding to the prevailing wind conditions [2]. This will produce a varying frequency output from the generator synchronized with the drive shaft rotation speed. This output can then be rectified in the generator side of an AC → DC → AC converter with grid frequency ready for synchronizing with the grid (see Figure 26.10).

The grid side converter can also be used to provide reactive power (VARs) to the grid for power factor control and voltage regulation by varying the firing angle of the thyristor switching in the inverter and thus the phase off the output current with respect to the voltage.

The range of wind speeds over which the system can be operated can be extended and mechanical safety controls can be incorporated by means of an optional speed control system based on pitch control of the rotor vanes as used in the fixed speed systems described earlier.

A major drawback of this system is that the components and the electronic control circuits in the frequency converter must be dimensioned to carry the full generator power.

26.10.4 Doubly Fed Induction (Asynchronous) Generator – DFIG

The doubly fed induction generator (DFIG) with asynchronous generator overcomes difficulties of the other models [2]. DFIG technology is currently the

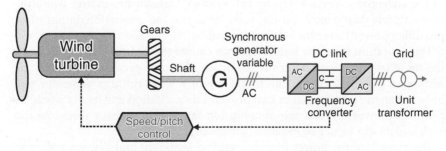

Figure 26.10 In-line conversion speed conversion.

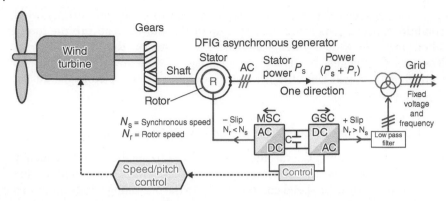

Figure 26.11 Asynchronous DFIG model.

preferred method of operation for the wind turbines and the most expensive (see Figure 26.11). The doubly fed asynchronous induction generator gets its excitation current from the grid through the stator windings and has limited control over its output voltage and frequency. It also permits a second excitation current input, through slip rings to a wound rotor permitting greater control over the generator output.

The DFIG system consists of a three-phase wound rotor generator with its stator windings fed from the grid and its rotor windings fed via a back-to-back converter system in a bidirectional feedback loop taking power either from the grid to the generator or from the generator to the grid.

DFIG generator operating principle is based on the feedback control system that monitors the stator output voltage and frequency and provides error signals if these are different from the grid standards. The frequency error is equal to the generator slip frequency (over the synchronizing frequency) and is equivalent to the difference between the synchronous speed and the actual shaft speed of the machine.

The generator flow diagram is typical for one WTG, gearbox driven a DFIG type asynchronous (induction) generator, with reactive power support feeding to the collector grid via a step-up 690 V/33 kV transformer. Active wind turbine controls (blade pitch, turbine yaw) maximize the generation output while providing power factor (or voltage) control.

The excitation from the stator windings causes the generator to act in much the same way as a basic squirrel cage or wound rotor generator. Without the additional rotor excitation, the frequency of a slow running generator will be less than the grid frequency that provides its excitation and its slip would be positive. Conversely, if it was running too fast, the frequency would be too high and its slip would be negative (generating).

The rotor absorbs power from the grid to speed up and delivers power to the grid in order to slow down. When the machine is running synchronously, the frequency of the combined stator and rotor excitation matches the grid frequency, there is no slip and the machine can be synchronized with the grid.

26.10.4.1 Converter – Grid Side Converter (GSC)

It carries current at the grid frequency. It is an AC → DC converter circuit used to provide a regulated DC voltage to the inverter in the machine side converter (MSC). It maintains a constant DC link voltage. A capacitor is connected across the DC link between the two converters and acts as an energy storage unit. In the opposite direction, the GSC inverter delivers power to the grid with the grid regulated frequency and voltage.

As with the in-line converter described earlier, by adjusting the timing of the GSC inverter switching, the GSC converter also provides variable reactive power output to counterbalance the reactive power drawn from the grid enabling power factor correction as in the in-line frequency control system described earlier.

26.10.4.2 Machine Side Converter (MSC)

It carries current at slip frequency. It is a DC → AC inverter that is used to provide variable AC voltage and frequency to the rotor to control the torque and speed of the machine.

When the generator is running too slowly, its frequency will be too low so that it is essentially motoring. The MSC takes DC power from the DC link and provides AC output power at the slip frequency to the rotor to eliminate its motoring slip and thus increase its speed. If the rotor is running too fast causing, the generator frequency to be too high, the MSC extracts AC power from the rotor at the slip frequency causing it to slow down, reducing the generator slip, and converts the rotor output to DC, passing it through the DC link to the GSC where it is converted to the fixed grid voltage and frequency and is fed to the grid.

26.10.4.3 DFIG Control, Frequency

The frequency of the rotor currents induced by transformer action from the stator is the same as the slip frequency and this is equivalent to the frequency error signal in the feedback loop.

The additional direct excitation of the rotor adds a second set of controlled currents to the currents already induced in the rotor by transformer action from the stator. These additional currents affect the rotation speed of the rotor in the same way as the stator-induced currents, producing an additional driving torque on the rotor except that the additional rotor currents are independent of the speed of the rotor. The frequency of the control current supplied by the MSC can be precisely controlled to match and thus neutralize the slip frequency so that, with zero slip, the generator rotates at the synchronous frequency determined by the grid. The greater the slip, the greater the compensating frequency required.

The control system has to respond to both positive (motor) slip and negative (generator) slip.

To increase the speed of a slow running rotor, the phase sequence of the rotor windings is set so that the rotor magnetic field is in the same direction as the generator rotor producing negative slip to counteract and thus neutralize the rotor's positive slip. To reduce the rotor speed, the phase sequence of the rotor windings is set in opposite direction from the generator's rotation producing positive slip to counteract the rotor's negative slip.

When operating at synchronous speed, the rotor current will be DC current and there will be no slip and no power flow through the rotor.

26.10.4.4 Voltage

The generator output voltage is determined by the magnitude of the excitation current supplied to the rotor and this can be adjusted by means of the rotor input voltage provided by the MSC. A chopper or pulse width modulator (PWM) is used to generate the variable DC control voltage necessary. The converter feedback controls thus enable the excitation current to be regulated by the MSC to neutralize the voltage error signal and thus obtain a constant bus voltage matching the grid voltage.

26.10.4.5 DFIG Performance

The DFIG system provides regulated power tied to the grid frequency and voltage when driven by varying levels of torque from the wind. Typical speed control range is ±30% of synchronous speed. For a greater speed control range, it may be necessary to implement IPC on the wind turbine's rotor vanes.

The generator power flow is shared by the stator and the rotor with 70% or more coming from the stator. The feedback loop only carries the slip power that is between 20 and 30% of the total.

Because of the reduced power flowing through the converters, compared with the in-line control system described earlier, the DFIG converters can be implemented with less expensive lower power components.

The DFIG machine can produce up to twice the power of a similar sized singly fed machine while incurring similar losses; however, the losses in the electronic controls must be added to this. Nevertheless, the DFIG machine efficiency is better than that of a singly fed machine.

26.11   Turbine Sizes

Wind turbines are available in a variety of sizes, and therefore power ratings. The largest machine has blades that span more than the length of a football field and 20 stories high. The turbines are getting larger in particular offshores (see Figure 26.12).

In year 2010, Enercon installed several 7.6 MW units in Belgium and Austria, 135 m tower, 126 rotor for a total blade height of 200 m [8, 9].

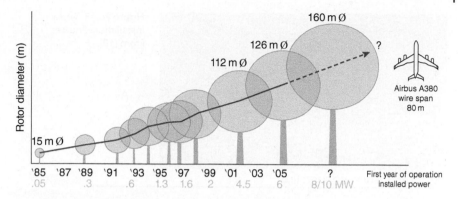

Figure 26.12 Historical growth. Source: Joe Beurskens.

1.5 MW GE1.5 sle Wind turbine for onshore

Tower 80, 85 m high.

Rotor 77 m diameter swept area 4657 m²

Assembly (blades and hub) weights 22 tons

Nominal wind: 12 m/s cut-in 3.5 m/s, cut-out 25 m/s

Active blade pitch

Braking – hydraulic control system: Programmable Logic Controller (PLC), remote control

Nacelle, with generator 52 tons).

Concrete base is constructed with 26 tons of reinforcing steel and 190 cubic meters of concrete.

The base is 15 m in diameter

Lightning protection: on blade tips

Cold weather package: cut-off at −30 °C.

2 MW Vestas for onshore (Figure 26.13)

Rated	1.8/2 MW
Wind class	IEC IIA
Speed	cut in–cut out 4–25 m/s, re-cut in 23 m/s
Operating temperature	−20 to 40 °C
With winter package	−30 °C
Sound	104 dbA
Tower height	80–95–105 m
Blade	44 m

Figure 26.13 Vesta installation. Source: From [10].

V100–1.8/2.0 MW™ IEC S/IEC IIIA

Rotor diameter	90 m
Swept area	6362 m^2
Generator	50 Hz, 4 pole, 60 Hz, 6 pole with slip rings
Transmission	Gearbox
Sales	>20 000 units installed

7.5–8 MW for offshore

Tower	135 m, tip: 198 m
Blade	80 m in two sections for land transport
Rotor diameter of 127 m	
Tower	35 tapering concrete rings
Base	14.5 m, top: 4.1 m

3.8 MW largest vertical-axis (Figure 26.14)

Le Nordais wind farm in Cap-Chat, Quebec
Tower height 110 m
Farm capacity 100 MW from 133 turbines
Single vertical axis, the rest horizontal

Mitsubishi: 2.4 MW, 12.5 m/s, IEC Class IIA

System	Variable speed, DFIG
Hub height	70, 80, 90, 100 m
Diameter/swept area	92m/6647 m^2
Rotor speed	14.5–16.3 rpm, variable

Figure 26.14 Vertical axis model. Source: From [10].

Pitch	Individual blades
Blade material	GFRP
Wind cut in/out	3/25 m/s
Survival speed	60 m/s (135 mi/h)
Gear box	three stages (one planetary, two parallel)
Gear box ratio	82.7
Generator speed	690 V, 1200/1340 rpm
Brake	Pitch control, shaft maintenance brake

26.12 Building a Wind Farm

There is more to selecting and installing a wind turbine. Typically, the WTGs operating in a farm are strung up to 10 units on one collector cable. Several collector cables are then connected to the collector switchgear to a common output transformer at 33 kV/grid voltage.

The wind farm will likely include 10–100 units placed in the locations to suit the wind flow and the roads. Typically, the units will generate power at 690 V. The output of each unit is then transformed to 33 kV in step up transformer at the tower base. The step up transformers are then connected by collector cables in several strings at 33 kV to the main switchyard, which is the point of interconnection to the power utility. Each unit may also have a capacitor bank at its location for supplying the reactive power for the generator and

Table 26.2 Cable ampacity.

WTG	At 690 V	At 33 kV
MW	A	A
1	850	18
2	1700	36
5	4250	90
7	6000	130

transformer magnetizing needs. The strings may also be looped together to allow for switching of the units on several multiple paths to the main substation.

Cables used for the wind generators are of large ampacity size as shown in Table 26.2 for individual units. The cable size will also depend on the location of the source of reactive power, imported from grid or received from the local capacitor banks.

Wind power has become a major component in the daily load cycle, thus a failure of the wind farm as a block of power may have a major cascading effect on the overall power system. To prevent this happening and mitigate the impact of the wind on the other generation, the wind farm must include additional equipment, which is discussed further in this chapter.

Additional capacitor banks are also installed at the main substation for MVAR requirements in the collector cables and main transformers. Furthermore, the main substation will include dynamic switching hardware to regulate the MVAR flows for continuous voltage control (LVRT) for line faults and ride through capability for voltage rises (HVRT).

26.12.1 Grid Integration Issues

26.12.1.1 Wind Farm Impacts on Utilities

Wind is an intermittent and highly variable generation resource. However, it is not sufficiently dependable to be considered as firm power. When available, it displaces the energy generated by fossil fuels, resulting in fuel savings and reduced emissions. Each type of generation has distinctive operating characteristics and responds differently to the power integration [9, 11] with the wind resource.

The following are some Impacts of wind farms on the utility generation, transmission and distribution systems:

- High VAR consumption (induction generators)
- Wind turbine – voltage sags due to inrush current during start/stop
- Voltage fluctuations and voltage regulations

- Possibility of tripping off due to either sudden low or high voltage
- Changes in wind speed can cause sudden power output changes
- Frequency issues
- Voltage flicker due to tower shadowing effects or power output changes
- Harmonics

26.12.1.2 Load Variability

Wind adds additional source of variability to the already variable nature of the power demand system. The wind resource affects the base generating plant operation like responding to a variable load, which may vary in a matter of few seconds, hours, and days.

Different types of generators respond faster or slower to load variations. Diesel generation, for instance, will respond faster than large thermal plants. This differential in response time affects the ratio of the generation mix for the base and peaking generation in operation.

Electric power operation requires that generation and consumption be balanced essentially instantaneously, and continuously. The fast load transients are mostly dealt with by the inertias of all the operating generating units. Therefore, having sufficient amount of running spinning reserve on the common bus is the key to adapting to the wind resource.

To analyze the additional variation caused by wind generators, every change in wind output does not need be matched one-for-one by a change in another generating unit moving in the opposite direction, but rather matched on an overall system basis. Wind simply adds to the existing system variation. Minute-to-minute fluctuations in individual loads are largely uncorrelated, providing an aggregation benefit. Similarly, minute-to-minute fluctuations in wind output are also uncorrelated with load. This implies that the additional variations that wind farms add to the system do not add linearly. Diversity of fluctuations happens in magnitude, time, and direction.

In an isolated power system that includes some wind generation, the installed generating thermal plant is likely less effective in drawing on the benefits of a large diverse interconnected power system which operates with multiple generators located in the various geographical locations and with plenty of spinning reserve available to counteract fluctuations in system voltage and frequency.

26.12.1.3 Wind Turbine Start

If a large wind turbine is switched on to the grid with a regular switch, the power users would experience a brownout due to the current flow required to magnetize the generator. This is followed by a power peak due to the generator current surging into the grid. Another unpleasant side effect would be extra wear on the gearbox, appearing like you slammed on the mechanical brake of the turbine.

To prevent this situation, modern wind turbines are soft starting, i.e. they connect and disconnect gradually to the grid using electronic converter contactors and circuit breakers. The electronics is automatically shorted with bypass switches as the turbine reaches 90% speed.

26.12.1.4 Voltage Requirements

In order to operate with the grid, the utilities have established specific interconnection requirements [9, 11]. Wind operators must

- be able to deal with the variations in voltage on transmission grid (±10%). A wind farm may require on load tap changers (OnLTCs) on power transformer(s), which adds cost to the collector grid design.
- handle large capacitor or reactor banks switching on the grid, which may cause gearbox damage due to sudden voltage changes on transmission grid.
- handle sudden voltage sag or rise on transmission grid. Utilities will demand the wind turbine owners provide the units with ride through capability for major system disturbances.

26.12.1.5 Reactive Power Requirements

Each WTG produces MW output, but almost no reactive MVARs needed for its magnetizing needs [9], and for the farm auxiliary distribution equipment. How much reactive power is needed to make the wind farm operate on the system? It depends on the length of the collector cables and the transformer impedances in the farm distribution system. On a typical farm installation, the collector cables tend to be long (see Figure 26.15).

A 159 MW wind farm in Australia connected on two collector cable strings operates satisfactorily with an addition of switchable 47 MVARs of capacitors located within the farm. The farm distribution system consumes $11 + 22 + 24 = 47$ MVARs for the unit transformers, collector cables, and two main transformers, respectively. This reactive power does not have to be imported on the HV line from the utility and cause further MW losses on the line. The switching system maintains a power factor at about 96%.

The MVAR requirement can be calculated as part of a power study for the worst-case scenario. Based on the aforementioned, a suitable MVAR number would be around 35–40% of the MW installed capacity, to also allow for a short time overloading capability.

The reactive power flow is extremely variable, dependent on the wind resource and MW production, thus greatly affecting the voltage profile of the circuit. Even on a calm day, the farm requires a significant amount of MVARs. For that reason, the switchable capacitors are located close to the source (Nacelle) to limit the flow of MVARs on the cables and to cause additional line MW losses. OnLTCs furnished on the main substation transformers cope with the system voltage variations.

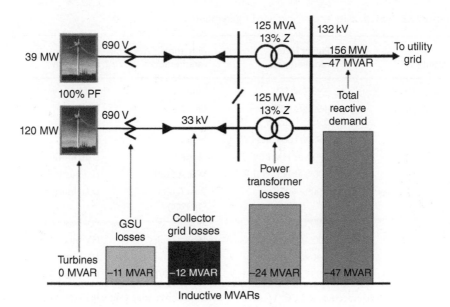

Figure 26.15 Typical farm project. MVARs required.

26.12.2 Utility Stability Requirements for Wind Farms

The fast development of wind power generation brings in stricter requirements for wind turbines for their integration into the large grids, to mitigate the impacts on the power quality, and dynamic stability of the grids. This generally applies for the units >1 MW. It is in the interest of everyone that wind farms stay connected to the grid and stabilize the grid voltage in case of grid faults.

The issues are the management of steady-state active and reactive power feed in capability, continuously acting voltage control and fault ride through capability (LVRT, HVRT).

Weak grid coupled with wind gust fluctuations, intermittent power production, lightning strikes can cause sudden momentary dips in voltage, resulting in fluctuating grid voltage and wind turbine trips, thus creating even greater fluctuations and loss of production.

In addition to the individual turbine controllers to adjust the speed to reduce the torque transients, control the blade pitch to capture more wind, adjusting jaw direction and MVAR resources, the farm must also have the energy integration solutions that allow power plants to stay on line. With respect to the grid faults, the voltage rise is less common than the voltage sag, but it is usually associated with a system fault conditions. The requirements by the utilities on the wind farms are significant, but not uniform as shown in Table 26.3 with

Table 26.3 Typical utility interconnection requirements.

Utility	Power factor requirements	Ride thru requirements	Voltage regulation
1. ERCOT	±95% at HV	None	HV bus
2. Alberta	+90%/−95% at disturbance V	Eon Netz	HV bus
3. Exelon	+95%/100% at HV	None	HV bus
4. Pacificorp	100% at HV POI	None	Distribution bus
5. Xcel	100% at HV POI R	Worst faults	HV bus R
6. Sask Power	+90%/−95%	Post fault recovery	Distribution bus
7. SDGE	+90%/−95%	WECC	HV bus
8. HELCO	100%/−88% at HV	Worst faults	HV bus
9. IESO	+90%/−95%	Worst fault	HV bus
10. South Australia	±93% at HV POI	HVRT/LVRT	HV bus

POI, point of interconnection; WECC, Western Electricity Coordinating Council; LVRT, low voltage ride through; HVRT, high voltage ride through.
Source: From [9].

respect to the following operational requirements at the main substation point of interface (POI):

(1) Power factor, generator magnetizing, MVARs need for the cables and transformers,
(2) Ride through capability: HVRT, High voltage ride through,
(3) LVRT: Low voltage ride through,
(4) Voltage regulations.

The requirements will vary depending on the location of the POI within the grid. Here are some of the recommendations by the USA – Joint NERC/FERC committee for the Interconnection Standards for Wind Energy – December 2005.

(1) Dynamic power factor control of ±95% control at the point of interconnection.
(2) Voltage regulation capability.
(3) HVRT for grid voltage rises at voltages up to 130–140%.
(4) LVRT down to zero remaining voltage
(5) Three phase faults at high side of power transformer cleared in four to nine cycles.
(6) Single line-ground faults with delayed clearing.
(7) Need to be supported by case-specific studies.

(8) Additional utility requirements.

(9) Wind farm dynamic compensation system to meet postfault voltage recovery targets.

(10) Remain on line during emergency conditions (voltage range 0.90–1.10 pu).

26.12.2.1 HVRT Capability

A voltage rise can occur due to a single line-to-ground fault on the system, which can result in a temporary voltage rise on the un-faulted phases. Voltage rise can also be generated after sudden load drops. The abrupt interruption of current can generate a large voltage change. Switching on large capacitor banks may also cause a voltage rise.

The wind turbines with DFIG are more affected during fault conditions because the stator is directly connected to the grid while the rotor is connected and fed through a converter and DC link, which enables its control. Without having a provision for mitigation of overvoltage due to grid faults there is a danger of a severe damage to the units. In order to avoid these problems, the DIFG units must be furnished with HVRT capability.

Traditionally, wind turbines were disconnected from the grid in case of an abnormal grid voltage. In some areas, the amount of wind energy has increased essentially to the point that disconnection of an entire wind farm would severely affect the overall system stability. Transmission system companies now demand that wind turbines must offer ride through capability under the abnormal conditions.

The HVRT hardware typically incorporates power electronics-based controllers to enhance the unit controllability and increase power transfer capability. This is accomplished by rapid exchange of active and reactive power with the power system independently. The equipment includes series and shunt-connected controllers placed between the unit transformer and the group transformer to inject reactive shunt current into the grid system to lower the voltage swell at the wind turbine.

26.12.2.2 LVRT Capability

The equipment is added to the turbine systems to enhance the transient stability of wind turbines during an external short-circuit fault and allow them to ride through the disturbance. Voltage collapse following a fault clearance is most pronounced in DFIG, which have two sets of powered magnetic windings. In a wind farm having many distributed generators, trip of one generator may cause a chain reaction and lead to a cascading failure effect.

Modern large-scale wind turbines, typically 1 MW and larger, are normally required to include LVRT systems that allow them to operate through such an event, and thereby "ride through" the voltage dip.

Depending on the code application the protection may, during and after the dip, be required to

- disconnect temporarily from the grid, but reconnect and continue operation after the dip
- stay operational and not disconnect from the grid
- remain connected and support the grid with reactive power

The DFIG stator is directly connected to a grid, while its rotor is connected to a grid through a back-to-back converter. When a fault occurs, currents of 3.0–5.0 times nominal are induced in the rotor and causing torque oscillations in the machine, which may damage the DFIG system.

The solution to the LVRT problems is to insert a resistance (called crowbar) to bypass the generator rotor circuit to significantly lower the fault current flowing through the generator and then remove it when the system voltage recovers. A disadvantage of this approach is that the DFIG loses its controllability when the crowbar is inserted. This may lead to the DFIG absorbing a large amount of reactive power from the grid, causing further grid voltage degradation.

To provide for the controllability while protecting the DFIG, other strategies employ the use of superconducting current limiter (SCL) in the rotor circuit. It reduces fault current level at the stator side and improves the fault ride-through capability of the system. SCL is a superconducting resistive limiter, presenting zero resistance during normal and high resistance during fault conditions.

The main advantages of proposed solutions using SCL in the rotor circuit compared with other technologies employed on DFIG LVRT improvement are

- The SCL has no influence on the power generation of the DFIG during normal operation;
- Higher rotor current limitation during critical period of voltage dip is achieved;
- The SCL effect automatically appears in the rotor circuit faster than other solutions;
- The device recovers itself to the superconducting state without any external command.

26.12.3 Managing Variability and Voltage Regulation at Wind Farms

For up to 10% wind penetration in a given hour of the day, there is generally no concern. There is sufficient flexibility built into the system including spinning reserve to cope with varying loads [12]. When wind is generating >10% of the electricity that the system is delivering in a given hour, its variability is an issue that has to be addressed. Some of it can be resolved with wind forecasting, and an appropriately flexible approach to generation and load management.

Depending on the time frame of the wind oscillations, experienced operating utilities deal with the wind issues by three methods: regulation, load following, and unit commitment. The big power plant operator can let the generators automatically handle the instantaneous load swings (regulation), manually adjust generation (fueling) in anticipation of daily load changes (load following), and prepare to add/remove additional units for an anticipated load/generation change over the next few hours or days (unit commitment).

In an isolated power system, the load following and unit commitment are of less value as tools for dealing with the load changes or wind resource. Since all the base load units are in operation at all times, and the load is relatively constant the plant is in operation with a substantial amount of spinning reserve.

26.12.3.1 Impact On Cost

Wind integration impacts the operating cost. The utilities, according to National Renewable Energy Laboratory (NREL) estimate these costs are relatively low, ranging from 0.1 to 0.4 cents/kWh of wind power, mostly due to overcompensating with controllable (spinning reserve) generators. This is mostly caused by inadequate wind forecasting.

26.12.3.2 Impact On Stability

Though at first glance, the impact of wind on system stability appears to be a concern, the following factors indicate the situation can be managed:

- Interconnected power systems routinely cope with varying and uncertain demand and handle unexpected system disturbances (loss of transmission or generation).
- The plant isolated power system will have to cope with the wind component with its own generators as there are no interconnected power resources.
- Ability of utilities to forecast wind power output in both hourly and day ahead time frames if there is a seasonal pattern of the wind.
- Multiple wind farm generators in different locations can provide a smoothing effect owing to the diversity of wind profiles across the wind farming areas.

26.12.4 Methods of Transient Regulation of Power Generation

26.12.4.1 AGC

The most important method of addressing the wind issues at the plant isolated power system is the regulation by automatic generator control (AGC). AGC generates signals that cause an increase/decrease in output to maintain balance between the load and generation. The regulation by AGC occurs in a time scale ranging from several seconds to 10 minutes. These movements in loads and generation are not typically predicted or scheduled in advance. To meet these fluctuations, AGC must be properly programmed to anticipate the rate of change in the operating conditions and sufficient generation must be online and synchronized to provide enough flexibility to respond. The regulation is done by fast acting governors initiating more or less fuel input as needed.

Regulation for fast responses is a capacity service and does not involve any net energy exchange, but simply the use of generator inertia.

26.12.4.2 Load Following

This technique covers approximately the time period of 10 minutes to several hours. In this time scale, economic dispatch decisions are made in response to the trend in demand to establish proper plant operating conditions. The operator is responsible for ensuring that sufficient capacity is always available to meet these large, relatively slow swings in load demand. If the wind generation is also increasing and can be forecast reliably, the system operator would need fewer load following thermal resources. Load following may go in the other direction if the wind output drops while the load increases. Having to operate base load units at lower power to be able to accommodate the unpredictable wind results in higher power losses for the thermal plant.

26.12.4.3 Unit Allocation

This activity can range from several hours to several days, depending on the load change and the type of generation available to absorb the change depending on the wind pattern. The coal-fired boilers have slower transient characteristics to respond to load changes. While diesel generators and gas turbines are fast responding, boilers fueled by coal are slow to respond, in start up or shut down. To economically operate the system, many forward-looking decisions must be made to ensure sufficient generation is available to meet the loads plus a spinning reserve margin, without scheduling more operating capacity than needed.

26.12.4.4 Reactive Power Charges

Most wind turbines operate with induction generators. These generators cannot self-excite. They require input current from the grid to magnetize the unit. Grid utilities usually apply a tariff charge for this reactive power. Furthermore, the grid companies require that wind turbines be equipped with switchable capacitor banks to partly compensate and reduce the current inflows.

26.12.4.5 Availability

The figures for annual energy output assume that wind turbines are operational and ready to run all the time. In practice, however, wind turbines need servicing and inspection once every six months to ensure they remain safe. In addition, component failures and accidents (such as lightning strikes) may disable wind turbines.

Statistics show that the reputable turbine manufacturers consistently achieve availability factors above 96%, i.e. the machines are ready to run more than 96% of the time. Total energy output is generally curtailed <2%, since wind turbines are never serviced during high winds.

The availability factor is therefore usually ignored when doing economic calculations, since other uncertainties like wind variability are far larger.

26.13   Wind Energy in Cold Climates

The operation of wind turbines in a cold climate [13] such as Canada's involves additional challenges not present in warmer locations, such as the following

- Accumulation of ice on wind turbine blades resulting in reduced power output and increased rotor loads.
- Cold weather shutdown to prevent equipment failure.
- Limited or reduced access for maintenance activities.

Wind turbine manufacturers are increasingly recognizing the impacts of cold climate operation and are building turbines better equipped to handle winter conditions. With the installation of "cold weather packages" that provide heating to turbine components such as the gearbox, yaw, and pitch motors and battery, some turbines can operate in temperatures down to −30 °C.

Various types of rotor blade de-icing and anti-icing mechanisms, such as heating and water-resistant coatings are currently being employed, as well as operational strategies to limit ice accumulation. Based on actual measurements, icing can occur up to 20% of the time between the months of November and April. Wind turbines must therefore be able to sustain at least limited icing without incurring damage that would prevent normal operation. However, it is not clear which cold climate solutions deliver the highest performance while still being cost-effective, making this an active area of research. The frequency, severity, and type of icing event varies significantly by region, meaning that mitigation methods need to be tailored to local conditions. Different manufacturers' turbines may also be more or less suited to particular solutions. Improving the accuracy of ice detection and forecasting systems represent another avenue for minimizing cold climate losses.

The goals of Canmet ENERGY-Ottawa (CE-O)'s cold climate research program are to analyze the impact of cold climate operation on Canadian wind energy generation and to support the development of targeted solutions to improve cold weather performance. For a study, CE-O analyzed production data from 23 wind farms across eight Canadian provinces with the objective of quantifying the degree to which cold climate operation affects wind energy production in Canada. Over the six-year study period from May 2010–April 2016, the average loss factor for the summer period from May to October was estimated to be 4.2%, compared to 8.1% for the winter period from November to April, resulting in an average cold climate loss factor of 3.9%. For individual wind farms, the 2010–2016 average cold climate loss factor ranged from −6% (higher losses in summer than winter) to 16%. Cold climate losses were estimated to total 959 GWh across the country each year, representing lost revenue of US$ 113 million annually.

26.13.1 Gaspe Region

One only need look at the cold snap of January 2013, when Quebec wind farms operated at full capacity for many days despite temperatures staying below −25 °C. In fact, wind turbine production at the Gaspé research site was almost twice the annual average during this period as well as in early January 2014. Data show that periods of extreme cold have always resulted in maximum yield every year since the commissioning. In fact wind turbines generate more electricity from November to April due to winter's strong winds and the greater density of cold air.

26.14 The Effect of Rain on the Wind Turbine Performance

Manufacturers are expected to develop wind turbines with a lifespan of 20–25 years. The individual components of a wind turbine, especially the rotor blades, are exposed to extreme environmental influences. This is a result of the continuous exposure to the elements and particularly of high rotor blade tip speeds. The tip speeds over a velocity of 90 m/s, result in leading edge erosion.

The damage mechanism due to droplet impingement [14–16] is based on enduring loads from individual impacts. The stress duration of an individual impact amounts to just a few milliseconds. The surface damage is induced by high pressure and the formation of a lateral jet that occurs at the burst of a droplet. The pre-stressed material is even more vulnerable so that existing cracks are widened and the material is eroded. The occurrence of jets leads to sheer stress on the surface material.

26.14.1 Conclusion

An examination of the water drop and particle erosion damage process concerns a variety of suppliers: rotor blade, wind turbine, and varnish manufacturers, as well as wind turbine operators and service companies. All of these organizations must be engaged regularly to inspect the turbine blades, foils, and coatings.

Based on a NASA study (3/1988), the rainfall rates experienced at the wind farm were generally below 8 mm/h, usually falling under the light rain classification. The study described did not show a decrease in performance from light rain; rather an increase of approximately 3% was indicated. The 3% increase may be attributable to a reduction of blade roughness, rain effects on the anemometer, or a combination of the two. The general conclusion was drawn for this study: for the ESI-54 wind turbine, for rainfall rates generally <8 mm/h, the rain had a beneficial effect on performance by slightly improving the aerodynamic properties (maintaining clean blades). The test model was applied to a wind-turbine blade airfoil and studied the effect of rain for different rainfall rates in addition to the effect of surface tension and surface property of the airfoil.

It was observed that, at low rainfall rates, the performance of the airfoil is highly sensitive to the rainfall rate. However, if the rainfall rate is high enough to immerse most of the airfoil surface under water, a further increase in the rainfall rate does not have a substantial effect on the performance of the airfoil.

26.15   Wind Turbines in the Desert Environment

Sand storms and blowing sand environment is definitely a challenge, not smaller than that of the turbines spinning in oceans. Certainly, this environment will demand equipment adjustments and a different maintenance program. Thousands of turbines already operating in deserts prove that this is a productive frontier for the wind resource (see Figure 26.16).

Note: This excerpt was taken from SeaWind/SNC Lavalin study for Riko Diq, Pakistan, a mining project, 2008 [12].

The existence of sand storms, which occur approximately 9 d/yr at this site, will affect the performance of the turbine, not only during the storm but also over their lifespan. An industry study was carried out to identify what experience exists with projects in desert conditions to demonstrate whether wind turbine technology can operate successfully in these environmental conditions. Short-term effects and system deterioration would be the result of for example

Figure 26.16 Desert environment. Source: From [10].

sand particles penetrating electronic systems, hydraulic, oil lubrication, or oil cooling systems.

Electronic systems could malfunction through short circuits and would cause the turbine to cease operating until the faulty components are replaced. Particles entering fluid systems would lead to rapid deterioration of pumps, control valves, motors, hydraulic cylinders, etc., and would require premature replacement to take place, possibly in a matter of months, if not weeks.

Longer-term effects could manifest themselves when particles enter critical items such as main bearings, blade bearings, or the gearbox and would result in increased turbine vibrations, element deterioration, and would ultimately require the repair, exchange, or reconditioning of the element.

Long-term effects (more than three to five years) are likely to be mainly observed through particle impact on the external structure causing the UV and corrosion protection to deteriorate and requiring repairs. Through the inherent robust design of modern state-of-the-art wind turbines the risks of a modern MW type turbine being affected by airborne particles are already considerable reduced, and the success of the measures to mitigate the impact of environmental conditions is especially demonstrated when considering that turbines have been successfully adapted for years now to operate in offshore wind farms, where environmental conditions are much more demanding and aggressive than onshore.

26.15.1 Blades

Internally a blade is hollow, but the inside structure is practically hermetically sealed from the outside environment and no risk of contamination is therefore possible. The outside area of the blade is at risk from impact erosion from airborne particles, and this is aggravated toward the blade tip, where the rotational speed is much higher than the actual wind speed and also through the aerodynamic acceleration caused by the blade profile, whereby the leading edge is not only exposed directly to the particles and must divert them but also accelerates the wind around the profile (see Figure 26.17).

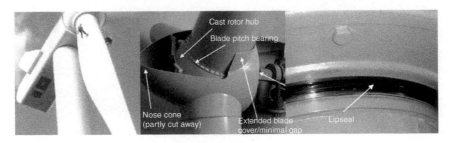

Figure 26.17 Nacelle and blade, hub, hub bearing.

As a result the areas of possible concern are the paint system that could erode extensively than UV damage to the underlying composite structure occurs. The most secure, but least efficient method of mitigating the risk of external erosion would be to stop the turbine each time a sand storm occurs. The blades would then only experience the actual wind speeds and with the blades turned into the wind, the exposed area would be limited to only the leading edge. In the case of this project, where sand storm occurrence is low 9 d/yr, this may be worth considering against the requirement to repaint certain areas of the blades more frequently. A more common method is however to protect the leading edge using anti-abrasion tape, so-called Helicopter tape. Finally, there are wind farm projects that have been operational in excess of five years in comparable or worse conditions.

They have been exposed to sand storm conditions without any necessity for blade repairs or repainting, even though the turbines are not switched off during sand storms.

26.15.2 Rotor Hub

The rotor hub is a cast structure that holds the blade pitching mechanisms, blade bearings, and which is fixed onto the main shaft. The rotor hub is protected from the environment by a nose cone, which is a glass fiber reinforced aero-dynamical structure which fits the contours of rotor hub, blade roots, and the front end of the nacelle cover (see Figure 26.17).

The nose cone has no need for vents, air intakes or exhaust and can therefore be made to be sealed tight against the environment. It usually has three access hatches that can be properly sealed and small gaps must only be maintained between the structure and the blades to allow free pitching of the blades and between the structure and the front end of the nacelle cover.

Turbine suppliers already minimize these gaps to 20–30 mm, to keep out water, for example, and it would be easily achievable to increase the protection even further by installing a better sealing system using, for example, brushes fixed to the nose cone structure. In addition, the nose cone can be slightly modified so that it overlaps the nacelle cover by 50 mm so that with the turbine facing the wind direction, sand particles would never find a direct path into the unit.

26.15.3 Air Intakes and Exhausts

Air intakes are necessary to provide ventilation and cooling, both at the bottom of the tower (where the power electrical equipment is usually located) and in the nacelle to cool the generator, oil coolers, etc. The cooling system in the nacelle does not usually draw air into the actual nacelle but consist of heat exchangers systems that channel the cooling air through the heat exchangers from the intake, pointed toward the wind direction, to the outlet, located toward the back of the nacelle. The nacelle therefore will not be exposed to the direct environmental conditions and only the heat exchangers must be able to deal with sand particles. This however is no different from any type of diesel generators or machinery that operates reliable in similar conditions and as such it is not likely that this will be an issue. The cooling system at the bottom of the tower does draw air into the internal environment and as such must be made to deal with sand particles. Current wind turbine tower design already includes weather protection grates and filter elements to keep out moisture and particles, see example mentioned later in the text and increased filtering can be easily achieved by adding another filtering stage, if necessary.

The cooling system at the bottom of the tower always use forced ventilation, whereby air is drawn into the internal environment through the filters of the air intakes. This offers another benefit in that a slight over dimensioning of the ventilation fans, or leaving the fans on full power during sand storms will create an overpressure in the internal environment and air will be forced out through any openings that may exist (such as the slight gap between the nose cone and the front end of the nacelle), ensuring no foreign particles enter the turbine internal environment. This is a principle that is applied successfully to achieve climate regulation in offshore turbine where the air drawn into the turbine must be treated for its corrosive nature by air conditioning units before entering the turbine internal areas.

26.15.4 Bearings

The main bearing or the blade pitch bearings are designed to withstand extreme occasional loads in addition to continuous fatigue loadings, both of which place high demands on the materials over the lifespan of the turbine. Particles could enter either through the greasing mechanism, or via the lip seal which seals the bearing at the shaft from the external environment.

When considering the likelihood of sand penetration into bearings, it must be considered that these lip seals are designed to keep much smaller dust particles from penetrating into the bearing and it is therefore unlikely that any sand particles would pose a concern (see Figure 26.17).

26.15.5 Electronic/Electrical Systems

The electrical systems, consisting of the generator, switchgear, converters, transformers, etc., are inherently well protected from outside influences, if only for operator safety reasons, and are unlikely to suffer any influence from sand particles inside the tower or nacelle. Electronic systems in wind turbines consist of a large range of industrial type control systems, which are not proprietary for turbine systems but are widely used in other, industrial environments, some of which are far more aggressive and potentially damaging than airborne dust and sand.

Electronic equipment is contained in secure steel cabinets and the only possibility for dust penetration would be when the cabinets are opened for service or repairs, via cable conduits or through ventilation openings. Contamination can be avoided by ensuring the area is suitably clean prior to accessing the electronic components, avoiding maintenance during dusty and wind conditions, sealing cable conduits properly during the assembly process and installing filter elements over air vents. Specific requirements depends very much on the turbine type and should be discussed with the turbine manufacturer, any necessary modifications will however be minor.

26.16 Cost, Component Percentage Share

An approximate cost percentage share of major components for a 5 MW, 100 m tower, wind turbine are

• Tower	26%
• Blades	23%
• Gearbox	13%
• Generator	4%
• Main shaft, hub, brake	7%
• AC/DC convertor	5%
• Unit transformer, cables	5%
• Miscellaneous turbine	21%

The additional costs of the wind farm infrastructure, 33 kV collector cables, switching stations, main transformers, regulating capacitor banks, and grid integration protection must be included in the overall farm design.

The comparative costs are shown in Figure 26.18 reproduced courtesy of IRENA Renewable Data Base.

US DOE 2008 – Renewable Energy Data Book [18], Figure 26.19 shows the levelized cents/kWh cost comparison, whereby the wind competes with the best renewable sources, like hydro power.

According to the European Wind Energy Association (EWEA), the cost of wind energy has fallen over the years as the technology has matured. Historically, the cost per produced kWh for new turbines has fallen by between 9% and 17% for each doubling of installed capacity.

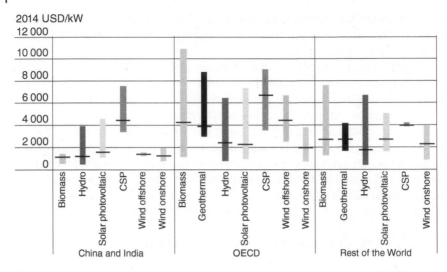

Figure 26.18 Renewable energy capital cost comparison. Source: Courtesy of IRENA Publications – International Renewable Energy Agency [17].

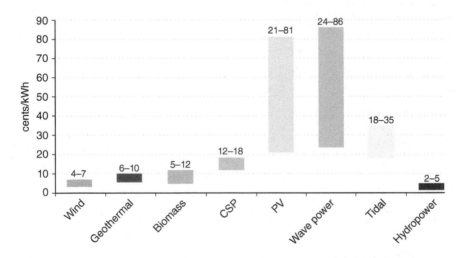

Figure 26.19 Energy cents/kWh levelized cost comparison. Source: From [18].

References

1 What is a wind Rose? https://Novalynx.com.
2 Electropaedia – Energy Technologies – www.mpoweruk.com.
3 Windpower Engineering & Development – www.windpowerengineering.com 2011.

4 Tudelft: mstudioblackboard – www.mstudioblackboard.tudelft.nl.

5 Magdi Ragheb and Adam Ragheb Betz Law and Optimal Rotor Tip Speed.

6 Lundquist, J. and Clifton, A. (2012). *How Turbulence Can Impact Power Performance. North American Windpower.*

7 Wikipedia – Wind Classifications; (IEC 61400-1) https://en.wikipedia.org/wiki/IEC_61400.

8 Enercon, E-126 Turbine – www.enercon.de/home.

9 John Diaz de Leon, P.E. Consulting T&D Planning Engineer "Renewable Energy – Connecting Wind Farms to the Grid".

10 https://www.awea.org/ – American Wind Energy Association.

11 NREL: Comparison of Integration Standards – https://www.nrel.gov/docs/fy16osti/64225.pdf.

12 Seawind (March 2008). Reko Diq Mining Project: The Impact of Sand Particles on Turbine Performance and Reliability.

13 Natural Resources Canada (2017): https://www.nrcan.gc.ca/energy/renewable-electricity/wind/7321.

14 Walker, S.N. Wade, J. NASA 3/1988 Effects of Precipitation on Wind Turbine Performance.

15 Liersch, L., Michael, J. (2014). Investigation of the impact of rain and particle erosion on rotor blade aerodynamics.

16 Cohana, A.C., Arastoopura, H.. (2016). Numerical simulation and analysis of the effect of rain and surface property on wind-turbine airfoil performance.

17 IRENA Report for 2014: Renewable Energy Capital Cost Comparison $/kWh

18 US DOE (2008). *Renewable Energy Data Book*. US DOE.

Index

a

ABC cable overhead distribution 205
Accelerating torque 305
Action of tap changers 39, 458
Affinity laws 331
Ambient derating factors 36
Analog cards 4-20mA, 0-20mA 387
Analog card scaling 388
Arc flash classifications 81, 104
Area lighting 159
Arrester
  classifications 226
  energy class 224
  protection margin 224
  rating 227
  selection 225
Autotransformers 131, 541
Availability 476, 490
  inherent 491
  operational 492
Availability factor 437
AVR droop 452

b

Bandwith 396
Baseband 396
Base load 36, 443
100BaseT ethernet 400
Basic insulation level BIL 72, 220, 540
Bathtub mortality 479

Battery
  Ah capacity 181
  chargers 180, 186
  float/ boost 184
  life 183
  room requirements 185
Below and above grounding 138, 148, 150
Betz's limit 603
Black start 434
Blade design 602, 611, 636
Break down torque 305
Breakers
  continuous ratings 54
  failure 366
  frame size rating 76
  HMCP 99
  interrupting rating 77, 78
  magnetic for motors 85, 96
  maximum loading 77
  MPCB 99
  status contacts 80
  thermal and magnetic for feeders 84, 96
Brushless exciter 430
Budget estimate 26
Bundled conductor 193
Burden 348
Buried installations 109
Bus duct 58, 279
Bus tie breaker switching 61

Practical Power Plant Engineering: A Guide for Early Career Engineers, First Edition. Zark Bedalov.
© 2020 John Wiley & Sons, Inc. Published 2020 by John Wiley & Sons, Inc.

c

Cable armor grounding   256, 260
Cable bending radius   278
Cablebus   58, 280
Cable color coding   268
Cable derating factors   258, 264
Cable insulation level   256
Cable list   13
Cable openings   115
Cables for fire protection   514
Cable shielding   254
Cable short circuit rating   259
Cable terminations, stress cones   267
Cable trays   273
    fill and classification   273
    loading   273
    support span   273
Capacitors   290
    impact on harmonics   297, 299
    voltage improvement   295
Capacity factor   438
Cathodic protection for buried
        pipelines   518
Circuit breaker, types   84
Class A wiring   497
Close and latch current   77
Coax RG cables   397
Codes and standards   4
Color rendering index (CRI)   172
Commissioning
    dry, wet   19
    report   25
Common mode failure   476, 489
Communication cables   268
Concentrated solar plants (CSP)   580,
        586
Concrete cable duct banks   109, 111
Conductor sag calculations   194. 204
Connected load, operating load   51
Constant kA rating   79
Constant MVA rating   79
Continuous UPS   188

Control cables   268
Control room fire protection   504
Core flux test   420
Corrosion protection for ships   525
Cost of change   8
Counter clockwise (CCW) and
        clockwise (CW)   243–249
Cross current compensation, CCC
        455
CT and PT polarities   342, 345
CTs and PTs   342, 345
CT saturation   350
Current limiting fuses, type E   88
Current source inverter (CSI)   325
Current, voltage THD   298, 339
Cut-in wind speed   613
Cut-out wind speed   614

d

Data sheets   553
Dead transfer   62
Delta tertiary winding   131
Design criteria   2, 7, 158, 542
Design limits   613
DeviceNet cables   270
Device numbering   11
DG waste heat recovery   469
Diesel engine plant   463
Differential protection   345, 347, 365
Digital governor   424
Digital I/O cards   386
Direct normal irradiance (DNI)   568
Distributed control system (DCS)
        385, 386
Diversity factor   52
DNP3   405
Double ended substation   92
Doubly fed induction generator (DFIG)
        617
Dry type transformers   90, 538, 564
Duty electrical and diesel pumps   511
Dynamic braking   333

e

Effective (solid) grounding   123
Effect of system grounding   226
Effect of voltage change   304
EHV cable termination   266
EHV power cables   265
ELCID test   420
Electrical rooms, control rooms   112,
      115, 119
Emergency lighting   159, 161
Enclosures
   IEC, NEMA   550
End of line (EOL) resistor   497
Energy heat balance and efficiency
      diagram   470
Engine ratings
   continuous, prime and standby   462
Engines for ships   473
Equipment grounding   122
Equipment withstand capability   219
Essential lighting   160
Ethernet 100BaseT   400
Ethernet cables   270
Ethernet TCP/IP   400
EXIT lighting   161
Exothermic connections   148, 236
Explosion proof motors   311
Exponential distribution   480

f

Failure rate   478
Feasibility study   5
Fence grounding   152
Fiber optic
   cables   271
   single, multi mode   398
Fieldbus cables   272
Fire control panels   499
Fire detectors   498
   heat   498
   ionization   498
   smoke   498
Fire hydrants   506

Fire integrity test for gas   516
Fire pump sizing   513
Fire wall ratings   116
Fire water storage tanks   510
Firm capacity   45
Firm power   462
Flexlogic   367
Flow diagrams   5, 34
Fluorescent lamp   174
Foot-candle   163
Foundations for grounding (Ufer)   150
Fresnel linear mirror   583
Fuel consumption   461, 464
Fuel storage requirements   464

g

Galvanic sacrificial anodes   519
Galvanized steel   519
Gas fire suppression   503
Generator
   capability and limits   413
   characteristics   415
   cooling   434
   neutral grounding   127
   special circuit breakers   431
   step up transformer   432
   transient conditions   419
   unit, system grounding   128
Glare factor   163
Governor deadband   446
Governor droop   446, 447
Grid
   integration issues   624
   resistance measurements   142
   tests   24
Groundbeds   521
Ground flash density   218
Grounding grid   138
Grounding of telecommunications
      152
Ground potential rise (GPR)   142
Guy wire   193

h

Harmonics 298
    caused by VFDs 335
    correction 338
    order limits 337
Heat transfer RHO factor 112
Hi-bay lighting 159
High and low pressure sodium lamp
        176
Hi-pot test 421
H$_2$S issues in Hydro power plants 527
Human machine interface, HMI 389
HV equipment testing 215

i

IEC 101 404
IEC 104 405
IEC 61850 404
IEEE device designation 343
IEEE 519 Standard 338
IGBT transistors 334
Illuminance level 164
Impressed current protection 521
Infinite bus 43
Instrumentation cables 268
Insulation coordination 221
Insulation temperature class: B, F, H
        537
Interlocks 21
Interposing relays 385
Isokeraunic lightning levels 212
Isolated phase bus (IPB) 549, 557

j

Jockey pump 511

k

Key interlocks 88
Key one line diagram 32, 50

l

Ladder logic 383
Large motor starting 311

LC filters, passive, active 337
Lead acid battery 184
LED 177
Levelized cost of energy 27
Light distribution charts 165
Lighting design criteria 158
Lighting in wet conditions 159
Lighting panels 170
Lightning strikes 211
Lightning wave phenomena 213
Line and cable charging current 206
Line compensation 300
Line inductance and capacitance 199,
        206
Line power transfers 456
Line reactors 301
Line reclosing 373
Line surge impedance 214
Lithium ion battery 185
Load break switches 91
Load factor 443
Load flow study 40
Load-frequency control 444
Load list 13, 51
Load scheduling and dispatch 442
Load shedding 374
Locked rotor torque 305
Lock out trip relays (86) 375
Low-bay lighting 159
Low rise/high rise building lightning
        protection 231
Lumen 162
Lux 162
LV motors, upto 200 kW 542, 557
LVRT, HVRT capability 628, 629

m

Maintainability 476
Make before break principle 62
Managing wind variability 625
Marginal cost of energy 29
Maximum continuous over voltage
        (MCOV), U_c 227, 228

Mercury lamp   174
Metal-clad enclosure   60, 76
Metal halide lamp   174
Metering accuracy class   347
Metering PTs and CTs   347
M of N systems   486
Moment of inertia H   445
Motor   558
    controls integration   388
    enclosures   310
    inrush current   311
    service factor   315
    starters   95, 103
    starting criteria   316
    starting on VFD   315
    start voltage drop   40, 64,
        258, 316
    surge protection   230, 300, 548
Motor control centers (MCC)   84,
    94–104, 560
MTTF, MTTR   478
Multi-function relays (MFR)   351
Multi-junction PV solar cells   571
Multy ratio CTs   350
MV controllers   59, 84–88
MV motors >200 kW   545, 558

n

NACE Air quality   529
National Electrical Manufacturers
    Association (NEMA)
    Class E2   59, 85
    guidelines for electrical rooms   112
    guidelines for transformers   116
    NEMA, IEC T frame sizes   307
    premium motors   316
    starter sizes   310
    TEFC motors   310
    TENV motors   310
    torque A, B, C, D, E   306
National Fire Protection Association
    (NFPA)
    class I and II materials   231

    guidelines for electrical rooms   112
    guidelines for transformers   116
Networking   393
    DeviceNet   395
    fieldbus   395
    profibus   395
Neutral and ground fault protection
    362
Ni-Cd battery   184
Nominal voltages   242
Non-linear loads   297, 335
Non-maintenance VRLA battery   184
Normal lighting   160

o

O/C instantaneous value   364
Outdoor lighting   159, 167
Overhead *vs.* underground   66, 198
Overload relays   100
Overvoltages   210

p

Parabolic mirror troughs   582
Parallel resonance   299
Parity   396
Pay back calculations   30
Peaking storage plants   593
Peak loading plants   443
Peer to peer   399
Permafrost in colder climates   108
Permanent magnet (PMG)   426, 430
Permissives   21
Personal protective equipment (PPE)
    81
Per unit calculations   41
Phase sequence   243
Photovoltaic solar plants (PV)   571
Pick up setting   353
P&ID diagrams   34
Plant control architecture   394
Plant design
    conceptual, detailed   5, 6
Plant phasing diagram   533

Polarization index (PI)   421
Pole foundation   194
Pole grounding   194
Polling (master/slave)   399
Portable hand extinguishers   507
Power agreement   38
Power cable selection   262, 277
Power distribution system   50
Power factor (PF)
  correction   288
  leading/ lagging   286
  penalties   285
Power loss calculation   203
Power system studies   6
Power tower   580
Power transformers   64, 531, 555
Pre-action sprinkler systems, wet, dry
    502
Pre-commissioning   17
Primary injection   21
Programmable logic controller (PLC)
    381
  or DCS?   391
  scan   384
Project development   3, 5
Protection elements   360
  for generators   354
  for motors   356
  for transformers   354
Pull stations   499
Pulse width modulation, PWM   324
PV panel technology   569, 576

r
Radial distribution   89
Raised floors   119
Rated wind speed   613
R class type fuses   60, 88
Reactive power requirements impact
    626
Rebar grounding   149
Reduced voltage starter   314
Regenerative VFDs   332

Relay accuracy class   347
Relay coordination   359
Relay logic   380
Relay setting groups   366
Reliability   477, 483
  criteria   37
  run   23
Remote reference earth   142
Remote site grounding   154
Removal of H_2S gas   527
Residual (discharge) voltage   221
Resistance grounding   125, 262
Restricted earth fault REF   362
Right of way   194
Rigid and EMT conduits   278
Ring main   89
Road clearances   201
Road lighting   167
Room numbering   115
RS-232C, RS-422, RS-485   400

s
Salt energy storage   591–594
SCADA transmission protocols   403
Schematic and wiring diagrams   14
Secondary injection   17
Self-reset relay (94)   375
Series and parallel systems   483
Shaft encoder   329
Shielded twisted pair STP   397
Shielding criteria, switchyard, line
    217, 229
Short circuit ratio (SCR)   416
Short time withstand   78
Single contingency outage   37
Single or double incomers   43
Siting a wind farm   600
  cold climate   633
  desert   635
  offshore   601
  onshore   601
Soil conditions   139
Soil resistance measurements   140

Soil resistivity   525
Soil thermal resistivity   112
Solar resource   567
Source impedance   40
Spinning reserve   444
Standpipes   506
Star/delta transformers   130
Star/star transformers   130
Static exciter   427–429
Structure: tangent, dead end, angle
   197
Sun trackers   574, 575
Supercharging vs turbocharging   465
Supply voltage   239
Switches and outlets   159, 171
Switchgear   57, 59, 75, 79, 562
Synchrocheck relay   357
Synchronizer relay   357
Synchronous condensers   292
Synchronous impedance   418
Synchronous motors   292, 315, 436
System grounding   122

t
Tap setting   353
Time-current grading   371
Time dial setting   354
Time grading   369
Torque, constant or variable   329
Touch and step potential   144, 145
Transformer
   cooling   44, 46, 536
   deluge system   496, 508
   inrush current   362, 541
   for lighting   168
   neutrals grounding   65, 130
   oil containment   118
   phase shift   532
   rectifiers   524
   tap changers   38, 537
   testing   539
   winding configurations
      46, 68, 532

Transmission
   duplex   397
   simplex   397
Trip supervisory relay   376
Turbine cavitation   433

u
Ungrounded systems   122
Uninterruptible power supply UPS
   71, 188
Unit substations   61, 89–94,
   563, 565
Unshielded twisted pair (UTP)   397
Utility interconnection requirements
   627
Utilization factor   52
Utilization voltage   239

v
Vector design for low speed operation
   328
Vector drive, sensorless   329
Ventilation, louvers, air dampers   507
Vesda fire protection   496, 504
VFD specification   548, 560
   constant V/Hz ratio   326
   critical cable length   334
   starting torque   326
VI characteristics   222
Voltage drop calculations   202
Voltage–reactive power control   452
Voltage regulation   64
Voltage source Inverter (VSI)   325

w
Water pumping stations   510
Web based HMI   402
Wind capacity factor CF   606
Wind energy captured   607
Wind energy distribution   607
Wind gusts   603
Wind power density   605
Wind rose   603

Wind turbine
  blade pitch control 611
  capital cost comparison 639
  effect of rain 634
  fixed speed 616
  IEC design classification 610
  sizes 620
  tip speed ratio 605
  tower 602
  variable speed 616
Yaw control 612

Wind turbulence 608
World plugs and sockets 250

X
Xd, Xd', Xd" impedances 418
XLPE and EPR cables 256

Z
Zg grounding transformers
    132
ZnO arresters 219